Interaction Effects in Linear and Generalized Linear Models

Advanced Quantitative Techniques
in the Social Sciences

Series Editor: Shenyang Guo, *Washington University in St. Louis*

VOLUMES IN THE SERIES

1. **HIERARCHICAL LINEAR MODELS: Applications and Data Analysis Methods, 2nd Edition**
   Stephen W. Raudenbush and Antony S. Bryk

2. **MULTIVARIATE ANALYSIS OF CATEGORICAL DATA: Theory**
   John P. van de Geer

3. **MULTIVARIATE ANALYSIS OF CATEGORICAL DATA: Applications**
   John P. van de Geer

4. **STATISTICAL MODELS FOR ORDINAL VARIABLES**
   Clifford C. Clogg and Edward S. Shihadeh

5. **FACET THEORY: Form and Content**
   Ingwer Borg and Samuel Shye

6. **LATENT CLASS AND DISCRETE LATENT TRAIT MODELS: Similarities and Differences**
   Ton Heinen

7. **REGRESSION MODELS FOR CATEGORICAL AND LIMITED DEPENDENT VARIABLES**
   J. Scott Long

8. **LOG-LINEAR MODELS FOR EVENT HISTORIES**
   Jeroen K. Vermunt

9. **MULTIVARIATE TAXOMETRIC PROCEDURES: Distinguishing Types From Continua**
   Niels G. Waller and Paul E. Meehl

10. **STRUCTURAL EQUATION MODELING: Foundations and Extensions, 2nd Edition**
    David Kaplan

11. **PROPENSITY SCORE ANALYSIS: Statistical Methods and Applications, 2nd Edition**
    Shenyang Guo and Mark W. Fraser

12. **INTERACTION EFFECTS IN LINEAR AND GENERALIZED LINEAR MODELS: Examples and Applications Using Stata**
    Robert L. Kaufman

Interaction Effects in Linear and Generalized Linear Models

Examples and Applications Using Stata®

Robert L. Kaufman

Temple University

Los Angeles | London | New Delhi
Singapore | Washington DC | Melbourne

FOR INFORMATION:

SAGE Publications, Inc.
2455 Teller Road
Thousand Oaks, California 91320
E-mail: order@sagepub.com

SAGE Publications Ltd.
1 Oliver's Yard
55 City Road
London EC1Y 1SP
United Kingdom

SAGE Publications India Pvt. Ltd.
B 1/I 1 Mohan Cooperative Industrial Area
Mathura Road, New Delhi 110 044
India

SAGE Publications Asia-Pacific Pte. Ltd.
3 Church Street
#10-04 Samsung Hub
Singapore 049483

Acquisitions Editor: Helen Salmon
Editorial Assistant: Megan O'Heffernan
Production Editor: Jane Martinez
Copy Editor: QuADS Prepress (P) Ltd.
Typesetter: C&M Digitals (P) Ltd.
Proofreader: Scott Oney
Indexer: Jeanne R. Busemeyer
Cover Designer: Janet Kiesel
Marketing Manager: Susannah Goldes

The trademarks pertaining to Stata software are the property of StataCorp LP, 4905 Lakeway Drive, College Station, TX 77845 USA. Their use in no way indicates any relationship between StataCorp LP and SAGE Publications, Inc.

Printed in the United States of America

Library of Congress Cataloging-in-Publication Data

Names: Kaufman, Robert L., author.

Title: Interaction effects in linear and generalized linear models : examples and applications using Stata / Robert L. Kaufman, Temple University.

Description: Thousand Oaks, California : SAGE Publications, Inc., [2019] | Includes bibliographical references and index.

Identifiers: LCCN 2018015288 | ISBN 9781506365374 (hardcover : alk. paper)

Subjects: LCSH: Regression analysis. | Regression analysis—Data processing. | Mathematical statistics. | Mathematical statistics—Data processing. | Linear models (Statistics) | Stata.

Classification: LCC QA278.2 .K38 2019 | DDC 519.5/36—dc23
LC record available at https://lccn.loc.gov/2018015288

This book is printed on acid-free paper.

MIX
Paper from
responsible sources
FSC® C008955

18 19 20 21 22 10 9 8 7 6 5 4 3 2 1

BRIEF CONTENTS

Series Editor's Introduction   xvii

Preface   xix

Acknowledgments   xxi

About the Author   xxiii

Chapter 1 • Introduction and Background   1

PART I • PRINCIPLES

Chapter 2 • Basics of Interpreting the Focal Variable's
              Effect in the Modeling Component   33

Chapter 3 • The Varying Significance of the Focal
              Variable's Effect   63

Chapter 4 • Linear (Identity Link) Models: Using the
              Predicted Outcome for Interpretation   99

Chapter 5 • Nonidentity Link Functions: Challenges of
              Interpreting Interactions in Nonlinear Models   143

PART II • APPLICATIONS

Chapter 6 • ICALC Toolkit: Syntax, Options, and Examples   185

Chapter 7 • Linear Regression Model Applications   245

Chapter 8 • Logistic Regression and Probit Applications   289

Chapter 9 • Multinomial Logistic Regression Applications   345

Chapter 10 • Ordinal Regression Models   411

Chapter 11 • Count Models   479

Chapter 12 • Extensions and Final Thoughts   525

Appendix: Data for Examples   555

References   573

Index   577

DETAILED CONTENTS

Series Editor's Introduction   xvii

Preface   xix

Acknowledgments   xxi

About the Author   xxiii

1. Introduction and Background   **1**

  Overview: Why Should You Read This Book?   1

  The Logic of Interaction Effects in Linear Regression Models   4

    What Is an Interaction Effect?   **4**

    Why Should You Consider Including an Interaction Effect in Your Analysis?   **5**

    How Do You Specify an Interaction Effect in the Prediction Function of a Linear Regression Model?   **6**

    When Is an Interaction Effect Statistically Significant?   **9**

    Common Errors in Specifying and Interpreting Interaction Effects   **11**

      Excluding Lower Order Terms   11

      Interpreting Coefficients as Unconditional Marginal Effects   13

      Interpreting Main Effect Coefficients When Not Meaningful and the Myth of Centering   13

      Not Interpreting the Moderated Effect of Each Predictor Constituting an Interaction   14

  The Logic of Interaction Effects in GLMs   15

    What Are GLMs?   **15**

    (Interaction) Effects in the Modeling Component   **16**

    (Interaction) Effects on the Observed Outcome   **16**

    Common Errors in Using Interaction Effects in GLMs   **18**

      Improperly Treating Product Terms for an Interaction   18

      Limited Range of Moderator Values Used to Probe Moderated Effect of Focal Variable   19

      Comparing Estimated Coefficients Across Nested Models (for Some GLMs)   19

  Diagnostic Testing and Consequences of Model Misspecification   20

    Diagnostic Testing   **20**

      Link Function Test   20

      Assessing Overall Model Fit/Departures   20

      Residual–Predictor Plots or Partial Residual–Predictor Plots   21

      Residual–Omitted Variable Plots   21

      Analysis of Influential Cases   22

    Consequences of Model Misspecifications   **23**

      Including an Interaction Specification When Not Needed   23

      Excluding an Interaction Specification That Is Needed   23

      Unspecified Heterogeneity and Group Comparisons   24

  Roadmap for the Rest of the Book   24

    Overview of Interpretive Tools and Techniques   **24**

      Defining the Moderated Effect of F With the GFI (Gather, Factor, and Inspect) Tool   25

Calculating the Varying Effect of F and Its Significance: SIGREG (Significance Regions) and EFFDISP (Effect Displays) Tools   25

The Predicted Outcome's Value Varying With the Interacting Predictors: The OUTDISP (Predicted Outcome Displays) Tool   25

Is the ICALC Toolkit Necessary?   25

Organization and Content of Chapters   **26**

Part I: Principles   26

Part II: Applications   27

Chapter 1 Notes   29

PART I • PRINCIPLES

2. Basics of Interpreting the Focal Variable's Effect in the Modeling Component   **33**

Mathematical (Geometric) Foundation for GFI   34

GFI Basics: Algebraic Regrouping, Point Estimates, and Sign Changes   37

One Moderating Variable   **37**

Case 1: b_1 and b_3 Have the Same Sign   37

Case 2: b_1 and b_3 Have Opposite Signs   39

Two (or More) Moderating Variables   **40**

A Three-Way Interaction   **42**

Wrap-Up   **44**

Plotting Effects   46

Overview   **46**

One-Moderator Effect Display Examples   **47**

Moderated Effect of Headship Type   47

Moderated Effect of Number of Children   48

Two-Moderator Effect Display Examples   **49**

Line Plot of Focal by Moderator 1, Repeated for Values of Moderator 2   49

Effect Displays for Two Interval Moderators   51

Effect Displays for a Three-Way Interaction   **54**

Summary   58

Special Topics   60

Derivation of Equation 2.17   **60**

Chapter 2 Notes   60

3. The Varying Significance of the Focal Variable's Effect   **63**

Test Statistics and Significance Levels   64

Wald Tests Versus LR Tests   **64**

Potential Adjustment of the Significance Level for Simultaneous Testing   **65**

JN Mathematically Derived Significance Region   66

JN for One Moderator   **69**

Interpretation of Boundary Value Analysis Results From ICALC   73

JN for Two Moderators (or Three-Way Interaction)   **74**

Interpretation of Boundary Value Analysis Results From ICALC   76

Summary   **79**

Empirically Defined Significance Region   79

One-Moderator Results and Interpretation   **81**

Application to the Two-Moderator Example   **82**

Application to the Three-Way Interaction Example   **82**

Confidence Bounds and Error Bar Plots   84

    One-Moderator Examples of Confidence Bounds and
       Error Bar Plots   85

    Confidence Bounds and Error Bar Plots for Two Moderators   88

    Plots for a Three-Way Interaction   90

Summary and Recommendations   96

Chapter 3 Notes   97

4. Linear (Identity Link) Models: Using the Predicted Outcome for Interpretation   99

Options for Display and Reference Values   100

    Focal Variable   100

    Moderator Variable(s)   101

Reference Values for the Other Predictors (Z)   102

    Define by Central Tendency Values   103

    Define by Representative (Substantively Interesting) Values   103

    Define by As-Observed Values   103

Constructing Tables of Predicted Outcome Values   105

    Single Moderator   106

    Two or More Moderators   109

       Isolating Each Moderator's Effect on the Focal Variable   113

    Three-Way Interaction   115

Charts and Plots of the Expected Value of the Outcome   119

    Bar Charts for Categorical Focal and Moderator Variables   121

       Example 1: Effect of Education Moderated by
          Household Headship Type   122

       Example 2: Two Moderators, Headship Type Moderated
          by Education and by Any Children   123

    Scatterplots for Interval Focal Variables   127

       Example 3: Interval-by-Interval Interaction, Age by SES Effects on
          Frequency of Sex   127

       Example 4: Interval-by-Categorical Interaction, Number of
          Children Predicted by the Interaction of Income and Birth Cohort   131

       Example 5: Three-Way Interaction of a Mix of Interval and
          Categorical Predictors, Voluntary Association Memberships
          Predicted by the Interaction of Sex, Age, and Education   134

Conclusion   140

Special Topics   141

    Equivalence of As-Observed and Central Tendency
       Options for Linear Models   141

Chapter 4 Notes   142

5. Nonidentity Link Functions: Challenges of Interpreting Interactions in Nonlinear Models   143

Identifying the Issues   143

    The Goal of Interpretation, a Caveat   148

Mathematically Defining the Confounded Sources of Nonlinearity   149

    Confounding in Comparing Predicted Values   149

    Confounding in Comparing Slopes   153

Revisiting Options for Display and Reference Values   156

Solutions   159

Example 1: Two-Way Nominal-by-Nominal Interaction 160

Example 2: Two-Way Interval-by-Nominal Interaction 164

Example 3: Two-Way Interval-by-Interval Interaction 167

Example 4: Three-Way Interval-by-Interval-by-Nominal Interaction 172

Summary and Recommendations 177

Derivations and Calculations 179

Equation 5.13 for Slope of Logistic Prediction Function,
Main Effects Model 179

Equation 5.14 for Slope of Logistic Prediction Function,
Interaction Model 180

Chapter 5 Notes 181

PART II • APPLICATIONS

6. ICALC **Toolkit: Syntax, Options, and Examples** 185

Overview 185

INTSPEC **Tool: Interaction Specification** 185

GFI **Tool: Gather, Factor, and Inspect** 185

SIGREG **Tool: Significance Regions** 186

EFFDISP **Tool: Graphic Displays of the Moderated Effect** 186

OUTDISP **Tool: Display of a Predicted Outcome
by the Interacting Variables** 187

INTSPEC: Syntax and Options 187

One-Moderator Example 189

Two-Moderator Example 190

Three-Way Interaction Example 191

Several Important Details in Specifying the Interaction 192

GFI Tool: Syntax and Options 193

One-Moderator Example 194

Two-Moderator Example 195

SIGREG Tool: Syntax and Options 199

One-Moderator Example 201

Three-Way Interaction Example 203

Advanced Options: Factor Change Coefficients,
Coefficients Scaled in Standard Deviations of $g(y)$,
and SPOST13 Marginal Effects 207

Factor Changes 207

Scaled by $g(y)$'s Standard Deviation 210

SPOST13 Marginal Effects 211

EFFDISP Tool: Syntax and Options 213

One-Moderator Example 215

Two-Moderator Example 218

Three-Way Interaction Example 222

OUTDISP Tool: Syntax and Options 227

One-Moderator Example 229

Two-Moderator Example 234

Three-Way Interaction Example 238

Next Steps 243

Chapter 6 Notes 244

7. Linear Regression Model Applications   245

Overview   245

Properties and Use of Linear Regression Model   **245**
Data and Circumstances When Commonly Used   245
GLM Properties   246
Diagnostic Tests and Procedures   247

Data Source for Examples   **247**

Single-Moderator Example   247

Data and Testing   **247**

The Effect of Age Moderated by SES   **248**
Setup With INTSPEC Tool   248
GFI Analysis   248
Significance Region Analyses: SIGREG and EFFDISP Tools   249
Outcome Displays: OUTDISP Tool   252
Recap   253

The Effect of SES Moderated by Age   **254**
Applying the ICALC Tools   254
Recap   260

Summary and Recommendations   **260**

Two-Moderator Example   261

Data and Testing   **261**
Strategy for Interpreting Two-Moderator Interaction Models   262

The Effect of Birth Cohort Moderated by Family Income   **263**
INTSPEC Setup and GFI Analysis   263
Significance Region Analyses: SIGREG and EFFDISP Tools   264
Outcome Displays: OUTDISP Tool   265

The Effect of Education Moderated by Family Income   **267**
INTSPEC Setup and GFI Analysis   267
Significance Region Analyses: SIGREG Tool   268
Outcome Displays: OUTDISP Tool   268

The Effect of Family Income Moderated by Birth Cohort and Education   **271**
INTSPEC Setup and GFI Analysis   271
Significance Region Analyses: SIGREG and EFFDISP Tools   273
Outcome Displays: OUTDISP Tool   278
What to Present and Interpret?   283

Special Topics   284

Customizing Plots With the pltopts( ) Option   **284**

**Aside on Using the Path Diagram for a Multicategory
Nominal Moderator**   **285**

**Testing Differences in the Predicted Outcome Among
Categories of a Nominal Variable**   **285**

Chapter 7 Notes   287

8. Logistic Regression and Probit Applications   289

Overview   289

Properties and Use of Logistic Regression and Probit Analysis   **289**
Data and Circumstances When Commonly Used   290
GLM Properties and Coefficient Interpretation
for Logistic Regression   291
GLM Properties and Coefficient Interpretation for Probit Analysis   291
Diagnostic Tests and Procedures   292

Data Source for Examples   **292**

One-Moderator Example (Nominal by Nominal)   292

Data and Testing   **292**

Part I: The Effect of Sex Moderated by Residential Location 293
    INTSPEC Setup and GFI Analysis 293
    Significance Region Analyses: SIGREG and EFFDISP Tools 295
Part II: The Effect of Residential Location Moderated by Sex 300
    INTSPEC Setup and GFI Analysis 301
    Significance Region Analyses: SIGREG With Varying effect( ) Options 301
Part III: Outcome Displays With the OUTDISP Tool 304
    Adding a Display of Predicted Values From a No
        Interaction Effects Model 306
    Dual-Axis Labeling 307
Wrap-Up 309
    What to Present to Interpret a One-Moderator
        Interaction Effect From Logistic Regression 309
    Comparison of Probit and Logistic Regression Results 310

Three-Way Interaction Example (Interval by Interval by Nominal) 313
    Data and Testing 313
        Strategies for Interpreting the Three-Way Interaction 314
    GFI Results for the Three Predictors 314
        Racial Contact 314
        Education 315
        Race 316
    Factor Change Interpretation 316
        Moderated Effect of Racial Contact 316
        Moderated Effect of Education 317
        Moderated Effect of Race 318
        Summary 320
    Standardized Latent Outcome Interpretation 320
    Using Significance Region Tables for Interpretation 321
        Moderated Effect of Racial Contact 321
        Moderated Effect of Education 321
        Moderated Effect of Race 323
        Summary of Moderated Effects 324
    Using Outcome Displays for Interpretation of the Latent Outcome 324
        Tabular Display 324
        Graphic Display 327
    Using Predicted Probabilities for Interpretation 330
        Significance Region Tables for the Discrete Change Effects 330
        Predicted Probability Plots 333
    What to Present for a Three-Way Interaction
    From a Logistic Regression 339

Special Topics 340
    Customizing Dual-Axis Scatterplots and Bar Charts 340
        Scatterplot Customization 340
        Bar Chart Customization 341
    Alternative Plot Comparing Additive and Interaction Model Predictions 341

Chapter 8 Notes 343

9. Multinomial Logistic Regression Applications 345

Overview 345
    Properties and Use of Multinomial Logistic Regression 345
        Data and Circumstances When Commonly Used 345
        GLM Properties and Coefficient Interpretation for MNLR 346
        Diagnostic Tests and Procedures 347
    Data Source for Examples 348

One-Moderator Example (Interval by Interval) 348
    Data and Testing 348
        INTSPEC Setup and GFI Analysis 349

Factor Change (Odds Ratio) Interpretation of Education Effect **353**

Discrete Change Interpretation of Attendance Effect **355**

Cautions 358

Interpretation Using Displays of Predicted Probabilities **359**

ICALC and Stata Command Sequence 359

Interpretation of Predicted Probability Displays 361

Cautions 365

Interpretation Using Displays of Predicted

Standardized Latent Outcomes **365**

Wrap-Up 369

Two-Moderator Example (Interval by Two Nominal) 369

Data and Testing **369**

Strategies for Interpreting a Multiple-Moderator Interaction 370

The Effect of Sex Moderated by Education **371**

INTSPEC Setup and GFI Analysis 371

Discrete Change Effects 373

Predicted Probability Interpretation 375

The Effect of Race/Ethnicity Moderated by Education **379**

INTSPEC Setup and GFI Analysis 380

Factor Change (Odds Ratio) Interpretation 383

Predicted Standardized Latent Outcome Interpretation 387

The Effect of Education Moderated by Race/Ethnicity and by Sex **391**

INTSPEC Setup and GFI Analysis 391

Factor Change Interpretation 392

Interpretation Using Predicted Probabilities 396

Special Topics 405

Getting the Base Probability for a Discrete Change

Effect From SPOST13 **405**

Finding the Standard Deviation and Mean of the

Latent Outcomes (Utilities) **405**

Creating a Stacked Area Chart **407**

Option 1: Using *mgen* in Stata 407

In Excel 408

Chapter 9 Notes 409

10. Ordinal Regression Models **411**

Overview 411

Properties and Use of Ordinal Regression Models **411**

Data and Circumstances When Commonly Used 412

GLM Properties and Coefficient Interpretation for

Ordinal Regression Models 413

Interpretation of Interaction Effects **414**

Diagnostic Tests and Procedures 415

Data Source for Examples **415**

One-Moderator Example (Interval by Nominal) 415

Data and Testing **415**

Education Moderated by Sex **417**

Standardized Change in the Latent Outcome 418

Predicted Change in the Odds of More Frequent Purchase 419

Discrete Change in the Probabilities of Each Purchase Category 419

Effect Displays 420

Sex Moderated by Education **421**

Standardized Change in the Latent Outcome 422

Predicted Change in the Odds of More Frequent Purchase 423

Discrete Change in the Probabilities of Each Purchase Category 423

Effect Displays 424

OUTDISP for the Effects of Education and Sex Simultaneously **426**

Displays of the Predicted Latent Outcome   427

Displays of the Predicted Outcome Category Probabilities   429

Displays of the Predicted Outcome Category Probabilities, With
Superimposed Main Effects   431

Ordinal Probit Results   **433**

Two-Moderator Interaction Example (Nominal by Two Interval)   435

Data and Testing   **435**

Approaches to Interpreting the Two-Moderator Interaction   **436**

**Moderated Effect of Education by Race on Standardized
Class Identification**   **437**

**The Moderated Effect of Log Income by Race as Factor
Changes in the Cumulative Odds**   **438**

**The Moderated Effect of Race by Education and
Race by Log Income**   **441**

Factor Change Effect of Race on the Cumulative Odds
(Higher Versus Lower Class)   442

Discrete Change Effects of Race on the Probability of
Each Class Category   445

**OUTDISP for the Effects of Race, Education, and
Income Simultaneously**   **450**

Predicted Values in the Model Metric (Standardized Latent Outcome)   **451**

Table of Predicted Values   451

Plots of Predicted Values   453

**Predicted Values in the Observed Metric (Probability of
Outcome Categories)**   **457**

Predicted Value Plots   457

Predicted Value Tables   469

Evaluating Confounded Nonlinearities in the Interactive
Effect Model's Predicted Probabilities   472

Special Topics   473

**Testing the Equality of Factor Change Effects for
Different Moderator Values**   **473**

Option 1: Stata *test* Command   473

Option 2: Stata *testnl* Command   475

**How to Calculate the Average Standardized Latent
Outcome by Race Group**   **476**

Chapter 10 Notes   477

11. Count Models   **479**

Overview   479

Properties and Use of Count Models   **479**

Data and Circumstances When Commonly Used   479

GLM Properties and Coefficient Interpretation for Count Models   480

Diagnostic Tests and Procedures   482

Data Source for Examples   **482**

One-Moderator Example (Interval by Nominal)   482

Data and Testing   **482**

Work–Family Conflict Moderated by Occupational Status   **483**

SIGREG Results   485

EFFDISP Results   486

Occupational Status Moderated by Work–Family Conflict   **489**

SIGREG Results   490

EFFDISP Results   491

OUTDISP for Work–Family Conflict and Occupational
    Status Simultaneously 492
Three-Way Interaction Example (Interval by Interval by Nominal) 497
    Data and Testing 497
    Approaches to Interpreting the Three-Way Interaction 498
        Moderated Effect of Age on Log Number of Memberships 499
        Moderated Effect of Education as a Factor Change in Number of
            Memberships 501
        Moderated Effect of Sex as a Discrete Change in
            Number of Memberships 502
    OUTDISP for the Effects of Age, Education, and
        Sex Simultaneously 504
        Predicted Values Tables 506
        Predicted Values Plots 509
Special Topics 516
    Using Predicted Probabilities for Interpretation 516
        Table of the Predicted Probability Distribution of Counts 517
        Plotting the Predicted Probability Distribution of Counts 519
    Working With Interaction Effects in the
        Zero-Inflated Model Component 521
    Standardized Log Count for Poisson and Negative Binomial Models 522
    Getting the Count Value Equivalent to a Standardized Log Count Value 523
Chapter 11 Notes 523

12. Extensions and Final Thoughts 525

Extensions 525
    Interaction of a Polynomial Function of a Predictor
        With Another Predictor 525
        Moderated Effect of Race 527
        Moderated Effect of Age 528
        Tables and Plots of Predicted Values of Education by Age and Race 533
    Models With Censored (Selected) Outcomes 535
    Models for Survival Analysis (Cox Proportional Hazards Example) 542
        GFI and SIGREG for the Effect of Age Moderated by Site 544
        GFI and EFFDISP for the Effect of Site Moderated by Age 545
        OUTDISP for the Interaction of Site and Age 546
        Survival Curves for the Interaction of Site and Age 546
Final Thoughts: Dos, Don'ts, and Cautions 549
    Specifying Terms in the Prediction Function 549
    Interpreting Effects Versus Interpreting Coefficients 550
    Consider the Totality of an Interaction Specification 551
    Comparing Effects 551
    Model Misspecification and Diagnostic Testing 553
Chapter 12 Notes 553

Appendix: Data for Examples 555
Chapter 2: One-Moderator Example 555
Chapter 2: Two-Moderator Mixed Example 556
Chapter 2: Two-Moderator Interval Example 557
Chapter 2: Three-Way Interaction Example 558
Chapter 3: One-Moderator Example 558

Chapter 3: Two-Moderator Example   559

Chapter 3: Three-Way Interaction Example   559

Chapter 4: Tables One-Moderator Example and Figures Example 3   559

Chapter 4: Tables Two-Moderator Example   560

Chapter 4: Figures Examples 1 and 2   560

Chapter 4: Figures Example 4   560

Chapter 4: Tables Three-Way Interaction Example and
    Figures Example 5   560

Chapter 5: Examples 1 and 2   560

Chapter 5: Example 3   560

Chapter 5: Example 4   561

Chapter 6: One-Moderator Example   561

Chapter 6: Two-Moderator Example   561

Chapter 6: Three-Way Interaction Example   562

Chapter 7: One-Moderator Example   562

Chapter 7: Two-Moderator Example   562

Chapter 8: One-Moderator Example   562

Chapter 8: Three-Way Interaction Example   563

Chapter 9: One-Moderator Example   563

Chapter 9: Two-Moderator Example   564

Chapter 10: One-Moderator Example   565

Chapter 10: Two-Moderator Example   567

Chapter 11: One-Moderator Example   568

Chapter 11: Three-Way Interaction Example   568

Chapter 12: Polynomial Example   568

Chapter 12: Heckman Example   569

Chapter 12: Survival Analysis Example   571

References   573

    Data Sources   576

Index   577

SERIES EDITOR'S INTRODUCTION

I am very pleased to introduce Robert Kaufman's volume *Interaction Effects in Linear and Generalized Linear Models: Examples and Applications Using Stata*. An interaction effect is an important statistical concept that shows how the impact of an independent variable on an outcome variable varies by another independent variable, which is a core statistical technique widely applied in today's social behavioral and health research. The primary rationale for specifying, estimating, and testing interaction effects is on theoretical and substantive grounds. Once researchers find statistically significant interaction effects, they must interpret them in a clear, effective, and efficient fashion, because often this is the core finding of current research that shows differential effects, also known as moderating or buffering or accelerating effects, of a key predictor variable among diverse groups. Interpreting interaction effects requires different strategies than those typically designed and employed for interpreting main effects, and it is from here that researchers face challenges.

This volume provides various strategies to ease the burden of interpreting and presenting interaction effects. It is designed to be a supplement for a graduate class in quantitative methods but is also a solid reference for practicing professionals. Although the volume rigorously presents fundamental concepts based on key mathematics, such as partial derivatives, it presents the information in a reader-friendly fashion. The overall strategy of the volume is to lay the foundation by conveying the basic ideas of testing moderating effects and principles of presenting the effects, and then extending to describe how to do it with detailed examples. The focus on interpreting interactive effects for both linear and generalized linear models is a unique feature. In addition to methods for ordinary-least-squares regression, the volume discusses critical approaches for interpreting interaction effects of complicated models, such as logistic, probit, ordinal regression, multinomial logit, Poisson regression, and negative binomial regression. The extension of discussing interaction effects to Tobit model, Heckman's sample selection, and Cox regression, helps readers develop new skills to move forward.

A very attractive feature of the volume is the provision of the ICALC Stata toolkit. With this package, researchers can practice the interpretation principles to define moderated effects (GFI), identify significance regions (SIGREG), create plots to display moderating effects (EFFDISP), and create plots and tables to display interaction effects on the predicted outcome (OUTDISP). Simply put, this to-do toolkit makes the complicated task of interpreting and presenting interaction effects less complicated and enjoyable.

Perhaps the most important feature of the volume is that it provides a unified approach to interpreting interaction effects which can be applied across a wide range of techniques of analysis. In addition, this unified approach is much more comprehensive in its coverage of interpretive tools than is typical of the books currently available for any given single-technique of analysis. For this reason, I believe students of all ages and stages as well as practicing professionals will benefit from this volume.

Shenyang Guo
Series Editor

PREFACE

This book is about how to understand, interpret, and present interaction effects from both linear and generalized linear regression models (GLMs) using four interpretive tools. Despite the commonplace publication of research using interaction, many graduate students, data analysts, and faculty researchers struggle to properly test and interpret interaction effects. The roots of my approach run back to the first graduate statistics classes I taught nearly 40 years ago at the University of Texas at Austin. One of my examples included an interaction effect (not a topic for my course). Students would ask me how to determine how the effect of one of the interacting predictors on the outcome varied with the values of the other interacting predictor(s). Few students in my class knew how to take derivatives, so I developed what I now call the Gather, Factor, and Inspect approach to use simple algebra to define the algebraic expression for how the effect of the first interacting predictor depends on the other predictors with which it interacts. I developed the other interpretive approaches—defining theoretical and empirical significance regions, creating effects displays and specialized outcome displays—over the years, largely as a result of discussion with faculty colleagues and graduate students about how to interpret interaction effects for a variety of statistical techniques, not just ordinary least squares (OLS) regression.

So why did I decide to write a book on interpreting interaction effects? The existing didactic treatments most often discuss only a single analytic technique and cover a limited range of interpretive tools which are illustrated with applications to just the simplest form of interactions. And they rarely provide software code or spreadsheet-friendly formulas to automate the calculations. My goal in writing this book is to provide a unified approach for interpreting interaction effects which is more comprehensive in its coverage of interpretive tools and which applies across a wide range of techniques of analysis. And I created ICALC—for Interaction CALCulator—a downloadable toolkit of Stata routines to produce the calculations, tables, and graphics for each interpretive tool. I provide detailed discussions of how to apply and interpret the results of these tools to both simple and complicated forms of interactions for linear regression models and for a set of commonly used GLMs: binomial logistic regression, probit analysis, multinomial logistic regression, ordinal regression models, and Poisson and negative binomial regression models (including zero-inflated variants).

For these GLMs I particularly focus on the challenges of interpreting tabular and graphical presentations of predicted outcomes. A little-recognized problem is that displays of predicted values confound two sources of nonlinearity simultaneously: the nonlinearity of the interaction effect and the inherent nonlinearity in how these GLMs model the relationship between the outcome and the predictors. It can be difficult to visually distinguish such a graph from a model with an interaction effect from one without an interaction effect in nonlinear models. This confounding hampers an analyst's ability to interpret visual or tabular displays of interaction effects effectively as well as a reader's ability to understand what is presented. I propose and illustrate the use of alternative visual displays designed to help disentangle these two forms of nonlinearity. A similar concern complicates the interpretation of discrete change and marginal change measures of the effects of interacting predictors for GLMs with nonlinear link functions.

Researchers and analysts in a variety of social science and related disciplines—such as Sociology, Political Science, Economics, Psychology, Criminology/Criminal Justice, Urban Studies, Public Health, Communications, Education, and Business—will, I hope, find that the book's development and application of the interpretive tools make this a valuable professional reference book. It should be a worthwhile supplement for graduate level statistics classes in these same disciplines, with subsequent value as a professional reference book as students continue their careers. Although the book and the ICALC Toolkit are written explicitly for Stata users, I provide readers the flexibility to apply these tools and/or to customize tables and graphics on other platforms. In the Principles chapters I present sufficient mathematical details for and application examples of the underlying formulas to enable readers to write their own spreadsheet formulas or software code for the platforms they use. Further, ICALC automatically stores graphics during a session as editable-in-Stata graphs (memory graphs) and provides options to save numeric results, tables, and the underlying data used to construct graphics into an Excel spreadsheet. These features give users the flexibility to customize graphics in Stata or to use the saved data to create their own graphics using other platforms.

A major advantage of the ICALC Toolkit for Stata users is that the examples and applications throughout the book use ICALC to produce the results, with annotated explanations of the use of the commands in the application chapters. Moreover, the data and Stata do-files for all the examples are available to download online at www.icalcrlk.com so users can confirm their understanding of how to use ICALC by reproducing results for the examples. Users working on other platforms can use the examples to validate their applications by replicating the results from the examples.

ACKNOWLEDGMENTS

I owe thanks to many people who have contributed to the process of writing and publishing this book and to its content. I am deeply grateful to Lauren Krivo for her advice, support, and encouragement—both professional and personal—as I have worked, often obsessively, on this project for the past two and a half years. Helen Salmon, my editor at SAGE, has been an enthusiastic supporter and cheerleader for this book since I first broached the idea. And it has been a pleasure to work with her and the other editorial staff at SAGE again, especially Jane Martinez and Megan O'Heffernan but also Bennie Allen, Danielle Janke, Shari Countryman, Susannah Goldes, and Andrew Lee. My copy editor from QuADS Prepress, Rajasree Ghosh, was always responsive to questions and concerns, and her careful work greatly improved the clarity and consistency of the content. I am also indebted to the reviewers from SAGE for their quite helpful criticism of the initial proposal and, subsequently, of draft chapters of the book and to the Advanced Quantitative Techniques in the Social Sciences series editor, Shenyang Guo, for similarly constructive feedback. I almost always agreed with and implemented their suggestions in some fashion, and I can honestly say that the end result is much better than it would have been otherwise. Robert DePhillips provided excellent proofing and copyediting of most of the chapters as I prepared them for the final review. Leslie Krivo-Kaufman gave valuable advice and set up the website to host the ICALC Stata add-on companion to this book. I am indirectly grateful to Scott Long and Jeremy Freese for two reasons. They created the SPOST13 suite of Stata add-on commands on which parts of my ICALC commands rely. In addition, the clear and accessible writing style of their book *Regression Models for Categorical Dependent Variables Using Stata* (third edition) inspired me in my approach to writing. Josh Klugman and Libby Luth provided useful feedback on very early versions of some ICALC routines. And last but not least, I am grateful to the Department of Sociology at Temple University for the sabbatical that gave me the time and resources to begin working on this book in earnest.

SAGE and the author would like to thank the following reviewers for their time and feedback:

Wen Fan, *Boston College*

Andrew Fullerton, *Oklahoma State University*

Jennifer Hayes Clark, *University of Houston*

Nicole Kalaf-Hughes, *Bowling Green State University*

To Laurie, Leslie, and Alana for their love and encouragement.

And to the many graduate students and colleagues who have sought my advice on interaction effects over the years.

ABOUT THE AUTHOR

Robert L. Kaufman (PhD, University of Wisconsin, 1981) is professor emeritus of sociology and past Chair of the Department of Sociology at Temple University. His substantive research focuses on economic structure and labor market inequality, especially with respect to race, ethnicity, and gender. He has also explored other realms of race/ethnic inequality, including research on wealth, home equity, residential segregation, traffic stops and treatment by police, and media portrayals of crime. Much of his research on applied statistics is motivated by specific substantive problems. But he also has a long-standing interest in the general topic of techniques for testing, interpreting, and presenting quantitative results in an appropriate and intuitive fashion, such as the detection and correction of heteroscedasticity and the interpretation of interaction effects. He has published a monograph in the SAGE University Series on *Quantitative Applications in the Social Sciences* and papers on quantitative methods in *American Sociological Review, American Journal of Sociology, Sociological Methodology, Sociological Methods and Research*, and *Social Science Quarterly*. He served on the editorial board of *Sociological Methods and Research* for 15 years and has taught graduate-level statistics courses almost every year for the past 38 years.

Sara Miller McCune founded SAGE Publishing in 1965 to support the dissemination of usable knowledge and educate a global community. SAGE publishes more than 1000 journals and over 800 new books each year, spanning a wide range of subject areas. Our growing selection of library products includes archives, data, case studies and video. SAGE remains majority owned by our founder and after her lifetime will become owned by a charitable trust that secures the company's continued independence.

Los Angeles | London | New Delhi | Singapore | Washington DC | Melbourne

INTRODUCTION
AND BACKGROUND

My goal is to correctly discuss an interaction term before I die.

—**Confidential Dissertator (ca. 2000)**

OVERVIEW: WHY SHOULD
YOU READ THIS BOOK?

The inevitable question that the author of a statistics book like this has to address is whether another book on interaction effects is necessary—why a reader should read it— given that analyses incorporating both simple and more complicated interaction effects are commonplace. My first answer is embodied in the chapter quote above, which is as true today as it was nearly 20 years ago. Many graduate students as well as post-PhD researchers continue to have difficulty properly testing, interpreting, or specifying interaction effects in linear regression, let alone for nonlinear regression techniques.

This is evident in the apparent short life cycle of publications on how to interpret interaction effects properly. Every 5 to 10 years, there is a renewed call for researchers to follow best practices to avoid common problems. Each explicitly argues that these points are important to continue to reiterate because they still do not consistently inform actual practice (e.g., Aiken & West, 1991; Brambor, Clark, & Golder, 2006; Braumoeller, 2004; Dawson, 2014; Hayes, Glynn, & Huge, 2012; Jaccard, 1998; Jaccard & Turisi, 2003; Kam & Franzese, 2007; Southwood, 1978). When quantitatively oriented colleagues and graduate students learn that I am writing a book on the interpretation of interaction effects, their typical reaction is along the lines of "Great. My students could really use it and so could I. When will it be out so we can read it?"

Second, I identify and discuss solutions to a little recognized problem with tabular and graphical presentations of predicted outcomes derived from analyses using many types of generalized linear models (GLMs). Look at the two plots in Figure 1.1 from a negative binomial regression (a count model) predicting the number of voluntary associations in which the respondent is a member. The only difference is that one model contains only linear terms for age, education, and sex (main effects) and the other also includes product terms between every pair of these predictors and a product of *Age × Education × Sex*—that is, a three-way interaction. Just from examining the plots, can you convince yourself which plot shows the main effect model's predictions and which portrays the interactive effect model's predictions? Is it Plot B

FIGURE 1.1 ● PREDICTED VALUE PLOTS FROM MAIN EFFECTS AND INTERACTIVE EFFECTS MODELS

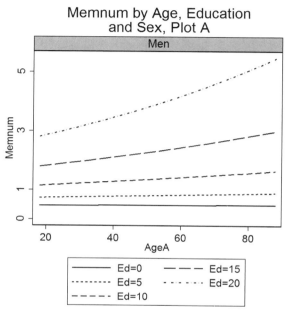

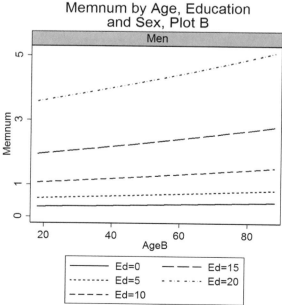

showing interaction effects in which education mutes the effect of age for men? Or is it Plot A representing education as enhancing the effect of age for men?

Moreover, if you were given only the main effects plot (whichever it is), can you honestly say you would identify it as representing a main effects model and not an interactive effects model? My point is that it can be difficult to visually distinguish a graph from a model with an interaction effect from one without an interaction effect in nonlinear models. And this hampers an analyst's ability to interpret visual or tabular displays of interaction effects effectively as well as a reader's ability to understand what is presented. The problem is that a plot or a table of predicted values incorporates and portrays two sources of nonlinearity simultaneously: the nonlinearity of the interaction effect and the inherent nonlinearity of many GLMs in how they model the relationship between the outcome and the predictors. This issue and recommended solutions are the topic of Chapter 5.

Moreover, four limitations of the current didactic literature (Aiken & West, 1991; Brambor et al., 2006; Braumoeller, 2004; Dawson, 2014; Hayes, 2013; Hayes et al., 2012; Jaccard, 2001; Jaccard & Turisi, 2003; Jaccard & Wan, 1996; Kam & Franzese, 2007; Southwood, 1978) motivated me to write this book and shaped its content to overcome and avoid these problems. Specifically, existing treatments are subject to at least one and usually more of the following shortcomings:

- They address a limited range of analytic models—often a single technique, most commonly ordinary least squares (OLS) regression.

- They provide details and examples only for the simplest interaction effect— a focal variable with a single moderator (often a two-category nominal predictor)—leaving readers to extend the approach to more complicated interaction effects themselves.

- They do not cover a wide range of tools for interpreting interaction effects.

- They provide limited, if any, assistance, with rare exceptions, for automating the calculations needed for many of the tools for interpreting interaction effects (i.e., software code/programs or spreadsheet-friendly formulas). And if they do so, it is specific to a particular technique of analysis.

These limitations result in piecemeal knowledge in which practitioners learn how to use some interpretive and calculating tools for one technique but different ones for another. Moreover, there is considerable (wasted) effort as researchers reinvent the wheel by creating their own specialized programs or spreadsheets for doing calculations and creating tables and graphics.

Consequently, my goal in writing this book is to provide a unified approach for interpreting interaction effects, which is more comprehensive in its coverage of interpretive tools as well as applicable across a wide range of techniques of analysis. I have also created a set of Stata routines (ado files) named ICALC—for Interaction CALculator— to apply the interpretive tools I discuss in this book. ICALC can be downloaded free of charge at **www.icalcrlk.com** as can the data sets and Stata syntax (do-files) for all the book's examples. Readers may find it helpful to follow along in the do-files as they read the application examples. I chose Stata for this platform primarily because I consider this book an intellectual companion to Long and Freese's (2014)

Stata is a registered trademark of StataCorp LLC.

book on interpretive techniques for common nonlinear models. And ICALC uses their SPOST13 suite of tools written for Stata for many of its calculations. Moreover, Stata is, by most indicators, one of the top four package platforms for data analysis, and unlike IBM® SPSS® Statistics and SAS®, its popularity is growing, not declining (Muenchen, 2017). You can download and install the ICALC Toolkit when you are running Stata with online access. In the Stata Command window, type *net search icalc*. The Stata Results window should show four packages. Click the link for *icalc_ado* to install the ICALC program and help files in your PERSONAL directory. Use the same process if you need to install SPOST13, which I would highly recommend. To provide readers the flexibility to apply these tools on other platforms, I strive to present sufficient mathematical details for and application examples of the underlying formulas to enable readers to write their own spreadsheet formulas or software code for the platform they use. Furthermore, ICALC automatically stores graphics during a session as editable-in-Stata graphs (memory graphs) and provides options to save numeric results, tables, and the underlying data used to construct graphics in an Excel spreadsheet. These features give users the flexibility to customize graphics in Stata or to use the saved data to create their own graphics using other platforms.

In the next several sections, I review basic background material about interaction effects, GLMs, and relevant statistical and diagnostic tests. I would recommend that readers well-versed in these topics at least skim through this material to ensure that nothing essential is missed. In particular, the section on confounded nonlinearities in GLMs is likely an unfamiliar issue. I conclude the chapter with a roadmap of the content and organization of the rest of the book.

THE LOGIC OF INTERACTION EFFECTS IN LINEAR REGRESSION MODELS

Let me start by answering four basic questions about interaction effects: (1) What is an interaction effect? (2) Why should you consider including an interaction effect in your analysis? (3) How do you specify an interaction effect in the prediction function of a linear regression model? (4) When is an interaction effect statistically significant?

What Is an Interaction Effect?

Conceptually, an interaction effect is a way of specifying that the relationship between the outcome and a first predictor, call it F, is contingent on the values of another predictor, call it M_1. Or, to put it a different way, the effect of F on the outcome varies with the values of M_1. For example, an economist might argue that the earnings return to work experience is greater for a worker with more education than for a worker with less education. That is, the earnings–experience relationship is different depending on the level of a worker's education; a worker with more years of education would receive a higher payoff to his or her work experience. Or a legal scholar might argue that the effect of education on approval of a legal ban against racial intermarriage differs by race. Specifically, education is more consequential for Whites' approval of a legal ban against racial intermarriage—it reduces their approval more—than it is for Blacks' approval.

The first predictor, F, is often labeled the focal variable in the interaction, and the second predictor, M_1, is often referred to as the moderator or the moderating variable. This corresponds to the fact that interaction effects are frequently developed with a primary theoretical or conceptual focus on one of the predictor's effects and how those effects are moderated by the second predictor. For this reason, interaction effects are sometimes called moderated effects (e.g., Hayes, 2013; Jaccard, 1998; Jaccard & Turisi, 2003).

This is a very useful heuristic device that I also adopt throughout the book, but the roles of moderating and focal variables should not be reified. They are arbitrary from a statistical point of view because the contingency of the *Outcome–F–M*$_1$ relationship works both ways. That is, specifying that the effect of *F* is contingent on *M*$_1$ also specifies that the effect of *M*$_1$ is contingent on *F*. Thus, in the first example, it is equally valid to argue that the earnings–education relationship is contingent on a worker's years of work experience. And the flipside in the second example is that the effect of race on approval varies with education.

Interaction effects are not limited to a single pair of predictors. You could specify that *F*'s effect on the outcome changes with two other predictors separately—a two-moderator model—or that it is dependent on the specific combination of values of the other two predictors—a three-way interaction. In the race–education–legal ban example, the legal scholar might extend the original hypothesis to argue that the effect of race varies not only by education but also by region of residence. A two-moderator model would specify that the race-by-education effect on approval is the same in each region, that the race-by-region effect is the same at each level of education, and that there is no interaction between education and region. In contrast, a three-way interaction model would specify that the race-by-education effect on approval differs across regions, that the race-by-region effect varies with education, and that the education-by-region effect varies by race.

Why Should You Consider Including an Interaction Effect in Your Analysis?

With some exceptions in practice, the primary rationale for estimating and testing interaction effects is on theoretical or substantive grounds. That is, you have developed new hypotheses or expectations that certain predictors should have contingent effects. But it is also conventional to include interaction effects to reflect current knowledge in the literature. Additionally, in many of the social sciences (certainly in my discipline of sociology), it is commonplace to propose and analyze outcome differences between groups defined by their social characteristics or statuses such as race/ethnicity, sex or gender, sexual orientation/identity, class, and so on. In the course of developing the rationale for group differences, it is not unusual (and often purposeful) to make an argument that groups diverge in outcome levels because different factors are important for some groups, or the same factors have varying effects. Both these arguments create an expectation that interaction effects exist and set the stage for testing for interactions between the groups and at least some of the predictors.

Diagnostic testing of model fit or for model misspecification sometimes provides the grounds for estimating and testing for interaction effects. For example, finding a significant diagnostic test for the presence of heteroscedasticity might instead indicate the presence of an interaction effect or some other misspecification of the model's functional form (Fox, 2008, p. 274; Greene, 2008, pp. 166–167; Kaufman, 2013, pp. 22–23). Similarly, a residuals analysis could find issues with the functional form of two predictors that might suggest an unspecified interaction effect.

Even the failure to find a significant effect for a predictor might lead you to test for interaction. When a predictor has opposite-signed effects for two groups—or changes sign across the range of its moderator—this can easily average out to a nonsignificant test of its effect. In such circumstances, I would recommend that you think seriously about the substantive sensibility of specifying an interaction effect before testing for its presence. Other specifications or corrective actions might be more conceptually appealing. And adding interaction effects as a result of data mining runs the risk of overfitting and hence misspecifying the model.

How Do You Specify an Interaction Effect in the Prediction Function of a Linear Regression Model?

To include an interaction effect, you model the outcome Y as a linear function of the focal and moderating variables (F and M_j), their product terms, and a set of other predictors. The coefficients for the focal and moderating variables are commonly referred to as the "main effects" of the predictors, while the coefficients for the product terms are called "interaction effects." Your model should, with rare exceptions, satisfy the principle of marginality (Fox, 2008; Nelder, 1977), also known as the criteria for a hierarchically well-formulated model (Jaccard, 2001; Kleinbaum, 1992):

> [This] specifies that a model including a *higher order term* (such as an interaction) should normally also include the "lower-order relatives" of that term (the main effects that "compose" the interaction). (Fox, 2008, p. 135)

Table 1.1 lists the predictors your model should contain for several forms of interaction effects to adhere to the principle of marginality. Specifically, you add a product term formed by multiplying together all the predictors in the highest order interaction as well as product terms for every lower order relative. For a one-moderator model, you add a predictor defined as the product of F and M_1. Notice that I said "add" a predictor because you keep the individual predictors in the model when you include the product term following the principle of marginality. Similarly, for a two-moderator model, you have three individual predictors (F, M_1, and M_2) plus two product terms ($F \times M_1$ and $F \times M_2$). For a three-way interaction model, you have three individual predictors (main effects), three pairs of product terms among the three predictors (lower order relatives), and a product term for $F \times M_1 \times M_2$ (highest order term).

TABLE 1.1 ● PREDICTORS INCLUDED FOR FORMS OF INTERACTION MODELS	
Interaction Form	**Predictors in Model for Interaction Specification**
One moderator	F, M_1, $F \times M_1$
Two moderators	F, M_1, M_2, $F \times M_1$, $F \times M_2$
Three-way interaction	F, M_1, M_2, $F \times M_1$, $F \times M_2$, $M_1 \times M_2$, $F \times M_1 \times M_2$
Two moderators, M_2 categorical	F, M_1, $D_{M_2=1}$, $D_{M_2=2}$, $D_{M_2=3}$ $F \times M_1$ $F \times D_{M_2=1}$, $F \times D_{M_2=2}$, $F \times D_{M_2=3}$

How does adding product terms work to create contingent effects of the interacting variables? For simplicity and specificity, let's work with a single-moderator specification:

$$Y = a + b_1 F + b_2 M_1 + b_3 F \times M_1 + \cdots \qquad (1.1)$$

Mathematically, the effect of a predictor on Y in a linear regression model is defined as the partial derivative of Y with respect to a predictor,[1] which is the slope of the regression plane. The partial derivative is equal to a predictor's estimated coefficient if the predictor is not part of an interaction specification (or a multiple-variable functional form, e.g., a parabolic effect). For an interaction specification, the partial derivative gives the effect of F as the main effect coefficient for F (b_1) plus the coefficient for the product term $F \times M_1$ (b_3) multiplied by the value of M_1:

$$\text{Effect of } F = \frac{\partial Y}{\partial F} = b_1 + b_3 M_1 \qquad (1.2)$$

This tells us very concretely how the effect of F on Y changes with the value of the moderator. Similarly, the effect of M_1 found by taking the partial derivative of Y with respect to M_1 is

$$\text{Effect of } M_1 = \frac{\partial Y}{\partial M_1} = b_2 + b_3 F \qquad (1.3)$$

Using these formulas, we can determine the nature and shape of the moderated effect of F (or M_1) on the outcome and how those change across different values of the moderator. By nature and shape of the effect, I mean the direction of the effect of F (positive or negative), whether it changes sign for different values of M_1, whether it changes significance across the values of M_1, and whether the moderated effect is ordinal or disordinal (see the Aside). How to probe the nature and shape of the interaction is covered in depth in Chapters 2 to 5.

ASIDE: ORDINAL AND DISORDINAL INTERACTIONS

Consider drawing a set of prediction lines for the value of Y plotted against F, each prediction line for a different value of M_1. These prediction lines will all cross at the same value of F (Jaccard & Turisi, 2003, p. 78). The interaction is ordinal if that value of F does not fall within the sample range of values of F. It is disordinal if the crossover value lies within the sample range of F. Conceptually, a disordinal interaction means that the outcome for a given value of the focal variable F and a specific value of the moderator M_1 is sometimes greater but is sometimes less than the outcome for the same value of F and a different value of M_1. And an ordinal interaction means that the outcome for a given value of the moderator M_1 is always greater (or always less) than the outcome for a different value of the moderator, for any value of F in the sample range.

What if one or all of your predictors are categorical? For each categorical predictor listed in Table 1.1, you would replace the single-variable expression with the set of binary indicators for your predictor. When you create product terms between two predictors, you multiply every term in the first predictor's expression by every term in the second predictor's expression. To illustrate, the last row in Table 1.1 shows the terms included in a two-moderator interaction specification if F and M_1 are interval variables but M_2 is a three-category construct represented by dummy variables for Categories 1 and 2 ($D_{M_2=1}$ and $D_{M_2=2}$; Category 3 is the reference category). To create the list of included predictors, you replace every occurrence of M_2 with its set of dummy indicators. For example, in the expression $F \times M_2$, you replace M_2 with $D_{M_2=1}$ and $D_{M_2=2}$ and multiply by F, which gives two product terms to include in the model: $F \times D_{M_2=1}$ and $F \times D_{M_2=2}$. The corresponding prediction function becomes

$$Y = a + b_1 F + b_2 M_1 + b_3 F \times M_1 + b_4 D_{M_2=1} + b_5 D_{M_2=2} + \\ b_6 F \times D_{M_2=1} + b_7 F \times D_{M_2=2} + \cdots \tag{1.4}$$

Taking the partial derivative of this prediction function, the moderated effect of F is

$$\text{Effect of } F = \frac{dy}{dF} = b_1 + b_3 M_1 + b_6 D_{M_2=1} + b_7 D_{M_2=2} + \cdots \tag{1.5}$$

Because this is a two-moderator interaction, the effect of F varies corresponding to both the values of M_1 and the categories of M_2. As I elaborate in Chapter 2, the expressions for the moderated effect of F like those in Equations 1.2, 1.3, and 1.5 are an important basis for understanding and interpreting interaction effects.

A path–style diagram can be a useful device to show succinctly the nature and form of your interaction specification, especially for more complicated interaction specifications or when some of your interacting predictors are categorical. Writing out the prediction function in terms of all the component predictors that constitute it is essential for running your analysis, and such an equation communicates the form of the interaction well to mathematically or formulaically inclined readers. But for many readers, a diagram like Figure 1.2 for the two-moderator example is much more comprehensible.

FIGURE 1.2 ● PATH–STYLE DIAGRAM OF INTERACTION EFFECT

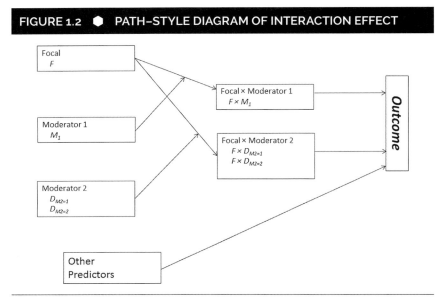

The path diagram has boxes for the focal and moderator variables in the leftmost column and for the two-way interaction terms in the second column, and the outcome is a vertical box on the right. There is a box for each construct in the interaction, and the box indicates whether the construct is measured by a single predictor or a set of indicators. The interaction terms are expressed as *Focal × Moderator* variable names to make the conceptual and mathematical relationships clearer. Lines connect main effect terms to the relevant two-way interaction terms and connect the two-way terms to the outcome. Intersecting lines—where one line stops with its arrowhead on the other line—indicate that the effects of the corresponding variables interact. This provides a visual and conceptual map of the interaction specification.

The diagram shows that F and M_1 interact because their lines intersect and lead to their corresponding two-way interaction term. Similarly, F and M_2 interact. But M_1 and M_2 do not interact because their lines do not intersect. The diagram shows a nonintersecting line from the "Other Predictors" box to the "Outcome" box as a reminder that a model usually has more than just the interacting predictors. In practice, the estimated coefficients could be shown in the diagram.

When Is an Interaction Effect Statistically Significant?

Part of the answer to why include an interaction specification is that it is statistically significant, but what does that mean in terms of what you specifically test? The essential statistical test is whether or not the coefficient(s) for the highest order term in the interaction are significant. If the coefficient(s) are not significantly different from zero, you would conclude that the lower order effects on Y do not vary at that highest level. The significance or lack of significance of any lower order term constituting the interaction is irrelevant to this decision (Aiken & West, 1991, p. 50).

For example, in a one-moderator model in which F and M_1 are interval level, the highest order term is $F \times M_1$, and if its coefficient is not significant then the effect of F does not depend on M_1 and you should use a model without the interaction term. If the $F \times M_1$ coefficient is significant, you would conclude that the effect of F varies with M_1 and use the interactive model results. The significance of the main effect of F is not relevant because this tests whether F's effect is significantly different from zero when $M_1 = 0$. To see why the main effect coefficient is the effect of F when $M_1 = 0$, look back at Equation 1.2, which defines the effect of F. Substitute 0 as the value for M_1, and this leaves b_1, the main effect coefficient for F:

$$\text{Effect of } F = b_1 + b_3 \times 0 = b_1 \tag{1.6}$$

This means that b_1 is the effect of F when M_1 takes on the value of zero and tells you nothing about the significance of F's effect at any other value of M_1. Its significance may or may not be a substantively interesting finding by itself, but it is not informative about the statistical grounds for including the interaction in your model. The same holds true for testing the main effect of M_1: This tests the effect of M_1 when $F = 0$.

What if you were testing a three-way interaction? You would test the coefficient(s) for the product of the three predictors ($F \times M_1 \times M_2$), and if it is significant keep all the interaction terms in your prediction equation. The significance of the two-way product terms, as well as the main effect terms, is not relevant to this decision because they are tests of effects when the moderating variables equal zero (Aiken & West, 1991, p. 50).

How you test the significance of the highest order term depends on whether or not you have directional hypotheses about the coefficients of the highest order term and on whether it consists of a single coefficient or multiple coefficients. Table 1.2 summarizes the commonly used test statistics for different combinations of the testing situations. When you have directional hypotheses, there is no choice; you use a z test or a t test, as appropriate for your estimation technique. If you are testing a single coefficient, then the decision rule is straightforward. Keep the highest order interaction term in your model if its coefficient is significant and exclude the term if it is not. If you are testing multiple coefficients, it is typically because the focal or one or more of the moderating variables is nominal with three or more categories. This creates multiple product terms in the highest order interaction, like the two-way interaction of F and M_2 in Equation 1.4. In this case, you need to decide before you conduct the statistical testing what results constitute sufficient grounds for keeping the interaction.

A stringent rule (rarely applied to my knowledge) would be that all of the directional tests are significant. A common approach is to keep the set of coefficients in the model if any one of them is significant, with a Bonferroni or other correction of the significance level for conducting multiple tests.[2] One complication is that a different choice of reference category (or an alternative parameterization) will on occasion lead to a different conclusion. For this reason, an alternative approach is to supplement the tests of individual coefficients with a global nondirectional test such as the likelihood ratio (LR) test or the Wald test. While a global test is nondirectional, it is unaffected by the choice of reference category and provides a guard against excluding a significant interaction by your choice of a reference category. The decision rule in this scenario is to keep the set of coefficients in the model if either the multiple coefficient tests or the global test yields a significant result.

For nondirectional hypotheses, more choices are possible when testing a single coefficient: a z or t test of the coefficient, a Wald test, or the LR test. However, note that the Wald test and the single-coefficient test will always give identical results.[3] Thus, it does not matter which one you use. For testing multiple coefficients, a global test (Wald or LR test) is generally preferred over conducting multiple t tests on the individual coefficients for the reason given earlier: The global test is robust against changing the reference category or other equivalent reparameterizations.

But using just a global test will occasionally find the set of coefficients not significantly different from zero when one or more of the individual coefficients is significant and of substantive interest. Thus, some analysts use the same either/or decision rule described earlier: Include the interaction if the highest order terms are significant

TABLE 1.2 ● **TESTS OF THE HIGHEST ORDER INTERACTION TERM(S)**

Test	Directional Hypotheses		Nondirectional Hypotheses	
	Single	Multiple	Single	Multiple
z test/t test	Preferred	Preferred	Yes	Supplement
Wald test		Supplement	Yes	Yes
LR test		Supplement	Preferred MLE	Preferred MLE

Note: LR = likelihood ratio; MLE = maximum-likelihood estimation.

using the global test or if at least one coefficient is significant from the multiple tests of single coefficients using a z or t test.

The LR test is preferred over the Wald test for nonlinear models and/or techniques of analysis using maximum-likelihood estimation, especially with a large sample size. The minor disadvantage is that you must run the model twice—once with all of the interaction terms and once excluding only the term(s) for the highest order interaction—and then test the change in the log likelihood between the two models. Some analysts prefer the Wald test because it is asymptotically equivalent to the LR test, and you do not have to make strong distributional assumptions as you do when using the LR test. Additionally, the Wald test does not require estimating two models, which is pragmatically an advantage only in those relatively rare instances in which model estimation takes a significant amount of time or if you are applying the test repeatedly.

What should you do if you test the highest order terms of, say, a three-way interaction ($F \times M_1 \times M_2$), and it is not significant? You would conclude that the effect of F does not vary across different combinations of the values of M_1 and M_2. But F's effect might vary with M_1 regardless of M_2's values and with M_2 regardless of M_1. You would test those possibilities by running a two-moderator model and testing each of those two-way terms ($F \times M_1$, $F \times M_2$, $M_1 \times M_2$) and keep a two-way term only if it is individually significant.

Common Errors in Specifying and Interpreting Interaction Effects

For concreteness in describing and discussing these errors, consider a single-moderator interaction specification in which number of children is predicted by the interaction of family income and a dummy indicator of birth cohort (1 = *Pre–Baby Boom*, 0 = *Baby Boom and younger*):

$$Children = a + b_1 Income + b_2 Cohort + b_3 Income \times Cohort + \cdots$$
$$Children = 4.4783 - 0.1075 Income - 1.0251 Cohort +$$
$$0.0327 Income \times Cohort + \cdots \qquad (1.7)$$

Excluding Lower Order Terms

The most frequent mistake analysts make in specifying an interaction effect is to not include in the prediction function all of the lower order terms, sometimes referred to as constitutive terms (Aiken & West, 1991; Brambor et al., 2006; Braumoeller, 2004; Jaccard, 2001; Kam & Franzese, 2007). Doing so violates the principle of marginality and fundamentally changes the meaning, estimated values, and/or statistical test results of the coefficients for the other terms in the interaction specification. Perhaps the most consequential issue is the potential for model misspecification (Kam & Franzese, 2007, pp. 100–101). Unless you have both a clear theoretical argument for this exclusion and a high certainty that empirically $b_2 = 0$, you run the risk of an omitted-variable bias affecting the estimates of any predictor that is correlated with the omitted variable.

Suppose the main effect of cohort was excluded from the prediction equation. Because *Income × Cohort* is very likely to be positively correlated with cohort, the effect of cohort will be partly attributed (added) to the *Income × Cohort* effect. Depending on the sign of the (omitted) effect of cohort and of *Income × Cohort*, this exclusion would either increase or decrease the estimated coefficient for *Income × Cohort*. Thus, the significance test of the interactive model of income and cohort versus a linear model of effects will be biased as well. Moreover, this exclusion constrains the prediction line for children plotted against income for each birth cohort to have the same intercept but a varying slope, which Fox (2008) aptly describes, for a different empirical example, as "a specification that is peculiar and of no substantive interest" (p. 138).

A related question that invariably comes up when I first introduce students (or colleagues) to interaction effects is whether you should exclude from the model nonsignificant lower order terms if the highest order term is significant. The proposed rationale for doing so is that keeping the lower order terms in the model unnecessarily inflates the standard errors of the other predictors (decreases the efficiency of estimates). However, the consequences of excluding the lower order terms when they should be included are much more consequential—biased coefficient estimates, as I just discussed.

The more fundamental problem with deciding to exclude nonsignificant lower order terms is that you could well make the opposite decision if you had a mathematically equivalent but different specification of your interaction model. In the example, suppose the main effect of income (b_1) is significant, so you don't think about excluding it from the prediction equation. Someone else does the same analysis except that instead of older cohort (1 = *older cohorts* and 0 = *most recent*) he or she uses recent (1 = *most recent cohorts* and 0 = *older cohorts*) and their prediction equation is

$$Children = a^* + b_1^* Income + b_2^* Recent + b_3^* Income \times Recent + \cdots \qquad (1.8)$$

But they find the main effect of income (b_1^*) is not significant, so they decide to exclude the main effect term for income from the prediction equation. The problem with these opposite decisions is that the two prediction equations are mathematically equivalent to each other—you can derive the coefficients of one from the coefficients of the other. To see this, realize that *Recent* = 1 − *Cohort* and rearrange the terms:

$$Children = a^* + b_1^* \, Income + b_2^* \left(1 - Cohort\right) + b_3^* \, Income \times \left(1 - Cohort\right) + \cdots$$
$$Children = \left(a^* + b_2^*\right) + \left(b_1^* + b_3^*\right) Income - b_2^* \, Cohort - b_3^* \, Income \times Cohort + \cdots \qquad (1.9)$$

We can now write the coefficients for the original parameterization in terms of the alternative parameterization:

$$a = a^* + b_2^* \qquad b_1 = b_1^* + b_3^* \qquad b_2 = -b_2^* \qquad b_3 = -b_3^*$$

The point is that the two analyses test different things when they test the main effect of income in their prediction equation. The first analysis tests the effect of income when *Cohort* = 0 and the second tests the effect of income when *Cohort* = 1 (i.e., *Recent* = 0). The underlying conceptual problem with excluding the main effect term is that it serves to anchor the moderated effect of income by setting a value for the effect when its moderator (whether cohort or recent) takes on the value of zero. When you change between equivalent parameterizations, the anchor value adjusts so that you get the same moderated effect of income. So you need to keep it in the model as long as the higher order term is in the model. (For a more detailed discussion, see Aiken & West, 1991; Allison, 1977; Brambor et al., 2006; Kam & Franzese, 2007.)

Interpreting Coefficients as Unconditional Marginal Effects

Because the lower order and higher order terms in the interaction specification are functionally related, you cannot interpret the coefficient for any interaction term without considering how it is related to the other interacting predictors (Aiken & West, 1991; Allison, 1977; Brambor et al., 2006; Braumoeller, 2004). Consider the numeric coefficient of −1.0251 for the main effect of cohort in Equation 1.7. The incorrect way to interpret this would be to say that the predicted number of children for the older cohorts is about one less child than for the younger cohorts. This depicts an unconditional relationship between cohort and children and ignores that the effect of cohort is functionally related to the interaction of *Cohort × Income*.

A correct interpretation would be that when *Income* = 0, the predicted number of children for the older cohorts is about one child less than for the younger cohorts. Notice how this statement makes clear that income values define a contingent relationship of cohort to children. But it does not tell the full story of that contingency. A better description would be that the predicted number of children for the older cohorts is about one child less than for the younger cohorts when *Income* = 0 and is predicted to increase by 0.0327 children for each $10K increase in income. Equivalently, and perhaps more informative, would be to replace the end of the sentence with "… *Income* = 0 and is predicted to increase by about one third of a child for a $100K increase in income."

Interpreting Main Effect Coefficients
When Not Meaningful and the Myth of Centering

It is always technically correct to interpret a main effect coefficient as the effect of the predictor when its moderator is equal to zero. But that technical interpretation is not always meaningful—in particular, when zero is not a possible value for the moderator or if zero is possible but is not within the sample range for your analysis. In those cases, the interpretation is not meaningful and may be confusing. For example, suppose you are analyzing how household size moderates the effect of rent on savings. The minimum household size is one, so it would not be sensible to interpret the effect of rent for a household with zero people.

This is one reason why some didactic works on interaction effects recommend centering an interval (continuous) predictor by subtracting its sample mean before running the analysis (Aiken & West, 1991; Dawson, 2014; Hayes, 2013; Jaccard & Turrisi, 2003). If a predictor's moderator is centered, then the main effect of the predictor is its effect when the centered moderator equals zero—which is to say, when the moderator equals its mean. In this case, the main effect term will have a meaningful interpretation. In the household size by rent example, the main effect coefficient for rent would now be meaningful because it would refer to the effect of rent when household size is equal to its sample mean. But the numeric value, interpretation, and statistical significance of the moderated effect (Equation 1.5) is the same whether or not you center your predictors (Kam & Franzese, 2007, p. 97).

The myth of centering refers to the second reason often given for centering; namely, that it supposedly reduces problems of collinearity between the components of the interaction specification. The reality is that centering makes no real difference in the estimation of the parameters of the model. Some coefficients and their meanings change because you changed the measurement of your predictors but not, as noted above, the overall moderated effect of the predictor. The minor and rare exception is that if the collinearity is so extreme that the parameters for the uncentered model cannot be estimated, it is possible but not likely that centering could make enough of a difference to estimate the parameters.

In any case, you can mathematically derive the uncentered coefficients and their standard errors from the centered coefficients, their standard errors, and the covariance between the coefficients, and vice versa (see Kam & Franzese, 2007, pp. 96–98). A point that is often overlooked is that collinearity among the components of an interaction specification is normal and expected. The predictors are functionally related in how they create the overall moderated effect and should be in many instances highly correlated. We should not expect to get precise estimates of the coefficients for the individual components of an interaction because there is by definition a mathematical relationship between the magnitude of the coefficients—each depends on the values of the other predictors in the interaction.

I want to clarify that I am not arguing against examining your data for potential problems of collinearity among your predictors by means of standard diagnostic tools, such as the presence of very high bivariate correlations, variance inflation indices, or the Belsley, Kuh, and Welsch (1980) collinearity diagnostics. Just do not be concerned about potential multicollinearity among the component predictors of an interaction specification.

In sum, there is neither any harm in centering your predictors nor any major advantage. The one gain is that centering can give a main effect coefficient a meaningful interpretation when a moderator value of zero is not possible. In general, the ability to meaningfully interpret the main effect can be useful in some analyses. This is invariably true when the moderator is a dummy variable indicator or a set of dummy variable indicators (which you would not center) because the value of zero corresponds to the reference category. For example, the main effect of income on children is the income effect for recent cohorts ($Cohort = 0$).

Not Interpreting the Moderated Effect of Each Predictor Constituting an Interaction

Brambor et al. (2006) document that many articles in top political science journals that report interaction effect analyses fail to provide an overall picture of the nature and significance of the moderated effect (Equation 1.5). In practice, many analyses only report and discuss the effect and statistical significance of a predictor when its moderator is equal to zero rather than reporting and describing how the effect and its significance changes across the range of the moderator (see the brief example discussed earlier in which I described the effect of cohort as it varies with income). This leads to incomplete and possibly misleading conclusions about the moderated relationship and the hypothesis it represents.

A related issue is a tendency to reify the heuristic device of designating one of the interacting predictors as the focal variable and the others as the moderating variables (also known as conditioning variables). This leads to a failure to reverse the roles of focal and moderating variables by only interpreting the moderated effect of the focal variable. This ignores that a higher order interaction term is symmetric and modifies the effects of all its constitutive predictors. As a result, authors provide and discuss partial and potentially incorrect descriptions of the nature of the complete interactive relationship (Berry, Golder, & Milton, 2012; Kam & Franzese, 2007). I think this tendency is reinforced by the fact that researchers sometimes propose and test hypotheses about contingent relationships based on asymmetric logic. That is, they initially develop hypotheses explicitly arguing how one factor creates contingent effects of the other and do not explicitly develop hypotheses or rationales for the interaction the other way around.

THE LOGIC OF INTERACTION EFFECTS IN GLMs

What Are GLMs?

They are a class of models that generalize linear regression by relaxing its assumptions to create a common statistical foundation for a wide range of specialized statistical models. To define a GLM, I follow the updated criteria described by Hardin and Hilbe (2012, chap. 2), which includes a wider set of models than the traditional formulation (Nelder & Wedderburn, 1972). Fox (2008, pp. 379–387) concisely frames the GLM approach as three questions whose answers define the type of GLM estimated:

1. What probability distribution characterizes the distribution of the outcome (Y) conditional on the predictors (X)? With limited recent exceptions (Hardin & Hilbe, 2012, p. 12), the choices are from the exponential distribution family and require that the variance be solely a function of the mean. Common choices include the normal, binomial, multinomial, Poisson, negative binomial, gamma, and inverse-Gaussian distributions.

2. What link function $g(\ )$ transforms the expected value of the outcome, $\mu = E(y)$, such that it is a linear function of the predictors? The link function $g(\ )$ must be monotonic, one-to-one, and differentiable. The most frequent options are the identity, log, inverse, inverse-square, square-root, logit, probit, log-log, and complementary log-log functions (see Fox, 2008, p. 380, table 15.1, for the mathematical definition of these link functions).

3. What variables X constitute the linear prediction function for η, the transformed expected value of the outcome—that is, for $\eta = g(\mu) = X\beta$? As for a linear model, you specify a linear and additive function of the predictors with coefficients β. Thus, the prediction function may include dummy variables, logged variables, polynomial functions, product terms for interactions, and the like.

For didactic and notational reasons, it is useful to think about a GLM as defined by two component equations. The first is the *linearizing (measurement) equation* that identifies the function $g(\ )$ that transforms the expected value of the observed outcome, $\mu = E(y)$, into the expected value of the modeled outcome (η):

$$\eta = g(\mu) \tag{1.10}$$

Equivalently, we can write μ as a function of η using the inverse[4] of the function g:

$$\mu = g^{-1}(\eta) = g^{-1}\left(\text{linear prediction function}\right) \tag{1.11}$$

For example, in a negative binomial model, $g(\ )$ is the natural log, $\ln(\ )$, and its inverse function is $g^{-1}(\ ) = e^{(\ )}$. That is, $\eta = \ln(\mu)$ and $\mu = e^{\eta}$.

The second component is the *modeling (structural) equation*, which specifies the linear prediction function for the modeled outcome (η). For interaction effects, the modeling equation specifies a linear function of the focal and moderating variables ($F, M_1, M_2, \ldots$), their interaction ($F \times M_1, F \times M_2, \ldots$), and a set of other predictors heuristically represented by Z:

$$\eta = a + b_1 F + b_2 M_1 + b_3 F \times M_1 + b_4 M_2 + b_5 F \times M_2 + \cdots + b_z Z \qquad (1.12)$$

Note that η can be the expected value of either an observed or an unobserved outcome depending on the link function for the specific GLM estimated. For example, for OLS, η is the observed outcome, Y; while for binomial logistic regression, η is the unobserved outcome, log odds(Y). This expression is sometimes called the "index function."

(Interaction) Effects in the Modeling Component

I labeled the second equation as the modeling or structural component to emphasize that you specify and test your conceptual model of effects in this expression. That is, when you test the coefficients, you do not directly test the effect of a predictor on the observed outcome, except when the link function $g(\ )$ is the identity link. As Equation 1.12 indicates, you directly test a predictor's effect on the modeled outcome η. When you perform a global test of multiple coefficients, you are thus testing whether or not those coefficients are needed in the model predicting the modeled outcome η. Thus, the prior material on how to test the statistical significance of an interaction applies to testing coefficients in the modeling component, not to testing the effects of coefficients on the observed outcome.

Equation 1.12 also means that you can interpret the coefficients for the predictors directly as effects on the modeled outcome. For example, take the partial derivative of η with respect to F in Equation 1.12 to find the effect of F on η:

$$\text{Effect of } F \text{ on } \eta = \frac{\partial \eta}{\partial F} = b_1 + b_3 M_1 + b_5 M_2 \qquad (1.13)$$

This is the counterpart to the moderated effect of F in a linear model shown in Equation 1.2. The difference is that Equation 1.13 is the effect of F on the modeled outcome and Equation 1.2 is the effect of F on the observed outcome. For a GLM with a nonidentity link function, these are not equivalent. As will be discussed throughout the book, this can sometimes be advantageous for interpreting interaction effects.

(Interaction) Effects on the Observed Outcome

This does not mean you cannot interpret the effect of predictors on the observed outcome in a GLM. Rather it means that the estimated coefficients by themselves do not describe that effect. To find the effect of a predictor on the observed outcome, you need to formulate the expression for the expected value of the observed outcome (μ)

as a function of the predictors. The effect of a predictor on the observed outcome is the partial derivative of that expression with respect to the predictor. Start by writing μ as a function of η from Equation 1.11 and then substitute for the linear prediction function from Equation 1.12:

$$\mu = g^{-1}(\eta) = g^{-1}\left(\text{linear prediction function}\right)$$
$$= g^{-1}(a + b_1 F + b_2 M_1 + b_3 F \times M_1 + b_4 M_2 + b_5 F \times M_2 + \cdots + b_z Z)$$

Using $XB = a + b_1 F + \cdots$ to make the equation more readable gives

$$\mu = g^{-1}(XB) \tag{1.14}$$

Taking the partial derivative with respect to F gives F's effect on the observed outcome as

$$\frac{\partial \mu}{\partial F} = \left(b_1 + b_3 M_1 + b_5 M_2\right) \times \frac{\partial}{\partial \eta} g^{-1}\left(XB\right)$$

This expression indicates how the effect of F on the observed outcome is related to the effect of F on the modeled outcome. The effect of F on the observed outcome is equal to F's effect on the modeled outcome—the first expression in parentheses—multiplied by a factor defined by the inverse link function and the coefficients and values of every predictor. This makes explicit that the effect of F on the observed outcome expresses two sources of nonlinearity—that specified by the interaction effect and that created by the inverse link function applied to the predictors' coefficients and values. When examining tables or plots of how the effect of F varies with its moderators or how predicted outcome values change with F, it is difficult to parse out how the patterns represent the nonlinearity of the interaction from how they represent nonlinearity induced by the link function. Chapter 5 explores in depth the interpretive complications of the presence of these conflated sources of nonlinearity and proposes solutions.

ASIDE: NONLINEAR EFFECT OF NONINTERACTING PREDICTOR

When F is not part of an interaction, its effect on the observed outcome is

$$\frac{\partial \mu}{\partial F} = b_1 \times \frac{\partial}{\partial \eta} g^{-1}\left(XB\right)$$

The nonconstant effect of F makes the obvious point that there is a nonlinear relationship between the observed outcome and F. Not quite as obvious is that the effect of F will vary with the values of the other predictors because the multiplicative factor calculated by $g^{-1}(XB)$ depends on those values. This is why the predicted value plot from a main-effects-only model seems to show an interaction effect (in case you were wondering, Plot B shows the main effects predictions). The coefficient for age is multiplied by a different factor at different levels of education yielding varying effects of F.

Common Errors in Using Interaction Effects in GLMs

The common errors concerning interaction effects in linear models all apply in some fashion when using them in GLMs. The errors in specifying interaction effects apply to the specification of the predictors in the modeling component and not directly to the observed outcome unless the link function is the identity link. But the common errors of interpretation apply to interpreting interaction effects on both the modeled outcome and the observed outcome. Beyond these errors, there are other pitfalls when interpreting interaction effects for GLMs with nonlinear link functions.

Improperly Treating Product Terms for an Interaction

The fact that a noninteracting predictor's effect gives the appearance of an interaction with the other predictors in nonlinear models has led some authors to propose that product terms for interaction are not always needed to find and interpret interaction effects (e.g., Berry, DeMerit, & Esarey, 2010). This conflates the nonlinear nature of the statistical model (link function)—what Nagler (1991, p. 1393) describes as an "artifact of the methodology"—with a substantively driven test and interpretation of the form of the relationship. Kam and Franzese (2007) describe this as the difference between implicit interaction resulting from the form of the link function and explicit interaction designed to model hypotheses or expectations. There is a clear and continuing agreement that product terms are necessary to model and test hypotheses about interaction effects in nonlinear models (Berry et al., 2012; Brambor et al., 2006; Braumoeller, 2004; Kam & Franzese, 2007).

A different error concerning the product terms is partly a software-driven error in interpreting the coefficient (and/or its marginal effect) for the product term in isolation from the other components of the interaction. Specifically, the problem occurs when statistical software treats the product term as a unique predictor without taking into account its functional relationship to the component predictors constituting the product term (Ai & Norton, 2003; Greene, 2010). It "mechanically computes a separate 'partial effect' for each variable that appears in the model" (Greene, 2010, p. 292), so the partial effect for the product term $F \times M_1$ is calculated as the partial derivative with respect to the quantity $(F \times M_1)$:

$$\frac{\partial \mu}{\partial (F \times M_1)} = b_3 \times \frac{\partial}{\partial \eta} g^{-1} (XB) \qquad (1.15)$$

But what you want is really two different partial effects that involve the product term, the partial effect of F when you describe the moderated effect of F, and the partial effect of M_1 when you interpret the moderated effect of M_1:

$$\frac{\partial \mu}{\partial F} = \left(b_1 + b_3 M_1 \right) \times \frac{\partial}{\partial \eta} g^{-1} (XB)$$

$$\frac{\partial \mu}{\partial M_1} = \left(b_2 + b_3 F \right) \times \frac{\partial}{\partial \eta} g^{-1} (XB) \qquad (1.16)$$

This has led some authors to interpret the computed partial effect in Equation 1.15 when it is in fact not a meaningful statistic. Note that the advent of the *margins* command in Stata is specifically intended to avoid this problem. But it does so only if the interactions are specified using Stata's factor-variable notation when estimating the GLM.

Limited Range of Moderator Values Used to Probe Moderated Effect of Focal Variable

Too often, the interpretation of the effect of F discusses its value calculated across a restricted range of the values of its moderators (see the review of published studies in Brambor et al., 2006). Doing so may conceal substantively important—and sometimes unexpected—variation in F's effect. Technically, this point applies to both linear and nonlinear models, but this limitation is much more consequential for the effect of F on the observed outcome in nonlinear models because the link function's nonlinearity affects the calculation of F's effect. With limited comparison points, it is more difficult to separate out how much of the difference in F's effect represents the explicit interaction by the moderator from how much is the nonlinearity of the link function. Thus, a recurring recommendation in the didactic literature is to examine how the effect of F changes with its moderator(s) across a substantively meaningful range of the moderators' values (Aiken & West, 1991; Brambor et al., 2006; Hayes, 2013; Kam & Franzese, 2007).

Comparing Estimated Coefficients Across Nested Models (for Some GLMs)

Typically, the reason to compare a predictor's coefficient across nested models for the same sample is to examine how other factors mediate (explain) the influence of that predictor by adding predictors in stages. In some GLMs, the model is identified by setting a fixed value for the error variance of a latent outcome. This results in coefficients that are identified only up to a multiplicative scaling factor (Long, 2009, pp. 47, 122–123; Maddala, 1983, p. 23). But the total variance of the (latent) outcome is not fixed as it would be for an observed outcome. Rather, it is the sum of the explained variance and the fixed value of the error variance. The problem is that the explained variance must increase as predictors are added to the model, even if minutely, which results in a larger total variance and hence a bigger scaling factor for the coefficients. Consequently, it is not unusual for the estimated coefficients to increase across stages of the nested model. Because the estimated coefficient can change solely due to the scaling factor changing, you cannot compare the coefficients across the nested models to examine how they are mediated by the other factors nor attribute such changes to suppressor effects.

The simplest and commonly recommended solution is to examine changes in y-standardized coefficients, defined as the coefficient divided by the estimated standard deviation of the latent outcome (Long, 1997; Mare & Winship, 1984; Mood, 2010; Williams, 2013). Because the coefficients at each stage are standardized by the latent outcome's changing standard deviation, it counters the differences in the scaling of the estimated coefficients on the condition that the sample analyzed must be the same at each stage. Other solutions include comparisons using predicted probabilities or functions of the predicted probabilities (Long, 2009, 2016) or techniques to decompose effects (Buis, 2010; Kohler, Karlson, & Holm, 2011). Note that this concern is not specific to interaction coefficients but applies to any coefficient in nested models.

ASIDE: MODERATED EFFECTS VERSUS MEDIATED EFFECTS

The discussion of comparing coefficients across models introduced the idea of the mediation of an effect that is sometimes confused with the moderation of an effect. A mediation analysis (also known as path analysis) seeks to understand the causal process through which a predictor X affects a penultimate outcome Y, drawing a distinction between its direct causal effect and its indirect causal effect. Conceptually, X's indirect (mediated) effect is how X affects Y through its causal effect on other causes of Y. That is, the indirect effect is how X affects intermediate outcomes (Z), which in turn have their own direct effects on Y. This is often described as Z mediates (explains) some or all of the total effect of X on Y.

In contrast, a moderation or interaction analysis does not necessarily adopt a causal analysis framework. Rather, it focuses on how the effect of X on Y is contingent on (varies with) other predictors (Z) in the analysis. This is commonly labeled as Z moderates the effect of X on Y. It is possible to have a model that specifies that X's effect on Y is both mediated and moderated by other predictors in the model. (For an excellent introduction to mediation and moderation analysis in OLS regression, see Hayes, 2013.)

DIAGNOSTIC TESTING AND CONSEQUENCES OF MODEL MISSPECIFICATION

Before deciding to include and interpret interaction effects, you should always conduct diagnostic tests of model fit and the validity of the modeling technique's assumptions for your data. Model misspecification can create the appearance of interaction, and vice versa (Aiken & West, 1991; Fox, 2008, p. 274; Greene, 2008, pp. 166–167; Kaufman, 2013, pp. 22–23). In this section, I first describe diagnostic tests that apply broadly to GLMs: testing the link function, assessing model fit/departures, residual analyses for model misspecification, and analysis of influential cases. I then discuss the consequences of misspecifying interaction effects. In each application chapter, I discuss additional diagnostic tests applicable to that specific GLM.

Diagnostic Testing

Link Function Test

The usual way to assess the appropriateness of the link function is by an added variable analysis (Hardin & Hilbe, 2012, p. 55). Specifically, you construct the predicted value of the index function and its square and then estimate your model with those two variables. The significance test for the squared term's coefficient provides an assessment of the adequacy of the link function. If the coefficient is not significant, you would conclude that the link function is appropriate. But if it is significant, you should consider alternative link functions. You can also directly compare two different link functions (for details, see Hardin & Hilbe, 2012, pp. 50–51), but to the best of my knowledge this is not often done.

Assessing Overall Model Fit/Departures

A plot of the standardized deviance residuals against the predicted value of the index function provides a nonspecific diagnostic for systematic departures from the model.

By nonspecific, I mean that it does not identify the potential source(s) of the problem, only that there is evidence that the model fit is not adequate. A well-fit model should show no pattern or trend of the residuals against the predicted index function. If the plot shows a pattern, this indicates a problem with the fit of the model but is not informative about why. Thus, if you find a pattern, you should follow up using the next two diagnostic tests to try to find and correct the sources of the lack of fit.

Residual–Predictor Plots or Partial Residual–Predictor Plots

A residual–predictor plot can help identify if a predictor is creating poor model fit because its functional form is not properly specified. If the plot shows a pattern of the residuals changing with a predictor, it indicates you may need to consider alternative functional forms for the predictor. The pattern you see is net of the predictor's effect in the current model, so it typically represents departures from a linear relationship and can help you decide what type of change to make. It is usually easier to determine an alternative functional form if you create a partial residual–predictor plot because the pattern you see is not net of the predictor's effect in the model; that is, the pattern you see is the pattern you want to reproduce with an alternative functional form. For a given predictor, you create the partial residual by subtracting the predictors' effect as specified in the index function. For instance, for a current model predictor with a

Linear effect, $Partial\ residual = Residual - b_{x_j} \times x_j$

Quadratic effect, $Partial\ residual = Residual - b_{x_j} \times x_j - b_{x_j^2} \times x_j^2$

$$(1.17)$$

In principle, such plots could provide information relevant to identifying interaction effects. If two noninteracting predictors each exhibit departures from good model fit, this might suggest a missing interaction effect, but other model misspecifications could also produce that result. If you create separate plots for selected values of one of the predictors in which you plot the residual or partial residual against the other predictor, you may be able to discern an interactive pattern—that is, different slopes. You could also try using a three-dimensional (3D) scatterplot of the residual against the two interacting variables. But in practice either of these plots can be difficult to examine and to discern interaction effects (Fox, 2008, pp. 284–286). They are useful primarily for a two-variable interaction when one of the variables is categorical or has very few interval values. Otherwise, you have to group an interval variable—potentially collapsing across important changes—and need a large sample size to have sufficient cases in the subsamples' separate plots. If your model has two moderators and/or a three-way interaction, you would have to create separate plots for each combination of values of the two moderators.

Residual–Omitted Variable Plots

In any analysis, it is always possible that you omitted a relevant predictor that, if correlated with predictors in the model, would manifest as poor model fit in the prior types of residual plots. A convenient way to determine if a predictor that is not in the model potentially should be included is to plot the residual against any omitted predictors; a systematic patterning of the residuals by the predictor would indicate its inclusion in your model. Alternatively, you could formally test this by adding potential omitted predictors to your model and conduct standard statistical tests of whether or not the added predictor(s) have significant effects. This diagnostic tool presumes that you have measures of the omitted predictors. In my experience, the main reason for an omitted–predictor bias is that you do not have measures to use in the analysis.

Analysis of Influential Cases

Influential cases are those observations in the estimation sample that have both high leverage values and large residual values. Leverage values measure the distance of an observation's values for the predictors from the typical values in the sample (Fox, 2008, pp. 245, 412; Long, 1997, p. 100; Pregibon, 1981, p. 706). Because cases with high leverage values are unusual relative to the average case in terms of the predictors, they have the potential to affect coefficient estimates especially if a high leverage values case has a large residual. This is particularly concerning for interaction analysis because such cases can "pull" coefficient estimates toward their effects on the outcome. In some instances, this can produce a significant interaction effect that is absent if the sample excludes the influential cases.

Although separate analyses of leverage values and of residuals can be somewhat informative about influential cases, the better choice is to use a direct measure of influence. Perhaps, the most common summary indicator of influence for linear models is Cook's distance measure (see Belsley et al., 1980, for an alternative summary indicator as well as coefficient-specific influence measures). For the ith observation, you calculate the sum of the squared change in the predicted value of every observation between the model estimated with the full sample and a model reestimated without case i, and normalize it by the product of the sample mean of y and the mean squared error of the regression (Fox, 2008, p. 250). Note that Cook's distance measure is computationally intensive since it requires reestimating the model for every observation in your estimation sample.

For GLMs, Pregibon's (1981) approximation of Cook's distance measure is the most commonly recommended diagnostic (e.g., Fox, 2008, pp. 245, 412; Hardin & Hilbe, 2012, p. 49; Long, 1997, p. 101). The approximation does not require reestimating the model for every observation; instead it estimates how much each coefficient would change without the ith case in the analysis:

$$C_i = \left(\Delta_i \, \underline{\hat{\beta}} \right)' \widehat{Var} \left(\hat{\beta} \right) \left(\Delta_i \, \underline{\hat{\beta}} \right) \tag{1.18}$$

where

$$\Delta_i \underline{\beta} = \widehat{Var} \left(\hat{\beta} \right) \underline{x}_i \frac{y_i - \hat{y}_i}{1 - h_{ii}}, \; \underline{x}_i = \text{column vector of predictor values for case } i,$$

$$\widehat{Var} \left(\hat{\beta} \right) = \text{variance} - \text{covariance matrix of } \hat{\beta}$$

and

$$h_{ii} = \hat{y}_i \left(1 - \hat{y}_i \right) \underline{x}_i' \widehat{Var} \left(\hat{\beta} \right) \underline{x}_i$$

The usual criterion for defining a case as problematic is $C_i > \dfrac{4}{n - k - 1}$. You should reestimate your model to decide if the excluded case(s) in fact change your results in a meaningful way and finalize your sample before interpreting the interaction effects.

Consequences of Model Misspecifications

Keep in mind that we never know with certainty the correct specification of the model. If the decision about whether or not to include an interaction specification in your model is not clear-cut, you need to make a decision balancing between two types of model misspecifications and their consequences. First, what happens to your model estimates and results if you include an interaction specification when in reality it is not needed, and second, what are the consequences when you exclude an interaction specification when in actuality it is needed. Last, an area of ongoing debate is the effect of unspecified heterogeneity on the estimation and interpretation of group differences—the relationship between the outcome and a categorical predictor—and how to deal with it. This issue is relevant to the interaction analysis if one or more of your interacting predictors is categorical, which is fairly common.

Including an Interaction Specification When Not Needed

This is a specific instance of the inclusion of irrelevant predictors of any kind in the model. In linear models, it is well-established that the consequence of including superfluous predictors is to increase the standard errors of the coefficients, but the coefficient estimates remain unbiased (Greene, 2008, p. 136). Thus, there is some loss of efficiency but not a problem of bias. The increase in coefficient standard errors also applies to GLMs. But if a GLM has a nonlinear link function, then the coefficient estimates are not necessarily unbiased. Many of the commonly used GLMs can identify the estimated coefficients only up to a multiplicative scale factor, usually the standard deviation of a latent interval outcome (e.g., logistic regression, probit analysis, ordinal regression models, multinomial logistic regression [MNLR]). As a number of authors have pointed out (Allison, 1999; Mood, 2010; Williams, 2009), the scaling factor increases as predictors are added to the model. Consequently, the estimates of all the coefficients are biased by the inclusion of a superfluous predictor—such as a product term for an interaction when it is not needed—because they are scaled by an incorrect factor. This situation is analogous to the problem of comparing coefficients across nested models discussed above, so the same solutions apply. For example, if you use y-standardized coefficients to interpret your results, then you remove the bias introduced by the scaling factor. The inefficiency of the estimates (increased standard errors) remains a concern regardless of the type of GLM model you estimate, but it is not always consequential in terms of significance testing and interpretation, particularly if you have a large sample size.

Excluding an Interaction Specification That Is Needed

This is a special case of the general problem of omitted variable bias that has a similar consequence for linear and nonlinear models. In linear models, the coefficients for the other predictors are biased and inconsistent unless the omitted variable is uncorrelated with the included predictors. (Note that a lack of correlation with the omitted variable would be unusual when the omitted variable is a product term of included predictors.) However, in GLMs with nonlinear link functions, the coefficients for the other predictors are always biased and inconsistent even if the omitted variable is uncorrelated with the included predictors (Greene, 2008, p. 787). And for those GLMs whose coefficients are multiplicatively scaled to an identifying factor, the scaling factor is also biased. Moreover, these biases affect the calculations used in a variety of techniques to calculate and probe the relationship of interacting or noninteracting predictors to the outcome, such as marginal and discrete effects, predicted outcomes, and odds ratios.

As with linear models, the potential consequences of excluding relevant interaction terms is worse than the consequences of including superfluous interaction terms for GLMs. That is, your main concern if you include an extraneous interaction term is that your standard errors will be too large, and you will have to be careful to properly take into account the scaling factor bias when interpreting coefficients—which is no easy task. In contrast, if you exclude a relevant interaction term, your coefficients and subsidiary calculations will definitely be biased in an unknown direction. The rational decision in this case would be to err on the side of inclusion of interaction terms when in doubt. The consequence of biased estimates of other coefficients is more certain if you incorrectly exclude the interaction terms than if you incorrectly include them.

Unspecified Heterogeneity and Group Comparisons

A key assumption for GLMs is that the variance function must be a function solely of the mean of the outcome (Hardin & Hilbe, 2012, p. 11). Implicit in this assumption is that there is no additional heterogeneity (heteroscedasticity) in the variance among the sample cases. For linear models, the presence of unspecified heteroscedasticity results in inefficient but unbiased (or consistent) coefficient estimates. The consequences are more serious in GLMs with nonlinear link functions if the unspecified heteroscedasticity is related to any of the model predictors, leading to biased and inconsistent coefficient estimates.

One situation in which heteroscedasticity related to the predictors is plausible, if not likely, is when comparing the outcome across social categories or social groups (Kaufman, 2013, p. 8); that is, when you use a categorical predictor, whether part of an interaction specification or not. This creates a problem akin to the comparison of coefficients across nested models (Allison, 1999; Mood, 2010; Williams, 2009). Making group comparisons is like comparing a coefficient between nested models; the comparison is confounded with the unspecified group heteroscedasticity. There is an ongoing debate over how to solve this problem and no resolution yet (Allison, 1999; Buis, 2010; Kohler et al., 2011; Kuha & Mills, 2018; Long, 2009, 2016; Mood, 2010; Williams, 2009, 2013). Some simulations have shown that using an incorrect heteroscedasticity specification is worse than not adjusting for heteroscedasticity at all (Keele & Park, 2005), while Kuha and Mills (2018) recently argued that the concerns are often irrelevant. Thus, as Williams (2013) put it, "At this point, it is probably fair to say that the descriptions of the problems with group comparisons may be better, or at least more clear-cut, than the various proposed solutions" (p. 11).

This issue reemphasizes the value of the diagnostic testing discussed above to check for problems in the fit of the model to the data and to make modeling changes accordingly. For example, Williams's (2010) reanalysis of Allison's (1999) example demonstrated that adding the square of a predictor corrected for the apparent heteroscedasticity and made moot any concern over confounding of group differences and heteroscedasticity in that analysis. This further points to the importance of models well-grounded in theory and formulated with careful consideration of the appropriate functional form of predictors.

ROADMAP FOR THE REST OF THE BOOK

Overview of Interpretive Tools and Techniques

My approach to understanding interaction effects involves producing and interpreting three types of information to understand the relationship between the outcome and the interacting predictors. I detail the principles of this approach in Part I and apply them to different GLMs in Part II. I briefly describe the three sets of information here, labeling them with the names of the ICALC tool (command) that you use to calculate and create that information. For a paper or a presentation, you would usually include only a fraction of this information. The complete set is intended to help you fully understand the nature of the interaction effects and to provide you with different options for what you present and discuss. I use the heuristic device of declaring one of the interacting predictors as the focal variable and the other(s) as the moderating variable(s), but a proper interpretation should treat each interacting predictor in turn as the focal variable.

Defining the Moderated Effect of *F* With the GFI (Gather, Factor, and Inspect) Tool

The building block for understanding the nature of the interaction effect is to find the algebraic expression for the effect of the focal variable on the "modeled" outcome. You use this to describe and probe the basic structure of the focal variable's relationship to the outcome. With the GFI tool, you can determine if and when the focal variable's effect changes sign (positive to negative or negative to positive) as it varies with the values of the moderators. And you can create a visual representation of the algebraic expression with a path diagram–like graphic.

Calculating the Varying Effect of *F* and Its Significance: SIGREG (Significance Regions) and EFFDISP (Effect Displays) Tools

In probing the nature of the interaction effect, we are invariably interested in more than whether the moderated effect of *F* changes sign. Typically, we want to know the magnitude of the effect at different values of the moderator(s) and where that effect is significantly different from zero. The SIGREG tool finds, where possible, an analytic solution to define the range of moderator values for which *F*'s effect on the modeled outcome is significant. It also produces an empirically derived significance region table for which it can calculate effect values and significance for alternative types of effects when applicable—factor changes or any of the SPOST13 marginal effects calculated by its *mchange* command (see Long & Freese, 2014, pp. 166–171).

These tables are optionally saved to an Excel file with cell formatting to identify and highlight sign and significance changes. The EFFDISP tool creates visual counterparts to the significance region tables produced by SIGREG, plotting information about the varying magnitude and optionally the significance of the focal variable's moderated effect. Multiple plots are produced if there is more than a single moderator or if the focal variable is categorical with more than two categories.

The Predicted Outcome's Value Varying With the Interacting Predictors: The OUTDISP (Predicted Outcome Displays) Tool

Tables or graphs that display the predicted values of the outcome are probably the most familiar, and in some ways the most understandable, way to present and interpret interaction effects. Earlier in this chapter, I introduced the problem that such displays confound the nonlinearity of the interaction with the nonlinearity of the GLM link function, and I discuss this in detail in Chapter 5. I propose alternative visual displays to deal with this confounding that you can create using the OUTDISP tool.

Is the ICALC Toolkit Necessary?

Stata users might wonder whether the ICALC toolkit is really necessary given the capabilities of the *margins* and *marginsplot* commands in Stata for producing information about interaction effects. Part of the reason I created ICALC is that the output created by two of the tools cannot be produced from the *margins* or *marginsplot* commands. GFI provides basic information about the pattern of the focal variable's effect, while SIGREG identifies significance regions for the focal variable's effect. Some of the basic output from ICALC for creating the content of tables of the moderated effect (SIGREG), graphics of the conditional effects (EFFDISP), or predicted outcome values (OUTDISP) can in some form be produced by the *margins* or *marginsplot* commands. But the ICALC toolkit provides types of tables, graphs, and options not available through the Stata commands.

Moreover, it reports results that are more compact and simpler to read and uses an interface that is hopefully more user-friendly in much the same way as are the post-estimation results produced by SPOST13.[5] A major advantage of the ICALC toolkit for readers of this book is that the examples and applications throughout the book use ICALC to produce the results, with annotated explanations of the use of the commands in the application chapters.

For analysts who prefer platforms other than Stata for producing tables and graphics, the ICALC toolkit includes options to save the necessary results to an Excel file that can be easily imported into other applications. It will save formatted tables that users can reformat directly in Excel or copy and paste the Excel table into another application. For graphics, ICALC saves the data values used to create the Stata figures in a rectangular data format—a column for the *x*-axis values/categories, one or more columns for the corresponding *y*-axis values for each graphed data series, and columns for the variables defining separate graphics for subsamples (if any).

Organization and Content of Chapters

Part I (four chapters) describes and explains the principles of interpretation. Part II (seven chapters) provides detailed applications of the principles and the ICALC toolkit to a set of commonly used GLMs and ends with a discussion of extensions.

Part I: Principles

Chapter 2: Basics of Interpreting the Focal Variable's Effect in the Modeling Component. This chapter focuses on the derivation and interpretation of the algebraic expression for the moderated effect of the focal variable on the modeled outcome; that is, F's effect contingent on one or more moderators. The emphasis is on a holistic interpretation of the moderated effect rather than interpreting the component coefficients. It introduces a simple mathematical analysis to determine if and under what conditions (values of the moderators) the sign of F's effect changes from positive to negative, or vice versa. And it demonstrates the use of plots of the moderated effect of F against the predictors to help understand and interpret the changing sign and magnitude of F's effect.

Chapter 3: The Varying Significance of the Focal Variable's Effect. In this chapter, I discuss how to define the significance region for the moderated effect of F. That is, three ways to determine the ranges of the moderators' values for which F's effect is statistically significant are as follows:

- Johnson-Neyman analytically defined boundary values, which mark the boundaries between significance and nonsignificance

- Empirically derived significance region tables reporting F's effect and its significance for user-specified moderator values

- Plots of F's effect against the moderators with confidence intervals demarcating significance regions

Depending on the specific GLM, the last two approaches can be applied to various effects calculated from the moderated effect of F, such as factor changes or discrete changes.

Chapter 4: Linear (Identity Link) Models: Using the Predicted Outcome for Interpretation. This chapter and the next cover the nitty-gritty of how to create and interpret tables and plots of the predicted outcome to understand and explain the pattern of the relationship of the interacting predictors with the outcome. I discuss the simpler case of GLMs with an identity link function (e.g., OLS regression) in this chapter to lay a foundation for the more complicated application to nonlinear link functions in Chapter 5. Topics include options for choosing display and reference values for the interacting predictors, reference value options for noninteracting predictors, and the use and interpretation of predicted value tables, charts, and plots.

Chapter 5: Nonidentity Link Functions: Challenges of Interpreting Interactions in Nonlinear Models. This chapter continues the discussion of the use of predicted outcome values. I detail why the approach to interpreting predicted values for identity link functions presented in Chapter 4 is potentially problematic for nonlinear link functions. I revisit the topic of how to choose reference values for other predictors, and then describe and demonstrate the use of three options for handling these problems when creating and interpreting a predicted values table or graphic.

Part II: Applications

Chapter 6: ICALC Toolkit: Syntax, Options, and Examples. In this chapter, I describe the ICALC toolkit, explain the syntax and options for using the tools, and provide brief examples of how to apply the five ICALC commands. The INTSPEC command must be run before any of the other commands to save and set up information about the nature of the interaction specification and display/reference values of the predictors. The other four commands do the calculations and create tables and graphics corresponding to the approaches to interpretation described in the Part I principles chapters: gather, factor, and interpret (GFI), significance regions (SIGREG), effect displays (EFFDISP), and outcome displays (OUTDISP).

Chapters 7 to 11: Applications to Frequently Used GLMs. Each technique-of-analysis chapter has the same basic structure. It begins with an overview of the types of data situations for which the technique is typically used and brief descriptions of published examples from multiple disciplines (sociology, political science, economics, criminology, and public health). I chose the examples to illustrate the variety of actual applications, concentrating on publications in disciplinary flagship journals. For the most part, I chose the first examples that I found using search engines, excluding those with egregious errors of estimation or interpretation. This overview is followed by a discussion of the technique's properties as a GLM, the relevant ways to interpret coefficients, technique-specific diagnostics tests, and a brief description of the data used for the application examples. The bulk of each chapter demonstrates the application of the ICALC commands to different specifications of interaction effects and how to interpret the results and output for that specification. Each chapter has at least one single-moderator empirical example and one multiple-moderator or three-way interaction empirical example. The interpretations always treat each interacting predictor in turn as the focal variable. Many of the chapters conclude with a special topics section. The organization of the application chapters is as follows:

- Chapter 7: Linear Regression Model Applications
- Chapter 8: Logistic Regression and Probit Applications
- Chapter 9: Multinomial Logistic Regression Applications
- Chapter 10: Ordinal Regression Models
- Chapter 11: Count Models

I have organized the application chapters in the order in which I think they should be read. Certainly, readers should first review Chapter 6 on the use of the ICALC toolkit, and then read Chapter 7 on OLS regression even if they do not intend to use it because there is foundational material woven into the chapter. In a similar vein, readers should be familiar with the applications in Chapter 8 before they read about multinomial logistic or ordinal probability models in Chapters 9 and 10.

Table 1.3 presents a simple rubric for knowing when to consider using each of the GLM models in Chapters 7 to 11. This rubric uses the level of measurement of the outcome—nominal/categorical, ordinal, and interval/ratio—and the number of categories (values) in the outcome measure to identify when to choose each technique. There is additional discussion in each chapter of the circumstances and diagnostic tests that might preclude using these recommendations of a technique.

Chapter 12: Extensions and Final Thoughts. In this chapter, I provide very brief extensions to the use of ICALC for the interpretation of interaction effects in three additional analytic techniques—Tobit analysis, the Heckman selection model, and the Cox proportional hazards model for survival analysis—and the interpretation when one of the interacting predictors is a quadratic function. I end with a brief reminder about important dos and don'ts and cautions in specifying and interpreting interaction effects.

TABLE 1.3 ● RUBRIC FOR CHOOSING GENERALIZED LINEAR MODEL TECHNIQUE

Technique	Type of Outcome Measure
Linear regression	Interval or ratio variable
Binomial logistic regression and probit analysis	Categorical with two categories
Multinomial logistic regression	Categorical with three or more unordered categories
Ordinal regression models	Categorical with three or more ordered categories
Count models	Nonnegative integers representing a count of events

CHAPTER 1 NOTES

1. See Section "Mathematical (Geometric) Foundation for GFI" in Chapter 2 for a more detailed discussion.

2. Suppose you are using a significance level of .05. The reason to correct the significance level is that the probability that one or more coefficients will be significant is greater than .05 and increases with the number of multiple tests.

3. For some estimation methods, the Wald test is an $F_{1,df}$ test statistic and the single-coefficient test is a t_{df} statistic. Because $t_{df}^2 = F_{1,df}$, you get the same test result from either. Similarly, for other estimation methods, the Wald test is a $\chi^2_{(1)}$ test statistic and the single-coefficient test is a z statistic. Because $z^2 = \chi^2_{(1)}$, you again get the same test result. Given this equivalence, some sources, including Stata, also identify the z and t tests of coefficients as Wald tests.

4. The Oswego City School District Regents Exam prep site has a great explanation of the inverse of a function (http://www.regentsprep.org/regents/math/algtrig/atp8/inverselesson .htm):

    "A function and its inverse function can be described as the 'DO' and the 'UNDO' functions. A function takes a starting value, performs some operation on this value, and creates an output answer. The inverse function takes the output answer, performs some operation on it, and arrives back at the original function's starting value."

5. Long and Freese (2014) provide a similar argument for a parallel issue raised concerning SPOST13 and give an example in which SPOST13 produces 50 lines of well-formatted output compared with more than 1,500 lines of output from the corresponding *margins* command (http://www.indiana.edu/~jslsoc/web_spost13/sp13_whymstar.htm).

PRINCIPLES

BASICS OF INTERPRETING THE FOCAL VARIABLE'S EFFECT IN THE MODELING COMPONENT

This chapter describes and illustrates the use of algebraic regrouping of the modeling equation component of a GLM—which for OLS is the same as the equation for the observed outcome—as a technique for interpreting the effect of the focal variable F and how it varies across the values of one or more moderating variables M. In Chapter 1, I described GLMs as consisting of two component equations:

1. A *modeling equation* specifying the modeled outcome η as a linear function of the focal and moderating variables $(F, M_1, M_2, \ldots)$, their interaction $(F \times M_1, F \times M_2, \ldots)$, and a set of other predictors heuristically represented by Z:

$$\eta = a + b_1 F + b_2 M_1 + b_3 F \times M_1 + b_4 M_2 + b_5 F \times M_2 + \cdots + b_z Z \qquad (2.1)$$

Note that η can be the expected value of either an observed or an unobserved outcome depending on the specific GLM estimated; for OLS, it is the observed outcome. The right-hand side of this equation is sometimes called the "index function."

2. A *linearizing equation* specifying the function $g(\ )$ that transforms the expected value of the observed outcome, $\mu = E(y)$, into the expected value of the modeled outcome (η):

$$\eta = g(\mu) \qquad (2.2)$$

Equivalently, we can write μ as a function of η using the inverse[1] of the function g, $\mu = g^{-1}(\eta)$. For example, in a negative binomial model, $\eta = \ln(\mu)$ and $\mu = e^{\eta}$.

You can algebraically rearrange the terms in the modeling component (Equation 2.1) to define and then interpret a mathematical expression of how the effect of the focal variable F on η varies with the values of the moderating variables M. The next section develops the justification for this process. In this chapter and the next, I describe how this expression can be used to explore the pattern and shape of the effect of F. This is a three-step process that I call GFI:

- *Gather* together all of the terms in Equation 2.1 that involve the focal variable F.

- *Factor* out F to identify the function of the other terms that defines the moderated effect of F.

- *Inspect* and interpret the moderated effect of F.

While the choice of which predictor to treat as the focal variable initially should be theory driven, a full understanding of an interaction specification requires exploring, in turn, the moderated effect of each of the component predictors of the interaction.

MATHEMATICAL (GEOMETRIC) FOUNDATION FOR GFI

Virtually every textbook on OLS regression contains a 3D graph, similar to the one shown in the top panel of Figure 2.1, for Y predicted by two independent variables, F and M, from a model without an interaction term. $\widehat{Y}$ is plotted on the vertical axis against F and M on the two dimensions on the floor. The 2D plotted surface portrays how the predicted outcome changes with F and M simultaneously, with other predictors (if any) set to fixed values. The 2D surface is flat because there is no interaction between F and M. The slope of the 2D surface with respect to F for any value of M describes how the outcome changes with F holding M constant. The value of the partial slope of any 2D surface with respect to F is given by the partial derivative with respect to F of the equation defining the surface— $\dfrac{\partial equation}{\partial F}$ —and vice versa for M. Geometrically, it describes the predicted amount of change in the outcome for a one-unit change in F at a fixed value of M.

ASIDE: TAKING PARTIAL DERIVATIVES

For those readers who may be rusty about (or unaware of) how to get the partial derivative of different algebraic expressions, I apply three principles below to take the partial derivative of the predicting equation with respect to a particular variable, say F:

1. If an expression is F multiplied by a constant (or multiplied by an expression not a function of F), its partial derivative is equal to the constant (expression).

2. If an expression does not contain F, its partial derivative is equal to 0.

3. The partial derivative of a sum of expressions is equal to the sum of the partial derivatives of each expression.

Consider the OLS prediction equation for $\widehat{Y}$ from a model without an interaction:

$$\widehat{Y} = a + b_1 F + b_2 M + \cdots + b_z Z \tag{2.3}$$

Only one term in Equation 2.3 includes F, so the remaining terms have partial derivatives of 0. That one term is F multiplied by b_1. So the partial derivative of $\widehat{Y}$ with respect to F is b_1. That is, $\widehat{Y}$ changes by a constant amount as F increases regardless of the value of the moderator. Similarly, the partial derivative of M is b_2, which does not depend on the value of F. This produces the flat regression plane in the top panel of Figure 2.1. Look at the edge running above the F axis (roughly left to right). Note

FIGURE 2.1 ● TWO-DIMENSIONAL SURFACE OF PREDICTED OUTCOME VALUES FROM MAIN EFFECT AND INTERACTION EFFECT MODELS

Main Effect Model, Y Predicted by F and M

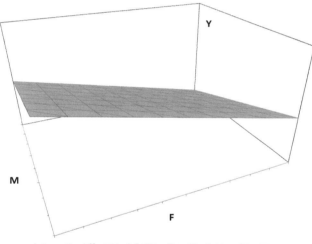

Interaction Effect Model, Y Predicted by F, M and F × M

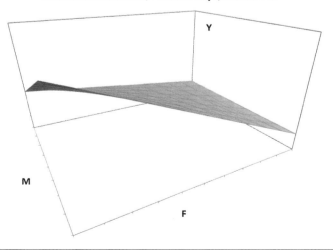

the lines moving away from the edge, which show F's effect for a specific value of M. Because the lines are parallel, they indicate a constant slope for F's effect.

Now let's look at the OLS prediction equation for a model with an interaction between F and a single moderator M:

$$\widehat{Y} = a + b_1 F + b_2 M + b_3 F \times M + \cdots + b_z Z \qquad (2.4a)$$

Rearrange the terms to bring together every term involving F:

$$\widehat{Y} = a + b_1 F + b_3 F \times M + b_2 M + \cdots + b_z Z \qquad (2.4b)$$

Rewrite the set of terms involving F to factor out F multiplied by an algebraic expression:

$$\widehat{Y} = a + \left(b_1 + b_3 \times M\right) F + b_2 M + \cdots + b_z Z \qquad (2.4c)$$

Only one term in Equation 2.4c includes F, so the remaining terms have partial derivatives of 0. That one term consists of a constant expression multiplied by F. The partial derivative of that one term then is equal to the constant expression and gives us the formula for the slope of how $\widehat{Y}$ changes when F increases by one unit:

$$\frac{\partial \widehat{Y}}{\partial F} = \frac{\partial \left(b_1 + b_3 \times M\right) F}{\partial F} = \left(b_1 + b_3 \times M\right) \qquad (2.5)$$

That is, the effect of F on $\widehat{Y}$ changes with the value of the moderator M, and an analogous expression holds true for the effect of M. This produces a 2D plotted surface that exhibits obvious curvature, such as that shown in the bottom panel of Figure 2.1. Now the lines moving away from the leading edge above the F axis are not parallel. That is, they show that the slope of the 2D surface with respect to F is different depending on the value of M.

The process of finding the partial derivative used above provides the justification for the GFI method for getting the effect of F moderated by M. The first two steps are precisely those used by GFI. Rearrange the terms in the modeling equation to bring together all the terms that involve F. Then factor out F to create a set of terms—call them ME—multiplied by F. The set of terms ME is exactly what GFI factors out from F to define the moderated effect of F. And as shown above, the algebraic expression in ME is what results from taking the partial derivative of $\widehat{Y}$ with respect to F, which is to say, it is the moderated effect of F.

The remainder of this chapter focuses on using GFI to probe the changing size, and possibly direction, of the effect of F, while Chapter 3 describes how to determine the varying statistical significance of the effect and of transformations of the effect. These provide a foundation for understanding how to use different approaches to the interpretation of an interaction specification: from the effect coefficients (b_j) directly, from transformations of the b_j (e.g., odds ratios), from predicted values of the observed outcome, or from predicted values of the modeled outcome. Chapter 4 starts with the simpler case of linear models and GLMs with an identity link function. I discuss and illustrate the interpretation of different types of interaction effects using tables and/or figures showing how the predicted outcome varies with the focal and modera-tor variables. This sets the stage for understanding the mathematically more compli-cated interpretation of such tools applied to GLMs with a nonidentity link function, discussed in Chapter 5, including the alternative of interpreting interaction effects using predicted values of the modeled outcome instead of the observed outcome—a mathematically straightforward but often conceptually less intuitive approach.

GFI BASICS: ALGEBRAIC REGROUPING, POINT ESTIMATES, AND SIGN CHANGES

One Moderating Variable

Let's start with the situation in which no variable other than M moderates the effect of F.

Gather the F terms from Equation 2.1:

$$\eta = b_1 F + b_3 F \times M + a + b_2 M + b_z Z \tag{2.6a}$$

Factor out F:

$$\eta = \left[b_1 + b_3 M\right] F + a + b_2 M + b_z Z \tag{2.6b}$$

The effect of F on η as it varies with M_1 is

$$b_1 + b_3 M \tag{2.6c}$$

You can use this formula to calculate point estimates for the effect of F. Notice that I wrote estimates in the plural to emphasize that there are multiple point estimates of the effect of F corresponding to the unique values of the moderating variable M. The signs of the coefficients b_1 and b_3 directly provide information about the effect of F on η, but correct interpretation must also take into account how F and M are measured. Consider how to interpret the effect of F in the case in which b_1 and b_3 have the same sign (both positive or both negative) and then when b_1 and b_3 have opposite signs.

Case 1: b_1 and b_3 Have the Same Sign

Suppose both coefficients are positive. The positive value of b_1 indicates a positive effect of F on η that increases in magnitude with higher positive values of M and decreases in magnitude for smaller values of M. If M takes on only nonnegative val-ues (which is the case for many social science indicators), then we can more simply interpret this as a positive effect of F that becomes increasingly large as the values of M increase. For example, in a regression of respondent's education (in years) on father's socioeconomic index (SEI, in standard deviations) and number of siblings (*Sibs*), the interaction of SEI by siblings and a few controls finds that[2]

$$\widehat{Education} = 14.511 + 0.892\,SEI - 0.275\,Sibs$$
$$+ 0.055\,SEI \times Sibs - 0.00027\,Age + \cdots \tag{2.7}$$

Treat SEI as focal and apply GFI.

Gather SEI terms:

$$\widehat{Education} = 0.892\, SEI + 0.055\, SEI \times Sibs + 14.511 - 0.275\, Sibs - 0.0003\, Age + \cdots$$

Factor out SEI:

$$\widehat{Education} = \left[0.892 + 0.055\, Sibs \right] SEI + 14.511 - 0.275\, Sibs - 0.0003\, Age + \cdots$$

The bracketed expression is the effect of SEI moderated by siblings. It tells us that adults who have no siblings are predicted to have 0.892 more years of education for each standard deviation difference in father's SEI, calculated by setting $Sibs = 0$ in the GFI expression:

$$0.892 + 0.055 \times 0 = 0.892$$

(*Aside:* Notice how this illustrates the interpretation of the main effect coefficient $b_1 = 0.892$; it is the effect of the focal variable, SEI, when the moderating variable, siblings, takes on a value of 0.) This baseline positive effect of SEI increases by 0.055 with each additional sibling. For example, if the respondent grew up with two siblings, the person could expect his or her education to be 1.002 years more for each standard deviation higher the father's SEI. That is, set $Sibs = 2$ in the GFI formula:

$$0.892 + 0.055 \times 2 = 1.002$$

However, if M includes negative values (e.g., when M is centered), then the effect of F could be negative for some values of M. It is simple to determine for what, if any, values of M the effect of F is positive or negative for any combination of signs of b_1 and b_3 and any measurement scale for M. F's effect changes sign when

$$M = -\frac{b_1}{b_3} \tag{2.8}$$

Specifically,

$$\text{Positive effect of } F \quad b_1 + b_3 M > 0 \quad \text{when } M > -\frac{b_1}{b_3}$$

$$\text{Negative effect of } F \quad b_1 + b_3 M < 0 \quad \text{when } M < -\frac{b_1}{b_3}$$

Applying this to the SEI effect moderated by siblings, for example, the effect of SEI would change to a negative effect when

$$Sibs < -\frac{0.892}{0.055} = -16.218$$

That is, the SEI effect is positive across the valid range of number of siblings (0–9 in the sample).

When both coefficients are negative, the interpretation follows the same process with analogous language as when both coefficients are positive, substituting positive for negative and decrease for increase, and vice versa. That is, if M has only nonnegative values, the effect of F is always negative and becomes increasingly more negative for higher values of M. Conversely, if M's range includes negative values, then Equation 2.8 can be used to determine when the effect of F is negative and when it is positive.

Case 2: b_1 and b_3 Have Opposite Signs

In this case, you start with the expectation that the effect of F may—but not necessarily will—change signs for different values of M. Consider the results from a negative binomial regression of how workers' socioeconomic status (*SEI*) and work–family conflict (*Conflict*) affect the experience of poor mental health (GSS Selected Data):

$$\ln Days = -2.99134 - 0.02612\ SEI - 0.03424\ Conflict$$
$$+ 0.00929\ SEI \times Conflict - 0.015259\ Age + \cdots \tag{2.9}$$

The observed outcome variable is the number of days of poor mental health reported during the prior month (0–30), which means that the modeled outcome is the log number of days. The focal variable is SEI—a worker's occupational socioeconomic index (ranging from 17.6 to 97.2)—and the moderator variable is work–family conflict (a scale from 1 = *never* to 4 = *often*). Applying GFI to the results in Equation 2.9, the moderated effect of SEI on ln (number of poor mental health days) is given by

$$-0.02612 + 0.00929\ Conflict \tag{2.10}$$

This indicates that the predicted log number of days decreases as workers' SEI increases $(-0.02612 < 0)$ but this negative effect diminishes in magnitude as work–family conflict increases $(0.00929\ Conflict)$. Using Equation 2.8, the sign changes at

$$Conflict = -\frac{-0.02612}{0.00929} = -(-2.81) = 2.81$$

This tells us that SEI's negative effect turns positive at a work–family conflict value higher than 2.81 on a 4-point scale. To put it another way, SEI appears to increase the number of poor mental health days for those workers with the highest levels of work–family conflict but appears to decrease it for workers with less work–family conflict. Note that the main effect of SEI $(b_1 = -0.02612)$ does not have a meaningful interpretation. It represents the effect of SEI when conflict equals 0, but the smallest valid value for conflict is 1. Readers should reverse the role of the focal and moderator variables for this example using the results for conflict in Equation 2.9 to understand why its moderated effect is always positive despite a negative main effect.[3] Also, notice the language I used to describe the changing effect of SEI: "appears" to increase/decrease reporting of poor mental health days. One reason for this qualifier is that we do not (yet) know for what values of conflict the SEI effect is significantly different from 0. (More on this in the next chapter.)

What if the main effect is positive and the interaction effect negative and/or the moderator had negative values? In these and most other cases, it is not possible to determine just from looking at the moderated effect equation from GFI whether the moderated effect changes sign or not. The only situation in which you can look at the effect equation and immediately know that the effect of F does not change sign is if the main effect coefficient and the interaction effect coefficient are of the same sign and the moderator variable can only have nonnegative values. In every other situation, you need to use Equation 2.8 to determine for what, if any, values of M the effect of F changes sign.

For the purpose of reading and evaluating the correctness of a published interpretation of an interaction effect, I think it is well worth knowing how to calculate the point, if any, at which the sign of the focal variable's effect changes from positive to negative. There are other tools available for determining if the sign changes when you are doing your own data analyses. For example, you can plot the effect of F against the range of values of M to empirically show if and where F's effect changes sign, as elaborated later in this chapter. From plots, it is easy to see sign changes and other aspects of the pattern of F's effect.

Two (or More) Moderating Variables

A slightly more complicated case is when there are two variables M_1 and M_2 in the modeling component, each of which independently moderates the effect of F (i.e., a pair of two-way interactions, not a three-way interaction). That is, there are terms for both $F \times M_1$ and $F \times M_2$:[4]

$$\eta = a + b_1 F + b_2 M_1 + b_3 F \times M_1 + b_4 M_2 + b_5 F \times M_2 + b_z Z \qquad (2.11)$$

Apply the same process of gathering terms involving F and then factoring out F.

Gather terms involving F:

$$\eta = \left[b_1 F + b_3 F \times M_1 + b_5 F \times M_2 \right] + a + b_2 M_1 + b_4 M_2 + b_z Z \qquad (2.12a)$$

Factor out F

$$\eta = \left[b_1 + b_3 M_1 + b_5 M_2 \right] F + a + b_2 M_1 + b_4 M_2 + b_z Z \qquad (2.12b)$$

The effect of F on η as it varies with M_1 and M_2 is

$$b_1 + b_3 M_1 + b_5 M_2 \qquad (2.12c)$$

To illustrate the interpretation, I use a logistic regression analysis predicting approval of a law banning interracial marriage (1 = *approve*, 0 = *disapprove*). The analysis includes two-way interactions of White (dummy coded 1 = *White*) with education (ranging from 0 to 20) and of White with whether the respondent grew up in the South (coded 1 = *South*, 0 = *non-South*). This analysis uses 1987 General Social Survey (GSS) data, which I have previously used for didactic purposes (see Kaufman, 1996, for a description of the coding of all the variables). The outcome in the modeling component for a logistic regression is the log odds of the outcome. The estimates from the full model are

$$
\begin{aligned}
\ln \widehat{odds}\left(approval\right) = {} & -1.5458 + 2.8351 \, White - 0.1054 \, Education \\
& -0.1424 \, White \times Education + 0.2623 \, South \\
& +1.1394 \, White \times South + 0.0291 \, Age - 0.1398 \, Class \\
& -0.3465 \, Contact
\end{aligned}
\qquad (2.13)
$$

Gathering the terms for White as the focal variable yields the expression

$$\ln \widehat{odds}\,(approval) = [2.8351\,White - 0.1424\,White \times Education \\ + 1.1394\,White \times South] - 1.5458 \\ - 0.1054\,Education + \cdots \tag{2.14}$$

Factoring out White,

$$\ln \widehat{odds}\,(approval) = \big[2.8351 - 0.1424\,Education + 1.1394\,South\big]White \\ - 1.5458 - 0.1054\,Education + \cdots$$

In this example, the positive main effect coefficient for White tentatively suggests that Whites are more likely than others to approve of an interracial marriage ban. This effect is larger in magnitude among respondents who grew up in the South because the interaction term for South is the same sign (positive) as the main effect of White. But the effect of White is smaller for respondents with higher levels of education because the education interaction term has an opposite signed (positive) effect. However, it is not immediately clear if the effect of White remains positive across the possible combinations of values of the moderating variables. An easy way to determine this (if there are no higher order interactions) is to calculate the effect of the focal variable at all of the combinations of the minimum and maximum values of the moderators[5] or for combinations of selected values across the range of each moderator. This also can be achieved by graphing the effect of the focal variable, as discussed later.

Table 2.1 presents the effect of White at the four combinations of minimum and maximum values of the two moderators. This shows that the effect of White is not uniformly positive and varies greatly in magnitude from −0.013 to more than 300 times that at 3.975. Note that the change in the White effect across the range of values for the moderators is greater for education (2.848 = 2.835 − (−0.013)) than it is for South (1.140 = 3.975 − 2.85).

TABLE 2.1 ● EFFECT OF WHITE VARYING BY SOUTH AND EDUCATION

South =	Education =	Effect of White
Min(0)	Min(0)	2.835
Min(0)	Max(20)	−0.013
Max(1)	Min(0)	3.975
Max(1)	Max(20)	1.127

ICALC command: gfi, ndigit(3). The ndigit( ) option specifies that the table's cells are formatted with three decimal places.

There is a more general approach that calculates the value of Moderator 1 at which the focal variable's effect changes sign contingent on the value of Moderator 2, and vice versa. It is a special case of the formulas for the three-way interaction example in the next section. You can apply the calculating formulas in Equation 2.17 to the two-moderator case by setting the value of b_7 to 0. You could then create and interpret sign changes using a table like Table 2.1.

A Three-Way Interaction

Last, let's complicate this further by specifying that two moderating variables M_1 and M_2 interact with each other in moderating the effect of F. The three-way interaction specifies that the change in the effect of F with M_1 also depends on the values of M_2. To model this, add terms for $F \times M_1 \times M_2$ and for $M_1 \times M_2$ to Equation 2.11:

$$\eta = a + b_1 F + b_2 M_1 + b_3 F \times M_1 + b_4 M_2 + b_5 F \times M_2 + \\ b_6 M_1 \times M_2 + b_7 F \times M_1 \times M_2 + b_z Z \tag{2.15}$$

Apply the same process of gathering terms involving F and then factoring out F.

Gather F terms:

$$\eta = \left[b_1 F + b_3 F \times M_1 + b_5 F \times M_2 + b_7 F \times M_1 \times M_2 \right] + \\ a + b_2 M_1 + b_4 M_2 + b_6 M_1 \times M_2 + b_z Z \tag{2.16a}$$

Factor out F:

$$\eta = \left[b_1 + b_3 M_1 + b_5 M_2 + b_7 M_1 \times M_2 \right] F \\ + a + b_2 M_1 + b_4 M_2 + b_6 M_1 \times M_2 + b_z Z \tag{2.16b}$$

The effect of F on η as it varies with M_1 and M_2 is

$$b_1 + b_3 M_1 + b_5 M_2 + b_7 M_1 \times M_2 \tag{2.16c}$$

The interpretation of the effect of F directly from this expression is straightforward in only one situation. If all four coefficients have the same sign and the moderator variables take on only nonnegative values, then the effect of F is consistently positive (negative), and this effect becomes larger in magnitude as each moderator variable increases in value. Moreover, the magnitude of the effect of F rises for M_1 (M_2) at an increasing rate for higher values of M_2 (M_1).

Otherwise, you can describe in the abstract what the initial effect of the focal variable is, how it varies with each of the moderator variables, and how the moderation of F by M_1 is itself moderated by M_2, and vice versa. However, such an interpretation rarely provides a clear picture of whether the effect of F is always the same sign or when and how it changes sign and by how much (see the example below).

Thus, we need a way to determine for what, if any, values of M_1 and M_2 the effect of F changes sign. For this purpose, I derived an expression for the value of M_1 at which the effect of F changes sign for a specified value of M_2, and vice versa (see Special Topics section at the end of the chapter):

$$M_1 = -\frac{b_1 + b_5 M_2}{\left[b_3 + b_7 M_2\right]} \quad \Leftrightarrow \quad M_2 = -\frac{b_1 + b_3 M_1}{\left[b_5 + b_7 M_1\right]} \tag{2.17}$$

You can apply these across the range of possible values of M_1 and M_2 to determine the values of M_1 and M_2 for which the effect of F is positive or negative. I find it useful to construct a 2D table, with rows defined by selected values of M_1 and columns by selected values of M_2. Each row reports the moderated effect of F as it changes with M_2 as well as the value of M_2 at which the sign changes (if it does). Analogously, the columns show the moderated effect of F varying by M_1 and the M_1 value at which the sign changes. (As I discuss in the next section, we can also graph the changing effect of F.) For an example, I estimate an OLS regression using the GSS 2010 extract sample data excluding three outliers. The outcome is the number of voluntary association memberships predicted by a three-way interaction among age, education, and sex plus controls for perceived social class and population size in millions. Age and education are coded in years, and sex is dummy coded (0 = *men*, 1 = *women*):

$$\begin{aligned} \textit{Memberships} = {} & -1.1548 + -2.6761 \, \textit{Sex} + 0.0072 \, \textit{Age} + 0.0399 \, \textit{Age} \times \textit{Sex} \\ & + 0.1580 \, \textit{Education} + 0.2231 \, \textit{Education} \times \textit{Sex} \\ & + 0.000527 \, \textit{Education} \times \textit{Age} \\ & - 0.00386 \, \textit{Education} \times \textit{Age} \times \textit{Sex} \\ & - 0.1169 \, \textit{PopSize} + 0.00765 \, \textit{Class} \end{aligned} \tag{2.18}$$

Applying GFI to the regression results, the effect of sex on memberships is

$$-2.6761 + 0.0399 \, \textit{Age} + 0.2231 \, \textit{Education} - 0.00386 \, \textit{Education} \times \textit{Age} \tag{2.19}$$

This indicates that there appears to be an initial negative effect of sex (−2.6761),[6] which diminishes in magnitude and possibly turns positive for higher age and education levels. But the positive moderation of the effect of sex by age (0.0399) is less at higher education levels (−0.00386), and similarly, the positive moderation of the effect of sex by education (0.2231) is less at higher ages (−0.00386). This description provides a sense of the pattern of how age and education moderate the effect of sex. But is the sex effect consistently negative, is it consistently positive, or does it switch between positive and negative?

To answer such questions, you can use Equation 2.17 to calculate at what, if any, value of age the education-moderated effect of sex changes sign and at what value of education the age-moderated effect of sex changes sign. I constructed Table 2.2

(and the GFI expression in Equation 2.19) from the output produced by the *gfi* command in the ICALC Stata toolkit shown in the table notes. The interior cells of the table report the effect of sex on membership for the corresponding values of age in the column and education in the row. Reading across a row shows how the effect of sex changes in magnitude and/or sign with age for the specified value of education. The last column reports the exact value of age (M_2) at which the sign of the effect changes if that change point is between the minimum and maximum age value, and it reports "never" if the sign does not change. For example, the row for *Education* = 0 indicates that at the youngest age (18) the effect of sex is −1.958, meaning that women are predicted to have two fewer memberships than men. But this difference diminishes in magnitude as age increases, and switches to a higher number of memberships for women than for men when *Age* ≥ 67.09 (see the last column in that row). Similarly, reading down a column indicates changes in the size and sign of the effect of sex with education for the given value of age, with the last row showing the value of education at which the sign changes or "never" if it does not. For instance, when *Age* = 38 (a little below mean age) the sex difference is one fewer memberships for women than for men with minimum education (0). As education increases, the sex gap diminishes in size and changes sign for *Education* ≥ 15.16 (as reported in the last row for that column).

Scanning across the entries in Table 2.2, the pattern of sign change and moderation of the effect of sex becomes clear. When education values are near the bottom of its distribution and age values near the top of its distribution, the effect of sex turns from negative to positive (look at the upper-right corner of the table). The sex effect also changes sign when age values are near the bottom of its distribution but education values are toward the top of its distribution (look at the lower left corner of the table). Moreover, looking diagonally from the top left corner to the bottom right corner, the effect of sex on memberships first diminishes in size and then increases.

Overall, the visual impression of the numbers in the table suggests that the effect of sex should be predominantly negative. But this format obscures the fact that there is usually an association between the multiple moderating variables that could result in a different distribution of positive and negative effects than is suggested by the distribution of effects across the cells. That is, the percentage of positive and negative effects empirically depends on the joint sample distribution of age and education. From the *gfi* command's output, the table note reports that 21% of the sample are predicted to have positive effects of sex based on their values for age and education. Consequently, almost 80% of the sample would have a negative effect of sex on the number of memberships, confirming the first visual impression.

Wrap-Up

This section provided a foundation on the algebra of interaction effects in the modeling equation component of a GLM, which is central to the interpretation of an interaction specification. I emphasized how to find and then interpret the algebraic expression that defines the moderated effect of the focal variable. And I focused on a holistic discussion of the coefficients in the expression to describe the conditional effect of the focal variable—the pattern of how the conditional effect changes with the moderators—rather than a piecemeal interpretation of the meaning of the main effect coefficients separate from the interaction coefficient(s). This foundation is useful not only for your own data analyses but also when you read and evaluate someone else's interpretation of his or her interaction effects model. To use it, you only need to know the information that is (or should be) always presented: the coefficients and how the predictors are measured.

TABLE 2.2 ● SIGN CHANGE ANALYSIS OF THE EFFECT OF SEX ON G(MEMBERSHIPS) MODERATED BY EDUCATION (M_1) AND AGE (M_2)

When Education =	When Age =								Sign Changes Given M_1
	18	28	38	48	58	68	78	88	
0	−1.958	−1.559	−1.160	−0.761	−0.363	0.036	0.435	0.834	When $M_2 = 67.089$
2	−1.651	−1.329	−1.007	−0.686	−0.364	−0.042	0.280	0.601	When $M_2 = 69.306$
4	−1.343	−1.099	−0.854	−0.610	−0.365	−0.120	0.124	0.369	When $M_2 = 72.921$
6	−1.036	−0.869	−0.701	−0.534	−0.366	−0.199	−0.031	0.136	When $M_2 = 79.867$
8	−0.729	−0.638	−0.548	−0.458	−0.367	−0.277	−0.187	−0.096	Never
10	−0.421	−0.408	−0.395	−0.382	−0.369	−0.355	−0.342	−0.329	Never
12	−0.114	−0.178	−0.242	−0.306	−0.370	−0.434	−0.498	−0.562	Never
14	0.193	0.052	−0.089	−0.230	−0.371	−0.512	−0.653	−0.794	When $M_2 = 31.709$
16	0.501	0.283	0.064	−0.154	−0.372	−0.590	−0.809	−1.027	When $M_2 = 40.946$
18	0.808	0.513	0.217	−0.078	−0.373	−0.669	−0.964	−1.260	When $M_2 = 45.359$
20	1.116	0.743	0.370	−0.002	−0.375	−0.747	−1.120	−1.492	When $M_2 = 47.944$
Sign changes given M_2	When $M_1 = 12.741$	When $M_1 = 13.545$	When $M_1 = 15.160$	Never	Never	When $M_1 = 0.928$	When $M_1 = 5.598$	When $M_1 = 7.171$	

Note: Percentage of in-sample cases with positive moderated effect of Women = 21.2.

ICALC command: gfi, ndig[3]. The ndig[ ] option again specifies cell formatting.

In this section, I showed how to use tables of the moderated effect of F to report how the effect of F varied with the moderator(s). In the next section, I discuss an alternative way to present the information in such tables—especially when there are multiple moderating variables and/or if there is a three-way interaction—graphs of the effect of F as it varies with the moderating variables.

PLOTTING EFFECTS

Overview

For many researchers and readers, the old adage—a picture is worth a thousand words—applies to understanding the pattern of how the effect of F is moderated. Different types of visual displays of the moderated effect of F can be constructed for this purpose, with the choice largely dictated by the level of measurement of the moderators. For a single-interval moderator, this can be achieved effectively with a line plot of the effect of F on the y-axis against the values of M on the x-axis. A drop-line plot works well for a categorical moderator; the moderator categories are equally spaced on the x-axis, with a marker symbol vertically above it at the value of the effect of F on the y-axis and a drop line connecting the marker symbol to a zero-effect horizontal reference line. Alternatively, you could use a bar chart, which replaces the marker and drop line with a rectangular bar.

For a model with two (or more) moderators or a three-way interaction, you have several options. If all the moderators are categorical, then drop-line plots or bar charts of F plotted against M_1, repeated for each category of M_2, are the best choice. For all interval moderators, you can use the following:

1. A 3D surface plot can be used in which the vertical axis (z) represents the size of F's effect and the two axes on the floor of the plot (y and x) correspond to the two moderators. Plotting F's effect against the moderators creates a 2D surface showing how the effect of F varies with M_1 and M_2.

2. Although Stata does not have 3D surface plot capabilities, ICALC can draw a contour plot, which is the 2D equivalent of a surface plot, albeit harder to interpret. Like a surface plot, it uses the two moderators to define a 2D surface, but it is flat on the floor and differentiates the magnitude of F's effect using shaded/colored areas (contours) of similar effect values. ICALC optionally exports the contour plot data points to an Excel spreadsheet, where you can create a 3D surface plot from the data or input it to another graphics program.

3. A set of line plots like that for a single-interval moderator, conditioned on and repeated for a selected set of values of M_2, can be used. Each graph plots F's effect on the y-axis against M_1 on the x-axis for a given value of M_2.

For a mixture of interval and categorical moderators, I would recommend line plots of F against the interval moderator conditioned on and repeated for the categories of the nominal moderator. Alternatively, you could use drop-line plots or bar charts, treating selected values of the interval moderator as if they were categorical.

I often find it useful to present the frequency distribution of the moderator(s) to indicate the relative frequency of effects of different sizes and/or signs as they actually exist in the data. Keep in mind when looking at these plots that they show the *effect* of F on η and how this effect changes, *not* how the expected value of η (or μ)

varies with F and the moderator(s). Such visual displays are frequently more useful for researchers as an aid to their understanding of the moderated relationship—so that they can interpret it well—than as a graphic to include in a journal article or book. The following sections illustrate the application and interpretation of these effect display options for one-moderator, two-moderator, and three-way interaction specifications.

But first let me draw the reader's attention to an issue of terminology. In this book, I refer to such visual displays as "effect displays" since they show how the focal variable's *effect* varies with the moderators. This is purposeful to distinguish them from what I call "outcome displays," which show how an *outcome* varies with the interaction variables. This unfortunately may create some confusion for readers familiar with Fox's (2008) well-known textbook, in which he uses the term *effect displays* to refer to plots of the fitted outcome values against predictors.

One-Moderator Effect Display Examples

For the one-moderator case, I use household wealth (measured as net worth in $10K) regressed on the interaction between household headship type (couple, single male, and single female) and the number of children (0–10). The interaction is an ideal didactic example because it demonstrates several situations. The effect of one of the interacting predictors changes sign while the other doesn't; the focal effect of children illustrates the use of a drop-line plot, while the focal effect of headship type applies a line plot; and it shows how to work with a focal variable with multiple effect parameters. (See the appendix for more data and measurement details.)

Moderated Effect of Headship Type

Because headship type is a three-category indicator, there are two moderated effects of headship type: the effect of single-male household versus couple household and the effect of single-female household versus couple household. For background information, the GFI results provide the following expressions for the effect of headship type on net worth:

$$\textit{Single-man}\left(\text{vs. couple household}\right): -1.699 + 0.062 \times \textit{Kids}$$
$$\textit{Single-woman}\left(\text{vs. couple household}\right): -3.770 + 1.677 \times \textit{Kids}$$

Thus, the effect of both types of single-headed households are negative when they have no children and move toward zero (possibly positive) as the number of children rises, but much more rapidly for single-woman than for single-man.

Figure 2.2 presents two line plots, one for the effect of single-male and one for the effect of single-female households. The x-axis is the number of children (0–10), and the y-axis is the magnitude of the effect of headship type on net worth. A horizontal *dashed gray line* provides a reference point for no effect (zero) of headship type and also serves to indicate when the headship type effects change sign. In the left-hand plot, the *solid black line* indicates that the effect of single-man is always negative but decreasing in magnitude—moving closer toward zero—as the number of children in the household increases. That is, single-male-headed households have a lower net worth than couple-headed households, but the difference between the household types diminishes from about $17K to $11K as the number of children increases. The right-hand plot shows a negative effect of single-woman, which changes more sharply

FIGURE 2.2 ● **ONE-MODERATOR EFFECT DISPLAY, LINE PLOT**

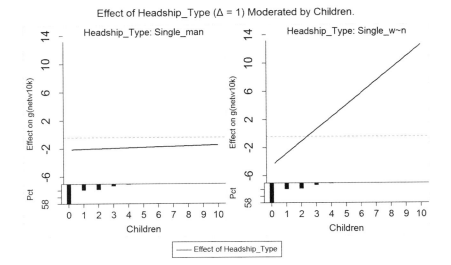

Effect of Headship_Type (Δ = 1) Moderated by Children.

ICALC command: **effdisp, plot(type(line) freq(tot)).** The type( ) option specifies that a line plot is drawn, and freq( ) requests that the total sample distribution of the moderator be shown.

with number of children and turns positive for 3 or more children (technically at 2.25 children according to the GFI results). This means that single-female-headed households are predicted to have a lower net worth than couple-headed households when they each have less than 3 children, with the difference rapidly falling from about $38K to $21K to only $400 as the number of children rises from 0 to 2. But for 3 or more children, single-female-headed households are predicted to have an increasingly higher net worth than couple-headed households. Note, however, that the percentage distribution of number of children shown in the spike plot at the bottom of the graph indicates a relatively small percentage of households with 3 or more children.[7]

Moderated Effect of Number of Children

Reversing the roles of the focal and moderator variables, the GFI expression for the effect of Number of Children as it varies with Headship Type is

$$-0.968 + 0.062 \times Single\text{-}man + 1.677 \times Single\text{-}woman$$

To calculate the effect of number of children for each headship type, substitute the appropriate 0s and 1s into the headship-type dummy variables:

Couple-headed household effect
$(Single\text{-}man = 0$ and $Single\text{-}woman = 0\,) - 0.968 = -0.968 + 0.062 \times 0 + 1.677 \times 0$
Single-male-headed household effect
$(Single\text{-}man = 1$ and $Single\text{-}woman = 0\,) - 0.902 = -0.968 + 0.062 \times 1 + 1.677 \times 0$
Single-female-headed household effect
$(Single\text{-}man = 0$ and $Single\text{-}woman = 1\,) - 0.709 = -0.968 + 0.062 \times 0 + 1.677 \times 1$

FIGURE 2.3 ● ONE-MODERATOR EFFECT DISPLAY, DROP-LINE PLOT

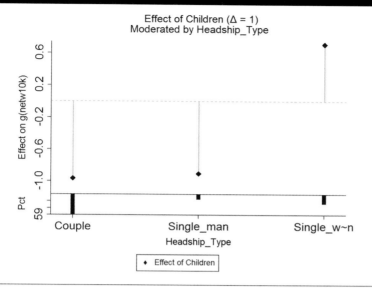

ICALC command: **effdisp, plot(type(drop) freq(tot))**. The type( ) option specifies that a drop-line plot is drawn, and freq( ) requests the moderator's sample distribution.

Thus, children have a negative effect on wealth for couple-headed households, a smaller but still negative effect for single-male-headed households, and a positive effect for single-female-headed households.

Figure 2.3 presents a drop-line plot portraying the size and direction of the effect of number of children for each household headship type. The horizontal dashed line marks a zero effect and serves as the dividing line between positive effects and negative effects. The plot makes clear that the effect of children on net worth is negative and similar in magnitude for couple-headed and single-male-headed households, predicting a decline in net worth of $9K to $9.7K per child, whereas for single-female-headed households the effect is positive, predicting a $7K increase in net worth per child. The headship type distribution at the bottom of the plot indicates that the effect of children on wealth is negative for about 70% of households, those headed by a couple or by a single man.

Two-Moderator Effect Display Examples

Line Plot of Focal by Moderator 1, Repeated for Values of Moderator 2

Choosing which variable to use as the first moderator and which as the second moderator in this situation is partly a substantive issue and partly an issue of ease of comprehension and interpretation. All else being equal, you should treat the moderator of greater substantive interest as the first moderator because the line plots focus your attention more on how F's effect varies with M_1. However, if one moderator is interval and the other categorical, it is easiest to understand the set of graphs if you plot F's effect against the interval moderator repeated for each category of the nominal moderator. That is, comparing the line plots across the second moderator's values lends itself a little better to categorical comparisons than to numerical comparisons.

Selecting the values of M_2 to use for creating repeated plots depends on its measurement and distribution. If M_2 is a categorical variable or an interval variable with a limited number of values (e.g., how many days per week someone rides public transportation), you should repeat the line plot of F against M_1 for each category of M_2. If M_2 is an interval variable, many authors recommend selecting values spanning the middle part of the distribution of the moderator or some other pick-a-point approach (e.g., Aiken & West, 1991; Hayes, 2013; Jaccard, 1998, 2001; Jaccard & Turisi, 2003; see also the discussion in Bauer & Curran, 2005). Often used are the mean, 1 standard deviation below the mean, and 1 standard deviation above the mean. Like Bauer and Curran (2005), I think it is important to consider the full range of the values of the moderator, at least initially, in case there are substantial differences in the effect of F outside the central part of M_2's distribution. Thus, I recommend trying five to seven values equally spaced from the minimum to the maximum of M_2. Equal spacing facilitates comparisons across the repeated line plots because adjacent plots always represent the same change in M_2's value. Otherwise, for a moderator that is not badly skewed, I would recommend selecting five values: (1) M_2's minimum, (2) its mean minus 1 standard deviation, (3) its mean, (4) its mean plus 1 standard deviation, and (5) its maximum. If the moderator is skewed, I would suggest instead using the minimum and maximum plus the first, second, and third quartiles of M_2 to define the repeated line plots.

The empirical illustration uses the modeling component prediction equation from a logistic regression analysis to illustrate the repeated line plots option. The modeled outcome is the log odds of Approval of a Law Banning Interracial Marriage, which is predicted by a pair of interactions—race by education and race by region—as well as controls for age, social class, and interracial contact. Race is dummy coded (1 = *White*, 0 = *Black*), education is years completed (0–20), and region of upbringing is dummy coded (1 = *South*, 0 = *non-South*). (See the appendix for more details on the data and measurement.) The GFI expression for the moderated effect of race is

$$2.8351 - 0.1424 \times Education + 1.1394 \times Region$$

This expression tells us that the effect of race (White–Black difference) starts as positive when *Education* = 0 and *South* = 0, decreases in magnitude as education levels rise above 0, but increases in magnitude as South changes from 0 to 1. In line with the advice above, I use the interval moderator (education) as the first moderator and the categorical moderator (region) as the second moderator.

Figure 2.4 portrays this pattern with line plots of the effect of race on the *y*-axis against education on the *x*-axis, separately for the two regions of upbringing. The *solid line* slanting from left to right shows the changing magnitude of the effect of White on the log odds of approval, and the *dotted horizontal line* provides a reference point of no effect (0). Because there is no three-way interaction, the slopes of the plotted solid lines are identical and indicate how the effect of White varies with education. The positive effect of White on the log odds of approval declines with increasing education but remains greater than zero except at the highest level of education (20) for those raised outside the South. Notice the difference between the two regional plots in the vertical placement of the solid line showing the effect of White. The plotted line is higher for *Region* = *South* than for *Region* = *non-South*, indicating that race has a larger positive effect on approval for those raised in the South than for those raised outside the South.

FIGURE 2.4 ● TWO-MODERATOR EFFECT DISPLAY, LINE PLOT FOR EACH VALUE OF SECOND MODERATOR

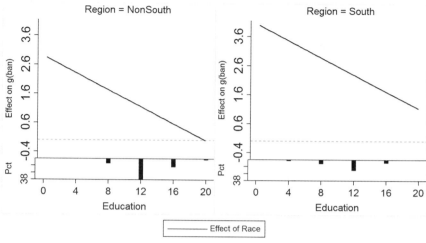

Effect of Race (Δ = 1)
Moderated by Education and Region.

ICALC command: **effdisp, plot(type(line) freq(tot)) ndig(1)**. The type( ) option specifies drawing a line plot, freq( ) requests showing the total sample distribution of the moderator, and ndig( ) formats the y-axis labels.

The spike plots at the bottom of the graph show the total sample percentage of cases for each combination of education (in 4-year categories) and region. The cases are primarily in the range of education from 12 to 16 years (especially 12 years), with more cases outside the South than in the South. This means that the effect of race shown by the plotted lines is primarily in the lower part of the effect range for those raised outside the South (about 1.3 and less) but in the middle part of the effect range for those raised in the South (between 2 and 2.6).

You can extend the repeated line plots approach to three moderators of the focal variable by creating a line plot of F by M_1 for every combination of the selected values of M_2 and M_3. This is particularly effective if M_2 and M_3 are both nominal. If two of the three moderators are interval, you could instead extend the use of 3D surface plots or 2D contour plots, which are discussed next.

Effect Displays for Two Interval Moderators

Both 3D surface plots and 2D contour plots require that both moderators be interval level; the focal variable can be either interval or nominal. I created the 3D plot in Figure 2.5 from the data saved by ICALC to an Excel file when it produced the 2D plot in Figure 2.6. I use the same OLS results to present and compare the use of these plot types. I regressed number of children in the household on the interactions of family income by education and family income by age, controlling for number of siblings, race, and religious intensity. Family income is measured in $10K, while age and education are given in years (see the appendix for more data and measurement details). Using family income as the focal variable, the GFI expression for the moderated effect of income is

$$0.0197 - 0.0017 \times Age + 0.0066 \times Educ$$

You might be tempted to interpret this as indicating that income has an initial positive effect on number of children (0.0197), which decreases with age and increases with education. But that would be correct only if *Age* = 0 were a possible sample value, which it is not.

Figure 2.5 presents the 3D surface plot of the income effect in the top plot and the joint distribution of the moderators in the bottom plot. After some experimentation,

FIGURE 2.5 ● TWO INTERVAL MODERATORS EFFECT DISPLAY, 3D SURFACE PLOT

Effect of Income on # of Children Moderated by Age and Education

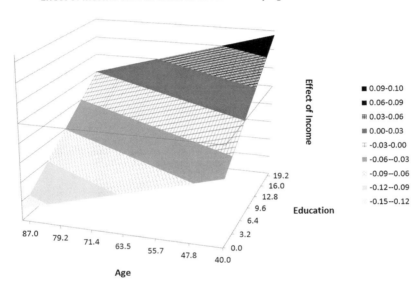

Frequency Distribution of Age & Education

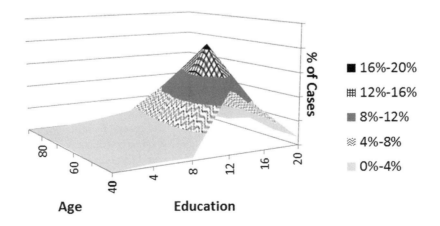

Note: Produced in Excel from plotting data saved by ICALC. See ICALC command for Figure 2.6.

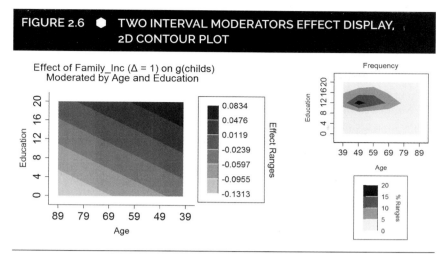

FIGURE 2.6 ● TWO INTERVAL MODERATORS EFFECT DISPLAY, 2D CONTOUR PLOT

ICALC command: effdisp plot(type(contour) freq(subtot) save(Output\Fig_2_5.xlsx)) pltopts(xsc(reverse)). The type( ) specifies a contour plot, freq( ) requests the joint frequency distribution of age and education as a percentage of the total sample, save( ) outputs the plotting data to an Excel file, and pltopts( ) customizes the plot's x-axis.

I found that the best 3D perspective is produced by setting the age axis values to increase from right to left and those on the education axis to increase from front to back. This surface plot shows the overall pattern of the effect of income in an accessible way; that is, the income effect moves in the negative direction as age rises (right to left) and in the positive direction as education increases (front to back). The stripes (contours) on the effect surface show the direction and magnitude of the income effect as it varies with age and education. Darker shades of gray show larger positive effects (and smaller negative effects). And I alternated solid fills with pattern fills to enhance the recognition of different effect sizes. At the right front corner of the surface (where *Age* = 40 and *Education* = 0), the effect of income is negative in the range of −0.03 to 0.00. Moving away from that point, as age increases from right to left, the income effect is a progressively larger-in-magnitude negative effect. You can see this by the downward slope of the surface that is apparent on its front edge as well as by the ever lighter shades of gray indexing larger negative (and smaller positive) effects at the edge. And along lines parallel to the edge—corresponding to different values of education—you can see the same pattern of changing effect size/direction. The opposite pattern characterizes how the effect of income changes as education rises when you examine the right-side edge of the surface moving from front to back. The edge slopes upward, and the negative income effect turns increasingly positive as the path moves into and past the darkest gray (not black) stripe. And at any line parallel to that edge—representing different age values—the income effect steadily changes from negative to positive, shown by the change from lighter to darker shades of gray.

The front edge of the darkest gray band marks the dividing line between negative and positive effects; negative income effects are in the region toward the front edge of the surface. It runs from the point defined by *Age* = 40 and *Education* = 7.32 to the point defined by *Age* = 89 and *Education* = 19.95 (these specifics come from the sign change analysis table produced by the GFI tool). Thus, income has a negative effect when years of education are low enough to be counterbalanced by high enough years of age—in particular, in the front left quadrant of the surface area. Impressionistically, the surface area mainly comprises negative effects (68%),[8] but this ignores the actual distribution of cases across age and education. Looking at the joint frequency distribution in the right-hand plot of Figure 2.5 shows the cases concentrated in the area where the effect of income is positive. Indeed, 68% of the sample cases are predicted to have positive effects of income on number of children according to the GFI analysis.

The 2D contour plot of the effect of income on the left side of Figure 2.6 presents the same information in a similar fashion on the surface area of the plot. It draws contour bands (*stripes*) showing areas of similar effect sizes that are differentiated by their coloration—the lighter gray the shading, the more negative (and less positive) the income effect, and the darker gray the shading, the more positive (and less negative) the effect. The same pattern of how the effect of income changes is apparent. At any given value of age on the horizontal axis, the income effect moves in the positive direction as education increases—the contours change from lighter gray to darker gray going from the bottom to the top of the plot. But from right to left, the contours become increasingly lighter gray as age increases, indicating that the income effect changes in the negative direction. However, the lack of a geometric portrayal of the changing magnitude of the effects makes it more difficult to appreciate the changes in the size of the effects. Similarly, the 2D contour plot of the distribution of cases across age and education on the right side of the figure shows where the cases are concentrated, but not nearly as dramatically as the 3D plot.

If your analysis has three moderators of the same focal variable, you have several choices for how to create multiple effect displays for the selected values of M_3. You could plot multiple effect surfaces on the same 3D surface plot. These surfaces would lie parallel to one another but would have different coloration patterns from one another to represent how the effect of F varies with the third moderator. Or you could produce 3D or 2D plots of F against M_1 and M_2 for selected values of M_3 or repeated line plots of F against M_1 for every combination of selected values of M_2 and M_3. Among these options, I would recommend the repeated line plots on the grounds of feasibility and relative ease of interpretation. 3D plots with multiple surfaces can be effective presentation devices, but you cannot create them in Excel or Stata; you need access to and expertise in software programs that can create such plots. 2D contour plots are difficult to compare because they lack a geometric representation of the changing size of F's effect. Having spent some amount of time staring at and trying to compare changes in the coloration patterns of 2D contour plots, I can attest to the difficulty of making sense of the results. Repeated 3D surface plots are a viable option, but they require some investment of time to produce, whereas ICALC can create repeated line plots using the *effdisp* command.

Effect Displays for a Three-Way Interaction

I discuss three options for portraying how the moderated effect of F varies with M_1, M_2, and $M_1 \times M_2$ when you specify a three-way interaction. A 3D surface plot is particularly effective in this situation, a 2D contour plot does not work as well as it did for the two-moderators case, and a repeated line plot can be a more than workable choice. I demonstrate these points by plotting OLS regression results for a three-way interaction between education (in years), age (in years), and Sex (dummy coded 1 = *female*) predicting the number of voluntary associations memberships (see the appendix for more data and measurement details). Technically, I should run these analyses from negative binomial regression results, but I use the OLS regression results for didactic reasons—namely, to simplify the language as I first introduce how to discuss the more complicated patterns of effects for a three-way interaction.

Figure 2.7 presents two 3D plots. The upper plot shows how the effect of sex on membership varies with the combination of education and age values, and the lower plot portrays the joint distribution of sample cases across education and age. The coloration of the effect surface area in the top plot is designed to facilitate telling the story by highlighting where the effect of sex is negative (*lighter gray shades/fills*) and where it is positive (*darker gray shades/fills*). Notice that the effect is not a flat plane but rather exhibits curvature because the sex effect becomes more positive (and less negative) as age increases when education is low—look at the leading edge of the surface along the age dimension and how the effect surface slopes upward from the "south" corner to the "east" corner. Similarly, the sex effect diminishes with age at higher years of education—take note of the far edge of the effect surface along the age dimension and how the surface slopes downward from "east" to "north." The same pattern characterizes how the sex effect varies with education: The sex effect moves in the positive direction with increasing education at younger ages (look at the leading edge on the education dimension) but moves in the negative direction at older ages (examine the far edge for the education dimension). This produces a saddle shape for the effect of sex. When age and education scores are both in the central part of their values, the effect of sex is more positive (and less negative) than it is when education and age are both low or both high—the south corner and the north corner of the surface. But that central area has a more negative (and less positive) sex effect than when education is high and age is low (west corner) or when education is low and age is high (east corner).

The bottom plot shows the relative frequency of cases, and hence effects, for the joint distribution of age and education. The height of the surface represents the relative frequency of the region, with darker shades/fills where there are more cases. This shows relatively few cases in areas for which the effect of sex on membership is uniformly negative—that is, in the west and especially the east corners of the surface. The sample cases are concentrated (57%) in an area roughly bounded by 11 to 17 years of education and 20 to 55 years of age, which is a mixture of negative (65%) and positive (35%) effects.

Figure 2.8 presents the 2D plot equivalents of the 3D plots in Figure 2.7. The coloration of the effect surface (darker gray for more positive effects and lighter gray for negative) in the left-hand plot is relatively effective in showing the varying magnitude of the sex effect and in mapping out the general pattern of the changes. Look at any age between 18 and 58 years, and trace a vertical path upward to determine how the effect of sex varies with education for the given value of age. The coloration changes from lighter to darker shades, showing that the sex effect changes from negative to positive, or at least to a smaller negative, toward the upper end of that age bracket. Beyond roughly age 58 years, the pattern reverses, and the sex effect becomes increasingly negative in magnitude as education rises. There is an analogous pattern of changes in the sex effect with age for a set value of education. For low education levels (around 0–9), the sex effect changes from negative to positive as age increases, while for higher years of education, the effect of sex becomes an ever-larger negative effect as age increases. But the 2D plot does not show these patterns nearly

FIGURE 2.7 ● THREE-WAY INTERACTION EFFECT DISPLAY, 3D SURFACE PLOT

Effect of Income on # of Children Moderated by Age and Education

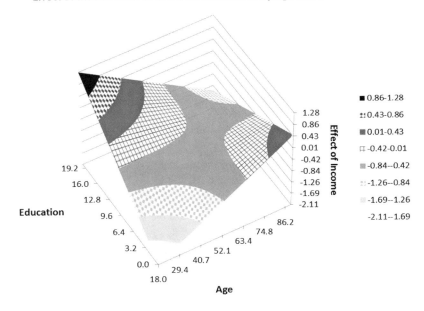

Frequency Distribution of Age & Education

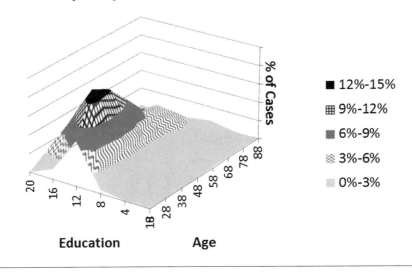

Note: Produced in Excel from plotting data saved by ICALC. See ICALC command for Figure 2.8.

as easily or accessibly as the 3D surface plot in Figure 2.7 because you have to rely on interpreting the changes in shading without a vertical dimension to visually portray the patterns. However, the 2D plot of the joint distribution of the sample cases across age and education is as easy to grasp as the corresponding 3D plot.

FIGURE 2.8 ● THREE-WAY INTERACTION EFFECT DISPLAY, 2D CONTOUR PLOT

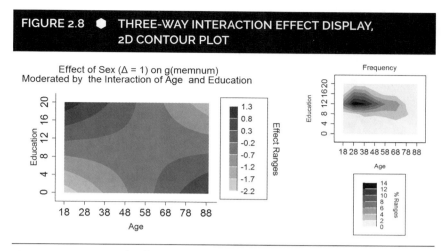

ICALC command: effdisp, plot(type(contour) save(Output\Fig2_7.xlsx) freq(subtot)) pltopts(ccuts(-2.2(.5)1.3)) ndig(1). The type( ) specifies a contour plot, freq( ) requests the joint frequency distribution of age and education as a percentage of the total sample, save( ) outputs the plotting data to an Excel file, pltopts( ) customizes the plot's cut points used to define contours, and ndig( ) formats the legend.

The last option is to produce and interpret two sets of repeated line plots: a set of line plots of the sex effect against age repeated for selected values of education and a second set of line plots of the sex effect against education repeated for selected values of age. After some experimentation to ensure that the pattern changes were well represented, I chose to repeat the sex effect by age line plot for *Education* = 0, 4, 8, 12, 16, 20 and to repeat the sex effect by education line plot for *Age* = 18, 41, 64, 87. Figure 2.9 presents the line plots for the sex effect by age in two panels. The upper panel clearly shows how the sex effect changes toward a positive (and less negative) effect with increasing age for low values of education. And the lower panel, just as plainly, portrays how the sex effect changes toward an ever-more negative effect with rising age when education is higher. The spike plot at the bottom of each line plot shows the joint distribution of sample cases across age and education, revealing their concentration in the vicinity of 12 years of education. This suggests that for much of the sample the sex effect on membership is negative (women belong to fewer voluntary associations than men) and that the gap between women and men widens as age increases. Results from the GFI tool indicate that 75% of the total sample would have a negative effect of sex on membership.

The second set of line plots of the sex effect by education in Figure 2.10 portrays a broadly similar pattern of how the second moderator (age) affects the change in the sex effect with the first moderator (education). That is, at younger and middle ages, the effect of sex on membership changes from negative to positive as education rises, but at older ages (about 60 and older), the sex effect becomes increasingly negative as education increases.

All in all, I think that using repeated line plots is often an optimal choice given how readily they show the patterns of moderation and because ICALC can produce them on request, unlike the 3D surface plot, which requires a fair degree of work in

FIGURE 2.9   ●   THREE-WAY INTERACTION EFFECT DISPLAY, REPEATED LINE PLOTS SET 1

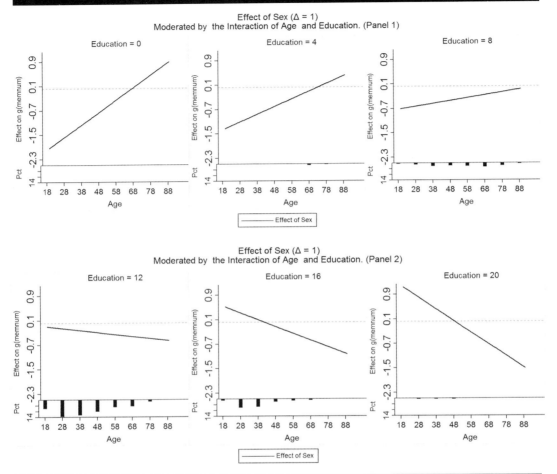

ICALC command: effdisp, plot(type(line) freq(subtot)) ndig(1). The type( ) sets line plot, freq( ) gets age-education distribution, and ndig( ) formats the *y*-axis.

Excel to create. On the other hand, this generates more plots to interpret and possibly to present.

Summary

In this chapter, I showed how to derive and interpret an algebraic expression for the effect of the focal variable on the modeled outcome (η) contingent on one or more moderators, including the case of a three-way interaction between the focal variable and two other moderators. This provides a basic building block for working with and understanding interaction effects. I also discussed how to mathematically determine if and when the moderated effect of F changes sign. And I illustrated the use of

FIGURE 2.10   ●   THREE-WAY INTERACTION EFFECT DISPLAY, REPEATED LINE PLOTS SET 2

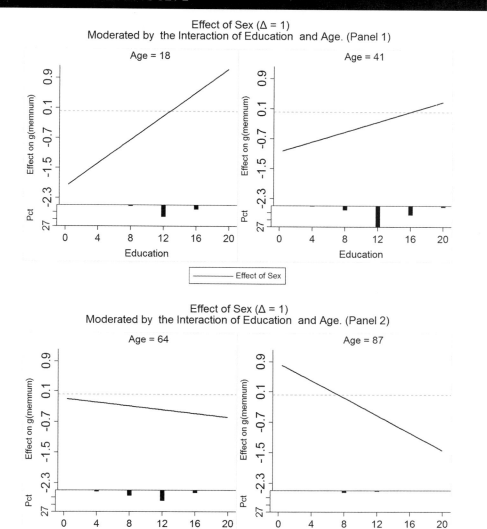

ICALC command: effdisp, plot(type(line) freq(subtot)) ndig(1). The type( ) sets line plot, freq( ) gets age-education distribution, and ndig( ) formats the *y*-axis.

plots to help understand and interpret the moderated effect of *F*, which can often be easier to comprehend than the expressions from the algebraic regrouping. Using plots forgoes the necessity of logically thinking through the implications of how the measurement scale of the moderators combined with the sign of the coefficients affects the effect of the focal variable. A crucial issue to consider at this point is whether the varying effect of *F* is always statistically different from zero, particularly if it changes sign. Chapter 3 covers three different ways to determine and present information about the potentially varying significance of the effect of F.

A point I purposefully deferred discussing is how to choose the set of values of the moderating variable(s) at which the moderated effect of the focal variable is calculated and displayed. Chapter 4 provides a relevant set of recommendations as part of a broader discussion of setting display and reference issues for the focal variable, the moderators, and the other predictors for creating tabular and graphic displays of the relationship between the interacting predictors and the predicted outcome.

Given the didactic purposes of this chapter, I reversed the roles of the focal and moderator variables only in the first effects display example because that served the purpose of illustrating different types of effects displays. Thus, I want to remind readers that in an actual analysis you should always treat each of the component predictors of an interaction in turn as the focal variable. You will see this principle at work in the application chapters in Part II.

Special Topics

Derivation of Equation 2.17

In a three-way interaction, for a given value of M_2, at what value of M_1 does the effect of F change sign (if it does)? That is, we want to find the value of M_1 for which the expression in Equation 2.16c equals 0, with M_1 expressed in terms of the value of M_2.

Set the expression in Equation 2.16c to 0

$$b_1 + b_3 M_1 + b_5 M_2 + b_7 M_1 \times M_2 = 0$$

Collect terms involving M_1 $\quad (b_3 + b_7 \times M_2) M_1 + b_1 + b_5 M_2 = 0$

Isolate M_1 terms on left-hand side $\quad (b_3 + b_7 \times M_2) M_1 = -(b_1 + b_5 M_2)$

Solve for M_1 $$M_1 = -\frac{b_1 + b_5 M_2}{(b_3 + b_7 M_2)}$$

This is the first result shown in Equation 2.14 for M_1. An analogous derivation solving for M_2 in terms of M_1 yields the second result.

CHAPTER 2 NOTES

1. The Oswego City School District Regents Exam prep site has a great explanation of the inverse of a function (http://www.regentsprep.org/regents/math/algtrig/atp8/inverselesson .htm): "A function and its inverse function can be described as the 'DO' and the 'UNDO' functions. A function takes a starting value, performs some operation on this value, and creates an output answer. The inverse function takes the output answer, performs some operation on it, and arrives back at the original function's starting value."
2. This example uses the GSS 2010 extract sample data.
3. The main effect for work–family conflict is −0.03424.
4. There might or might not be an interaction between M_1 and M_2. But whether or not it is present and is part of the other predictors' specification in $b_z Z$ does not affect the interpretation of the moderated effect of F.
5. The number of combinations is 2^m, where m is the number of moderating variables.

6. I say there "appears" to be an initial negative effect because the main effect coefficient for Sex is its effect when Age and Education are both equal to zero, but zero is not a valid value for Age.

7. Given that, we might suspect that the positive effect for 3 or more children, which rises above a $100,000 advantage, is an artifact of the sharp decline in the absolute magnitude of the effect of children at the bottom of the distribution.

8. Norming each variable by its range, the total area is 1 ($49/49 \times 20/20$), and the positive effect area is a right triangle with an area of 0.316 ($\frac{1}{2} \times 49/49 \times (19.95 - 7.32)/20$). The negative effect area then is the total area minus the positive effect area, $0.684 = 1 - 0.316$.

THE VARYING SIGNIFICANCE OF THE FOCAL VARIABLE'S EFFECT

Having established how the focal variable's effect varies with the moderators, a related and often important question is whether the moderated effect is always significantly different from zero across the range of values of the moderating variable(s). Or are there parts of the range for which the effect of F is not significant? For example, if the effect of work–family conflict on mental health is moderated by sex, there would be considerable substantive interest in knowing whether or not work–family conflict has a significant effect for both women and men, or for only one of the sexes. In this chapter, I discuss three somewhat interrelated approaches to determining the significance region for F's effect and how to apply them to single-moderator, multiple-moderator, and three-way interaction specifications.

First, I build on Bauer and Curran's (2005) extension of the Johnson-Neyman (JN) method to analytically define the regions of significance and nonsignificance for the moderated effect of F. I extend it to interaction effects in the modeling component of GLMs. I demonstrate how to write and solve the analytic equations by hand to lay a foundation for readers to understand how to apply the equations and use them in any software. I then use the SIGREG tool in ICALC to calculate and translate the solutions into tables and plots to facilitate interpretation. Readers who do not intend to write their own software code or spreadsheet formulas may want to skim the first part of this material on developing and applying the formulas and focus on the discussion of how to interpret the results, beginning with the section "Interpretation of Boundary Value Analysis Results From ICALC."

Second, I show how to empirically define regions of significance by testing the significance of F across the range of values of a single moderator or at combinations of selected values of multiple moderators. I suggest creating a table whose rows and/or columns are defined by the moderators' values and whose cells report the moderated effects. I illustrate the use and interpretation of tables created by ICALC, which formats the cells to visually identify the moderated effect's sign and significance.

A related approach is to calculate and then plot confidence bounds or confidence intervals around the effect of F as it varies with the first moderator, with the plot repeated for selected values of additional moderators. You would plot confidence bounds for an interval first moderator and confidence intervals for a categorical first moderator. Confidence bounds are continuous lines marking the upper and lower limits of a specified confidence interval level (e.g., 95%) for the effect of F; the bounds and F's effect are continuously plotted across the range of the interval moderator. Analogously, for a categorical first moderator, you calculate the confidence interval for the effect of F at each category of the first moderator. You then plot a marker showing the moderated effect of F and draw error bars above and below the marker to show the confidence interval limits. If the confidence bounds/intervals enclose the value of 0 for a given value of the first moderator, this indicates that the effect of F is not significantly different from 0 at that value of the moderator(s). I discuss the interpretation of such confidence bounds plots and error bar plots using plots created by ICALC.

Before turning to the elaboration of these three approaches, in the next section, I address two issues concerning the choice of an appropriate test statistic and of an appropriate significance level (confidence interval level) to explore the changing significance levels of the moderated effect of F. I also want to remind the reader that the discussion and presentation in this chapter assumes that you have previously determined that the interaction specification is significant, as discussed in Chapter 1. Defining significance regions is not a substitute for demonstrating that the interaction specification as a whole is statistically significant. There is no reason to explore for which values of the moderator(s) the effect of F is significant if your prior conclusion was that the effect of F does not significantly vary with the moderator(s).

TEST STATISTICS AND SIGNIFICANCE LEVELS

Wald Tests Versus LR Tests

Either a Wald test or an LR test is commonly used to test hypotheses about linear or nonlinear functions/restrictions of the coefficients from GLMs, including tests of single coefficients (Greene, 2008, pp. 498–506; Long & Freese, 2014, p. 115). These test statistics are asymptotically equivalent, converging to a $\chi^2(J)$ distribution for a test with J degrees of freedom. But their results can differ in small samples for which their sample properties are not generally known (Greene, 2008, p. 498). There is a general preference to test GLM hypotheses with an LR test when feasible (Greene, 2008, pp. 498–506; Long & Freese, 2014, p. 205), especially for models estimated by maximum likelihood. One reason is that the Wald test result is not necessarily invariant to equivalent specifications of the tested restriction if one of the specifications is nonlinear and the other linear but the LR test result is invariant in such cases (Greene, 2008, p. 502).

On the other hand, you must make stronger distributional assumptions to use the LR test than to use the Wald test (Greene, 2008, p. 502). And on practical grounds, to perform an LR test on a single restriction, you must estimate your model twice, once as a full (unrestricted) model and once with the restriction, while the Wald test can be applied without reestimating the full model. If you are performing a small number of tests, the repeated estimation is not an onerous task unless you are analyzing a very large data set with a computationally intensive algorithm. But this is a concern for two of the three methods for determining the significance region because it could require numerous reestimations of different restricted models. For example, to empirically derive the significance region for the effect of F interacted with a single moderator would require reestimating the model 11 times if you divided the range of the moderator into 10 equal intervals for testing. For a two-moderator or a three-way interaction, it might require 121 reestimations for all the combinations of selected values of the two moderators. And calculating confidence intervals, especially confidence bounds, would necessitate even more reestimations. Moreover, the LR test cannot be used with the JN method to solve for significance boundaries. For these reasons, I develop the different approaches to significance regions in this chapter using the Wald test given its asymptotic equivalence to the LR test, that it does not require model reestimation, and that it can be used with the JN approach.

Potential Adjustment of the Significance Level for Simultaneous Testing

There is an unresolved debate about how to treat significance levels in performing multiple tests of the moderated effect of F, with a standard Bonferroni adjustment as well as several alternatives (Aitken, 1973; Gafarian, 1964; Huitema, 2011; Potthoff, 1964; Rogosa, 1980, 1981). Bauer and Curran (2005, pp. 382–383) argue that such methods provide valid corrections, but they also note that in many instances their use can be tantamount to throwing out the baby with the bathwater. The criteria may be so stringent that they create extremely narrow significance regions or, equivalently, overly broad confidence bounds, which could limit our ability to distinguish where the moderated effect is and is not significant, particularly in small samples.

A different approach to this issue is to consider that you have already conducted an inferential test of significance to determine that the effect of F does change significantly with the moderator(s). The multiple significance tests are not being used inferentially per se but rather to descriptively mark the boundaries between significant and nonsignificant effects. As such, there is no need to adjust the significance level when you define the significance region.[1]

Given the lack of consensus on whether to do such corrections, I will follow the common practice of not making a correction for multiple testing and acknowledge that the resulting significance regions are primarily of heuristic value (see also Bauer & Curran, 2005; Hayes, 2013). But I do illustrate in Chapter 6 how to make the simultaneous testing corrections using ICALC, which provides three adjustment options. A standard Bonferroni adjustment or a Sidak adjustment of the nominal α level can be used if the focal variable is categorical with more than two categories (C) or if a moderator is categorical or otherwise has a limited number of values (C). For the Bonferroni, use $\alpha^* = \dfrac{\alpha}{\left(\dfrac{C(C-1)}{2}\right)}$ instead of α when calculating the critical value

for your test statistic (Huitema, 2011, p. 280). The Sidak correction gives a somewhat larger adjusted value for the significance level, calculated as $\alpha^{**} = \left(1 - (1-\alpha)^{\frac{1}{C}}\right)$.

It is often preferred because the Bonferroni gives an estimate that is frequently very conservative. Potthoff (1964) developed a critical value that can be applied when testing across the range of values of one or more moderators. In the general case of an interaction of F with Q moderators, you would use alternative critical values as follows. If using an F test, as discussed below, instead of a critical value of $F_{\alpha,1,df}$, use $(Q+1)F_{\alpha,Q+1,df}$. And if using a Wald test, replace $\chi_\alpha^2(1)$ with $\chi_\alpha^2(Q+1)$ as the critical value.[2] In the case of a multicategory focal variable, either a Bonferroni or a Sidak adjustment can be done simultaneously with a Potthoff correction (Huitema, 2011, p. 280) by using α^* or α^{**} to determine the critical value for the Potthoff correction.

JN MATHEMATICALLY DERIVED SIGNIFICANCE REGION

The original JN method analytically derived expressions for the boundary values of the significance region, the pair of points at which the moderated effect of F changes from significant to nonsignificant, and vice versa (Johnson & Fay, 1950; Johnson & Neyman, 1936). Prior to Bauer and Curran (2005), the JN method was limited to a dummy focal variable by continuous moderator–variable interaction in analysis of covariance (ANCOVA) models. Bauer and Curran (2005) extended the JN method approach to apply to any combination of measurement types of F and a single moderator in ANCOVA, linear regression models, and multilevel regression models (Hayes, 2013, has written macros for SPSS and SAS that apply Bauer & Curran's [2005] extension to OLS models). I apply the logic of their extension to derive equivalent procedures to apply to interaction effects in the modeling component of GLMs, tested with either an F or a χ^2 test (see Lazar & Zerbe, 2011, for a parallel development for single-moderator interactions). I begin with the single-moderator case and then extend the procedures to multiple moderators and three-way interactions. Readers not interested in the technical details should skim the material until the "Interpretation of Boundary Value Analysis Results From ICALC" section and then begin reading more carefully.

Bauer and Curran (2005) rely on a fundamental principle of random variables to develop their formulas for the JN calculations—namely, that the variance of a linear function of random variables is a linear function of the variances and covariances of those random variables. Thus, we can construct an estimate of the sampling variance for $\hat{\omega}$—a weighted linear composite of a model's coefficients ($\underline{b}$)—if we know the fixed weights for the linear composite and the estimated sampling variances ($s_{b_i}^2$) and covariances (s_{b_i,b_j}) for the modeled coefficients $\underline{b}$:[3]

$$\text{Let } \underline{b} = \begin{bmatrix} b_1 & b_2 & \cdots & b_k & a \end{bmatrix}' \text{ and}$$

$$\widehat{Var}(\underline{b}) = \begin{bmatrix} s_{b_1}^2 & s_{b_1,b_2} & \cdots & s_{b_1,b_k} & s_{b_1,a} \\ s_{b_2,b_1} & s_{b_2}^2 & \ddots & s_{b_2,b_k} & s_{b_2,a} \\ \vdots & \ddots & \ddots & \vdots & \vdots \\ s_{b_k,b_1} & s_{b_k,b_2} & \cdots & s_{b_k}^2 & s_{b_k,a} \\ s_{a,b_1} & s_{a,b_2} & \cdots & s_{a,b_k} & s_a^2 \end{bmatrix} \text{ where } s_{b_i,b_j} = s_{b_j,b_i} \text{ for } i \neq j \qquad (3.1)$$

and $\hat{\omega} = \underline{c}'\underline{b}$, where $\underline{c}'$ is a set of $k+1$ fixed weights that define the effect of F;[4] then,

$$\widehat{Var}(\hat{\omega}) = \underline{c}'\,\widehat{Var}(\underline{b})\,\underline{c} \Rightarrow se(\hat{\omega}) = \sqrt{\underline{c}'\,\widehat{Var}(\underline{b})\,\underline{c}}$$

We can use this principle to estimate the sampling variance for the effect of F at varying values of the moderating variables and hence to calculate the regions of significance and confidence bounds for the effect of F. The starting point is to specify the weights (c') that define the effect of F as a linear composite of the model's coefficients and that specify the formula for the sampling variance. The contents of c' for the effect of the focal variable contingent on the moderator(s) should be a 1 in the position corresponding to the coefficient for F, a value of each moderator variable in the coefficient position corresponding to the interaction of F and that moderator (e.g., m_1), and, if there is a three-way interaction, the product of the values of the two moderators in the coefficient position corresponding to the three-way interaction (e.g., $m_1 \times m_2$). All the other weights should be zero. Table 3.1 (rows 3 and 4) shows the specification of c' and the resulting formula for the sampling variance of the moderated effect of F for the interaction of F with a single moderator, with multiple moderators, and as part of a three-way interaction. It uses lowercase ms to represent a specific value of a moderating variable.

We can calculate the region of significance using the point estimate at particular values of the moderator variables $\hat{\omega}$ and the formula for the sampling variance of F's effect—as it varies with the values of the moderator(s)—starting from the formula for the appropriate significance test for F's effect. Bauer and Curran (2005) develop their formulas starting from a t test that $\hat{\omega} = 0$. However, we can only use the t test when we know the finite sample properties of an estimator, which we do for only a limited number of estimators (Gould, 2013). An approach that is appropriate when we know either the finite sample properties or the asymptotic distribution of an estimator is to use a Wald test that $\hat{\omega} = 0$. The test statistic is $Wald = \dfrac{\hat{\omega}^2}{\widehat{Var}(\hat{\omega})}$; which has a $\chi^2(1)$ asymptotic distribution for maximum-likelihood estimators or an $F_{1,df}$ distribution for estimators for which we know finite sample properties. Note that the test result using the χ^2 distribution is asymptotically equivalent to using an F test or a t test.[5] To determine the formula for the region of significance for an interaction effect, we set the Wald critical value for the α level of significance (W_α) equal to the expression for the Wald statistic and then solve for those values of the moderator variable(s) that yield that critical value. These values of the moderator variables mark the boundary where the effect of F changes between significance and nonsignificance.

So for a specified critical value W_α, we need to find the values of the moderators such that

$$\frac{\hat{\omega}^2}{\widehat{Var}(\hat{\omega})} = W_\alpha \Leftrightarrow W_\alpha \, \widehat{Var}(\hat{\omega}) - \hat{\omega}^2 = 0 \tag{3.2}$$

We can apply this formula to different forms of interaction effects by rewriting this generic equation in terms of the expressions for $\hat{\omega}$—the effect of F at specific values of the moderators—and $\widehat{Var}(\hat{\omega})$ shown in Table 3.1 (rows 2 and 4). The result will always be a quadratic expression in terms of the unknown values of the moderator variable (m_1) with coefficients $\tilde{a}$, $\tilde{b}$, and $\tilde{c}$:

$$\tilde{a}\, m_1^2 + \tilde{b} m_1 + \tilde{c} = 0 \tag{3.3}$$

whose solution according to the quadratic formula is

$$m_1 = \frac{-\tilde{b} \pm \sqrt{\tilde{b}^2 - 4\tilde{a}\tilde{c}}}{2\tilde{a}} \tag{3.4}$$

TABLE 3.1 ● SPECIFICATIONS FOR ESTIMATING STANDARD ERRORS, SIGNIFICANCE REGIONS, AND CONFIDENCE INTERVALS FOR THE EFFECT OF *F* AT PARTICULAR VALUES OF MODERATING VARIABLES FOR COMMON FORMS OF INTERACTION EFFECTS

One Moderator		**Two Moderators**
Modeling equation[a]	$\eta = b_1 F + b_2 M_1 + b_3 F \times M_1 + b_z Z + a$	$\eta = b_1 F + b_2 M_1 + b_3 F \times M_1 + b_4 M_2 + b_5 F \times M_2 + b_z Z + a$
$\hat{\omega}$ = Effect of F at $M_1 = m_1$, $M_2 = m_2$	$\hat{\omega} = b_1 + b_3 m_1$	$\hat{\omega} = b_1 + b_3 m_1 + b_5 m_2$
$\underline{c}$ = Weights for calculating $Var(\hat{\omega})$	$[1\ 0\ m_1\ 0 \dots 0]$	$[1\ 0\ m_1\ 0\ m_2\ 0 \dots 0]$
$Var(\hat{\omega}) = $ Sampling variability of $\hat{\omega}$	$s_{b_1}^2 + 2m_1 s_{b_1,b_3} + m_1^2 s_{b_3}^2$	$s_{b_1}^2 + 2m_1 s_{b_1,b_3} + 2m_2 s_{b_1,b_5} + 2m_1 m_2 s_{b_3,b_5} + m_1^2 s_{b_3}^2 + m_2^2 s_{b_5}^2$
Significance region coefficients for M_1 (M_q)	$\tilde{a} = W_\alpha s_{b_3}^2 - b_3^2$ $\tilde{b} = W_\alpha 2 s_{b_1,b_3} - 2 b_1 b_3$ $\tilde{c} = W_\alpha s_{b_1}^2 - b_1^2$	$\tilde{a} = W_\alpha s_{b_3}^2 - b_3^2$ $\tilde{b} = W_\alpha \left(2 s_{b_1,b_3} + 2m_2 s_{b_3,b_5}\right) - 2 b_1 b_3 - 2 b_3 b_5 m_2$ $\tilde{c} = W_\alpha \left(s_{b_1}^2 + 2m_2 s_{b_1,b_5} + m_2^2 s_{b_5}^2\right) - \left(b_1^2 + 2 b_1 b_5 m_2 + b_5^2 m_2^2\right)$
Significance region derivatives for M_1	$\dfrac{d\hat{\omega}}{dm_1} = b_3$ $\dfrac{dVar(\hat{\omega})}{dm_1} = 2 s_{b_1,b_3} + 2m_1 s_{b_3}^2$	$\dfrac{d\hat{\omega}}{dm_1} = b_3$ $\dfrac{dVar(\hat{\omega})}{dm_1} = 2 s_{b_1,b_3} + 2m_2 s_{b_3,b_5} + 2m_1 s_{b_3}^2$
Three or More Moderators (J)		**Three-Way Interaction**
Modeling equation[b]	$\eta = b_1 F + b_2 M_1 + b_3 F \times M_1 + b_4 M_2 + b_5 F \times M_2 + \cdots$ $+ b_{2J} M_J + b_{2J+1} F \times M_J + b_z Z + a$	$\eta = b_1 F + b_2 M_1 + b_3 F \times M_1 + b_4 M_2 + b_5 F \times M_2$ $+ b_6 M_1 \times M_2 + b_7 F \times M_1 \times M_2 + b_z Z + a$
$\hat{\omega}$ Effect of F at $M_J = m_J$	$\hat{\omega} = b_1 + b_3 m_1 + b_5 m_2 + \cdots + b_{2J+1} m_J$	$\hat{\omega} = b_1 + b_3 m_1 + b_5 m_2 + b_7 m_1 m_2$
c = Weights for calculating $Var(\hat{\omega})$	$[1\ 0\ m_1\ 0\ m_2 \dots 0\ m_J\ 0 \dots 0]$	$[1\ 0\ m_1\ 0\ m_2\ 0\ m_1 m_2\ 0 \dots 0]$
$Var(\hat{\omega}) = $ Sampling variability of $\hat{\omega}$	$s_{b_1}^2 + \displaystyle\sum_{j=1}^{J} m_j^2 s_{2j+1}^2 + \sum_{j=1}^{J} 2 m_j s_{b_1, b_{2j+1}}$ $+ \displaystyle\sum_{j=1}^{J-1} \sum_{k=j+1}^{J} 2 m_j m_k s_{b_{2j+1}, b_{2k+1}}$	$s_{b_1}^2 + 2m_1 s_{b_1,b_3} + 2m_2 s_{b_1,b_5} + 2m_1 m_2 s_{b_1,b_7} + 2m_1 m_2 s_{b_3,b_5}$ $+ 2m_1^2 m_2 s_{b_3,b_7} + 2m_1 m_2^2 s_{b_5,b_7} + m_1^2 s_{b_3}^2 + m_2^2 s_{b_5}^2 + m_1^2 m_2^2 s_{b_7}^2$

Three or More Moderators (J)	Three-Way Interaction
Significance region coefficients for M_1 (M_q) $\tilde{a} = W_\alpha s_{b_{2q+1}}^2 - b_{2q+1}^2$ $\tilde{b} = W_\alpha \left(2s_{b_1,b_{2q+1}} + \sum_{j\neq q}^{J} 2m_j s_{b_{2q+1},b_{2j+1}} \right)$ $\quad - 2b_1 b_{2q+1} - \sum_{j\neq q}^{J} 2b_{2q+1} b_{2j+1} m_j$ $\tilde{c} = W_\alpha \left(\begin{array}{l} s_{b_1}^2 + \sum_{j\neq q}^{J} \left[m_j^2 s_{b_{2j+1}}^2 + 2m_j s_{b_1,b_{2j+1}} \right] \\[2mm] + \sum_{j\neq q}^{J-1} \sum_{k=j+1}^{J} 2m_j m_k s_{b_{2j+1},b_{2k+1}} \end{array} \right)$ $\quad - b_1^2 - \sum_{j\neq q}^{J} m_j^2 b_{2j+1}^2 - \sum_{j\neq q}^{J-1} \sum_{k=j+1}^{J} 2m_j m_k b_{2j+1} b_{2k+1}$	$\tilde{a} = W_\alpha \left(2m_2 s_{b_3,b_7} + s_{b_3}^2 + m_2^2 s_{b_7}^2 \right) - 2b_3 b_7 m_2 - b_3^2 - b_7^2 m_2^2$ $\tilde{b} = W_\alpha \left(2s_{b_1,b_3} + 2m_2 s_{b_1,b_7} + 2m_2 s_{b_3,b_5} + 2m_2^2 s_{b_5,b_7} \right)$ $\quad - 2b_1 b_3 - 2b_1 b_7 m_2 - 2b_3 b_5 m_2 - 2b_5 b_7 m_2^2$ $\tilde{c} = W_\alpha \left(s_{b_1}^2 + 2m_2 s_{b_1,b_5} + m_2^2 s_{b_5}^2 \right) - \left(b_1^2 + 2b_1 b_5 m_2 + b_5^2 m_2^2 \right)$
Significance region derivatives for M_1 (M_q) $\dfrac{d\hat{\omega}}{dm_1} = b_{2q+1}$ $\dfrac{dVar(\hat{\omega})}{dm_1} = 2m_q s_{b_{2q+1}}^2 + 2s_{b_1,b_{2q+1}} + \sum_{j\neq q}^{J} 2m_j s_{b_{2q+1},b_{2j+1}}$	$\dfrac{d\hat{\omega}}{dm_1} = b_3 + b_7 m_2$ $\dfrac{dVar(\hat{\omega})}{dm_1} = 2s_{b_1,b_3} + 2m_2 s_{b_1,b_7} + 2m_2 s_{b_3,b_5} + 4m_1 m_2 s_{b_3,b_7}$ $\quad + 2m_2^2 s_{b_5,b_7} + 2m_1 s_{b_3}^2 + 2m_1 m_2^2 s_{b_7}^2$

a. Because the term $b_z Z$ represents a set of additional predictors, the rightmost entries in $\underline{c}$ would in practice consist of a set of zeros as designated by the 0 … 0 notation in the last column.

b. The term $b_z Z$ is a set of other predictors; thus, the last entries in $\underline{c}$ would be a set of zeros as designated by the 0 … 0 notation in the last column.

To apply this, we need to specify the content of $\tilde{a}$, $\tilde{b}$, and $\tilde{c}$ for the different forms of interaction effects that are shown in row 5 of Table 3.1. Note that the solutions (roots) in a particular empirical application could be undefined if $\tilde{a} = 0$, or the solution could be a complex number if $\tilde{b}^2 - 4\tilde{a}\tilde{c} < 0$. Either situation indicates that the boundaries do not exist; that is, there are no values of the moderators at which the significance of the effect of F changes. If there are multiple moderators, we have to solve this equation for one moderator at a time, contingent on the values of the other moderators that will be incorporated into the expressions for $\tilde{a}$, $\tilde{b}$, and $\tilde{c}$. As a practical matter, the easiest approach is to calculate $\tilde{a}$, $\tilde{b}$, and $\tilde{c}$ for an empirical application and then substitute those calculated values into Equation 3.3 to find the boundary values of m_1.

JN for One Moderator

To illustrate this, let's start with the single-moderator case. For this case only, I show how to derive the content of $\tilde{a}$, $\tilde{b}$, and $\tilde{c}$ as diagrammed in Equation 3.5. You substitute the specific content from rows 2 and 4 in Table 3.1 for $\hat{\omega}$ and $Var(\hat{\omega})$ into Equation 3.3. Then collect the terms by which m_1^2 is multiplied to define $\tilde{a}$ (*solid arrows*), the terms by which m_1 is multiplied to define $\tilde{b}$ (*dashed arrows*), and the remaining terms to define $\tilde{c}$ (*dotted arrows*).

$$W_\alpha \left(s_{b_1}^2 + 2s_{b_1,b_3}m_1 + s_{b_3}^2 m_1^2 \right) - \left(b_1 + b_3 m_1 \right)^2 = 0 \qquad \text{or} \qquad \text{expanding} \left(b_1 + b_3 m_1 \right)^2$$

$$W_\alpha \left(s_{b_1}^2 + 2s_{b_1,b_3}m_1 + s_{b_3}^2 m_1^2 \right) - \left(b_1^2 + 2b_1 b_3 m_1 + b_3^2 m_1^2 \right) = 0$$

$$\tilde{a} = W_\alpha s_{b_3}^2 - b_3^2 \qquad \tilde{b} = W_\alpha 2s_{b_1,b_3} - 2b_1 b_3 \qquad \tilde{c} = W_\alpha s_{b_1}^2 - b_1^2 \qquad\qquad (3.5)$$

Calculating the values of $\tilde{a}$, $\tilde{b}$, and $\tilde{c}$ from your analysis and plugging them into Equation 3.4 yields the values of m_1 at which the effect of F changes from significant to nonsignificant.

We can determine whether these values of m_1 mark changes in F's effect from significant to nonsignificant or the other way around by evaluating the derivative of W with respect to m_1 at the value of the roots (solutions) calculated from Equation 3.4. If the derivative is positive, then the value of the Wald statistic is increasing as m_1 increases past the boundary value, and the effect of F is becoming significant at that point. Conversely, if the derivative is negative, then the test statistic is decreasing as m_1 increases, so the effect of F is turning nonsignificant. Taking the derivative of W in Equation 3.2 provides a calculating formula whose content varies with the form of the interaction effect:

$$\frac{dW}{dm_1} = \frac{2\hat{\omega}}{Var(\hat{\omega})} \times \frac{d\hat{\omega}}{dm_1} - \frac{dVar(\hat{\omega})}{dm_1} \times \left(\frac{\hat{\omega}}{Var(\hat{\omega})} \right)^2 \qquad (3.6)$$

Table 3.1 lists the specific content for each expression for the four forms of interaction effects. To compute the value for the derivative at those values of m_1, we calculate the values of $\hat{\omega}$ and $Var(\hat{\omega})$ and of their derivatives as shown in Table 3.1 (rows 2, 4, and 6) and use those values in Equation 3.6.

Let's apply this to the OLS regression example from Chapter 2 of the interaction between headship type and number of children as it affects net worth, using headship type as the focal variable and children as the moderator. I illustrate the JN boundary value calculations on the moderated effect of the single-female-head category in Table 3.2. To use Equations 3.2 and 3.3, we need the values of the coefficients for F (−3.7695913) and for the interaction of $F \times M$ (1.6774052), as well as the sampling variance—also known as the squared standard error—for each coefficient (0.19176000 and 0.15242387, respectively) and the covariance between the two coefficients (−0.08331753).[6] And we need to select a critical value for the Wald statistic (W_α). Because OLS regression uses least squares estimation with known finite sample properties, we specify an $F_{1,4221}$ critical value; for $\alpha = .05$, the critical value is 3.8421128 (if these had been maximum-likelihood estimated results, we would specify a critical value of $\chi^2(1) = 3.84146$). Putting these values into the equations produces the results shown in Table 3.2. I show intermediate results in the calculation of the boundary values (roots) and the derivatives in this table so that readers

Significance Region Boundary (Equation 3.2)	Derivative at Boundary Value (Equation 3.3)
$\tilde{a} = 3.8421128 \times 0.15242387 - (1.6774052)^2 = -2.2280585$	$\hat{\omega} = -3.7695913 + 1.6774052\, m_1$
$\tilde{b} = 3.8421128 \times 2 \times (-0.08331753) - 2 \times (-3.7695913)$ $\times 1.6774052 = 12.006033$	$Var(\hat{\omega}) = 0.1917600 + 2m_1(-0.08331753)$ $+ m_1^2(0.15242387)$
$\tilde{c} = 3.8421128 \times 0.191760001 - (-3.7695913)^2$ $= -13.473055$	$\dfrac{d\hat{\omega}}{dm_1} = 1.6774052$
	$\dfrac{dVar(\hat{\omega})}{dm_1} = 2(-0.08331753) + 2m_1(0.15242387)$
Boundary 1 $m_1 = 1.59330$	Derivative at Boundary 1 $= -15.66421$
$\dfrac{-12.006033 + \sqrt{(12.006033)^2 - 4(-2.2280585)(-13.473055)}}{2(-2.2280585)}$	$\dfrac{2(-3.7695913 + 1.6774052 \times 3.79562)}{0.1917600 + 2 \times 3.79562(-0.08331753) + 3.79562^2(0.15242387)} \times 1.6774052$
$= \dfrac{-12.006033 + 4.90609922}{-4.45611699} = \dfrac{-7.09993418}{-4.45611699} = 1.5933$	$-[2(-0.08331753) + 2 \times 3.79562(0.15242387)] \times \left(\dfrac{-3.7695913 + 1.6774052 \times 3.79562}{0.1917600 + 2 \times 3.79562(-0.08331753) + 3.79562^2(0.15242387)}\right)^2$
	$= \dfrac{-2.1939617}{0.31320449} \times (1.6774052) - (0.03190790)\left(\dfrac{-1.09698085}{0.31320449}\right)^2$
	$= -7.0048858 \times (1.6774052) - (0.03190790) \times (-3.5024429)^2$
	$= -11.7500319 - 3.9141759 = -15.66421$

(Continued)

TABLE 3.2 ⬡ (CONTINUED)

Significance Region Boundary (Equation 3.2)	Derivative at Boundary Value (Equation 3.3)
Boundary 2 $m_1 = 3.79526$ $$\frac{-12.006033 - \sqrt{(12.006033)^2 - 4(-2.2280585)(-13.473055)}}{2(-2.2280585)}$$ $$= \frac{-12.006033 - 4.90609922}{-4.45611699} = \frac{-16.91213262}{-4.45611699} = 3.79526$$	Derivative at Boundary 2 = 2.79573 $$\frac{2(-3.7695913 + 1.6774052 \times 3.79562)}{0.1917600 + 2 \times 3.79562(-0.08331753) + 3.79562^2(0.15242387)} \times 1.6774052$$ $$- \left[2(-0.08331753) + 2 \times 3.79562(0.15242387)\right] \times \left(\frac{-3.7695913 + 1.6774052 \times 3.79562 + 3.79562^2(0.15242387)}{0.1917600 + 2 \times 3.79562(-0.08331753) + 3.79562^2(0.15242387)}\right)^2$$ $$= \frac{5.1932034}{1.7548523} \times (1.6774052) - (0.9903421)\left(\frac{2.59660172}{1.75485230}\right)^2$$ $$= 2.95933933 \times (1.6774052) - (0.9903421) \times (1.47796697)^2$$ $$= 4.9640112 - 2.1682772 = 2.79573$$

Note: OLS = ordinary least squares; JN = Johnson-Neyman.

can verify their understanding of how to apply the formulas. For those who want to implement this approach in a spreadsheet or as code for their preferred statistical software, this also provides a way to validate their application. The SIGREG tool in ICALC produces a table of boundary value results, which I use in the next section to discuss their interpretation.

Interpretation of Boundary Value Analysis Results From ICALC

```
Boundary Values for Significance of Effect of Headship_Type on g(Netw10K)
Moderated by Children
   Critical value F = 3.842 set with p = 0.0500


+------------------------------------------------------------------------------+
| Effect of     | When        | Sig Changes         | When         | Sig Changes |
| Headship_T    | Children >= |                     | Children >=  |             |
|---------------+-------------+---------------------+--------------+-------------|
|   Single_man  |    1.21     | to Not Sig [-5.69]  | -0.84 (< min)| to Sig[4.90]|
| Single_woman  |    1.59     | to Not Sig [-15.66] |    3.80      | to Sig[2.80]|
+------------------------------------------------------------------------------+
     Note: Derivatives of Boundary Values in [ ]
```

ICALC command: sigreg, ndig(2). The ndig( ) option specifies the number of decimals in table results.

The results from the Stata output window show the boundary value analysis for both headship type categories. There is a pair of columns for each boundary value: The column labeled "When Children >=" reports a boundary value, and its adjacent column, "Sig Changes," notes whether passing that boundary results in a change to a nonsignificant effect or a change to a significant effect for the focal variable. The number in square brackets is the value of the derivative at the boundary value; a positive value indicates a change to significance and a negative value a change to nonsignificance. Recall that the sign change analysis of this example in Chapter 2 showed that the moderated effect of single-man is always negative but the moderated effect of single-woman changed from negative to positive at *Children* = 2.25.

For the moderated effect of single-man, the boundary values occur at *Children* $\geq$ −0.84 and *Children* $\geq$ 1.21. The boundary at −0.84 marks a change to a significant effect according to its "Sig Changes" column, while the boundary at 1.21 defines a transition to a nonsignificant effect type on wealth. Since the effect of single-man is always negative, this tells us that the net worth of single-man households is predicted to be less than that of married-couple households for any number of children, but the difference is only significant in the three quarters of households with zero or one child (i.e., the real-world correspondence to the criterion *Children* $\leq$ 1.21).

Turning to the moderated effect of single-woman, the boundary value at *Children* = 1.59 indicates that the single-woman effect becomes nonsignificant in households with more than 1 child; note the negative derivative [−15.66]. The other boundary at *Children* = 3.7953 has a positive derivative [2.80], meaning that the effect of single-female-head changes back to significant at that number of children. The prior sign change analysis shows that the effect of single-woman turns from negative to positive between the two boundary values, at *Children* $\geq$ 2.25. The sign change and boundary value results together provide a clear picture of the changes in the moderated effect of single-female-head:

- When the number of children is between 0 and 1.59, the effect of single-woman is negative and significant. Thus, for households with less than

2 children—a little more than three quarters of households—single-woman households are predicted to have a significantly lower net worth than married-couple households.

- Between 1.59 and 3.80 children, the effect of single-woman is not significant (22% of households), and the effect changes from negative to positive at 2.24 children. That is, the difference between single-woman and married-couple households is not significant when they have 2 or 3 children, changing from a "deficit" if have 2 children and an "advantage" if have 3 children.

- For the 2% of households with more than 3.80 children ((i.e., 4 or more children), single-woman households are predicted to have a significantly higher net worth than married-couple households.

This is a good illustration of the potential value of defining the significance region. It lets the researcher provide an informative summary of how the effect of the focal variable changes with the moderator as well as when the focal variable's effects are and are not significant. In practice, I would recommend only presenting to your audience the boundary values as part of your discussion of what they tell you about the moderated effect of F, without mentioning the derivatives, much as I did in the bullet points above.

JN for Two Moderators (or Three-Way Interaction)

Starting with a model with two moderators, we again apply the generic expression in Equation 3.2 to find the boundary values for the moderators using the specifications for this type of interaction effect from Table 3.1. This is more complicated because the boundary values for M_1 depend on the values of M_2, and vice versa. Note that rows 5 and 6 of the "Two Moderators" column include m_2 in the expressions for the significance region coefficients and derivatives. And as we will see, interpreting these boundary values is not straightforward at first sight. The example is the logistic regression analysis of a marriage ban approval predicted by the effect of race moderated by education and race by region. Let's begin with the boundary values for education (M_1) contingent on region (M_2)—that is, the values of education at which the effect of Race changes significance and how those boundary values vary by region. For readers interested in implementing this procedure on their own, I present a walk-through of the steps and some calculations for one of the boundary values to make clear how to implement the formulas in a spreadsheet or other software. I would strongly advise against hand calculations as they are very error-prone.

The first step is to calculate $\tilde{a}, \tilde{b},$ and $\tilde{c}$ using the formulas in Table 3.1, the values of the coefficients and their sampling variances and covariances, and a $\chi^2(1)$ critical value of 3.84146 because the coefficients are maximum-likelihood estimates. The top panel of Table 3.3 shows the information needed for these calculations, and the bottom two panels show the results of a sample calculation. The middle panel shows how calculating $\tilde{a}, \tilde{b},$ and $\tilde{c}$ gives formulas that depend on the value of the second moderator variable (m_2). The bottom panel illustrates how to use these formulas to

TABLE 3.3 ● CALCULATING SIGNIFICANCE REGION BOUNDARY VALUES FOR EDUCATION CONTINGENT ON REGION

Variance/Covariance of	With			Coefficient
	b_1	b_3	b_5	
Race b_1	0.793785			2.835118
Race × Education b_3	−0.055509	0.0044859		−0.142396
Race × Region b_5	−0.223369	0.006098	0.25100725	1.139440

Plugging this information into the equations in row 5 for the "Two Moderators" column of Table 3.1

$$\tilde{a} = 3.84146 \times 0.004859 - \left(-0.142396^2\right) = 0.001611$$

$$\tilde{b} = 3.84146\left(2 \times (-0.055509) + 2 \times m_2 \times 0.006098\right) - 2 \times 2.835118 \times (-0.142396) - 2 \times (-0.142396) \times 1.139440 \times m_2$$
$$= -0.426471 + 0.046850 \times m_2 + 0.807419 + 0.3245034 \times m_2$$
$$= 0.380948 + 0.371354 \times m_2$$

$$\tilde{c} = 3.84146\left(0.793785 + 2 \times m_2 \times (-0.223369) + m_2^2 \times 0.251007\right)$$
$$\quad -\left(2.835118^2 + 2 \times 2.835118 \times 1.139440 \times m_2 + 1.139440^2 \times m_2^2\right)$$
$$= 3.049293 - 1.716126m_2 + 0.964319m_2^2$$
$$\quad -\left(8.037894 + 6.460893m_2 + 1.298324m_2^2\right)$$
$$= -4.988601 - 8.177020m_2 - 0.334090m_2^2$$

Example: Boundary values when *Region* = 1 using Equation 3.3

$$\tilde{a} = -0.001611 \quad \tilde{b} = 0.380948 + 0.371354 \times 1 = 0.752302$$

$$\tilde{c} = -4.988601 - 8.177020 \times 1 - 0.334090 \times 1^2 = -13.49971$$

$$m_1 = \frac{-\tilde{b} \pm \sqrt{\tilde{b}^2 - 4\tilde{a}\tilde{c}}}{2\tilde{a}} = \frac{-0.752302 \pm \sqrt{0.752302^2 - 4 \times (-0.001611) \times (-13.499711)}}{2 \times (-0.001611)}$$

$$\frac{-0.752306 \pm 0.692075}{-0.003222} = 18.7, \ 448.3$$

find the boundary values for education for a specific value of region (1 = *South*). One of the boundary values (448.4) is well outside the possible range of education values of 0 to 20. The other boundary is at *Education* = 18.7.

Interpretation of Boundary Value Analysis Results From ICALC

Race Effect Significance, Boundary Values for Education on g(Bwlaw)
Given Region
    Critical value Chi_sq = 3.841 set with p = 0.0500

Effect of Race	When Education>=	Sig Changes	When Education >=	Sig Changes
At Region =				
NonSouth	13.9136	to Not Sig [-1.7710]	222.6202 (> max)	to Sig [0.0015]
South	18.6925	to Not Sig [-1.5426]	448.4186 (> max)	to Sig [0.0007]

Note: Derivatives of Boundary Values in [ ]

Race Effect Significance, Boundary Values for Region on g(Bwlaw) Given Education
Critical value Chi_sq = 3.841 set with p = 0.0500

Effect of Race	When Region >=	Sig Changes	When Region >=	Sig Changes
At Education =				
0.00	-23.8495 (< min)	to Not Sig [-0.0503]	-0.6261 (< min)	to Sig [6.6207]
2.00	-21.6677 (< min)	to Not Sig [-0.0552]	-0.5848 (< min)	to Sig [7.6229]
4.00	-19.4934 (< min)	to Not Sig [-0.0612]	-0.5360 (< min)	to Sig [8.8846]
6.00	-17.3294 (< min)	to Not Sig [-0.0685]	-0.4769 (< min)	to Sig [10.4685]
8.00	-15.1802 (< min)	to Not Sig [-0.0779]	-0.4031 (< min)	to Sig [12.4000]
10.00	-13.0528 (< min)	to Not Sig [-0.0903]	-0.3073 (< min)	to Sig [14.5311]
12.00	-10.9597 (< min)	to Not Sig [-0.1072]	-0.1773 (< min)	to Sig [16.1955]
14.00	-8.9236 (< min)	to Not Sig [-0.1318]	0.0097 (< min)	to Sig [15.7695]
16.00	-6.9882 (< min)	to Not Sig [-0.1605]	0.2974	to Sig [11.6564]
18.00	-5.2373 (< min)	to Not Sig [-0.2376]	0.7696	to Sig [5.8400]
20.00	-3.8064 (< min)	to Not Sig [-0.3641]	1.5618 (> max)	to Sig [12.2070]

Note: Derivatives of Boundary Values in [ ]

ICALC command: sigreg, plot(White) ndig(4). The ndig( ) option specifies number of decimals in table results.

The ICALC analysis creates a boundary value table for each moderator, with the row showing how the boundary values for that moderator are contingent on the values of the other moderator. Note that boundary values are annotated to indicate when the boundary value lies below the valid range of the moderator (<min) or above the valid range (>max). The first table describes how the significance of race's effect changes with education dependent on the value of region, and the second reports the changing significance of race's effect across the categories of region contingent on the value of education. Let me detail the interpretation for the first row of the first table and then provide a holistic interpretation of each set of results. Keep in mind that the sign change analysis in Chapter 2 found that the effect of race is virtually always positive. For those raised outside the South (*Region* = 0), the boundary values for the significance of the effect of race are at *Education* ≥ 13.8 and at *Education* ≥ 222.6. The boundary at 13.8 marks the transition to a not significant effect because the derivative is a negative value (−1.7710). That is, the effect of race becomes non-significant as education crosses this boundary for those raised outside the South. The other boundary value (222.6) shows that the effect of race remains nonsignificant until education increases beyond 222 years, at which point it becomes significant, as denoted by the positive value of the derivative (0.0015).

More simply put, for those raised outside the South, the effect of race is positive and significant from education's minimum value (0) through almost 14 years of education and is not significant when *Education* > 13.8. In contrast, for those raised in the South, the effect of race is significant for *Education* < 19, and it is not significant only for those with 19 or more years of education.

The boundary values table reports if and when the effect of race changes significance across the categories of region, depending on the value of education. When *Education* = 0 to 12 years, you see the same pattern of boundary values and derivatives for region. Using the row for *Education* = 0 as an example, the boundary value at *Region* ≥ −23.85 marks a transition to a nonsignificant effect of race because the derivative is negative. The second boundary value at *Region* ≥ −0.6261 denotes a transition to a significant effect of race because the derivative is positive. However, the valid range of values for *Region* are 0 to 1, which lies above the second boundary value, meaning that the effect of race is significant for both southern upbringing (*Region* = 1) and nonsouthern upbringing (*Region* = 0). For *education* = 14 to 18, the second boundary value lies between 0 and 1 (0.01, 0.30, and 0.77), which means that the effect of race is significant when region passes that boundary; that is, it is significant for southern upbringing (*Region* = 1) but not for nonsouthern upbringing (*Region* = 0). Last, for *Education* = 20, the second boundary value (1.56) is beyond the valid range of region; thus the effect of race is not significant in either region at this value of education.

This is sometimes a little easier to visualize if you create a plot of the boundary values of M_1 as they vary with M_2 and on the same graph plot the boundary values of M_2 as they vary with M_1. To do this for the effect of race, you draw two curves: one marking where the race effect becomes significant with increasing region values and the other one marking where the race effect turns nonsignificant with increasing region values. Both curves are plotted with region boundary values on the *y*-axis as they vary with education values on the *x*-axis. On the same graph, you plot education boundary values by drawing two curves for the education boundary values—one for change to significance and the other for change to nonsignificance. The education boundary values define the *x*-axis, and the curves show how they vary with the region values on the *y*-axis. Figure 3.1 provides such a simultaneous plot. The lines plotted with no markers represent the region boundaries: a solid line marking where the race effect changes to not significant and a dashed line representing where the race effect changes to significant. And the markers plotted without a line show the education boundaries; the × symbol marks where the race effect changes to not significant, and a gray • indicates where the race effect becomes significant. The y-coordinate and x-coordinate values of the plotted points are truncated to a value slightly lower than the minimum or slightly higher than the maximum of the moderator defining the axis. The area within the curves and markers shows where the race effect is significant. Specifically, the race effect is significant when the region value lies above the dashed line and, simultaneously, the education value is to the left of the × boundary.

The way to understand how the figure shows this is either to pick a point on the education axis and determine what happens to the significance of the race effect as the values of region increase from the bottom of the plot area to the top or to pick a point on the region axis and examine what happens as education values increase from left to right. Consider first the point on the x-axis where *Education* = 10, and follow a vertical line from the x-axis at the bottom of the plot to the top of the plot. If you read straight up from that point—region changes and education stays at 10—the path quickly passes two boundary lines in succession, the first (*solid*) line marking a change to a nonsignificant effect of race, almost immediately followed by the second (*dashed*) line indicating a change to a significant effect of race. Because the upward

---

FIGURE 3.1 ● JOHNSON-NEYMAN BOUNDARY VALUE PLOT FOR TWO MODERATORS

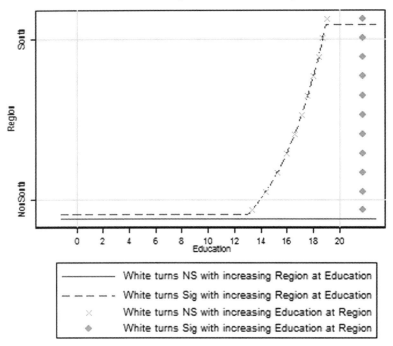

Boundary Value Plot for Significance of White on g(ban)
Moderated by Education and Region

————	White turns NS with increasing Region at Education
– – – – ·	White turns Sig with increasing Region at Education
×	White turns NS with increasing Education at Region
•	White turns Sig with increasing Education at Region

ICALC command: sigreg, plot(White) ndig(4). The plot( ) names the graph, and ndig( ) formats table cells.

path crosses no more boundary lines, the effect of race is significant for both categories of region. If we instead examine the path where *Education* = 16, the upward path crosses the line of nonsignificance first, then moves past the Region's minimum value before it crosses the line of significance. This means that the effect of Race is not significant when *Region* = *non-South* and *Education* = 16 but is significant for *Region* = *South* because the upward path then crosses the line of significance before it reaches the value for *Region* = *South*.

Now start on the *y*-axis where region equals non-South (bottom of the *y*-axis), and trace a horizontal path as the education values increase. The path crosses the line of gray × markers showing where the race effect turns nonsignificant after it reaches the minimum value of education. This means that the effect of race when *Region* = non-South is significant for *Education* ≥ 0 until the path crosses the line of × symbols marking the change to a not significant effect, which it does for *Education* ≥ 13.8 years. Now trace the horizontal path when *Region* = *South* (top of the *y*-axis) to see that the effect of race for this region of upbringing is significant when *Education* ≤ 18.9, the value at which it crosses the line of nonsignificance marked by the × symbols.

Summary

I think it is debatable whether the boundary value plot in Figure 3.1 is easier to comprehend than the results in the boundary value analysis tables. For a two-moderator interaction, both take some degree of concentration to understand and require more than a brief explanation to an audience. Applying the JN method to a three-way interaction is only different from the two-moderator application in terms of the content substituted into the formulas for calculating the boundary values and their derivatives and involves more terms (see rows 5 and 6 in Table 3.1). But the patterns defining the significance region are usually more complex. Let us very briefly consider the boundary values plot in Figure 3.2 for the three-way interaction between sex, age, and education as they predict the number of voluntary association memberships. This treats sex as the focal variable. At the youngest ages (≤40), the sex effect is significant for education values in the lower part of its distribution (below the solid line), is not significant for education values in the middle of its distribution (between the solid and dashed lines), and then turns significant again for the youngest ages at the top of the education distribution (above the *dashed line*). Conversely, at most of the older ages, the effect of sex is not significant at the bottom of the education distribution (below the *dashed line*), significant at the middle of the distribution (between the dashed and solid lines), and then not significant again at the top of the education distribution (above the *solid line*) except for the very oldest ages.

In sum, the JN boundary value approach provides accessible results for a single-moderator interaction specification, and I would highly recommend its use for such models. But for models with multiple moderators or a three-way interaction, we need a more readily understood and explained method for exploring significance regions. I describe two such approaches in the following sections, empirically derived significance region tables followed by plots showing confidence bounds or confidence intervals.

EMPIRICALLY DEFINED SIGNIFICANCE REGION

A mathematically less elegant but more practical and intuitive solution is to calculate the Wald statistic $\left(\dfrac{\hat{\omega}^2}{\widehat{Var(\hat{\omega})}} \right)$ and its significance level for combinations of selected

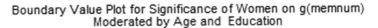

FIGURE 3.2   ●   JOHNSON-NEYMAN BOUNDARY VALUE PLOT FOR THREE-WAY INTERACTION

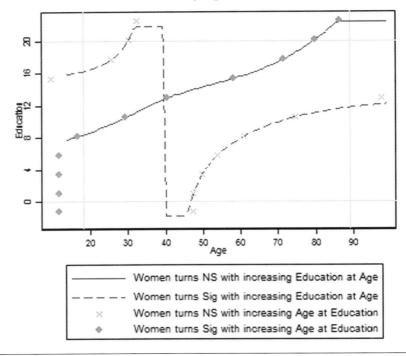

ICALC command: sigreg, plot(female) pltopts(xlab(20(10)90) ylab(0(4)20)). The plot( ) option names the graph, and pltopts( ) customizes the labels on the *x*- and *y*-axes.

values of the moderator(s). For an interval moderator, I recommend a first pass using around 8 to 10 evaluation points evenly spaced from its minimum to its maximum; use all the categories of a categorical moderator. For two moderators (or a three-way interaction), imagine a crosstab-like table where the rows and columns represent M_1 and M_2, respectively, and the cell entries report the sign and significance of the effect of F for the selected values of the moderators. For a single moderator, the table would have a single row. For three (or more) moderators, there would be multiple such 2D tables, one for each selected value of the remaining moderators. If the focal variable is nominal with more than two categories, you need a set of tables for each effect of F. Such calculations can be easily implemented in a spreadsheet or a statistical software program relying on the formulas for $\hat{\omega}$ and $\widehat{Var}(\hat{\omega})$ specified in Table 3.1 for the different forms of interaction effects. To illustrate for the one-moderator example, let's calculate the moderated effect of single-woman and its significance when *Children* = 4. The calculating formulas for this example's moderated effect and its variance when the moderator value is m_1 are shown in the top right-hand box of Table 3.2. Substituting $m_1 = 4$ into these formulas gives

$$\hat{\omega} = -3.7695913 + 1.6774052 \ m_1 = -3.7695913 + 1.6774052 \times 4 = 2.94003$$

$$Var\left(\hat{\omega}\right) = 0.1917600 + 2m_1\left(-0.08331753\right) + m_1^2\left(0.15242387\right)$$

$$= 0.1917600 + 2 \times 4\left(-0.08331753\right) + 4^2 \times\left(0.15242387\right) = 1.964002$$

$$\text{Wald} = \frac{\hat{\omega}^2}{\widehat{Var}(\hat{\omega})} = \frac{2.94003^2}{1.964002} = \frac{8.643773}{1.964002} = 4.401103$$

Because the Wald test statistic (4.401) is greater than the critical value ($F_{1,14221}$ = 3.842), we mark the moderated effect of 2.940 as significant (see the output results in the next section).

As the applications below illustrate, highlighting cells with significant effects of F and using different font styles for positive and negative effects can facilitate interpreting and explaining the regions of significance. These significance region tables were all produced using the SIGREG tool in ICALC.

One-Moderator Results and Interpretation

```
Significance Region for Effect of Headship_Type (1 unit difference)
    on g(Netw10K) at Selected Values of Children
   ------------------------------------------------------------------------
                                        At Children=
   Effect of  |  0    1    2    3    4    5    6    7    8    9    10
   ---------+----------------------------------------------------------------
   Single_man  | -1.7* -1.6* -1.6  -1.5  -1.5  -1.4  -1.3  -1.3  -1.2  -1.1  -1.1
   Single_woman | -3.8* -2.1* -0.4   1.3   2.9*  4.6*  6.3*  8.0*  9.6* 11.3* 13.0*
   ------------------------------------------------------------------------
     Key: Plain font  = Pos, Not Sig    Bold font*   = Pos, Sig
          Italic font = Neg, Not Sig    Italic font* = Neg, Sig
```

ICALC command: sigreg, nobva ndig(1). The nobva option suppresses boundary value analysis, and ndig( ) specifies 1 decimal format for table results.

This significance region table reports results for the moderated effect of two headship types, each effect embodying a contrast to married-couple households. Because the moderator (children) has only 11 values, I evaluated the moderated effect's significance at every value. The key for the table tells us that negative effects are in italics with an * added if the effect is significant. Positive effects are in a plain font if not significant or a bold font with an * if the effect is significant. Visually, this makes it fairly easy to look at the table and immediately realize that the effects for both categories of headship type are negative and significant for households with few children but positive and significant for single-woman households with four or more children.

Specifically, the effect of single-man household on net worth (in $10K) is always negative; that is, the predicted net worth of single-man households is less than that of married-couple households. The differences range from −$17K to −$11K, decreasing as children increases, and are large enough to be significant in households with zero to one child. The predicted net worth for single-woman households is also significantly less than that for married-couple households with zero to one child but larger in magnitude than for single-man households. Surprisingly, this effect becomes significant and increasingly positive—a larger net worth for single-woman than for married-couple households—for households with four or more children.

Although the boundary value analysis of the moderated effect of household type also clearly shows the values at which the moderated effect changes significance, the advantage of the significance region table is its reporting of the effect sizes.

Furthermore, the subsequent application chapters for specific GLMs will demonstrate how to create significance region tables in an effect metric other than the modeled metric of η, such as the factor changes or discrete effect changes calculated by SPOST13 (Long & Freese, 2014). The one minor disadvantage to this alternative is that it does not calculate the exact boundary values, leaving you to infer approximately where the change in significance occurs. But you can pair the results with a JN boundary value analysis to calculate the exact boundary values.

Application to the Two-Moderator Example

```
Significance Region for Effect of White (1 unit difference)
on g(Bwlaw) at Selected Values of Region and Education
------------------------------------------------------
                                 At Region=
       At Education= |     NonSouth       South
    ---------------+----------------------------------
               0   |       2.84*         3.97*
               2   |       2.55*         3.69*
               4   |       2.27*         3.41*
               6   |       1.98*         3.12*
               8   |       1.70*         2.83*
              10   |       1.41*         2.55*
              12   |       1.13*         2.27*
              14   |       0.84*         1.98*
              16   |       0.56          1.70*
              18   |       0.27          1.41*
              20   |      -0.01          1.13
    ------------------------------------------------------
Key: Plain font   = Pos, Not Sig    **Bold font***   = Pos, Sig
     *Italic font*  = Neg, Not Sig    *Italic font**  = Neg, Sig
```

ICALC command: sigreg, nobva ndig(2). The nobva option suppresses boundary value analysis, and ndig( ) specifies 2 decimals to format table results.

These results for the two-moderator example demonstrate how simple it can be to use a significance region table to communicate the patterns of moderation and statistical significance of the focal variable's effect. It reports the moderated effect of White—note that ICALC uses the category label for the dummy variable race—with the significant effects highlighted in a bold font with an *, while the nonsignificant effects are in plain text. Other than telling your audience how the cells are formatted to identify the sign and statistical significance of the effects, such a table would require very little discussion as the conclusions are then fairly obvious. The effect of White (race) on approval is almost always positive (Whites are more likely to approve of a marriage ban than Blacks), the effect is larger in magnitude for those raised in the South, and it decreases in magnitude as education increases. Moreover, the White effect is significant except at the upper end of the education distribution. For those raised outside the South, the effect of being White is not significant for those with more than 14 years of education. But for those raised in the South, it is significant for everyone except those with 19 or 20 years of education.

Application to the Three-Way Interaction Example

Table 3.4 presents a different version of a formatted significance region table written to an Excel file by ICALC, designed to better highlight regions of significance versus regions of nonsignificance of the effect of sex. The Key beneath the table describes the meaning of the formatting. Nonsignificant effects are in plain text if positive and in italicized text if negative. Significant positive effects are in bolded text with an *, and the cell is shaded with a medium gray fill. Significant negative effects are in bolded and italicized text with an *, and the cell is shaded with a lighter gray fill. When you look at the table, the regions of significance jump out at you with obvious differentiation of positive and negative effects, making the patterns of moderation easier to discern and describe. The shaded area running on the diagonal from the top left to the lower right represents significant negative effects of sex; that is, women's predicted number of memberships is less than men's. This area is defined essentially by the combination of education and age values, both in the bottom half of their respective ranges or both in the top half of their ranges. Women are predicted to have

---

TABLE 3.4 ● EXCEL FORMATTED SIGNIFICANCE REGION TABLE, THREE-WAY INTERACTION EXAMPLE

Effect of Women (One-Unit Difference) Moderated by the Interaction of Age and Education on g(Memnum), Formatted to Highlight Sign and Significance

Education	Age							
	18	28	38	48	58	68	78	88
0	*−1.958**	*−1.559**	*−1.160**	*−0.761*	*−0.363*	0.036	0.435	0.834
2	*−1.651**	*−1.329**	*−1.007**	*−0.686**	*−0.364*	*−0.042*	*−0.280*	0.602
4	*−1.343**	*−1.099**	*−0.854**	*−0.610**	*−0.365*	*−0.120*	*−0.124*	0.369
6	*−1.036**	*−0.869**	*−0.701**	*−0.534**	*−0.366*	*−0.199*	*−0.031*	*−0.136*
8	*−0.729**	*−0.638**	*−0.548**	*−0.458**	*−0.367**	*−0.277*	*−0.187*	*−0.096*
10	*−0.421*	*−0.408**	*−0.395**	*−0.382**	*−0.369**	*−0.355**	*−0.342*	*−0.329*
12	*−0.114*	*−0.178*	*−0.242**	*−0.306**	*−0.370**	*−0.434**	*−0.498**	*−0.562**
14	0.193	0.052	*−0.089*	*−0.230**	*−0.371**	*−0.512**	*−0.653**	*−0.794**
16	0.501	0.283	0.064	*−0.154*	*−0.372*	*−0.590**	*−0.809**	*−1.027**
18	**0.808***	0.513	0.217	*−0.078*	*−0.373*	*−0.669*	*−0.964**	*−1.260**
20	**1.116***	**0.743***	0.370	*−0.002*	*−0.375*	*−0.747*	*−1.120*	*−1.492**

Key	
Plain font, no fill	Pos, Not Sig
Bold*, filled	Pos, Sig
Italic, no fill	Neg, Not Sig
Bold italic*, filled	Neg, Sig

Note: Memnum = number of memberships.

ICALC command: sigreg, nobva save(Output\Table3_4.xlsx tab) ndig(3). The nobva option suppresses boundary value analysis, ndig( ) specifies 1 decimal format for table results, and save( ) outputs the formatted table to an Excel file.

significantly more memberships than men only near the top of the education range (roughly 18–20 years) for those with ages near the bottom of the age range (approximately 18–35 years). This example exemplifies how effective this approach can be for the more complex pattern of moderation and statistical significance that is typical of a three-way interaction. If you look back at the corresponding boundary value plot in Figure 3.2, I am sure you will agree.

In the next section, I describe a different but related method for identifying and portraying information about patterns of moderation and statistical significance. It elaborates the effect displays discussed in Chapter 2 by calculating confidence bounds/confidence intervals for every plotted value of the moderated effect and adding them to the plot. In general, many researchers and audiences prefer such visual displays to the tabular displays just discussed.

CONFIDENCE BOUNDS AND ERROR BAR PLOTS

An alternative way to think about and present significance regions relies on the idea that we can calculate a confidence interval for $\hat{\omega}$—the effect of F at specified values of the moderators $(m_1, m_2, \ldots, m_q)$. If the confidence interval includes the value of 0, then the effect of F is not significant for those values of the moderators. I focus on two types of plots to show how the effect of F and its statistical significance change with the values of its moderator(s). Each is an elaboration of a type of effect display plot described in Chapter 2, so similar guidelines apply for choosing a plot type. For an interval first moderator, you add two plotted lines to the line plot of the moderated effect of F, representing the upper and lower bounds of the calculated confidence interval. That is, construct a confidence bounds plot by graphing three lines, $\hat{\omega}$ and the upper and lower bounds of its confidence interval on the y-axis drawn across the moderator's range of values on the x-axis. If the first moderator is categorical, then we can adapt a drop-line effect display plot by replacing the drop line with up-and-down vertical lines to portray the confidence interval limits. Specifically, create an error bar plot with the categories of the moderator on the x-axis and a marker placed vertically corresponding to the moderated effect of F and placed horizontally above the moderator category on the x-axis. Show the confidence interval by drawing a vertical line upward from the marker to the point representing the upper limit of the confidence interval—marked with a short horizontal line—and a second vertical line downward from the marker to the point for the confidence interval's lower limit, also marked with a short horizontal line. Both confidence bounds plots and error bar plots provide a visual portrayal of the sign and magnitude of F's effect as well as its significance region, which are typically easy to interpret. I briefly discuss the usage of a 2D contour plot with significance markers at the end of the three-way interaction example. I do not consider 3D surface plots because it is very difficult to create an effective 3D graph that also displays significance markings.

I follow the common practice of determining the upper and lower limits of the confidence interval using the sampling variability of $\hat{\omega} : \widehat{Var}(\hat{\omega})$ as specified in Table 3.1. This has good large-sample properties and is easy to apply, hence its predominance in practice that I will adopt. As I discussed earlier, using the LR statistic to calculate confidence intervals can be very computationally intensive, and the same applies to constructing confidence intervals using score tests. Thus, I rely on the Wald test statistic to calculate confidence interval limits using either a t-critical value or a z-critical value. The calculations use the t-statistic critical values if we know the finite sample properties of the coefficients (e.g., OLS regression), but they use the z-statistic critical values if we rely on asymptotic results (e.g., maximum-likelihood estimation). I use the symbol tz in the formula for the $100(1 - \alpha)\%$ confidence interval (CI) to represent this choice of test statistic:

$$CI = \hat{\omega} \pm tz_{\frac{\alpha}{2}} \times \sqrt{\widehat{Var}\left(\hat{\omega}\right)} \quad \text{or}$$

$$CI_{Upper} = \hat{\omega} + tz_{\frac{\alpha}{2}} \times \sqrt{\widehat{Var}\left(\hat{\omega}\right)} \qquad CI_{Lower} = \hat{\omega} - tz_{\frac{\alpha}{2}} \times \sqrt{\widehat{Var}\left(\hat{\omega}\right)}$$

(3.7)

For a single moderator (M_1), we can plot $\hat{\omega}$, CI_{Upper}, and CI_{Lower} against M_1, which defines the x-axis to construct confidence bounds plots or error bar plots. If there is more than one moderator, we can construct multiple plots of $\hat{\omega}$, CI_{Upper}, and CI_{Lower} against one of the moderators, with a separate plot for each combination of (selected) values of the other moderators. As I argued above, it is important to examine the results for the full range of values of the other moderators to best understand the boundaries of the significance regions. I construct and discuss such plots for different forms of interaction models to illustrate their use. I suspect that most readers will find these plots the most accessible and informative of the techniques for understanding significance regions and the pattern of the moderated effect of F.

One-Moderator Examples of Confidence Bounds and Error Bar Plots

I use the OLS regression example of household wealth predicted by the interaction between household headship type and number of children to illustrate creating and interpreting these plots. I begin with an error bar plot for the effect of children on wealth moderated by headship type (Figure 3.3) and then discuss a confidence bounds plot for the effect of headship type on wealth moderated by children (Figure 3.4). Both use a 95% confidence level ($\alpha = 1 - 0.95 = 0.05$) and—because these are OLS results—a t-critical value ($t_{\frac{\alpha}{2}} = 1.96013$) to define the upper and lower limits in Equation 3.7.

As an example, let's apply Equation 3.7 to calculate the confidence interval for the moderated effect of children when headship type is single-woman shown in Figure 3.3. First, we need to substitute the results from the OLS regression (coefficients and their sampling variances/covariances) into the expressions for $\hat{\omega}$ and $\widehat{Var}(\hat{\omega})$ shown in rows 2 and 4 of the column labeled "One Moderator" in Table 3.1:[7]

$$\hat{\omega} = -0.96769 + 1.67741 \times 1 = 0.70972$$

$$\widehat{Var}(\hat{\omega}) = 0.030003 + 2 \times 1(-0.024806) + 1^2(0.15242) = 0.13281$$

Substituting these values into Equation 3.7 gives the upper and lower confidence interval limits plotted in Figure 3.3 for the effect of children for headship type = single-woman:

$$CI = 0.70972 \pm 1.96013 \times \sqrt{0.13281} = 0.70972 \pm 0.71432$$

$$= -0.00461, 1.42405$$

(3.8)

FIGURE 3.3 ● ERROR BAR PLOT, ONE MODERATOR

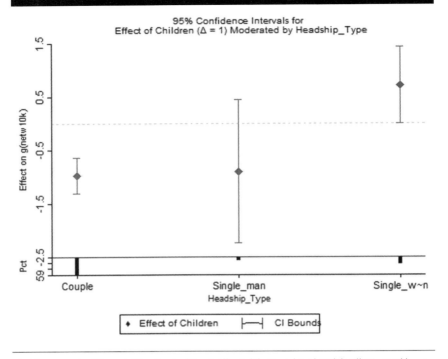

ICALC command: effdisp, plot(type(errbar) freq(tot)) ndig(1) pltopts(ylab(-2.5(1)1.5)). The plot( ) option specifies an error bar chart showing the frequency distribution of the moderator, headship type; ndig( ) formats the *y*-axis labels; and pltopts( ) customizes the *x*-axis labels.

FIGURE 3.4 ● CONFIDENCE BOUNDS PLOT, ONE MODERATOR

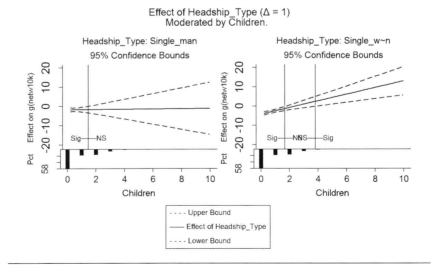

ICALC command: effdisp, plot(type(cbound) freq(tot)) ndig(0) pltopts(xlab(0(2)10)). The plot( ) option specifies a confidence bounds plot showing the frequency distribution of the moderator, headship type; ndig( ) formats the *y*-axis labels; and pltopts( ) customizes the *x*-axis labels.

The error bar plot provides a concise and accessible picture of the differences in the effect of children on wealth contingent on headship type. The effect of children is significant only for married-couple households (almost 60% of the sample) because the horizontal dotted line for a zero effect lies outside the error bars (confidence interval) for this headship type. The error bars for the other two headship types enclose an effect value of 0, indicating no significant effect of children for single-headed households, whether headed by women or men. Notice that the lower error bar for the single-woman headship type barely includes the "no effect" line within the confidence interval. It is so close that I would be uncomfortable drawing the conclusion of no effect without a numeric calculation to verify the visual display. Not by coincidence, we have such verification from the sample calculations shown in Equation 3.8, that the lower limit of the confidence interval is −0.00461. This demonstrates the importance of knowing how to calculate the confidence interval limits by hand, although they might also be simple to do with your statistical software.[8]

Turning to the other side of this interaction effect, Figure 3.4 includes a confidence bounds plot for both headship type effect categories that displays a lot of information. Let me start by identifying the different lines and what they tell you. The effect of the focal variable is the *solid thick line* and its upper and lower confidence bounds are the *thick dashed lines*. The *dashed light gray horizontal line* shows an effect value of 0. Thus, for any value of children where the *dashed gray line* falls between the *larger dashed black lines*, the headship type effect is not significant. That is, there is no significant difference in wealth between that headship type and married-couple-headed households. There can be one or more *vertical solid gray lines* that mark transitions from statistical significance (Sig) to nonsignificance (NS), or vice versa. The spike plot beneath the confidence bounds plot shows the distribution of number of children in the sample. Notice that the scales on both the *x*- and *y*-axes are identical in each plot to allow accurate visual comparisons.

The plot for the single-man effect on the left shows that the effect is always negative and its magnitude hardly changes as the number of children increases. The significance region is also clearly demarcated: The effect of single-man is significant only for households with zero to one child—note the vertical bar marking the transition to nonsignificant between one and two children. This represents more than 70% of the sample. The right-hand plot clearly exhibits a different pattern for the moderated effect of Single-woman-headed households, with *vertical bars* marking the transition from significance to nonsignificance between one and two children and back to significance at slightly less than four children. Thus, like the effect of single-man, the single-woman effect is significant and negative when the number of children is less than two, but it is larger in magnitude than the single-man effect. This means that single-woman-headed households have even less wealth compared with married-couple households than single-man households. The effect of single-woman changes to positive and significant at a point where there are very few cases (about 2% of the sample have four or more children) and could be an artifact of the steeply declining negative effect of headship type from two children to zero child, which encompasses a vast majority of cases.

Like the corresponding significance region table in the prior section, error bar plots and confidence bounds plots effectively communicate the changing sign, size, and significance of the effect of a focal variable as these properties vary with its moderator. They are similarly effective for two-moderator models, to which I turn next.

Confidence Bounds and Error Bar Plots for Two Moderators

We can similarly use these plots for two (or more) moderators by creating multiple plots: For selected values of M_2, plot the moderated effect of F as it varies with M_1 (and M_2), along with the confidence bounds or error bars for the effect of F. The moderators' level of measurement can help determine the type of plot to create as well as which moderator to set as the x-axis and which moderator should define the multiple plots. If one moderator is categorical and the other interval, then the most informative plot is typically a confidence bounds plot with the interval moderator defining the x-axis, with the plot repeated for each category of the other moderator. If you want to focus attention instead on the categorical moderator's influence on the effect of F, you could create an error bar plot using the categorical moderator for the x-axis and repeating the plot for selected values of the interval moderator. For two nominal-level moderators, the decision is about which moderator defines the x-axis for an error bar plot and which defines the repeated plots. All else equal, I would use the moderator with the most categories to define the x-axis to minimize how many different plots you have to consider. If both moderators are interval level, there are no clear grounds for deciding which one defines the x-axis (M_1) and which one defines the repeated plots (M_2). However, it is important to select values spanning the entire range of M_2 to fully understand the significance region. I suggest dividing M_2's range into 6 to 10 equal intervals initially to help you understand the significance regions in the patterns of moderation, and then you could pick a smaller number of representative plots to present to your audience. This is demonstrated for a three-way interaction in the next section.

I use the logistic regression results for approval of a marriage ban to illustrate the creation and interpretation of a confidence bounds plot for a categorical and an interval moderator. The focal variable is race, and the moderators are region (categorical, southern vs. nonsouthern upbringing) and education in years (interval). Because region is categorical, I constructed confidence bounds plots of the moderated effect of race plotted against education, with separate plots for the two categories of region. Using education as M_1 and region as M_2, we can substitute into Equation 3.7 for $\hat{\omega}$ and $\widehat{Var}(\hat{\omega})$ to specify the calculating equation for the upper and lower bounds for the 95% confidence interval (see Table 3.1, rows 2 and 4 of the "Two Moderators" column, for the formulas for F's effect and its sampling variability for two moderators and Table 3.3 for the numbers for this example). I specify a z-critical value $\left(z_{\frac{.05}{2}} = 1.9600 \right)$ because the logistic regression results are maximum-likelihood estimates:

$$CI = \left(2.8351 - 0.1424 \times Education + 1.1394 \times South \right) \pm 1.9600 \times$$
$$\sqrt{\begin{array}{l} 0.7938 + 2 \times Education \times \left(-0.0555 \right) + 2 \times South \times \left(-0.2234 \right) + \\ 2 \times Education \times South \times 0.0061 + Education^2 \times 0.0045 + \\ South^2 \times 0.2510 \end{array}} \quad (3.9)$$

Figure 3.5 presents the separate confidence bounds plots for southern upbringing and for nonsouthern upbringing based on the expressions in Equation 3.9. Each figure plots the moderated effect of race (*thick solid line*) and the upper and lower bounds of the confidence interval for the effect of race (*dashed lines*) against the values of

FIGURE 3.5 ● CONFIDENCE BOUNDS PLOT, TWO MODERATORS

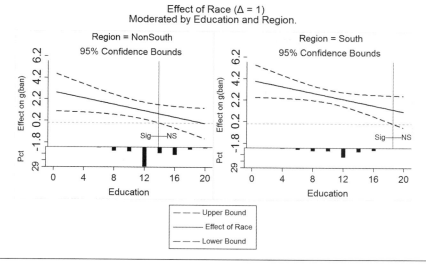

Effect of Race (Δ = 1)
Moderated by Education and Region.

ICALC command: effdisp, plot(type(cbound) freq(subtot)) ndig(1) pltopts(xlab(0(4)20)). The plot( ) option specifies a confidence bounds plot with the joint sample frequency distribution of the two moderators, ndig(1) sets the format of the *y*-axis labels, and pltopts( ) customizes the *x*-axis labels.

education. The *dotted horizontal line* provides a reference line of a zero effect, and the *gray vertical line* indicates at what value of education the moderated effect of race changes significance.

Both plots portray the same general story. The effect of race on approval is positive for almost all combinations of values of education and region of upbringing—the solid line for the race effect only crosses the zero effect reference line when *Education* = 20 and *Region* = non-South. Furthermore, the race effect declines in magnitude as education increases and becomes nonsignificant at a relatively high level of education. The effect of race is significant when the confidence bounds for a given value of education do not include the value 0; visually, this means that the zero effect dotted line lies below the lower bound curve. The *vertical gray line* in each plot line denotes the single significance change (to not significant) of the effect of race and marks the education value at which the lower bound of the confidence interval goes below 0 while the upper bound remains above 0. For nonsouthern upbringing, the effect of race becomes not significant when education is 15 years or higher, while for southern upbringing, the effect becomes nonsignificant only for those with 20 years of education. Note that the amount of change in the effect of race with education—the slope of the solid line—is the same in both plots because education and region do not interact in moderating the effect of race. Also notice that the effect of race is higher for southern upbringing than for nonsouthern upbringing at any level of education.

To provide a quick illustration of an error bar plot for a two-moderator interaction, I created such a plot for the same results (Figure 3.6), drawing error bars for *Education* = 0, 5, 10, 15, 20. There is a separate error bar plot for *Region* = *South* and *Region* = *non-South*. If the pair of error bars for a given education category include the zero effect line between the upper and lower error bar limits, it indicates a nonsignificant

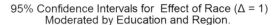

FIGURE 3.6 ● ERROR BAR PLOT, TWO MODERATORS

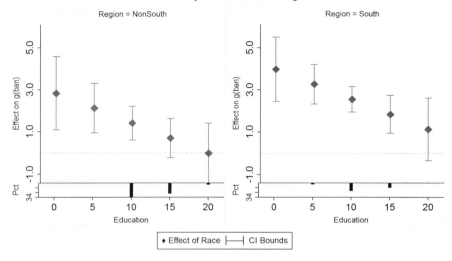

ICALC command: effdisp, plot(type(errbar) freq(subtot)). The plot( ) option specifies an error bar chart with the joint sample frequency distribution of the two moderators.

race effect for that category of the moderator. It is as straightforward to see the patterns in these plots as in the confidence bounds plots. The effect of race is positive except for *Education* = 20 and *Region* = *non-South* (the other diamond markers all lie above the zero effect line), the size of the effect declines from left to right (lower to higher education), and the race effect becomes nonsignificant for one or two of the rightmost categories (highest education levels). The one difference in the information provided is that error bar charts do not have vertical lines marking the boundaries between significant and nonsignificant effects. Such boundary markers only make sense for a numeric *x*-axis moderator, not a categorical one, and education is treated categorically for this illustration.

Overall, these plots provide what I think is an easily interpretable way to understand and present information about the significance region—much more so than the corresponding boundary value plots in Figure 3.1—while also providing information about the pattern of change in the sign and magnitude of the moderated effect of *F*. Depending on individual preferences for visual presentation versus numeric presentation, a significance region or effects-formatted table like those discussed in the previous section is also a very good, readily accessible option.

Plots for a Three-Way Interaction

For this last example, I focus on the creation and interpretation of confidence bounds plots for a three-way interaction and then briefly consider the use of 2D contour plots. I plot the results from an OLS regression of number of voluntary association memberships on the three-way interaction of sex, education, and age. Sex is a dummy variable (1 = *female*, 0 = *male*), and education and age are measured in years at the interval level. As before, sex is the focal variable, and I use education to define the x-axis of each plot, which is repeated for age from 18 to 88 years in 10-year intervals. To get the calculating equation for the upper and lower bounds for the 95% confidence interval for the moderated effect of sex, treat education as M_1 and age as M_2 and substitute the necessary information about $\hat{\omega}$ and $\widehat{Var}(\hat{\omega})$ into Equation 3.7 (rows 2 and 4 of the "Three-Way Interaction" column in Table 3.1 provide the formulas for F's effect and its sampling variability and Table 3.5 shows the numbers for this example). For a 95% confidence level, I specify a t-critical value $\left(t_{\frac{0.05}{2}} = 1.9617 \right)$ in the equation because the results are from an OLS regression:

$$
CI = \left(\begin{array}{c} -2.676088 + 0.223111 \times Education + 0.039889 \times Age \\ -0.00385716 \times Education \times Age \end{array} \right) \pm 1.9617 \times
$$

$$
\sqrt{ \begin{array}{l} 1.455585 + 2 \times Education \times (-0.109092) + \\ 2 \times Age \times (-0.248192 \times 10^{-1}) + \\ 2 \times Education^2 \times Age \times (-0.154280 \times 10^{-3}) + \\ 2 \times Education \times Age^2 \times (-0.370659 \times 10^{-4}) + \\ Education^2 \times 0.862388 \times 10^{-2} + Age^2 \times 0.474415 \times 10^{-3} + \\ Education^2 \times Age^2 \times 0.312088 \times 10^{-5} \end{array} }
$$

(3.10)

Figure 3.7 reports 95% confidence bounds plots of the moderated effect of sex as it varies with education for selected values of age. The age value increases between adjacent plots by 10 years, starting at $Age = 18$ (the sample minimum) up to $Age = 88$ (almost its maximum value of 89). The eight plots are vertically ordered by age increasing from the top to the bottom. To facilitate seeing the overall picture, the plots have the same scaling for the x- and y-axes to preserve the visual differences in the actual magnitude of the sex effect and its slope of change with education across plots. The *thick solid line* shows the moderated effect of sex, and the *thick dashed lines* represent the upper and lower confidence bounds. The *dotted horizontal line* provides a zero-effect reference line

TABLE 3.5 ● INFORMATION FOR CALCULATING CONFIDENCE BOUNDS FOR EFFECT OF SEX MODERATED BY THE INTERACTION OF EDUCATION AND AGE

Variance/ Covariance of	With				Coefficients
	b_1	b_3	b_5	b_7	
Sex b_1	1.455585				−2.676088
Sex × Education b_3	−0.109092	0.862388 × 10⁻²			0.223111
Sex × Age b_5	−0.248192 × 10⁻¹	0.189229 × 10⁻²	0.474415 × 10⁻³		0.039889
Sex × Education × Age b_7	−0.189957 × 10⁻²	−0.154280 × 10⁻³	−0.370659 × 10⁻⁴	0.312088 × 10⁻⁵	−0.385716 × 10⁻²

to help identify changes in the sign or the statistical significance of the sex effect. If the solid line crosses the zero-effect line, it indicates that the sex effect changes sign at the corresponding education value on the *x*-axis and the age value shown above each plot. Whenever a dashed boundary line crosses the zero-effect line, it marks a combination of the values of education and age at which the effect of sex changes significance. A *vertical gray line* marks these points with a label to indicate if the boundary marks a change to a significant effect (Sig) or to a nonsignificant effect (NS) as years of education increase.

Several patterns are apparent from a close study of these plots, echoing what we have seen from the other ways of portraying the significance region for this three-way interaction:

- The sex effect is mainly negative across combinations of education and age. Only when education and age are at essentially opposite parts of their ranges

FIGURE 3.7   ●   CONFIDENCE BOUNDS PLOTS, THREE-WAY INTERACTION, EIGHT SUBPLOTS

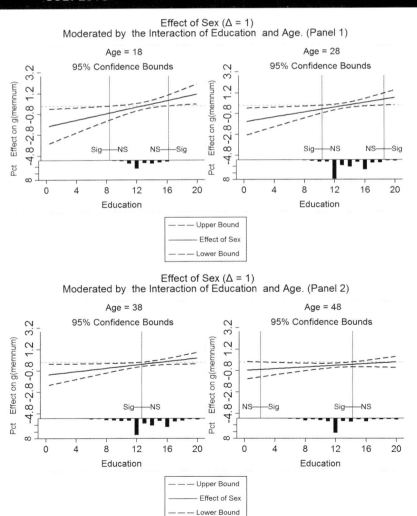

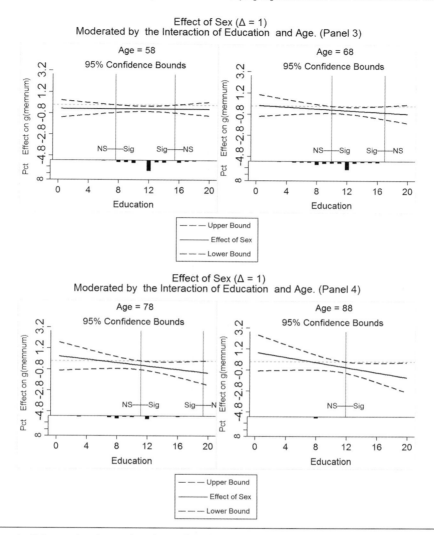

ICALC command: effdisp, plot(type(cbound) freq(subtot)) ndig(1) pltopts(xlab(0(4)20)). The plot( ) option gets a confidence bounds plot with a joint sample frequency distribution of the two moderators, and ndig( ) and pltopts( ) format the *y*- and *x*-axes, respectively.

is the sex effect positive—that is, the solid line crosses the dotted zero-effect line. In the plots for ages 18 to 38 years, the sex effect turns positive at around 13 years of education, and in the plots for ages greater than 68, the sex effect is positive only for very low values of education (6 years or less).

- The positive sex effects are statistically significant only in the plots for the two youngest ages (18 and 28 years) and only if education levels are relatively high (15 or more years). You can see this from both the vertical gray lines marking the change to significance and also the fact that the dashed confidence bounds lines do not include the value 0.

- The plots highlight how the combinations of education and age that produce significant negative effects of sex change systematically, loosely corresponding to similar positions in the distribution of age and education. That is, in the younger parts of the age distribution (≤48), the sex effect is negative and significant when education levels are below a boundary varying from about 8 years for the *Age* = 18 group to 13 years for the *Age* = 48 group. In the middle part of the age distribution (between 58 and 78), the sex effect is negative and significant when education values are also in the middle part of its distribution (a lower boundary varying from 7 to 11 years and an upper boundary between 15 and 18 years). At the top of the age distribution (*Age* > 78), the negative effect of sex is significant only for *Education* ≥ 12 years.

- The spike plot at the bottom of each confidence bounds plot shows the percentage of the total sample with the corresponding values of education and in an age bracket centered on the age value shown at the top of the plot. This shows that in the younger age brackets most of the sample have education values that represent nonsignificant effects of sex on membership. Conversely, in the middle age brackets, sample cases are more concentrated in the middle part of the education distribution, which yields significant effects of sex on membership.

I would argue that at least this number of plots is necessary in this case for a researcher to understand the pattern of the moderated effects and their significance. But this is quite likely too many to present to a reader or to expect a journal to publish. Having developed for yourself a set of bullet points like the ones above, you could select and present a smaller number of plots to illustrate the key points. In this case, I would recommend a plot from the younger part of the distribution, a plot from the older part, and two plots representing the middle ages. Figure 3.8 presents confidence bounds plots for *Age* = 18, 41, 64, and 87 (increments of 23 years from the minimum age to almost its maximum). The *Age* = 18 and *Age* = 87 plots illustrate that positive sex effects are produced by "opposite" education and age values, but they are only significant for the combination of younger ages and higher years of education. The younger age plot also represents the point that negative sex effects are mainly not significant in younger ages. The two middle age plots demonstrate the concentration of significant negative sex effects when education and age values are similarly located in their respective distributions.

Although confidence bounds plots are usually accessible and informative about interaction specifications with one or two moderators, I find them more difficult to use for three-way interactions (and I suspect most readers would agree). In contrast, I find that significance region tables, such as Table 3.4, can work well for three-way interactions. A visual counterpart to the significance region table is a 2D contour plot with significance markers overlaid on the contour plot. The markers are placed on a grid defined by combinations of the values of the two moderators. It takes some experimentation to determine how many values, and at what spacing, to select for each moderator. I would recommend trying between 4 and 7 grid points on each dimension.

Figure 3.9 presents a contour plot for this example, with a 6 × 6 grid and significance markers. *Large diamonds* mark a significant sex effect, and *small circles* denote a nonsignificant effect of sex; the location of the markers is slightly jittered to avoid drawing the eye to a straight-line grid. The four lightest gray shades of the contours represent negative effects of sex, with the lighter shades showing a more negative effect. The two darkest shades signify positive effects, with the darker shade showing a larger positive effect. Looking first at the positive effects, we can see that they are in

FIGURE 3.8 ● CONFIDENCE BOUNDS PLOTS, THREE-WAY INTERACTION, FOUR SUBPLOTS

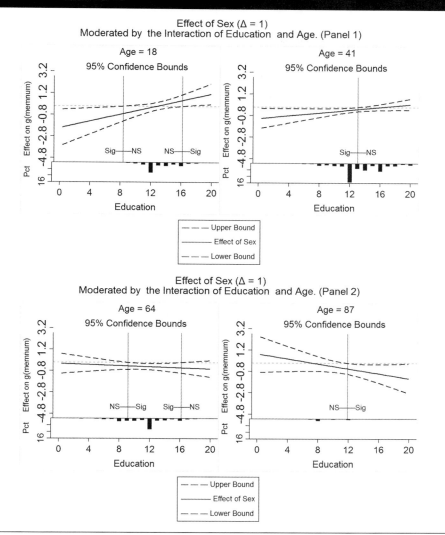

ICALC command: effdisp, plot(type(cbound) freq(subtot)) ndig(1) pltopts(xlab(0(4)20)). The plot( ) option gets a confidence bounds plot with a joint sample frequency distribution of the two moderators, and ndig( ) and pltopts( ) format the *y*- and *x*-axes, respectively.

the lower-right and upper-left corners of the plot, corresponding to "opposite" values of education and age. Moreover, positive effects are significant only for younger ages at higher years of education (lower-right corner). The statistically significant negative effects correspond to "matched" distributional locations for the values of education and age. The negative effect is largest in magnitude for low education matched to younger ages and high education matched to older ages (*lighter gray shades* in the lower-left and upper-right corners). The accompanying contour plot of the distribution of cases across education and age suggests that the sample cases are concentrated in an area centered on *Education* = 12 and *Age* = 32, which shows negative effects of sex with a mix of significant and not significant effects.

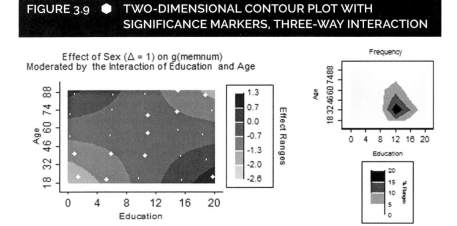

FIGURE 3.9 ● TWO-DIMENSIONAL CONTOUR PLOT WITH SIGNIFICANCE MARKERS, THREE-WAY INTERACTION

ICALC command: effdisp, plot(type(contour) freq(subtot)) ndig(1) sigmark pltopts(ccuts(-2.6(.65)1.30)). The plot( ) option specifies a contour plot with an accompanying joint sample frequency distribution of the two moderators, the sigmark option adds significance markers to the plot area, and ndig(1) and pltopts( ) customize the label format and cut points to define the contours.

SUMMARY AND RECOMMENDATIONS

This chapter focused on how to determine the pattern of changes in sign and magnitude as well as the significance region for the focal variable's moderated effect on the modeled outcome (η). I described three different options for defining and presenting the significance region, which differ somewhat in their ease of calculation and more so in the accessibility of the resulting definition of the significance region to audiences and researchers alike. Even though all of them can be produced using the ICALC toolkit, I have tried to provide enough detail on the formulas and calculations so that readers can implement these in other platforms. Given the extensive calculations required, hand calculations are not a realistic choice for any of these approaches.

The JN approach requires solving a quadratic equation to calculate the values of the moderator(s) at which the significance of the effect of F changes, either from significance to nonsignificance or vice versa. I would recommend this as a possible option if you have a single moderator of F or, in the case of multiple moderators, if at most one of the multiple moderators is measured at the interval level. In these instances, tables of boundary values or plots of the boundary values usually provide a clear and easy-to-understand description of the significance region. Otherwise, as I think the three-way interaction example illustrated, the resulting tables or plots of boundary values can be time-consuming if not difficult to understand and explain.

A second option is to directly calculate the significance of F's moderated effect at multiple values of a moderator spanning its range and, for multiple moderators, all the combinations of the selected values of different moderators. The results for Q moderators can be presented in a Q-dimensional table, where each dimension is defined by one of the moderators. The cell entries report the value of the moderated effect of F and information about its statistical significance. By highlighting cells with significant effects and formatting cells with positive and negative effects differently, the resulting table is eminently readable and interpretable. I would note that if you have three or more moderators, it can be more challenging to interpret such tables, but I think they are still the easiest to interpret of the three options.

The last option is to plot confidence bounds or confidence intervals for the effect of F as it varies with the moderators. That is, calculate the upper and lower bounds of the confidence interval for the effect of F across all of the values of its first moderator and selected values across the range of any other moderators. For an interval first moderator, you would create a confidence bounds plot, but would use an error bar plot for a categorical first moderator. For a single moderator, either of these plots is straightforward to interpret and explain. If your analysis has two or more moderators, you would need multiple such plots for each combination of selected values of the other moderators. For two moderators, if one of them is categorical, plotting confidence bounds can be a very effective way to present and understand the significance region. If both moderators are interval, the resulting set of plots can take some time and effort to comprehend, as the three-way interaction example demonstrated. In contrast, the formatted table approach to defining significance regions is much easier to understand and to present to audiences. The 2D contour plot with significance markers is also a viable option when you have two interval moderators, especially in comparison with a series of confidence bounds plots. In the end, I think that the choice between presenting plots of confidence bounds/confidence intervals versus the formatted table definition of significance regions is a matter of personal preference for a more visual versus a more numeric portrayal, respectively, of the information.

This chapter and Chapter 2 focused implicitly on working in the metric of the modeled outcome (η) and on how to understand and present the pattern of the sign and significance of the moderated effect of F on η. The next two chapters turn to the interrelated topics of how to present and interpret the moderated effect of F in terms of the predicted value of η and/or the predicted value of the observed outcome (μ). Chapter 4 covers models with linear link functions and interpretation using predicted values of the observed outcome. This provides a foundation for the more challenging application of these techniques to GLMs with nonlinear link functions in Chapter 5.

CHAPTER 3 NOTES

1. I would like to thank one of the reviewers for identifying this point of view.
2. The corresponding adjustment for calculations using a t or a z critical value would be to use the square root of the adjusted F or adjusted χ^2 critical value, respectively.
3. The intercept is listed as the last coefficient for convenience.
4. As we will see below, some of these entries can be values of the moderator variable(s), which we are assuming to be fixed rather than stochastic. Rogosa (1977, pp. 94–95) finds that the JN technique is robust to the use of stochastic predictors in OLS regression models. Bauer and Curran (2005, p. 398) speculate that such robustness is likely to apply to other types of modeled linear analyses.
5. The limiting distribution as df increases of the $F_{1, df}$ statistic is a χ^2 distribution with 1 df. Because the square of a two-tailed t with df degrees of freedom is an F statistic with degrees of freedom $(1, df)$, the t statistic test result is also asymptotically the same as the Wald statistic.
6. The standard errors for the coefficients are routinely reported and can be squared to get the sampling variance of the coefficients. Getting the covariance between the coefficients

for a GLM usually requires specifying an option or an additional command to report the variance–covariance matrix for the coefficients. For example, in Stata the command "matrix list e(V)" will display the variance–covariance matrix for the coefficients of the last model run.

7. Technically, we should use the expressions in the column labeled "2 Moderators" because our single conceptual moderator includes two predictors, M_1 = Single-woman and M_2 = Single-man. But when we use the expressions to calculate the moderated effect and its sampling variance for headship type equals single-woman, the value for $m_2 = 0$ and the expressions in the "2 Moderators" column simplify to the expressions in the "1 Moderator" column.

8. In Stata, you can use the *test* command to perform a Wald test that the moderated effect of F is significant at a specific value of M_1. The generic Stata command for this is as follows:

test _b[focal_varname] + m_1 *_b[focal_moderator_varname] = 0

In this example, it would be test _b[kids] + 1*_b[2.hoh#c.kids] = 0. The test statistic value is 3.79, with $p = .0515$.

LINEAR (IDENTITY LINK) MODELS

Using the Predicted Outcome for Interpretation

The last two chapters focused on the pattern of the focal variable's effect, in essence treating the focal variable's effect as an outcome. Understanding that pattern is a necessary first step to fully understand the nature of the interaction effect. Some researchers find that sufficient information, but many, if not most, researchers find it more intuitive to present and write up the results in terms of the pattern of change in the predicted (expected) value of the outcome corresponding to changes in the variables constituting the interaction specification. This chapter and the next turn attention to interpreting an interaction effect using tables, plots, or bar charts of the predicted outcome as it changes across the values of the focal and moderating variables. I applied the OUTDISP tool in ICALC to construct the tables and figures shown for the examples. This chapter considers the simpler case of models in which the observed outcome is a linear function of the predictors. In GLM terms, this is a model with an identity link function specifying the transformation of the expected value of the observed outcome into the expected value of the modeled outcome. In simpler terms, this means that the modeled outcome is the same as the observed outcome. This chapter lays the foundation for the next chapter, which considers the challenges of interpreting interaction effects in GLMs with nonlinear link functions.

For a linear model, the expected value of the outcome (μ) is a linear function of the predictors; that is, it is conditional on the focal and moderator variables as well as on the remaining predictors in the model. Thus, the predicted (expected) value of the observed outcome, $\widehat{Y}$, is calculated as a linear function of all the predictors:

$$\widehat{Y} = a + b_1 f + b_2 m_1 + b_3 f \times m_1 + \cdots + b_Z z \tag{4.1}$$

where f, m_1, and z are selected values of the focal, moderator, and other variables respectively.

To portray how $\widehat{Y}$ changes with the focal and moderator variables, you vary the values of the focal and moderator variables used in Equation 4.1—what I call the display values—but use nonvarying reference values for the remaining predictors in the model (Z). You then construct tables or various types of visual displays to show the pattern of change in $\hat{Y}$ across the display values of F and M. In the next section, I discuss options for choosing the display values for the focal and moderator variables and the reference values for the other predictors, depending on whether they are categorical or interval-level measures. You should make decisions about display values keeping in mind the distribution of the sample cases. You want to avoid giving too much credence in the presentation and discussion of results to regions in the joint distribution of the focal and moderating variables in which the sample is very sparse.

OPTIONS FOR DISPLAY AND REFERENCE VALUES

Focal Variable

For the focal variable, you should show how the predicted outcome varies across a wide range—usually the full range—of F's values. For a nominally measured focal variable, report the value of $\hat{Y}$ for each category of F and the chosen values of M in a table or a graphic.

For an interval-level focal variable, I recommend reporting $\hat{Y}$'s value at equal intervals from F's minimum to its maximum value, or at least close to those endpoints. For constructing tables, you may want to restrict how many display values you use for the focal variable's dimension. I would suggest about 7 to 10 display points, including the minimum and the maximum. In some cases, you might have as many display points as actual values—for example, preference for family size measured in whole numbers from zero to eight children. For others types of measures, there may be an intuitive ("natural") choice of intervals, such as every 5 or 10 years for age or every $1,000 for monthly earnings or every 0.5 units for a standardized scale. If not, to define D display points, calculate your interval as

$$\frac{Maximum - Minimum}{D - 1} \tag{4.2}$$

For plots, there is no reason to restrict how many values of F you plot against (unless you intend to show markers for each plotted point), so you should pick what makes sense to you given the scale of the focal variable. Or use the display point formula to define 20 or more plotting values (you will want fewer labels on the axis, of course). I do not recommend using equal-percentile intervals to pick display values, such as the values at every 10th percentile. Doing so would distort the linear relationship between F and $\hat{Y}$ because it is a nonlinear transformation of F.

You may want to modify these recommendations and restrict the display to less than F's full range if F is a badly skewed interval measure or if the data are otherwise sparse near the extremes of the value range. In that situation, I suggest setting display points by truncating the display values in the direction of the skew to avoid losing details of the pattern of the relationship at the other end of the distribution. For a positively skewed focal variable (or one with a sparse upper tail of the distribution), try truncating your display values at the 90th or 95th percentile in constructing plots. For tables, use that percentile value instead of the maximum value if you are calculating a fixed number of display values (e.g., $\frac{Maximum - Minimum}{D-1}$). If F is negatively skewed (very sparse near the minimum), then truncate display values at the 5th or 10th percentile and use the corresponding percentile value of F in place of the minimum to calculate a display interval.

Moderator Variable(s)

For the moderators, I would recommend about five to eight display values for the first moderator and even fewer, four to six, for any additional moderators. I draw this distinction given how you organize the information in tables, charts, and plots of the predicted outcome, discussed more fully below. Tables will have rows and columns defined by F and M_1, and the cells report the predicted outcome for the values of F and M_1 in the corresponding row and column. Plots theoretically will have three dimensions corresponding to F, M_1, and $\hat{Y}$. With two or more moderators, I suggest creating a series of tables or graphics for selected values of M_{2+} (combinations of values of M_2, M_3, etc.). If a moderator is nominal, then make the obvious choice of conditioning on each category of M_2. For an interval-level M, the common recommendation in the literature is to use the mean ± 1 standard deviation. As I have argued above, I think that it is important to start by examining how the interaction process plays out across the full range of a moderator's values. You may miss something important if you prematurely restrict your attention to the central part of the moderator's distribution. Consequently, I suggest using one of three options for choosing between five and eight display values:

1. *Mean, mean ± 1 standard deviation, mean ± 2 standard deviations, minimum, and maximum:* This choice set generally works well unless the moderator has a very skewed distribution. If the data are sparse in the tails, a reasonable variation would be to replace the minimum and maximum with values around the 5th and 95th percentiles, which are equally distant above and below the mean.

2. Median, median ± D, median ± 2D, median ± 3D, where

$$D = \frac{\text{The smaller of } \left(Median - Minimum\right) \text{ and } \left(Maximum - Median\right)}{3}$$

This option works well for a moderator with a skewed or otherwise sparse-tailed distribution. For a positively skewed distribution, this divides the distribution between the minimum and the median into three equal display intervals and then applies the same spacing of display intervals above the median. This will truncate the display at a point below the maximum because the distance between the median and the maximum is larger than the distance between the median and the minimum. Analogously, for a negatively skewed distribution, this truncates the display to a point above the minimum. A variation of this could also be used for a moderator that is not badly skewed, calculating equidistant intervals above and below the mean rather than the median.

Some authors (e.g., Hayes, 2013) suggest an option of using display values chosen at equal-percentile distances from the median (e.g., values at the 10th, 25th, 50th, 75th, and 90th percentiles). As I have argued above, this distorts the linear relationship between the moderator and the predicted value of the outcome and consequently may distort how you understand the relationship. I will show an example of this later.

3. *Equal intervals from the moderator's minimum to its maximum:* Use 5 to 10 evaluation points for the first moderator and 4 to 8 for subsequent moderators, defined by the display interval formula above, by the truncated display interval formulas for a skewed moderator, or by the choice of an intuitive interval. I prefer this option because it forces attention to a number of different points in the moderator's distribution.

In practice, presenting and discussing more than six sets of tables or figures is likely to create information overload and potentially focus attention on results where the sample distribution is sparse. Hence, my recommendation is for fewer display points for the M_{2+} moderators. But starting with more display points than that may help you as a researcher better understand the nature of the interaction effect. It can also help you decide which display points to actually present to best tell a coherent story, as was the case for the three-way confidence bounds plots at the end of Chapter 3.

REFERENCE VALUES FOR THE OTHER PREDICTORS (Z)[1]

At the outset, it is important to recognize that—for a linear link function—the only consequence of using different sets of reference values for Z is to shift the intercept in Equation 4.1. Remember that the term $b_z z$ in this equation symbolically represents the modeled effects for the remaining predictors at specified reference values $z_k : b_z z = \sum_{k=1}^{K} b_{z,k} z_k =$ a constant. Thus, using different reference values for the variables in Z would increase or decrease each of the calculated $\hat{Y}$ values by the same amount. On any visual representation, this shifts the entirety of any bar chart height, plotted line, or plotted surface upward or downward along the vertical dimension by the same increment. With that in mind, there are three common rubrics for picking reference values for Z: (1) central tendency values, (2) representative (substantively interesting) values, and (3) "as-observed" values. But below I show that only the first two are relevant for a linear (identity link) model in the sense that the as-observed values rubric is a special case of the central tendency values rubric for a linear model (but not for a nonlinear model) unless the Z predictors include polynomials or other multiplicative functional forms.

A second important point is how to specify properly the reference values for functionally related predictors among the variables in Z—for example, polynomial forms, sets of dummy variables for a nominal measure, or additional interaction forms not part of the interaction effect being analyzed. For such variables, the reference values across the related variables must be consistent with their functional relationship (see the excellent discussion with examples in Williams, 2012). For example, suppose your model includes a quadratic expression for the effect of work experience on earnings— $b_{z,1} Experience + b_{z,2} Experience^2$—and you are using the mean of each Z variable as its reference value. You should not use the mean of $Experience^2$ as the reference value for $Experience^2$. Rather, you should use the square of the mean of $Experience$ to preserve how $Experience^2$ is functionally related to $Experience$ in the model. Analogously, for a set of dummy variables, you should specify reference values to ensure that the sum of the reference values across the set of dummy variables is equal to 1. For an interaction form, the reference values for the higher order terms must be consistent with the reference values for the lowest order terms. Consider the three-way interaction between Z_1, Z_2, and Z_3 for which the model would include three first-order terms (one for each Z_k), three second-order terms (one for each pair of product terms $Z_k \times Z_{k'}$), and one third-order term ($Z_1 \times Z_2 \times Z_3$). The reference value for a second-order term should be the product of the reference values you specified for its first-order terms. Similarly, the third-order term's reference value should be the product of the three first-order terms' reference values.

Define by Central Tendency Values

This option was historically the most common—often the default—choice for specifying reference values and continues to be frequently used. The underlying logic is to portray how the predicted level of the outcome variable changes with the interaction variables, holding constant the other predictors at values that are typical of the sample. Measures of central tendency are one way to represent what is a typical value in the sample (and in the population if the sample is representative). A predictor's mean is the usual choice for an interval-level measure. If a predictor's distribution is badly skewed, the median is a better choice than the mean to represent what is typical of the sample values. Equivalently, for dummy-coded variables, specify the proportion of cases coded 1 in each category. A second, rarely if ever used, possibility for a dummy-coded variable would be to use a reference value of 1 for the modal category (i.e., the category with the most cases) and a reference value of 0 for the remaining categories.

Define by Representative (Substantively Interesting) Values

Rather than showing the pattern of the relationship between the predicted outcome and the interaction variables at values typical of the other predictors, this option portrays the pattern at one or more researcher-selected sets of values for each predictor in Z. These values are chosen to represent either a range of values (e.g., *Age* = 20, 30, . . . , 70) or specific values deemed of conceptual interest (e.g., *Age* of voting = 18, *Age* of majority = 21, *Age* eligible for Medicare = 65). For linear models, this option is not often used when interpreting interaction effects, in large part because different choices uniformly shift all the predicted values of the outcome (either larger or smaller). It is somewhat more commonly used to make comparisons of the expected value of the outcome across different scenarios, such as the difference in earnings between groups defined by age, race/ethnicity, and gender, even when there are no interaction effects for these variables in the model (e.g., earnings of young Black women vs. young Hispanic women vs. . . . vs. old White men). As I discuss in the next chapter on nonlinear models, the choice of reference values differentially affects the calculation of the predicted outcome's value for these models. As a result, this option is somewhat more commonly chosen when interpreting nonlinear models.

Define by As-Observed Values

There is a growing divide between those researchers who prefer this option and those who prefer the central tendency option (Williams, 2012). Both methods seek to construct values of $\widehat{Y}$ as it varies with the interaction variables, using values for the other predictors that are typical of the sample but differing in principle on how "typical" is defined. The as-observed option defines *typical* as the average value of the outcome for each combination of values of the interaction variables that we would observe in the sample if every sample member were hypothetically given that combination of interaction variables' values but kept their own values on the remaining predictors. This is a very abstract concept of *typical*, but it should become more understandable and sensible in the context of a specific example.

Consider the results from a weighted least squares (WLS)[2] regression of the frequency of sexual intercourse (per month) on the interaction between age in years and a socioeconomic status (SES) scale with controls for two interval predictors (number of children and religious attendance) and two nominal measures, each represented by a dummy variable (sex and marital status) (see the appendix for more data and measurement details). For the combination *Age* = 40 and *SES* = 45, we would calculate the expected value of sexual frequency by the following process:[3]

1. Assign each case the values *Age* = 40 and *SES* = 45, and then calculate their predicted frequency of sex using those values for age and SES but their own values for the other predictors.

2. Average the predicted frequency of sex across all the cases in the sample.

This value of frequency of intercourse is what we would hypothetically observe in a sample characterized by the actual (as-observed) values of the other predictors but the specified values of age and SES. This process would then be repeated for every combination of values of age and SES to construct the full pattern of how frequency varies with age and SES.

I noted above that for a linear model this is a special case of the central tendency option, specifically using the mean as the reference value for the variables in Z (see Special Topics at the end of this chapter). To demonstrate, let's see what the results would be of calculating the predicted value in this example for values *Age* = 40 and *SES* = 45. The estimated regression equation is

$$\hat{Y}_i = 14.912 - 0.201\, Age - 0.078\, SES + 0.00134\, Age \times SES - 0.743\, Female$$
$$- 1.038\, Never\ marry + 0.229\, Children - 0.121\, Attend \tag{4.3}$$

So for each case i in the sample, we set *Age* = 40 and *SES* = 45 (and *Age* × *SES* = 40 × 45), but leave the other predictors at their values for case i and calculate the case's predicted outcome:

$$\hat{Y}_{i,\,Age\,40,\,SES\,45} = 14.912 - 0.201 \times 40 - 0.078 \times 45 + 0.00134 \times 40 \times 45$$
$$- 0.743\, Female_i - 1.0384\, Never\ marry_i + 0.229\, Children_i \tag{4.4}$$
$$- 0.121\, Attend_i$$

We next average these predicted outcome values across the sample:

$$\overline{\hat{Y}}_{Age\,40,SES\,45} = \frac{\sum \hat{Y}_{i,Age\,40,SES\,45}}{n}$$

$$= \frac{\sum \begin{pmatrix} 14.912 - 0.201 \times 40 - 0.078 \times 45 + 0.0134 \times 40 \times \\ 45 - 0.743\ Female - 1.0384\ Never\ marry + 0.229 \\ Children - 0.121\ Attend \end{pmatrix}}{n}$$

$$= \frac{n \times (14.912 - 8.04 - 3.51 + 0.24)}{n} - 0.743 \frac{\sum Female_i}{n} - 1.0384 \tag{4.5}$$

$$\frac{\sum Never\ marry_i}{n} + 0.229 \frac{\sum Children_i}{n} - 0.121 \frac{\sum Attend_i}{n}$$

$$= 3.603 - 0.743\ \overline{Female} - 1.0384\ \overline{Never\ marry} + 0.229\ \overline{Children}$$

$$-0.121\ \overline{Attend}$$

The first term in Equation 4.5 (3.603) will change as we calculate the predicted outcome with different values for age and SES. But the last four terms for the other predictors in the model, calculated using "as observed" to define the reference values, will always be identical to using each predictor's mean as its reference value. This illustrates that for a linear model the as-observed option for reference values is equivalent to the central tendency option specifying the mean to define the reference value. Consequently, I will consider only the use of the central tendency option in the remainder of this chapter. But keep in mind that this equivalence is true only for a model with a linear link function, such as OLS regression. We will revisit this issue in the next chapter on GLMs with nonlinear link functions, where specifying the mean as a reference value does not give the same value for the predicted observed outcome as specifying as-observed values to define reference values.

CONSTRUCTING TABLES OF PREDICTED OUTCOME VALUES

The building block—not only for tables but also for figures—is the information presented in a 2D table whose columns correspond to the display values of the focal variable, whose rows represent the display values of the (first) moderator, and whose cells report the predicted outcome value using the focal variable's row value (r), the moderator's column value (c), and the chosen (nonvarying) reference values for the other predictors. Reading across a column will show how the predicted outcome changes with the focal variable for that row's value of the moderator, and we can compare the rows to see how the focal variable's effect changes with the values of the moderator. Analogously, reading down a column will reveal how the predicted outcome changes with the moderator for that column's value of the focal variable, and contrasting columns indicate how the moderator's effect changes with the focal variable. For more than one moderator, you would create sets of such 2D tables in which each table corresponds to a combination of the display values you specified for the 2nd, 3rd, . . . , *n*th moderators. I think it is often useful also to present (or bring into your discussion) information about the distribution of cases across the values of the focal and moderator variables. This could be either the joint or the marginal distribution of the variables. This helps avoid overinterpreting the import of results that rarely occur in the sample.

Single Moderator

For a single moderator, we construct a basic 2D table with the predicted outcome values calculated using Equation 4.1. That is, we plug in the column value for *F*, the row value for *M*, and the reference values for *Z* to get the predicted values in the cells. Let's apply this to the example we just used. We start by choosing display values for SES (the focal variable) and age (the moderator) and reference values for the other predictors. Both age (25–89 years) and SES (17.1–97.1 index points) have many distinct values, so we want roughly 10 display values of each for a first pass. For SES, I would specify display values every 10 points from 17 to 97, yielding 10 display values. For age, using 5- or 10-year intervals would give either too many or too few display values, so I instead use Equation 4.2 to calculate the interval for 10 display values. This gives a value of 7.11, which I round to 7 and slightly adjust its maximum display value from 89 to 88. The other predictors are either interval measures (not badly skewed) or dummy variables for a dichotomous construct, so it makes sense to use each one's mean as its reference value.[4] We already constructed the formula to predict the level of sexual frequency in Equation 4.3. We apply that (with an additional digit for accuracy) to calculate the predicted value in a cell with a row display value *Age = r* and a column display value *SES = c*:

$$
\begin{aligned}
\widetilde{Frequency}_{r,c} &= 14.9124 - 0.2014r - 0.07813c + 0.001343r \times c \\
&\quad - 0.7434 \, \overline{Female} - 1.0384 \, \overline{Never\ marry} + 0.2287 \, \overline{Children} \\
&\quad - 0.1213 \, \overline{Attend} \\
&= 14.9124 - 0.2014r - 0.07813c + 0.001343r \times c - 0.7434 \times 0.5458 \\
&\quad - 1.0384 \times 0.2337 + 0.2287 \times 1.9090 - 0.1213 \times 1.3729 \\
&= 14.9124 - 0.2014r - 0.07813c + 0.001343r \times c - 0.3784 \\
&= 14.5340 - 0.2014r - 0.07813c + 0.001343r \times c
\end{aligned}
$$

For example, for the cell in Table 4.1 corresponding to *Age* = 46 and *SES* = 37, the predicted frequency of sexual activity is

$$
\begin{aligned}
\widetilde{Frequency}_{46,37} &= 14.5340 - 0.2014 \times 46 - 0.07813 \times 37 \\
&\quad + 0.001343 \times 46 \times 37 = 4.66
\end{aligned}
$$

To make it easier to discern the patterns of how frequency of intercourse varies with age and SES, I recommend formatting the cell values with a font size proportional to the magnitude of the expected value.[5] The OUTDISP tool in ICALC can save such a formatted table to an Excel file.

Reading across any row in Table 4.1 shows the effect of SES for that row's value of age. For the first five rows (ages 25–53 years), the SES effect on sexual frequency is negative but diminishing in magnitude across the rows for older ages. For the youngest

TABLE 4.1 ● PREDICTED SEXUAL FREQUENCY BY THE INTERACTION OF AGE WITH SES[a]

Age	SES								
	17	27	37	47	57	67	77	87	97
25	8.74	8.30	7.85	7.40	6.96	6.51	6.07	5.62	5.18
32	7.49	7.14	6.79	6.44	6.08	5.73	5.38	5.03	4.68
39	6.24	5.98	5.73	5.47	5.21	4.95	4.70	4.44	4.18
46	4.99	4.83	4.66	4.50	4.34	4.17	4.01	3.85	3.68
53	3.74	3.67	3.60	3.53	3.46	3.39	3.32	3.25	3.18
60	2.49	2.52	2.54	2.56	2.59	2.61	2.64	2.66	2.69
67	1.24	1.36	1.48	1.60	1.71	1.83	1.95	2.07	2.19
74	−0.01	0.20	0.42	0.63	0.84	1.05	1.27	1.48	1.69
81	−1.26	−0.95	−0.65	−0.34	−0.03	0.27	0.58	0.89	1.19
88	−2.51	−2.11	−1.71	−1.31	−0.91	−0.51	−0.11	0.29	0.69

Note: SES = socioeconomic status.

a. Other predictors were held constant at their mean values.

ICALC command: outdisp, out(atopt((means) _all)) tab(save(Output\Table_4_1.xlsx) rowvar(mod)) ndig(2). The out( ) option sets predictors' reference values to their means, tab( ) saves a formatted table to an Excel file with rows defined by the moderator, and ndig( ) sets cell formats to 2 decimals.

age (25), the frequency of sexual activity decreases by 3.6 times per month (from 8.74 to 5.18) across the range of SES, whereas for a 53-year-old adult, frequency decreases by only 0.6 times per month (from 3.74 to 3.18). This effect turns positive (more frequent sex as SES increases) for ages 60 to 88, increasing in magnitude with age from a change in predicted frequency of 0.2 times per month at $Age = 60$ to a change of 3.2 times at $Age = 88$. The marginal distribution of age indicates that the effect of SES is negative for about two thirds of the sample and positive for about one third.

We can—and should—also examine the pattern in the columns to understand how the effect of age is contingent on SES. Looking at the change in predicted frequency down each column, age obviously has a negative effect at all values of SES, indicating a lower frequency of sex among older compared with younger adults. Age's effect decreases in magnitude as SES increases—compare the columns moving from left to right in the table. At the lowest level of SES (17), the predicted change in frequency across age's range is 11.25, diminishing to 7.87 when $SES = 57$ and further dropping to 4.47 at the highest SES (97).

Table 4.1 clearly portrays the overall picture of the relationship of frequency of sexual activity with age and SES. Older versus younger ages uniformly predict a lower frequency of sex, but that rate of decline in frequency is buffered by higher SES. On the other hand, the effect of SES on frequency is negative and diminishes in magnitude from ages 25 to 58 and then turns positive and increases in magnitude for older ages.

I argued above against choosing display values using equal-percentile intervals because it can visually distort the pattern of the relationship unless the predictors have close to a uniform frequency distribution. To illustrate this, Table 4.2 presents the predicted value of frequency using display values for age and SES at their respective deciles (10th, 20th, . . . , 90th percentile). If you read across the first row (10th percentile, $Age = 29$), you will see that the amount of change in the predicted value of frequency is not constant. Across the set of adjacent deciles, frequency decreases by 0.13, 0.20, 0.07, 0.44, 0.49, 0.04, 0.36, and 0.30, respectively. This fluctuating degree of change gives a very different impression from what the interaction model specifies. For a given value of age, there should be a constant effect of SES. This is what we see in the first row in Table 4.1, where the change in frequency between adjacent columns of SES is always a decrease of 0.44 (within rounding error). Comparing rows in Table 4.2 does still show that the negative effect of SES first diminishes in magnitude as age increases and then turns positive and increases in magnitude as age increases further. But depending on which pair of columns you read down to see the changing effect of SES, you get a different sense of the amount of change. For example, if you

TABLE 4.2   ●   PREDICTED SEXUAL FREQUENCY BY AGE WITH SES,[a] DECILE VALUES

Age Decile (Value)	SES Decile (Value)								
	10 (29)	20 (32)	30 (38)	40 (39)	50 (51)	60 (63)	70 (64)	80 (73)	90 (81)
10 (29)	7.55	7.42	7.22	7.15	6.71	6.22	6.18	5.82	5.52
20 (34)	6.73	6.63	6.46	6.41	6.04	5.63	5.60	5.31	5.06
30 (38)	6.09	6.00	5.86	5.81	5.51	5.17	5.14	4.89	4.69
40 (43)	5.28	5.21	5.10	5.07	4.84	4.59	4.57	4.38	4.22
50 (49)	4.30	4.27	4.20	4.18	4.04	3.89	3.87	3.76	3.67
60 (53)	3.66	3.63	3.60	3.59	3.51	3.42	3.41	3.35	3.30
70 (58)	2.85	2.84	2.84	2.84	2.84	2.84	2.84	2.83	2.83
80 (63)	2.04	2.06	2.09	2.10	2.17	2.25	2.26	2.32	2.37
90 (70)	0.90	0.95	1.03	1.06	1.24	1.44	1.45	1.60	1.72

Note: SES = socioeconomic status.

a. Other predictors are held constant at their mean values.

ICALC command: outdisp, out(atopt((means)_all)) tab(save(Table_4_2.xlsx) rowvar(mod)) ndig(2). The out( ) option sets predictors' reference values to their means, tab( ) saves a formatted table to an Excel file with rows defined by the moderator, and ndig( ) sets cell formats to 2 decimals.

compare the change between the 30th and 40th percentiles of SES, the sequence of SES effects changing with age is −0.07, −0.05, −0.05, −0.03, −0.02, −0.01, 0.0, 0.01, and 0.03. But between the 40th and 50th percentiles, the sequence of changing SES effects with age is −0.49, −0.37, −0.30, −0.23, −0.14, −0.08, 0.0, 0.07, 0.18. In contrast, comparing any two adjacent columns in Table 4.2 will show the same sequence of values for how the SES effect changes with age. Last, note that the age decile values are relatively uniformly spaced (4–6 years apart), while the SES decile values are not (1–12 index points apart). I leave it to the reader to do comparisons of the effect of age and how it changes with SES to see analogous problems.

Two or More Moderators

When you have multiple moderators of the focal variable, there are two approaches you can take to create a table of expected values, depending on whether you want to illustrate the joint effects of all the moderators or to isolate how particular moderators change the focal variable's effect. I begin with the joint effects case and then briefly consider how to get moderator-by-moderator expected value tables. When you have two moderators of the same focal variable, you construct a set of 2D tables with the focal variable defining the rows and the first moderator defining the columns. Each subtable in the set corresponds to a different display value of the second moderator. The expected value of the outcome ($\widehat{Y}$) is reported in the cells, calculated using the row value of the focal variable (r), the column value of the first moderator (c), the display value of the second moderator (s) for the subtable, and the reference values of the other predictors (z_{ref}). Applying Equation 4.1,

$$\widehat{Y} = a + b_1 r + b_2 c + b_3 r \times c + b_4 s + b_5 r \times s + \cdots + b_Z z_{ref} \tag{4.6}$$

This can be extended to additional moderators by creating a subtable for each combination of selected display values of the 2nd, 3rd, . . . , nth moderators and incorporating the relevant coefficients multiplied by the product of the row value of the focal variable (r) and the selected display value of the additional moderator into Equation 4.6. It is sometimes possible to format this as a fairly compact table with a set of panels, if one or more of your three interacting variables are nominal-level measures with a limited number of categories.

The example I present is for an interval-level focal variable interacting with two different interval-level moderators. The outcome is number of children ever born, the focal variable is household income (in $10K), the first moderator is age, the second moderator is education, and the other predictors in the model are race (two dummy variables), number of siblings, and religious intensity. The sample is limited to adults 40 years and older to reduce uncertainty about the completion of childbearing. Using the other predictors' OLS coefficients and their means to define reference values, you calculate the expected number of children ever born as

$$\begin{aligned}\widehat{Childs} = {}& 0.01966r + 0.04413c - 0.00170r \times c - 0.18682s \\ & + 0.00658r \times s + 2.12098\end{aligned} \tag{4.7}$$

where r = income display value, c = age display value, and s = education display value.

For income, I chose display values starting at its minimum ($500), with $15K increments. This interval size is a reasonable compromise between providing sufficient detail at the lower income levels ($10K would be better) and limiting how many display values are needed across the distribution of income ($20K would be better). I capped income's display values at $150,500 or more—between the 90th and 95th percentiles—because it has a very skewed distribution. The display values for age range from 40 to 80 in 10-year intervals, defining five evaluation points; an initial analysis with 5-year intervals suggested that the 10-year intervals are sufficient to show the pattern. The education values range from 5 to 20 years with a 5-year interval, defining four subtables presented in Table 4.3.[6] Each subtable reports the expected number of children corresponding to the display value of education, with the rows and columns defined by income and age, respectively. The font size of the cells is proportional to the magnitude of the expected number of children to facilitate seeing how this outcome changes with income, age, and education. Reading down a column shows the income effect: how the predicted number of children changes with income for a given age and level of education. Comparing a column with the other columns in the subtable reveals how the income effect changes with age, while comparing a column across the subtables indicates how the income effect changes with education.

Table 4.3 shows that the effect of income changes not only in magnitude but also in sign as age and education separately increase. At a given level of education, the income effect becomes more negative (less positive) in magnitude for older ages. In the subtables for *Education* = 5, the income effect is negative at any age: The expected number of children declines from the lowest income to the highest income in every column. But the magnitude becomes greater as age increases. At *Age* = 40, the expected number of children decreases by 0.023 for each $15K—compare any two adjacent rows in the same column—and at successive ages (columns) this predicted decline in the number of children is increasingly large: 0.048, 0.074, 0.099, and 0.125. Visually, we can see in this subtable that the expected number of children is smallest in the lower left-hand corner for the youngest age and highest income and increases as income decreases or as age increases. (Note how the differential font size makes this pattern obvious.) Inspecting the table for *Education* = 20 at the other end of the education distribution, we can see that the expected number of children is smallest for the youngest age and the lowest income—in the upper left-hand corner—and that the number of children increases as either age increases or income increases. Moreover, at this education level, the income effect is positive at any age but lessens in magnitude as age increases. The income effect at *Age* = 40 is to increase the number of children by 0.125 (per $15K), but that increase declines successively with age to 0.100, 0.074, 0.049, and 0.023. For the other levels of education, the income effect changes from a positive effect at younger ages to a negative effect at older ages. The age at which the income effect changes sign increases as education increases.

The moderating effect of education works in the opposite direction to that of age. For any given age, the income effect changes from a negative effect at the lowest level of education to an increasingly positive effect at a higher level of education—compare the column for an age across the subtables. At which education level the income effect changes sign depends on age. For *Age* = 40, the income effect changes from negative to a small positive between 5 and 10 years of education[7] and is increasingly positive at higher levels of education. Starting with an income effect of −0.023 at 5 years of education, the effects are 0.026, 0.076, and 0.125 for the subsequent levels of education. For *Age* = 80, the income effect does not switch from negative to

TABLE 4.3 ● CHILDS BY THE TWO-WAY INTERACTIONS OF FAMILY_INC WITH AGE AND WITH EDUCATION

Education	Family_Inc	Age				
		40	50	60	70	80
5	500	2.951	3.392	3.832	4.273	4.713
	15,500	2.928	3.343	3.758	4.173	4.589
	30,500	2.906	3.295	3.685	4.074	4.464
	45,500	2.883	3.247	3.611	3.975	4.339
	60,500	2.860	3.198	3.537	3.876	4.215
	75,500	2.837	3.150	3.463	3.777	4.090
	90,500	2.814	3.102	3.390	3.677	3.965
	105,500	2.791	3.053	3.316	3.578	3.841
	120,500	2.768	3.005	3.242	3.479	3.716
	135,500	2.745	2.957	3.168	3.380	3.591
	150,500	2.722	2.908	3.094	3.281	3.467
10	500	2.019	2.459	2.900	3.340	3.781
	15,500	2.045	2.460	2.875	3.290	3.705
	30,500	2.072	2.461	2.851	3.241	3.630
	45,500	2.098	2.462	2.826	3.191	3.555
	60,500	2.125	2.463	2.802	3.141	3.479
	75,500	2.151	2.464	2.778	3.091	3.404
	90,500	2.177	2.465	2.753	3.041	3.329
	105,500	2.204	2.466	2.729	2.991	3.254
	120,500	2.230	2.467	2.704	2.941	3.178
	135,500	2.257	2.468	2.680	2.891	3.103
	150,500	2.283	2.469	2.655	2.842	3.028

(Continued)

TABLE 4.3   ◆   (CONTINUED)

Education	Family_Inc	Age				
		40	50	60	70	80
15	500	1.086	1.527	1.967	2.408	2.848
	15,500	1.162	1.577	1.992	2.407	2.822
	30,500	1.238	1.628	2.017	2.407	2.796
	45,500	1.314	1.678	2.042	2.406	2.770
	60,500	1.390	1.728	2.067	2.406	2.744
	75,500	1.465	1.779	2.092	2.405	2.718
	90,500	1.541	1.829	2.117	2.405	2.692
	105,500	1.617	1.879	2.142	2.404	2.667
	120,500	1.693	1.930	2.167	2.404	2.641
	135,500	1.768	1.980	2.192	2.403	2.615
	150,500	1.844	2.030	2.216	2.403	2.589
20	500	0.154	0.594	1.035	1.475	1.916
	15,500	0.279	0.694	1.109	1.524	1.939
	30,500	0.404	0.794	1.183	1.573	1.963
	45,500	0.529	0.894	1.258	1.622	1.986
	60,500	0.654	0.993	1.332	1.671	2.009
	75,500	0.780	1.093	1.406	1.719	2.033
	90,500	0.905	1.193	1.480	1.768	2.056
	105,500	1.030	1.292	1.555	1.817	2.079
	120,500	1.155	1.392	1.629	1.866	2.103
	135,500	1.280	1.492	1.703	1.915	2.126
	150,500	1.405	1.591	1.777	1.964	2.150

ICALC command: outdisp, out(atopt((means) _all)) tab(save(Table_4_3.xlsx)) ndig(3). The out( ) option sets the predictors' reference values as their means, tab( ) saves the formatted table to an Excel file, and ndig(3) sets the cell format to 3 decimals.

positive until beyond 16 years of education, and the income effects from the lowest to the highest levels of education are −0.125, −0.075, −0.026, and 0.023, respectively. Visually, it is very easy to see that the expected number of children declines across increasing levels of education, as denoted by increasingly smaller font sizes from the first to the last subtable. It is only slightly harder to visually compare a given column across subtables to see that the income effect changes sign and eventually increases in magnitude in the positive direction.

However, what I think is not apparent is that the underlying childs–income–age relationship does not change with education and that the childs–income–education relationship does not change with age. That is, there is no three-way interaction. Compare the income effect on the predicted number of children for two different ages in each subtable to confirm that the effect of income changes with age in the same way at each level of education. For example, in the subtable for *Education* = 5, the income effect at *Age* = 40 is −0.023 (2.928 − 2.951) per \$15K, and the income effect at *Age* = 50 is −0.048. Thus, the income effect changes between those two ages by −0.025 (−0.048 − (−0.023)). In the subtable for *Education* = 10, the income effect at *Age* = 40 is 0.026, the income effect at *Age* = 50 is 0.001, and the change in the income effect with age is −0.025, the same as we just calculated for *Education* = 5. I leave it as an exercise for the reader to verify that the change in the income effect between *Education* = 5 and *Education* = 10 is 0.049 at *Age* = 40, *Age* = 50, and so on. As we will see shortly, constructing figures to portray how an outcome's expected values vary with the interacting variables often provides a more accessible and readily interpretable presentation of the relationship that should make this clear.

Isolating Each Moderator's Effect on the Focal Variable

To separate out the effect of each moderator on the focal variable, you can analyze each two-way interaction (F and M_1, F and M_2, . . .) individually as if it were single moderator. But you must set any term in the prediction equation involving the other moderators to a reference value, just as you would for the other predictors. Thus, to isolate how age moderates the effect of income on the number of children, we would modify the prediction equation developed above for age and education as moderators (Equation 4.7). Specifically, we would set the value of s for education to a reference value—such as its mean of 13.6175—everywhere it occurs in the equation:

$$\widehat{Childs} = 0.01966r + 0.04413c - 0.00170r \times c - 0.18682 \times 13.6175$$
$$+ 0.00658r \times 13.6175 + 2.12098$$
$$= (0.01966 + 0.00658 \times 13.6175)r + 0.04413c - 0.00170r \times c \quad (4.8)$$
$$+ (-0.18682 \times 13.6175 + 2.12098)$$
$$= 0.10926r + 0.04413c - 0.00170r \times c - 0.42304$$

It is important to recognize that for the term above in Equation 4.8 that included both education and income (0.00658 $r \times s$), you substitute the reference value of education for s but the display value of income for r. Using this expression, we plug in the display values for income and age to construct the predicted values reported in Table 4.4. It is easy to pick out the patterns (especially with the font size formatting):

- At *Age* = 40 to 60, the income effect is positive—the predicted number of children rises as income increases from the top to the bottom of the column—but also decreases in magnitude as age increases.

- But at *Age* = 70 and 80, the income effect is negative and increases in size with age.

- The effect of age on the number of children is always positive—within each row, it rises from left to right. But the age effect decreases in magnitude as income increases—compare the change across a row with the change across the next row.

Remember that these exact numbers would change if you picked a different reference value for education but the overall patterns would not. That is, the income effect will change toward a more negative effect (in some cases a smaller positive effect) as age increases, and the positive effect of age will diminish as income rises. The same process could be used to isolate the moderating effect of education.

In general, I do not recommend isolating the effect of each moderator when presenting or describing the interaction effects. That said, it is sometimes useful to isolate each moderator to help you understand what the patterns are so that you can better interpret the expected values table, which shows the effects of the focal variable and all the moderators together.

TABLE 4.4 ● PREDICTED CHILDS BY THE INTERACTION OF FAMILY_INC WITH AGE AND EDUCATION AT ITS MEAN

	Age				
Family_Inc	**40**	**50**	**60**	**70**	**80**
500	1.344	1.785	2.225	2.666	3.106
15,500	1.406	1.821	2.236	2.651	3.066
30,500	1.469	1.858	2.248	2.637	3.027
45,500	1.531	1.895	2.259	2.623	2.987
60,500	1.593	1.932	2.270	2.609	2.948
75,500	1.655	1.968	2.281	2.595	2.908
90,500	1.717	2.005	2.293	2.581	2.868
105,500	1.779	2.042	2.304	2.566	2.829
120,500	1.841	2.078	2.315	2.552	2.789
135,500	1.903	2.115	2.327	2.538	2.750
150,500	1.966	2.152	2.338	2.524	2.710

ICALC command: outdisp, out(atopt((means)_all)) tab(save(Table_4_4.xlsx)) ndig(3). The out( ) option sets the predictors' reference values as their means, tab( ) saves the formatted table to an Excel file, and ndig(3) sets the cell format to 3 decimals.

Three-Way Interaction

Constructing an expected value table for a three-way interaction uses the same process as for a focal variable with two moderators: Create a set of subtables for the selected display values of the second moderator. The only difference is in the formula to calculate the expected value reported in the subtables: Add the terms (boxed below) for the two-way interaction between the moderator variables and for the three-way interaction of the focal and moderator variables:

$$\widehat{Y} = a + b_1 r + b_2 c + b_3 r \times c + b_4 s + b_5 r \times s + \boxed{b_6 c \times s + b_7 r \times c \times s} + \cdots + b_Z z_{ref} \qquad (4.9)$$

where r, c, and s are again the display values for the row, column, and subtable indexes, respectively.

I use the prior OLS analysis predicting voluntary association memberships by the three-way interaction of sex, age, and education to illustrate the creation and interpretation of the expected values table. Sex defines the columns, age indexes the rows, and education defines the subtables. Putting the coefficients from that analysis into Equation 4.9 provides our calculating formula:

$$\begin{aligned}
\widehat{Y} = &-1.21184 + 0.0080918r - 2.76460c + 0.036413r \times c + 0.15505s \\
&+ 0.00053851r \times s + 0.23645c \times s - 0.0037381r \times c \times s \\
&- 0.14224\,\overline{Size} + 0.086893\,\overline{Class}
\end{aligned} \qquad (4.10)$$

where r, c, and s are the display values for age in the rows, sex in the columns, and education in the subtable headings, respectively.

Although I typically use the focal variable as the row variable, I think the table is more compact and readable with the focal variable as the column when it has very few categories. The display values for sex are its two categories (women and men), whereas for age and education I chose display values evenly spaced across their ranges—every 10 years from 18 to 88 for age and every 4 years from 0 to 20 for education. Table 4.5 consists of the six subtables corresponding to the six display values of education. Each subtable reports the expected values of memberships—with font size proportional to their magnitude—as it varies with sex and age for the given value of education.

What stands out and indeed dominates the first impression from Table 4.5 is the increasing font size—and predicted number of memberships—across the education subtables from low to high education. Less obvious is that the changing font sizes suggest that the predicted number of memberships mainly appears to increase with age. But a systematic examination provides a fuller and more complex understanding. There are three sets of comparisons to make to understand the interactive effect of sex, age, and education, focusing in turn on how the effect of each predictor is moderated by the interaction of the other two. Starting with the effect of sex, look across a row to see how predicted number of memberships differs between women and men at that row's age for the given years of education, and then scan down a column to find out how the sex difference changes with age. For example, look at the first subtable for *Education* = 0, and compare the sex difference in predicted number of memberships as age increases. Men's predicted memberships starts higher than women's, with the gap declining until *Age* = 68. After that point, women have a higher predicted number of memberships than men, with the gap now increasing

TABLE 4.5 ⬢ EXPECTED VALUE TABLE , THREE-WAY INTERACTION

Membership by the Three-Way Interaction of Sex, Age, and Education			
Education	Age	Sex	
		Men	Women
0	18	−0.631	−2.589
	28	−0.558	−2.117
	38	−0.486	−1.646
	48	−0.413	−1.174
	58	−0.341	−0.703
	68	−0.268	−0.232
	78	−0.196	0.240
	88	−0.123	0.711
4	18	0.039	−1.304
	28	0.133	−0.966
	38	0.226	−0.628
	48	0.320	−0.290
	58	0.414	0.049
	68	0.507	0.387
	78	0.601	0.725
	88	0.694	1.063
8	18	0.709	−0.019
	28	0.824	0.186
	38	0.939	0.391
	48	1.053	0.595
	58	1.168	0.800
	68	1.282	1.005
	78	1.397	1.210
	88	1.512	1.415

Membership by the Three-Way Interaction of Sex, Age, and Education			
Education	Age	Sex	
		Men	Women
12	18	1.379	1.265
	28	1.515	1.337
	38	1.651	1.409
	48	1.786	1.480
	58	1.922	1.552
	68	2.058	1.624
	78	2.193	1.696
	88	2.329	1.767
16	18	2.049	2.550
	28	2.206	2.488
	38	2.363	2.427
	48	2.519	2.365
	58	2.676	2.304
	68	2.833	2.242
	78	2.990	2.181
	88	3.146	2.119
20	18	2.719	3.835
	28	2.897	3.640
	38	3.075	3.445
	48	3.252	3.250
	58	3.430	3.056
	68	3.608	2.861
	78	3.786	2.666
	88	3.964	2.471

ICALC command: outdisp, out(atopt((means)_all)) tab(row(mod) save(Output\Table_4_5.xlsx)) ndig(3). The out( ) option sets the predictors' reference values as their means, tab( ) saves the formatted table to an Excel file with age defining subtable rows, and ndig(3) sets the cell format to 3 decimals.

with age. More abstractly, the effect of sex moves in the positive direction as age increases. This abstract pattern also characterizes the subtables for *Education* = 4 and 8, but note that the sex difference never turns positive for *Education* = 8.

A different pattern characterizes how age moderates the effect of sex at higher education levels. The gap between men and women moves in the negative direction, with a larger increase in the number of memberships for men than for women. At *Education* = 16 and 20, the sex effect changes from positive at the youngest ages (higher number of memberships for women than for men) to increasingly negative at older ages.

In a similar fashion, we can observe how the effect of age on memberships changes by sex and education. Again, start with the subtable for *Education* = 0. Note that in the column for men, the number of memberships increases by 0.072 for each 10 years' difference in age but for women, it increases by 0.471. Comparing the age effect for men and women across the education subtables shows that the age effect for men increases slightly as education rises—by 0.021 for each additional 4 years of education. In contrast, the effect of age for women decreases as education increases—by 0.133 per 4 years of education—and becomes a negative effect around 14 years of education.

Last, we can determine the pattern of the education effect as it is moderated by sex and age by comparing the predicted memberships in the same cells of adjacent education subtables. For example, the education effect for men aged 18 years is 0.659 per 4 years of education; that is, the number of memberships increases by almost 0.7 with an additional 4 years of education. I calculated this from the predicted memberships in the *Education* = 4 subtable minus the corresponding predicted value in the *Education* = 0 subtable (0.039 − (−0.631)). Note that you would get the exact same value by using any two adjacent subtables. Repeating this calculation across the age display values reveals that the education effect for men increases slightly with age. At *Age* = 88, the education effect for men increases only to 0.817. Conversely for women, the effect of education is larger at any age, but it declines substantially at older ages. At *Age* = 18, number of memberships is predicted to increase by 1.285 per 4 years of education, while at *Age* = 88, the education effect has diminished to 0.352 more memberships per 4 years of education.

Unearthing these patterns from just examining the results in Table 4.5 is not an easy task. I find it useful to do side calculations of the change in the predicted value with changes in the focal variable of the moment and how that changes across the display values of the two moderators. Table 4.6 shows the side calculation I did for how much membership changes with age for each combination of display values for sex and education. For example, the upper-rightmost cell (*Education* = 0 and *Sex* = women) is calculated from the first subtable of Table 4.5 (*Education* = 0) and the column for *Sex* = women. It is the difference between the number of predicted memberships for *Age* = 28 and *Age* = 18 (calculated with four decimals then rounded to three for Table 4.5):

$$-2.1172 - (-2.5886) = -2.1172 + 2.5886 = 0.4714$$

Notice how much easier it is to see the patterns in this table than in Table 4.5. Sometimes it is even better to forgo tables altogether and turn to visual representations of how the expected value of the outcome changes with the interacting variables, which is the next topic.

| TABLE 4.6 ● SIDE CALCULATION FOR EXPECTED VALUES TABLE, THREE-WAY INTERACTION |||

Change in Membership for a 10-Year Increase in Age		
Education	Men	Women
0	0.072	0.471
4	0.094	0.338
8	0.115	0.205
12	0.136	0.072
16	0.157	−0.062
20	0.178	−0.195

CHARTS AND PLOTS OF THE EXPECTED VALUE OF THE OUTCOME

Although bar charts, scatterplots, contour plots, or 3D surface plots can be used for any type of interaction effect, the most effective choice of a graphic display usually depends on the level of measurement of your focal and moderator variables and the number of moderating variables. Figure 4.1 presents a rubric for choosing among these options and includes a brief description of how the expected outcome, the focal variable, and the moderators are used to construct the visual display. A key criterion is whether the focal variable is measured as categorical or interval. For an interval focal variable, the change in the expected outcome across the values of the focal variable follows a set, numeric pattern. Using a bar chart in that case not only can result in sacrificing information in the visual display (from categorizing the focal variable) but also can make comparisons of the relationship across the values of the moderating variables more unwieldy and difficult to see compared with a scatterplot. Conversely, a bar chart is ideal for displaying the effect of a categorical focal variable, especially when its moderators are also categorical, because it emphasizes group differences rather than linear patterns. A partial exception for a categorical focal variable is when at least one of the moderators is interval. In that case, a scatterplot of the expected outcome against the first moderator, with separate lines for each category of the focal variable, allows you to easily see the group differences, which are represented by the vertical divergence between the lines.

When the focal variable is interval, I strongly recommend a scatterplot drawing connected lines for the expected outcome on the y-axis against the focal variable on the x-axis, with separate lines plotted for the first moderator's display values. The repeated lines visually highlight how the relationship between the outcome and the focal variable is conditioned by the moderator. In some cases, you can also easily see the contingent relationship between the outcome and the moderator, especially if the moderator is categorical. But in many cases, it is better to do a second visual display reversing the roles of the focal and moderator variables. (Note that such scatterplots are also known as line plots or connected-line plots.)

FIGURE 4.1 ● RUBRIC FOR CHOOSING A VISUAL DISPLAY OF THE EXPECTED OUTCOME VARYING WITH INTERACTING PREDICTORS

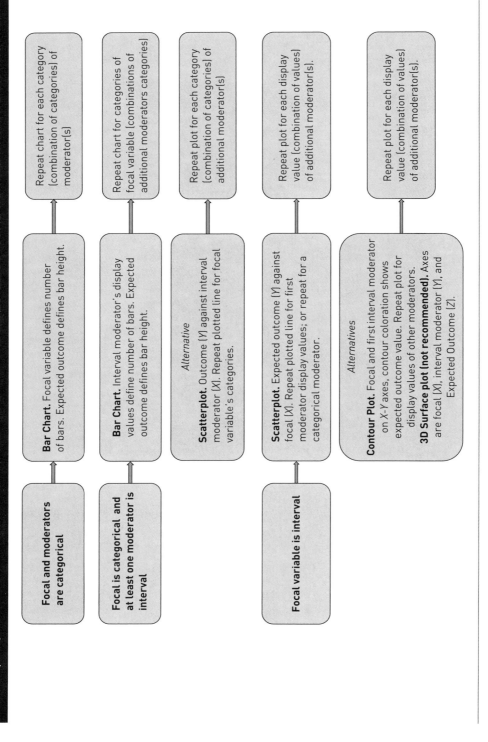

Focal and moderators are categorical

Bar Chart. Focal variable defines number of bars. Expected outcome defines bar height.

Repeat chart for each category (combination of categories) of moderator(s)

Focal is categorical and at least one moderator is interval

Bar Chart. Interval moderator's display values define number of bars. Expected outcome defines bar height.

Repeat chart for categories of focal variable (combinations of additional moderators categories)

Alternative

Scatterplot. Outcome (Y) against interval moderator (X). Repeat plotted line for focal variable's categories.

Repeat plot for each category (combination of categories) of additional moderator(s)

Focal variable is interval

Scatterplot. Expected outcome (Y) against focal (X). Repeat plotted line for first moderator display values; or repeat for a categorical moderator.

Repeat plot for each display value (combination of values) of additional moderator(s).

Alternatives

Contour Plot. Focal and first interval moderator on X-Y axes, contour coloration shows expected outcome value. Repeat plot for display values of other moderators.
3D Surface plot (not recommended). Axes are focal (X), interval moderator (Y), and Expected Outcome (Z).

Repeat plot for each display value (combination of values) of additional moderator(s).

For a two-way interaction in which both the focal and the moderator variables are interval, a 2D contour plot can be a good alternative. The focal and moderating variables define the two dimensions of the plot, and the expected value is shown by the coloration of the contours. But a 2D contour plot does not work well if you have more than one moderator; it is difficult to compare the coloration and boundaries of the contours across the repeated plots for the display values of the other moderator(s).

The other alternative is a 3D surface plot that defines the "floor" of the plot (the *x*- and *y*-axes) by the focal variable and the first moderator and the vertical dimension (*z*-axis) by the expected values. But I do not recommend using a 3D surface plot. As I noted in the discussion of effect displays, the issue is that the point of perspective from which you see the surface affects how you perceive and interpret the relationship between the outcome and the interacting variables. By definition, the projection of the 3D surface onto a 2D representation requires distorting the dimensions defining the 3D surface, so different choices of perspectives create different distortions. I provide a brief illustration of this at the end of the first scatterplot example of an interval-by-interval interaction. Note that the ICALC toolkit cannot produce a 3D surface plot directly in Stata (Stata graphics do not offer this capability), but it can write the data needed to create a 3D surface plot to an Excel spreadsheet if you create a 2D contour plot. Chapter 6, on the application of the ICALC toolkit, includes brief instructions on how to produce a 3D surface plot in Excel.

I want to emphasize that this rubric is only a guideline. Sometimes, a different choice will work better than that recommended by the rubric, so try alternatives if the first choice does not display the information well. In the next section, I describe the application of bar charts to a single-moderator example and a two-moderator example. In the following section on scatterplots, I illustrate their use for single-moderator models with either categorical or interval moderators, for a two-moderator example with interval focal and moderating variables, and for a three-way interaction with a categorical focal variable interacted with two interval-level moderators. For the three-way interaction example, I compare the use of scatterplots and bar charts for that example. I conclude with a brief illustration of how to apply and understand a 2D contour plot.

Bar Charts for Categorical Focal and Moderator Variables

The basic structure of the figure is a series of bar charts, one chart for each display value of the (first) moderator. Each individual bar chart has a bar for each category of the focal variable, where the height of the bar represents the expected value of the outcome variable and the coloration of the bar identifies the category of the focal variable. If you have more than one moderator, then you construct and present bar charts corresponding to each combination of display values of the multiple moderators. Throughout the book, I use a grayscale coloration scheme to distinguish bars, areas on graphs, and so on, to correspond to the types of figures that can most typically be published. Similarly, the default in the ICALC toolkit is a grayscale color scheme, with an option to specify a different color scheme.

The examples for a one-moderator interaction effect and for a two-moderator interaction effect both use the results from reanalyzing the wealth data used previously. The single-moderator model specifies an interaction of headship type (three categories) with education (five categories). The two-moderator example also includes the interaction of headship type with a measure of having any children (two categories).

Example 1: Effect of Education
Moderated by Household Headship Type

For this example, we need to construct a set of three bar charts, one panel for each category of headship type. The panels present the predicted net worth for the five categories of education, with the predicted values calculated contingent on the categories of headship type. These calculations follow the same procedure used above for creating expected value tables. That is, you use the regression results (Equation 4.1) to predict the expected outcome value for each combination of a focal variable category (fc) and a moderator category (mc) by setting the focal and moderating variables equal to their respective display values and setting all other predictors to their reference values:

$$
\begin{aligned}
\widehat{Wealth}_{fc,mc} = & -21.3 + 4.0\,Ed1_{fc} + 5.9\,Ed2_{fc} + 9.0\,Ed3_{fc} + 10.9\,Ed4_{fc} \\
& -1.2\,Head1_{mc} - 3.1\,Head2_{mc} + 1.4\,Head1_{mc} \times Ed1_{fc} \\
& +1.1\,Head2_{mc} \times Ed1_{fc} - 0.3\,Head1_{mc} \times Ed2_{fc} \\
& +1.3\,Head2_{mc} \times Ed2_{fc} - 2.7\,Head1_{mc} \times Ed3_{fc} \\
& -0.8\,Head2_{mc} \times Ed3_{fc} - 3.8\,Head1_{mc} \times Ed4_{fc} \\
& +3.5\,Head2_{mc} \times Ed4_{fc} + 0.4\,\overline{Age} - 0.0005\left(\overline{Age}\right)^{2} \\
& +1.2\,\overline{Retired} + 6.0\,\overline{NILF} + 2.8\,\overline{Income}
\end{aligned}
$$

Education is represented by four dummy variables (Edi_{fc}), so you set the display value to a specific category by specifying a value of 1 for the dummy variable for that category and a value of 0 for all other education dummy variables. That is, to set the display value to "High School"—the category represented by $Ed1$—you set $Ed1_{fc} = 1$ and $Ed2_{fc} = Ed3_{fc} = Ed4_{fc} = 0$. And you use these values in the calculations for the main effect terms and for the interaction effect terms involving Edi_{fc}. The same process applies to setting the display value for headship type. As an aside, note that age is specified as a quadratic function and the reference value for the square term is properly set to the squared value of mean age (not the mean of the squared value of age).

The bars in Figure 4.2 are labeled with the predicted net worth for the given categories of education and headship type to supplement the visual display with the exact predicted net worth. There are several patterns apparent from the bar charts. First, the magnitude of the education effect (the differences in the height of the bars across the range of education categories) is greater for households headed by a couple than those headed by a single man or a single woman. And the pattern of the differences between education categories also changes with headship type. For couple-headed households, there is a consistent stair-step increase in predicted wealth with education. For a single-woman-headed household, the stair-step increase holds true except between the two highest education categories, where wealth declines slightly. While the stair-step increase is consistent for single-man-headed households, the wealth differences among the middle three education categories are negligible. The disadvantage of not having at least a high school diploma is also greater for single-headed than for couple-headed households. Visually, compare the difference across headship types in the heights of the first education category with the subsequent categories. For example, the disparity between the first two education categories increases from 4.00 in couple-headed households to 5.41 and 5.15 in the single-male- and single-female-headed households, respectively. The takeaway message from the bar charts

FIGURE 4.2 ● EXPECTED VALUE BAR CHART, ONE MODERATOR

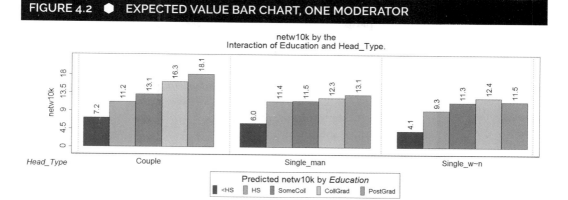

ICALC command: outdisp, out(atopt((means) _all)) plot(type(bar)) ndig(1) pltopts(plotregion(ma(t +4))). The out( ) option sets the predictors' reference values as their means, the plot( ) option specifies a bar chart, ndig( ) sets the bar label format to 1 decimal, and pltopts( ) customizes plot margins.

is that the predicted value of wealth consistently rises with education in all headship types. But the gains are almost always largest for married-couple-headed households. And for single-headed households, the education effect is primarily the advantage of having at least a high school degree versus not having a high school degree (note how small the stair-step differences are among their last four education categories).

Example 2: Two Moderators, Headship Type Moderated by Education and by Any Children

This example from the wealth reanalysis illustrates the use of bar charts for a focal variable that is moderated additively (not interactively) by two other predictors. That is, the model specifies an interaction between headship type and education and an interaction between headship type and any children. In this situation, you need to examine and explain three different relationships:

1. How the effect of headship type on wealth differs by education and by any children.

2. How the effect of education on wealth is contingent on headship type.

3. How the effect of any children on wealth varies by headship type.

As I discussed above for expected value tables, you can choose to examine the joint effect of the three variables on the outcome, or you can isolate the effect of each moderator on the relationship between a focal variable and the outcome. For presenting and explaining the results to your audience, I would recommend the joint effects approach for two reasons. Conceptually, it corresponds to the "reality" of the model you have specified. In practical terms, you can most often use a single set of bar charts to explain all three relationships. That said, if you are having difficulty seeing the patterns from the joint effects bar charts, then you should isolate the effects of each focal–moderator pair to help you recognize its patterns. You can then use those insights to help you understand the patterns in the joint effects bar charts.

In this situation, we construct the predicted value of wealth as it varies with headship type for every combination of categories of the two moderators—any children and headship type. All other predictors are set to their reference values (means). Figure 4.3 presents the resulting bar charts, where a bar's height represents the predicted value of wealth. These results are organized as separate panels for the two categories of any children. Each panel consists of five bar charts—separated by dotted vertical lines—for the five education categories. Each set of three bars reports the predicted wealth for one of the headship types. Comparing the predicted wealth within any bar chart shows the effect of headship type on wealth for a particular combination of any children and education. Keep in mind that wealth is in units of $10K. For example, in a household with no children and an education level of less than high school (the leftmost bar chart in the left panel), the net worth of a couple-headed household is about $16K more than for a single-man-headed (8.6 vs. 7.0) and about $43K more than for a single-woman-headed household (8.6 vs. 4.3).

To understand how any children moderates the headship type effect on wealth, compare the pattern within a bar chart between the left panel (no children) and the right panel (one or more children). Similarly, to examine how education moderates the effect of headship type from this figure, compare the pattern within a bar chart across the other categories of education in the same panel. Before you

---

FIGURE 4.3 ● EXPECTED VALUES BAR CHART, TWO CATEGORICAL MODERATORS

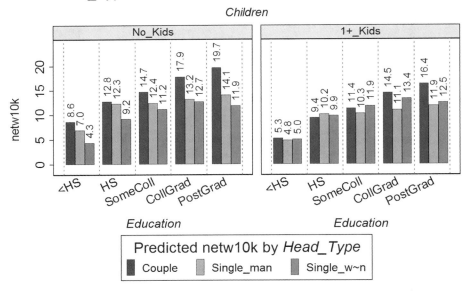

ICALC command: outdisp, out(atopt((means)_all)) plot(type(bar)) ndig(1) pltopts(plotregion(ma(t +2))). The out( ) option sets the predictors' reference values as their means, the plot( ) option specifies a bar chart, ndig( ) sets the bar label format to 1 decimal, and pltopts( ) customizes the plot margins.

read on, take some time to examine the figure to determine for yourself how you think children and education moderate the relationship between headship type and wealth.

To work out the moderating effect of any children, pick a bar chart for any education category, and compare it between the panel for no kids and the panel for one or more kids. For concreteness, let's compare the Some College bar charts between the panels. The panel for no kids evidences more substantial differences in predicted wealth among headship types than the panel for one or more kids:

No kids: *Couple − Single man* = $22,260, *Couple − Single woman* = $35,200

Single man − Single woman = $12,260

One or more kids: *Couple − Single man* = $10,560, *Couple − Single woman* = −$5,000

Single man − Single woman = −$15,560

Moreover, the ordering of headship types by wealth differs by the presence of children. Single-man households have higher predicted wealth than single-woman households if they have no children, but single-woman households have higher predicted wealth if the household has children except when *Education* = some college. No matter which education category you pick for comparison, you will see the same moderating effect of any children because this is not a three-way interaction, so the moderation of headship type by any children is unaffected by education.

Similarly, to understand how education moderates the effect of headship type on wealth, pick a category of any children—that is, one of the two panels—and evaluate how the differences in predicted wealth among headship types change across the education categories. Let's start with the left panel for households without any children and compare the wealth gap between couple-headed and single-headed headship types. From less than high school to high school, the wealth advantage of couple-headed households diminishes (negative direction) and then increases (positive direction) across the higher categories of education. The differences between single-man and single-woman households evidence a mainly opposite pattern. The wealth advantage of single-man households increases (positive direction) between the less than high school and high school categories, decreases (negative direction) across the next two education categories, and then increases again (positive direction) between the two highest education categories.

From a first look at the panel of results for households with children, the pattern of moderation of headship type by education might appear to be different. In this panel, some of the wealth differences between household types change sign, but in the panel for no kids, the sign of the wealth gap is always the same. The key to seeing the parallelism is to think about the direction in which the wealth gap changes rather than whether the gap changes from positive to negative. That was the reason for the parentheticals about positive or negative direction in the discussion above. For example, between the first two education categories, the wealth gap between couple-headed and single-man-headed households with one or more children changes from a $5K advantage for married couples to an $8K advantage for single-man-headed households. That is, the gap moves in the negative direction, just as it did in the panel for no kids. Readers should take another look at the patterns in the two panels to convince themselves of the parallelism.

Reversing the Roles of Focal and Moderator Variables. As an alternative to constructing and presenting two additional bar charts, you can use the results from the two-moderator bar chart to interpret how headship type moderates the effect of any children on wealth and how it moderates the effect of education on wealth. Let's start with the effect of any children. Pick the bar charts for one of the education categories, and for each headship type, compare its bar in the panel for no kids with its bar in the panel for one or more kids. The difference in predicted wealth is the effect of having any children on wealth for that headship type. Let's use the bar chart for High School and calculate the differences in predicted wealth:

Couple-headed: $9.45 - 12.79 = -3.34$ or $-\$33,400$

Single-man-headed: $10.2 - 12.34 = -2.14$ or $-\$21,400$

Single-woman-headed: $9.91 - 9.25 = 0.66$ or $\$6,600$

This readily shows that the presence of children predicts a fairly substantial loss of net worth for couple-headed households and single-man-headed households. I call this substantial given that median net worth in the sample is $54,745. Conversely, having children predicts a somewhat higher net worth for single-woman-headed households.

Turning to the education effect, pick one of the any children panels and examine how predicted wealth changes across education categories for each of the headship types in turn. Using the panel for one or more kids, I list the predicted wealth values from lowest to highest education for each headship type and describe its pattern:

Couple-headed: $5.28 \rightarrow 9.45 \rightarrow 11.36 \rightarrow 14.52 \rightarrow 16.40$

                         Stair-step increases of similar step sizes

Single-man-headed: $4.81 \rightarrow 10.20 \rightarrow 10.31 \rightarrow 11.07 \rightarrow 11.94$

                         Initial increase and then a plateau of slight increases

Single-woman-headed $5.01 \rightarrow 9.91 \rightarrow 11.86 \rightarrow 13.37 \rightarrow 12.54$

                         Initial larger increase and then an inclined plateau of smaller increases and final drop

The education effect on wealth obviously differs by headship type. Higher levels of education consistently predict more substantial increases in wealth for couple-headed than for single-headed households. For single-headed households, the effect of education on wealth is dominated by the difference between those with less than a high school degree versus all higher levels, with minimal additional gains in wealth with education beyond a high school degree.

There are three take-away messages from this detailed description of the relationship of wealth with headship type, children, and education from which you can frame a succinct interpretation of the relationship. First, headship type not surprisingly predicts a substantial wealth advantage for couple-headed compared with single-headed households. But this advantage is greater in households that never had any children than in those that have children. And it is greater in those households with the least education and those with the most education. Second, the disadvantage of having children is amplified for couple-headed and single-man-headed households. Last, the wealth advantages of higher education levels are most pronounced in magnitude for couple-headed households, while the wealth advantage of education among single-headed households is largely confined to the distinction between less than a high school diploma and all higher levels of education.

Scatterplots for Interval Focal Variables

My primary recommendation for visually representing the interaction between two interval variables is a scatterplot, which is also a good choice for the interaction between a categorical and an interval variable. The structure of a scatterplot should be very familiar. You plot a predicted outcome, whose values define the y-axis, against a predictor, whose values are represented on the x-axis. What is different in the case of interaction effects is that there is no single line that can show the relationship between the outcome and the predictor because that relationship changes with its moderator(s). Thus, you plot a separate line of the outcome–focal relationship for each of the display values of the first moderating variable. If you have additional moderators or a three-way interaction, you can construct a series of such scatterplots showing how M_1 moderates the effect of F on the outcome, one scatterplot for each combination of values of the other moderating variables. A contour plot is also a viable option for an interval-by-interval interaction, but it does not work as well for three-way interactions or multiple-moderator models.

Example 3: Interval-by-Interval Interaction, Age by SES Effects on Frequency of Sex

This first application uses the same regression results as did the first expected values table example to allow readers to assess whether they find the table or the scatterplot easier to work with and interpret. Figure 4.4 presents the scatterplot of predicted frequency of intercourse (y-axis) against SES (x-axis) with multiple lines drawn for the frequency–SES relationship, each calculated for one of the display values of age. The selected ages are evenly spaced in increments of 16 years from the minimum sample age (25 years) to the maximum (89 years). I initially created a plot with nearly twice that many plotted lines, from which I could determine that a scatterplot that had a plotted line at the middle of the age range with two lines for younger and two lines for older ages would capture the pattern of how age moderates the frequency–SES relationship.

The scatterplot makes quite clear that the effect of SES on frequency of sexual activity changes considerably with age from a negative effect at younger ages to a positive effect for older ages. The plotted lines are visually ordered by age from the top to the bottom of the figure, with the top line for the youngest age and the bottom line for the oldest age. The *solid line* for *Age* = 25 shows the steepest negative slope, indicating that the predicted frequency of sex declines with SES the most at the youngest sample age. The rate of decline is less for *Age* = 41 (*medium dashed line*) and is nearly flat for *Age* = 57 (*large dashed line*). The exact age at which the SES effect switches sign is 58.2 according to a GFI analysis. The next two lines, for *Age* = 73 and 89, show that the effect of SES on frequency of sex is increasingly positive at older ages.

Although Figure 4.4 portrays the frequency–SES relationship clearly and accessibly, it is more difficult to determine and explain the exact nature of the frequency–age relationship. At a given value of SES, the vertical distance between a pair of plotted lines represents the effect of age. This shows that the predicted value for the older age is always lower than the predicted value for the younger age; that is, there is a

FIGURE 4.4 ● INTERVAL-BY-INTERVAL SCATTERPLOT, SES MODERATED BY AGE

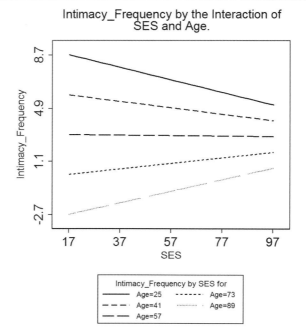

Intimacy_Frequency by the Interaction of SES and Age.

Intimacy_Frequency by SES for
Age=25   Age=73
Age=41   Age=89
Age=57

Note: SES = socioeconomic status.

ICALC command: outdisp, out(atopt((means)_all)) plot(type(scat)). The out( ) option sets the predictors' reference values as their means, and the plot( ) option specifies a scatterplot.

negative relationship between age and frequency of sex. Moreover, when *SES* = 17 the vertical gap between the successive ages is larger than it is when *SES* = 37, which is larger than the vertical gap when *SES* = 57, and so on. That is, as SES increases, the negative effect of age diminishes in magnitude. A better choice is to present a second scatterplot of frequency of sex plotted against age on the *x*-axis, with multiple lines drawn to represent how the effect of age is contingent on SES. Examining this scatterplot (Figure 4.5) immediately reveals that the effect of age is negative and that the steepest slopes for age occur at lower SES—compare the *solid line* for *SES* = 17 with the *medium dashed line* for *SES* = 37. Notice that the lines cross around *Age* = 57, signifying that at older ages frequency of sex declines more rapidly with age for people with low SES than for those with higher SES. More important, notice how much easier it is to interpret the moderation of the effect of age from Figure 4.5 than to try to reconstruct it from Figure 4.4.

This example demonstrates the potential drawback of using a scatterplot, that it is often necessary to present and interpret two scatterplots rather than a single visual display of the relationship between the outcome and the two interacting predictors. A 2D contour plot can be an effective alternative when your focal and moderating variables are both interval level. Let me suggest a device to help visualize how a 2D contour plot represents what is 3D information—the predicted outcome and its two interacting predictors. Imagine yourself suspended above a floor looking straight down. The bottom border of the floor (*x*-axis) has tick marks and values representing

FIGURE 4.5 ● INTERVAL-BY-INTERVAL INTERACTION, THE EFFECT OF AGE MODERATED BY SES

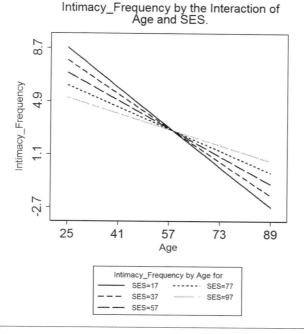

Intimacy_Frequency by the Interaction of Age and SES.

Intimacy_Frequency by Age for
—— SES=17   · · · · · · SES=77
– – – SES=37   · · · · · · · · · SES=97
— — SES=57

Note: SES = socioeconomic status.

ICALC command: outdisp, out(atopt((means) _all))) plot(type(scat)) ndig(2). The out( ) option sets the predictors' reference values as their means, plot( ) specifies a scatterplot, and ndig( ) formats the *y*-axis.

the first of your interacting predictors, and the left border (*y*-axis) has tick marks and values representing the other interacting predictor. For each combination of possible values of the predictors (*x*–*y* coordinates), you calculate the expected value of the predictor and paint a colored dot on the floor corresponding to its *x*–*y* coordinates. The coloration of the dot represents the magnitude of the value of the outcome within a range of similar values. This creates a floor painted with stripes of each color (contours), which show where on the floor there are similarly low and similarly high expected values; that is, they show how the magnitude of the predicted outcome changes corresponding to the values of the interacting predictors.

Figure 4.6 presents the 2D contour plot for this example. The *x*-axis is age, the *y*-axis is SES, and the contours for the predicted frequency of sex are colored in shades of gray, with a lighter shade representing a smaller predicted frequency of sex and a darker shade representing a higher predicted frequency. You interpret a contour plot in much the same way as a table of predicted values. To describe the effect of age at a given SES value, examine the changing colors of the contours from left to right, starting in the bottom-left corner at the lowest value of SES. The contours change from the darkest gray shade (highest frequency of sex) for the youngest age to the lightest gray shade (lowest frequency) for the oldest age. That is, the frequency of sex declines with increasing age. Around the next marked level of SES (37), the left-to-right path of the age effect crosses a smaller portion of the most frequent contour and of the least frequent contour, indicating a smaller change in predicted frequency across the range of age as SES increases. In the region of the next SES level (57), the left-to-right path does not include either the highest frequency contour or the lowest frequency contour. The path thus includes a narrower range of changes in predicted frequency as age increases, and the amount of change continues to grow smaller at higher levels of SES. The same pattern is evident at the top two levels of SES.

You interpret the effect of SES conditioned by age in an analogous way. Examine a path from the bottom of the contour plot to the top, starting in the bottom left-hand corner at the youngest age. As you scan upward (SES increases), the path crosses from the darkest shade through the next two lighter shades. That is, the predicted frequency of intercourse decreases as SES increases at the youngest age. At the next marked age (41 years), the path of increasing SES only crosses two contours, the third darkest and the next lighter shade, and the path is almost exclusively in the darker of the two shades. This still represents a negative effect, but the magnitude of the change in predicted frequency is smaller than at younger ages. At *Age* = 57, the SES path first runs along the border between two contours, indicating that the predicted frequency barely differs with SES at this age. Moving on to the next marked age (73 years), the path crosses only two contours, but it crosses from lighter to darker (by only a little), meaning that the predicted frequency now increases with rising SES at older ages. Similarly, at the oldest age (89), the path begins in the lightest contour (lowest frequency of sex) and then crosses the next two darker-shaded contours, showing that the positive effect of SES on predicted frequency continues to increase in magnitude at older ages.

In sum, the contour plot shows the same patterns we saw in the pair of scatterplots that each highlighted how one of the interacting variables is moderated by the other. The advantage of a contour plot is that you need to present and interpret only one figure, not two. On the other hand, the contour plot is a more abstract representation

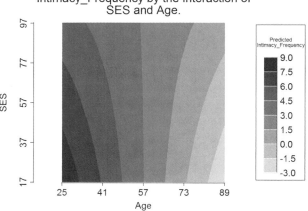

FIGURE 4.6 ● **TWO-DIMENSIONAL CONTOUR PLOT, INTERVAL-BY-INTERVAL INTERACTION**

Intimacy_Frequency by the Interaction of SES and Age.

ICALC command: outdisp, out(atopt((means)_all)) ndig(1) plot(type(contour)) pltopts(ccuts(-3(1.5)9) zlab(-3(1.5)9) cleg(tit(, size(*.625))))). The out( ) option sets predictors' reference values as their means, the plot( ) option specifies a contour plot, and ndig( ) and pltopts( ) customize contour cut points and legend labels and title.

of the magnitude of the predicted outcome and how that magnitude varies with the predictors. As such, it is much less accessible and requires more explanation than visual patterns of the plotted lines in the scatterplots.

A Note on 3D Surface Plots. To illustrate the issue of distortion in 3D surface plots, Figure 4.7 presents two versions of the 3D surface plot for this example. Each has the same number of contours and coloration as the 2D contour plot. The only difference between the versions are the angles of rotation used to define the projection of the 3D surface into two dimensions. Note how the contour lines are curved on the top plot, which suggests curvature of the surface. But the contours appear to follow nearly parallel straight lines on the bottom plot, giving an impression of a flat surface. Also, notice how the relative length of opposite sides of the 3D surface differs between the two plots, appearing more unequal for one dimension in one plot but the opposite dimension in the other plot. I think that most analysts and audiences would find their interpretation of the plots affected, at a minimum in subtle ways, by the differences in the plots.

Example 4: Interval-by-Categorical Interaction, Number of Children Predicted by the Interaction of Income and Birth Cohort

In the previous examples, the interacting variables were either all categorical or all interval level. If they are a mix of categorical and interval measures, you can choose to construct either a scatterplot with repeated plotted lines or repeated bar charts. The scatterplots use the interval predictor as the *x*-axis, and you create a plotted line of the predicted outcome by *X* for each category of the categorical predictor. For bar charts, you use the categorical predictor to define the set of bars, and you create a separate bar chart for each display value of the interval predictor. I apply both options to an analysis of the number of children predicted by the interaction of family income (in $10K) with respondent's birth cohort (Depression Era, World War II [WWII], Baby Boom, and Post-Boom).[8] (This is a reanalysis of an earlier example that used age as an interval measure (in years) and included a second interaction, between family income and education.)

The scatterplot in Figure 4.8 portrays the predicted relationship of number of children by family income contingent on birth cohort. The predicted effect of income on children is negative for the Depression Era cohort (*small dashed line*) but positive for the other cohorts. The slope is steepest for the Post-Boom cohort (*solid line*) and roughly equal but shallower for the WWII and Baby Boom cohorts. You can also interpret the relationship between birth cohort and the predicted number of children as it varies with income by examining the vertical distances among the plotted lines. The predicted differences in children among the cohorts generally diminish as family income rises;[9] the distance between each pair of lines grows smaller except for a slight increase between the prediction lines for the WWII and Baby Boom cohorts. Moreover, until income surpasses a little more than $150K, there is a consistent ranking of the cohorts according to the predicted number of children: the Depression Era cohort has the highest predicted number of children, followed by the WWII, then the Baby Boom, and last the Post-Boom cohorts. At the higher incomes, the ordering of cohorts is always WWII, Depression Era, Post-Boom, and Baby Boom.

How does a bar chart compare with the scatterplot in terms of ease of comparison? Let's look at the bar charts reporting the predicted number of children for the four birth cohorts, with a separate set of bars for display values of income ranging from $0 to $240K in $60K increments (Figure 4.9). Bar charts initially focus your eye on

FIGURE 4.7 ● THREE-DIMENSIONAL SURFACE PLOTS, INTERVAL-BY-INTERVAL INTERACTION

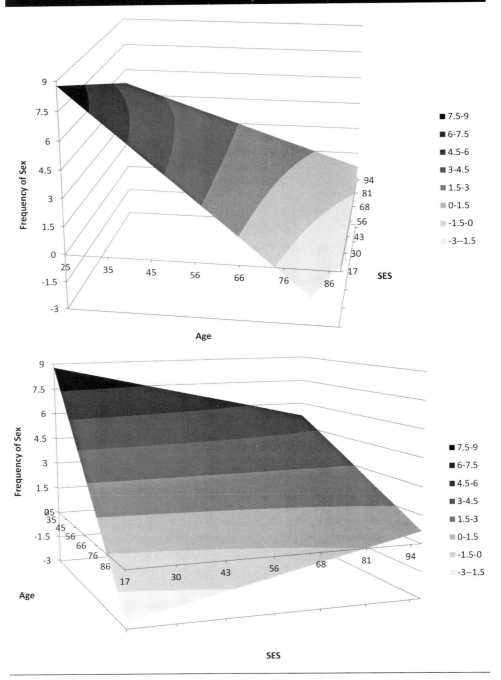

ICALC command saves plot data to Excel: outdisp, out(atopt((means) _all)) plot(type(contour) save(Fig_4_7.xlsx)) ndig(1) pltopts(ccuts(-3(1.5)9)). The out( ) option sets the predictors' reference values as their means, the plot( ) option specifies a contour plot and saves the plotting data to Excel, and the ndig( ) and pltopts( ) options customize the contour cut points and legend labels.

FIGURE 4.8 ●   SCATTERPLOT OF PREDICTED OUTCOME FOR AN INTERVAL-BY-CATEGORICAL INTERACTION

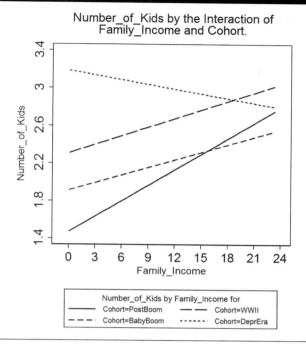

Number_of_Kids by the Interaction of Family_Income and Cohort.

Number_of_Kids by Family_Income for
- Cohort=PostBoom
- Cohort=BabyBoom
- Cohort=WWII
- Cohort=DeprEra

ICALC command: outdisp, plot(type(scat)) out(atopt((means)_all)) pltopts(ylab(1.4(.4)3.4)). The out( ) option sets the predictors' reference values as their means, the plot( ) option specifies a scatterplot, and the pltopts( ) option customizes the *y*-axis labels.

the differences among the bars in each panel, that is, the effect of birth cohort. This clearly shows the convergence of heights among the bars as income rises; that is, the differences among the cohorts diminish. And it is also easy to see that the ranking of the cohorts by the height of the bars is consistent through the third panel (*Income* = $120K), then switches to a different ordering for the higher income levels, just as the scatterplot showed. With a small amount of effort, you can read the income effect from the bar chart as well. If the differences are clear enough, as they are in this case, you can visually examine the heights and, if not, do side calculations on the reported predicted values. For each birth cohort, examine the change in its bar's height as income increases. The *black bar* for the Post-Boom cohort increases in height, and the difference in its height relative to the bars for the next two cohorts grows smaller, showing that income has a steeper slope for the Post-Boom cohort. The next two bars—for the Baby Boom and WWII cohorts—also grow taller as income increases, with the gap between the pair of bars slowly growing, indicating a somewhat larger slope for the WWII cohort (*darker gray bar*). Last, the *lightest gray bar* for the Depression Era cohort diminishes in height with increasing income, reflecting a negative income effect for members of this cohort.

Although my personal preference would be to present a scatterplot, I think that scatterplots and bar charts both effectively communicate the relationship between an outcome and the interaction between an interval and a categorical predictor. The

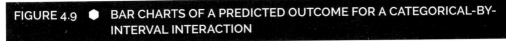

FIGURE 4.9 ● BAR CHARTS OF A PREDICTED OUTCOME FOR A CATEGORICAL-BY-INTERVAL INTERACTION

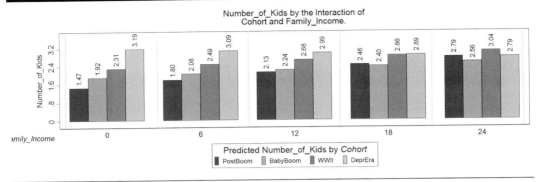

ICALC command: outdisp, plot(type(bar)) ndig(2) out(atopt((means)_all)) pltopts(plotreg(ma(t +2))). The out( ) option sets the predictors' reference values as their means, the plot( ) option specifies a bar chart, ndig( ) formats the bar labels with 2 decimals, and the pltopts( ) option customizes the region margins.

next example of a three-way interaction also has a mixture of categorical and interval interacting predictors, and readers can judge whether both types of visual display are equally accessible for this more complicated form of interaction.

Example 5: Three-Way Interaction of a Mix of Interval and Categorical Predictors, Voluntary Association Memberships Predicted by the Interaction of Sex, Age, and Education

This last example describes the creation and interpretation of visual displays for a three-way interaction, in this case a mixture of interval and categorical interacting predictors. For creating the scatterplots, the best approach is typically to plot the predicted outcome against one of the interval predictors, with a prediction line on the same plot calculated for each value of the categorical predictor. (If all your interacting predictors are interval, then you use the display values of one of them to define the repeated prediction lines.) Moreover, the scatterplot (and the calculation of the prediction lines) is repeated for each display value of the other interval predictor. You, thus, have as many scatterplots as display values of the third predictor. In most cases, you can use that set of scatterplots to understand the relationship of the outcome and the categorical predictor contingent on the two interval predictors. But you would almost certainly have to repeat this process, switching the roles of the two interval predictors to most easily determine the pattern of the effect of the third predictor.

The application example uses the results from the prior example of memberships predicted by the three-way interaction of sex (two categories) with education and age (both interval). The first figure consists of six scatterplots of predicted membership against education, with separate prediction lines for women and for men (Figure 4.10). The six scatterplots correspond to the display values I chose for age, spanning its range from 18 to 89 in increments of 14 years.

FIGURE 4.10 ● SCATTERPLOTS FOR THREE-WAY INTERACTION, EDUCATION ON X-AXIS

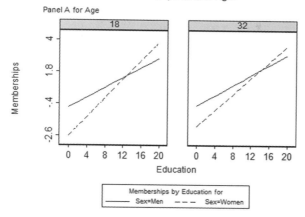

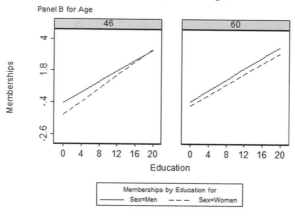

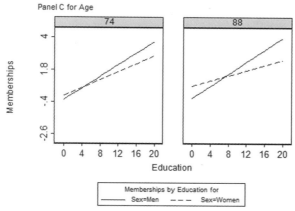

ICALC command: outdisp, plot(type(scat)) out(atopt((means)_all)) ndig(2). The out( ) option sets the predictors' reference values, the plot( ) option specifies a scatterplot, and ndig( ) formats the *y*-axis labels.

It is immediately obvious that the effect of education on memberships is positive for both men and women at all ages—each prediction line has a positive slope—but also that the effect of education varies by both sex and age. Note that the slope of the memberships–education prediction line decreases across the scatterplots as age increases for women (*dashed line*), but it hardly changes for men (*solid line*). You can also see how age and education moderate the effect of sex on memberships by examining the vertical distance between the prediction lines and how it changes within and between scatterplots. When *Age* < 60 years (see Panel B), women have fewer predicted memberships than men at lower levels of education (the dashed line for women lies below the dashed line for men) but more memberships at higher levels of education. And the transition point in the education distribution increases with age, from around 12 years at *Age* = 18, to 14 years at *Age* = 32, to 17 years at *Age* = 46. And as age continues to rise, men's advantage changes from larger at low levels of education to larger at higher levels of education.

The set of scatterplots in Figure 4.11 lets you easily determine the changing effect of age on memberships, which would be hard to tease out from Figure 4.10. Each scatterplot shows two prediction lines for the memberships–age relationship, one for women and one for men. The scatterplots correspond to the display values for education, every 4 years from no education to 20 years. The age effect is positive for men across education's range (*solid line*), and its slope hardly changes. In contrast, the age effect for women switches from a positive, steeper than men's, slope at low education to an increasingly negative effect of age at higher years of education. I leave it as an exercise for readers to examine the pattern of the effect of sex in these scatterplots to confirm that it corresponds to the description above.

Bar charts are also a sensible option given that one of the interacting predictors is a nominal measure. You use the nominal predictor's categories to define the bars, construct a panel of bar charts corresponding to the display values of one of the interval predictors, and repeat the panel for each of the display values of the other interval predictor. (If all of your interacting predictors are categorical, then you use all the categories of each predictor rather than choosing display values to construct the bar charts.) In much the same way that I interpreted the scatterplots above, you can make comparisons within and between the bar charts and the panels to describe how the three interacting predictors affect the outcome. Figure 4.12 provides the series of bar charts for this example. There are a pair of bars for the predicted memberships for men (*black bar*) and for women (*gray bar*). Each panel corresponds to a display value for age (from 18 to 89 in steps of 14 years), and within each panel there are six pairs of bars for the display values of education (every 4 years from 0 to 20). This organization of the information is parallel to the organization of the first set of scatterplots in Figure 4.10. Notice that a few bars drop downward, representing a negative value of the predicted memberships, but most bars rise upward to show a positive predicted number of memberships.

Within every panel, the positive heights of all the bars increase from left to right— keep in mind that the decreasing length of a negative bar means an increasing predicted value—indicating that the predicted number of memberships increases with education at all ages for both women and men. But the effect of education differs by age. At younger ages (≤32), the heights of the bars change more rapidly for women (*gray bars*) than for men (*black bars*), while at older ages, the bar heights change less for women than for men. In fact, if you compare the *black bar* for men's predicted memberships at a given level of education across the panels, you will notice relatively little change in the height (predicted memberships). To work out the pattern

FIGURE 4.11 ⬢ SCATTERPLOTS FOR THREE-WAY INTERACTION, AGE ON X-AXIS

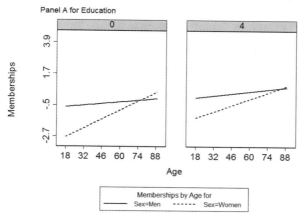

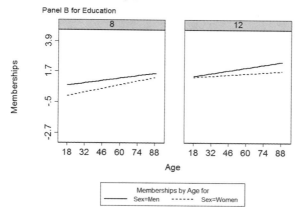

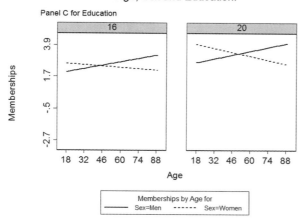

ICALC command: outdisp, plot(type(scat)) out(atopt((means)_all)) ndig(2). The out( ) option sets the predictors' reference values, the plot( ) option specifies a scatterplot, and the ndig( ) option formats the y-axis labels.

FIGURE 4.12 ● BAR CHARTS FOR A THREE-WAY INTERACTION

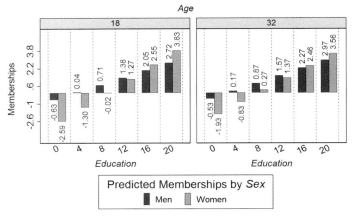

Memberships by the Three-way Interaction of Sex, Education and Age.

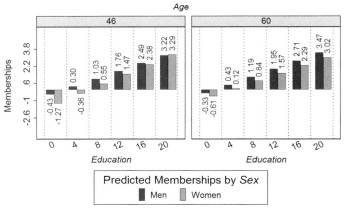

Memberships by the Three-way Interaction of Sex, Education and Age.

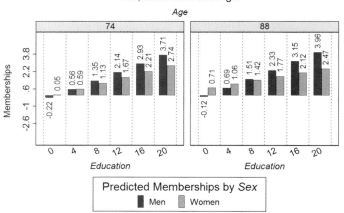

Memberships by the Three-way Interaction of Sex, Education and Age.

ICALC command: outdisp, plot(type(bar)) out(atopt((means)_all)) ndig(2) pltopts(plotreg(ma(t +3 b +3))). The out( ) option sets the predictors' reference values as their means, the plot( ) option specifies a bar chart, ndig( ) formats bar labels with 2 decimals, and the pltopts( ) option customizes the plot margins.

of moderation of the effect of sex on memberships, compare the difference in height between each pair of bars (i.e., the effect of sex) as it changes within and between panels. This shows the same pattern as described above: At younger ages, women have a higher predicted number of memberships than men at high, but not low, levels of education. And as age increases, the education value demarcating the transition from men's advantage to women's advantage also increases. Broadly speaking, the sex gap favoring men changes from largest when education is low to largest when education is high as age increases.

For a complete interpretation using bar charts, I would typically present a second set, switching the roles of education and age. I do not do so here because my purpose in presenting the bar charts is to demonstrate the parallelism in the kind of comparisons you make to interpret scatterplots and bar charts and to allow readers to compare the accessibility of the information presented in bar charts versus scatterplots. Personally, I find it slightly easier to see and describe the patterns of the prediction lines in scatterplots than to work out the patterns of the changing heights of the bars in bar charts. Nonetheless, I think the bar charts are a quite reasonable and effective alternative.

While you can construct a 2D contour plot in this case, I do not recommend it because it is difficult to get a good sense of the magnitudes of the effects as they are changing and its results are less accessible. Nonetheless, I discuss how to create and interpret a 2D contour plot. You use the two interval predictors to define the x- and y-axes of the contour plot, and the outcome's predicted value to define the contours and coloration, and you calculate the predicted outcome and construct a plot for each category of the nominal predictor. So in this example, education and age define the x- and y-axes, the contours provide information about the predicted number of memberships, and we create one contour plot to get the predicted number of memberships for women and another contour plot to get the predicted number of memberships for men (Figure 4.13).

FIGURE 4.13 ● CONTOUR PLOTS FOR A THREE-WAY INTERACTION

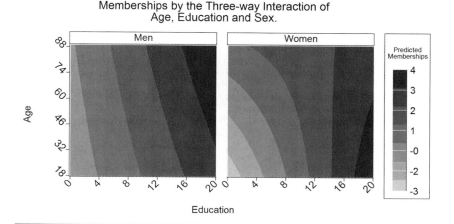

Memberships by the Three-way Interaction of Age, Education and Sex.

ICALC command: outdisp, plot(type(contour)) out(atopt((means)_all)) pltopts(ccuts(−3(1)4) zlab(−3(1)4) cleg(title(, size(2.8)))) ndig(0). The out( ) option sets the predictors' reference values as their means, the plot( ) option specifies a contour plot, and the ndig( ) and pltopts( ) options customize the contour cut points, the legend labels, and the legend title.

To interpret the effect of education, trace a path from left to right for a given age on the *y*-axis, starting at the youngest age and working up. For both men (left-hand plot) and women (right-hand plot), the education path always crosses from the *lighter gray contours* (fewer memberships) to *darker gray* (more memberships). The path for men crosses four, sometimes five contours, while the path for women crosses more contours at younger ages (six to seven contours) and progressively fewer at older ages (three to four). This suggests that the slope of the education effect for women changes from much steeper at young ages to shallower at older ages, whereas the moderated education effect for men does not change slope nearly as much.

Start at a given level of education on the *x*-axis, and trace the path from the bottom to the top of the contour plot to see the effect of age. For men, the age paths indicate a positive slope that does not change much across education because it usually crosses almost all of a single contour and a small part of a second contour, always moving from fewer to more memberships (lighter gray to darker gray). In contrast, the age paths for women cross more contours, going from lighter to darker, than for men at younger ages (a steeper positive effect), but at the oldest ages the age paths for women cross fewer contours, but now going from darker to lighter gray (a negative effect).

Finally, to interpret the effect of sex (the difference in predicted memberships between women and men), you have to determine for which combinations of age and education women have more memberships than men (*darker contours*) and for which they have fewer memberships (*lighter contours*). There are several areas that can be easily identified: More memberships for women at younger ages and higher education (compare the lower-right corners) and at the lowest education and highest ages (contrast the upper-left corners), and fewer memberships for women when education and age are both near the bottom of the distribution (see the lower-left corners) and when they are both near the top of their distribution (note the upper-right corners). But it is difficult elsewhere to see the differences and changes.

I suspect that most readers would agree with my assessment that the 2D contour plots are not as accessible or easy to explain as either scatterplots or bar charts and that they do not provide as good a numeric sense of the magnitude of a predictor's effect and how it changes with the other predictors. The primary advantage of the contour plot is that it requires fewer physical diagrams to present and interpret. For the three types of plots for this example, I used the following:

- *For bar charts:* six panels containing six bar charts, each bar chart with two bars

- *For scatterplots:* six scatterplots, each with two prediction lines

- *For 2D contour plots:* two contour plots, one for each sex

CONCLUSION

This chapter provided a grounding in the use and interpretation of tables and figures of the patterning of the predicted outcome variable to aid in the interpretation of interaction effects in GLMs with a linear identity link. Although I provided a rubric in Figure 4.1 for choosing which type of visual display to use for different types of interaction effects, I want to reiterate that the recommended options are simply suggestions of what to try first. Your own preferences and style play a role in determining which ones you will find the most effective, as do the actual patterns in the data. And as the later application example chapters illustrate, the best interpretations of such tables and figures also rely on other information about the nature of the interaction effect. Specifically, you should incorporate what you learn from using GFI (including sign changes) and from determining the significance regions for effects to understand the patterns of the effects.

This chapter also laid the foundation for the next chapter, which considers the additional challenges of interpreting interaction effects in GLMs with nonlinear link functions and discusses how to apply and/or adapt the interpretive tools discussed in this chapter.

SPECIAL TOPICS

Equivalence of As-Observed and Central Tendency Options for Linear Models

To show the equivalence, I apply Equation 4.1 to find the predicted outcome value for each option for a model with an interaction between F and M_1 with two additional predictors, Z_1 and Z_2. For specified values of the focal and moderating variables, $F = f$ and $M_1 = m_1$, the prediction equation is

$$\widehat{Y} = a + b_1 f + b_2 m_1 + b_3 f \times m_1 + b_{Z_1} Z_1 + b_{Z_2} Z_2 \tag{4.11}$$

Option 1: Central Tendency (Mean)

Substitute the means of Z_1 and Z_2, $\bar{z}_1$ and $\bar{z}_2$, into Equation 4.11.

$$\widehat{Y}_{Central\ tendency} = a + b_1 f + b_2 m_1 + b_3 f \times m_1 + b_{Z_1} \bar{z}_1 + b_{Z_2} \bar{z}_2 \tag{4.12}$$

Option 2: As Observed

Calculate the predicted outcome value for each case i using $F = f$ and $M_1 = m_1$ as the values for all cases but use each case's own values for Z_1 and Z_2, z_{1i} and z_{2i}:

$$\widehat{Y}_i = a + b_1 f + b_2 m_1 + b_3 f \times m_1 + b_{Z_1} z_{1i} + b_{Z_2} z_{2i} \tag{4.13}$$

To get the as-observed prediction, calculate the average across the cases of their predicted outcome in Equation 4.13:

$$\widehat{Y}_{As\ observed} = \frac{1}{N} \sum_{i=1}^{N} \left(a + b_1 f + b_2 m_1 + b_3 f \times m_1 + b_{Z_1} z_{1i} + b_{Z_2} z_{2i} \right) \tag{4.14}$$

The first three terms are constant across cases, so their average yields the same three terms, while the average of each of the last two terms gives the coefficient multiplied by the mean of the corresponding predictor:

$$\widehat{Y}_{As\ observed} = a + b_1 f + b_2 m_1 + b_3 f \times m_1 + b_{Z_1} \bar{z}_1 + b_{Z_2} \bar{z}_2 \tag{4.15}$$

This is the same result as calculated using the central tendency (mean) option. This proof generalizes, with limited exceptions, to models with additional moderators of F and/or additional predictors Z_k because these additions either expand the first set of terms that do not vary across cases (additional moderators) or the second set of terms that vary across cases (additional Z_k).

However, the two options are not equivalent when the additional predictors include polynomials or other multiplicative functional forms because these will be treated differently by the two options. Suppose the model includes Z_3 and its square:

$$\hat{Y} = a + b_1 f + b_2 m_1 + b_3 f \times m_1 + b_{Z_1} Z_1 + b_{Z_2} Z_2 + b_{Z_3} Z_3 + b_{Z_4} Z_3^2$$

Following the same steps as above shows that the central tendency option uses the square of $\bar{z}_3$ as the reference value but the as-observed option uses the mean of Z_3^2.

CHAPTER 4 NOTES

1. This discussion focuses on the outcome's expected value, but this set of issues and rubrics applies more broadly to selecting reference points for calculating a large variety of predicted or "marginal" statistics (Williams, 2012).
2. The WLS in this analysis applies a correction for heteroscedasticity and is mathematically equivalent to an estimated generalized least squares analysis (Kaufman, 2013).
3. My description of this process draws heavily on Williams's (2012) clear and succinct description in the context of calculating the marginal effect of a predictor.
4. Using the mean for a dummy variable is equivalent to specifying the average proportional representation in the sample of the category it represents.
5. Sorting the expected values into quintiles, the font sizes assigned to the top to the bottom quintiles are 16, 15, 14, 13, and 12.
6. Although the minimum value of education is 0, there are so few cases at or near this minimum (1.2%) that I chose 4 for the smallest display value for education.
7. Applying the GFI tool, the sign changes at *Education* = 7.32.
8. The cohorts are defined as those born in 1940 or earlier, 1941 to 1945, 1946 to 1964, and 1965 to 1980, respectively.
9. If you take the absolute values of the difference between each pair of lines and add them together, this sum uniformly declines across the range of income.

NONIDENTITY LINK FUNCTIONS

Challenges of Interpreting Interactions in Nonlinear Models

IDENTIFYING THE ISSUES

GLMs with nonlinear link functions have two functionally related outcomes measured in different metrics: (1) an observed outcome measured in its original (μ) metric in the measurement component of the GLM and (2) a transformed (and sometimes unobserved) outcome in the modeling component measured in the transformed (η) metric—that is, $\eta = g(\mu)$. This chapter is motivated by what I would describe as a "linear perception bias," when you use the observed (μ) metric to construct tabular or graphical displays of the expected value of the outcome as it varies across the categories or values of predictors. Consider a figure with a set of curves in a scatterplot, or bars in a bar chart, portraying the relationship between the predicted outcome (*y*-axis) and the focal variable (*x*-axis), with each curve (or set of bars) representing a different value of the moderator (see the four plots in Figure 5.1). If the plot of the predicted outcome against the focal variable is not a set of parallel straight lines or perfectly parallel curves,[1] then we tend to automatically perceive it as indicating an interaction between the predictors as they affect the outcome. Study the four graphics (one bar graph and three scatterplots) in Figure 5.1. Each portrays results from a GLM with a nonlinear link function: the joint effect of a pair of variables on the predicted outcome in its observed metric. Panels A and B are from a logistic regression (the outcome is approval of an interracial marriage ban), Panel C is from a negative binomial regression (the outcome is number of poor mental health days), and Panel D is from an ordinal logistic regression (the outcome is the frequency of purchasing chemical-free produce). Before you read on, think about how you would interpret what each figure shows about the relationship between the predicted outcome and the joint effect of the two variables and whether or not the graphic indicates an interaction effect between the two variables.

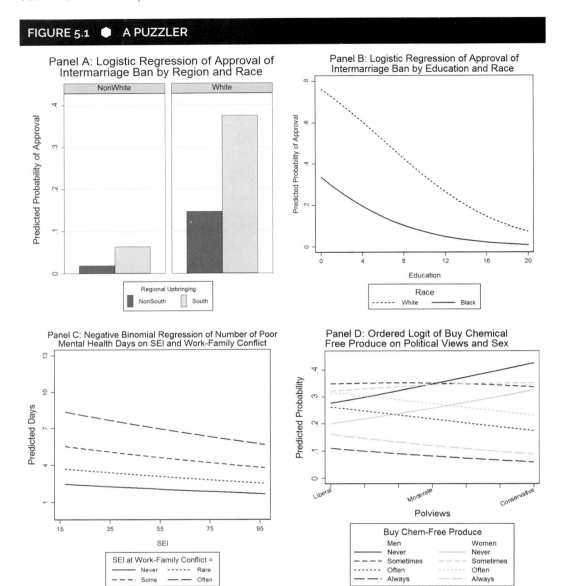

FIGURE 5.1 ● A PUZZLER

Panel A: Logistic Regression of Approval of Intermarriage Ban by Region and Race

Panel B: Logistic Regression of Approval of Intermarriage Ban by Education and Race

Panel C: Negative Binomial Regression of Number of Poor Mental Health Days on SEI and Work-Family Conflict

Panel D: Ordered Logit of Buy Chemical Free Produce on Political Views and Sex

Putting aside the question of interaction effects for the moment, the following is how I would explain these visual displays in purely descriptive terms. The bar graph in Panel A shows that the difference in the predicted probability of approval between southern and nonsouthern upbringing is much larger among Whites than among Blacks (compare the *gray bar* with the *black bar* in the race chart) and the difference in approval between Whites and Blacks is larger in the South than in the non-South (compare the *gray bars* with each other and the *black bars* with each other). The plot in Panel B indicates a higher probability of approval by Whites (*dashed line*) than by Blacks (*solid line*), with the gap declining as education rises. And among Whites, there is a steeper decline in the probability of approval as education increases than among Blacks. The plot in Panel C shows that the predicted number of poor mental health days decreases most rapidly with SEI for those with the highest levels of work–family conflict (often = *long dashes*) and declines less as conflict drops (sometimes = *medium dashes*, rare = *short dashes*, and never = *solid line*). The vertical gaps between the lines indicate more days of poor health for higher levels of work–family conflict, but these gaps decline at higher levels of SEI. Finally, Panel D takes a little more work to describe since it has a pair of curves (men in *black*, women in *gray*) for each of the four ordered values of the outcome (purchase never, rarely, sometimes, often). Each curve shows how the predicted probability of one of these choices changes with political views. The probability of outcome value 1 (never = *solid line*) increases as conservatism increases, with the gap between men and women growing slightly, while the probability of outcome value 2 (sometimes = *medium dashes*) is relatively constant with men initially having higher probability than women, which reverses as conservatism increases. Outcome values 3 and 4 (often = *short dashes* and always = *long dashes*) show similar patterns; the predicted probability of each outcome decreases as conservatism increases, and the gap between men and women decreases slightly as conservatism increases.

As you might have guessed, the question about interaction effects was a trick question. All four of these figures were produced from analyses with no interaction between the pair of variables in the modeling component of the GLM. What visually and descriptively appear to be interaction effects are in fact due to the nonlinearity of the relationship between the observed outcome in the μ-metric and the predictors. If I instead had made these figures using the η-metric, the result would have been perfectly parallel straight lines in the scatterplots, with no implication of an interaction effect. This is one part of the linear perception bias: The nonlinearity of the link function creates the appearance of an interaction effect when there is none.[2]

The second part of the perception bias is that it, consequently, can be difficult to determine the extent to which the moderator changes the effect of the focal variable solely by looking at a visual display of the outcome drawn in the observed μ-metric. In a linear link model, there is an implicit visual reference point—the degree to which the plotted lines diverge from being parallel. With nonlinear link models, the visual display of an interaction effect combines the nonlinearity of the interaction with that due to the nonlinearity of the model, which is what we saw in Figure 5.1. And we lack any inherent reference point to distinguish between the two sources. To illustrate this, Figure 5.2 presents graphics from models by adding an interaction effect to the main effects models that produced Panels A, B, and C of Figure 5.1 (for Panel D, there is no corresponding model because the interaction term is not significant).

Consider what the scatterplot in Panel B tells us about how the Black–White difference in the probability of approval—the vertical gap between the two curves—is moderated by education. It seems apparent that education moderates the Black–White difference to a considerable degree, with Whites having a much higher probability of approval than Blacks (about 60%) at the lowest levels of education but only a small gap (3%) at the highest level of education. What is not obvious is that much of this difference results from the nonlinearity of the logistic regression analysis, which we cannot judge looking only at Figure 5.2.

We can see the extent of confounding if we compare figures for models with and without the interaction effect. Rather than flipping pages to compare Panel B in

FIGURE 5.2 ◆ MODEL WITH INTERACTION EFFECTS

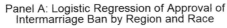

Panel A: Logistic Regression of Approval of
Intermarriage Ban by Region and Race

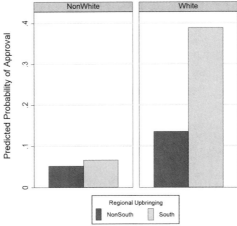

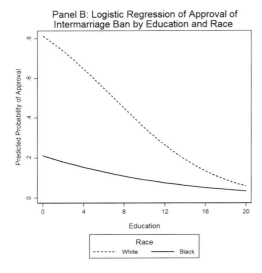

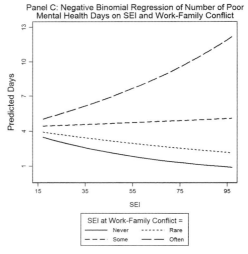

Figure 5.1 with Figure 5.2, look at Figure 5.3, which presents the results from the main effects analyses superimposed on the visual presentation of the interaction effects for each example. The black curves in Panel B represent the predicted probabilities from the interaction effects model, while the gray curves are from the main effects model. The moderating effect of education is the difference between the vertical gaps between the *black curves* (the interaction effect plus the nonlinear link effect) and the vertical gaps between the *gray curves* (only the nonlinear link effect). *This shows that a very substantial portion of what seemed due to an interaction effect is actually due to the nonlinear relationship between the probability of approval and the*

FIGURE 5.3 ● INTERACTION EFFECTS WITH MAIN EFFECTS SUPERIMPOSED

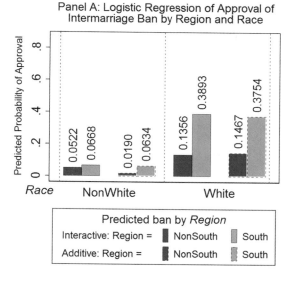

Panel A: Logistic Regression of Approval of Intermarriage Ban by Region and Race

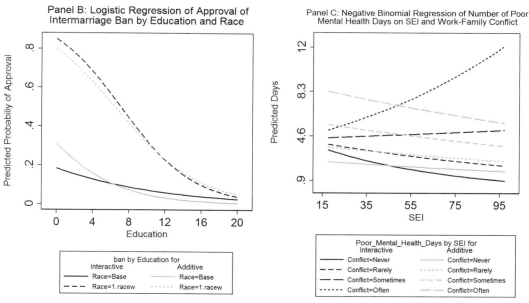

Panel B: Logistic Regression of Approval of Intermarriage Ban by Education and Race

Panel C: Negative Binomial Regression of Number of Poor Mental Health Days on SEI and Work-Family Conflict

predictors. This exemplifies the problem of overinterpreting the magnitude of the interaction that can result from relying on the predicted values of the outcome in the observed μ-metric.

The confounding of the analytic model's nonlinearity with the interaction effect's nonlinearity can also lead to understating the magnitude of the interaction. The negative binomial regression results in Panel C of Figure 5.3 demonstrate this possibility. The *black curves* show how the effect of SEI on the number of poor mental health days is moderated by work–family conflict. The number of poor health days decreases with SEI among workers with a low frequency of work–family conflict (*solid* and *medium dashed lines*) but increases if a worker has more frequent work–family conflicts (*small* and *long dashed lines*). Comparing this with the main effects scatterplot (*gray curves*) shows that work–family conflict moderates the SEI effect to such a degree that it changes what had been the steepest negative relationship between poor health days and SEI in the main effects scatterplot (work–family conflict = often = *long dashes*) to a positive relationship in the interaction effects scatterplot. That is, the moderating effect is even greater than is apparent from the interaction effects scatterplot (*black curves*).

The Goal of Interpretation, a Caveat

Although these examples are (I hope) persuasive about the problem of confounding that contributes to the linear perception bias, they are illustrations, not proof of the problem. In the next section, I briefly develop the mathematical foundation for the problem of confounding. And subsequent sections present solutions/alternatives to using visual displays of the predicted outcome in its observed μ-metric *when you want to determine the degree to which the focal variable's effect is moderated by other predictors*. This caveat about the purpose of creating tables and figures is crucial for avoiding a very likely, and somewhat subtle, misconception about my critique of the value of such displays using the observed variable's μ-metric. They do provide an accurate and appropriate representation of the pattern of the relationship between the observed outcome and the focal and moderator variables. It is only the attempt to use them to describe the pattern or degree of moderation embodied in the interaction effect that could be problematic. It is the difference between the following pair of statements' description of Panel B in Figure 5.2:

1. The probability of approval is consistently higher for Whites than for Blacks, but this gap is substantially reduced at higher levels of education, falling from a 63% difference at the lowest level of education to a 3% gap at the highest level. Moreover, the probability of approval decreases with education, at a faster rate for Whites than for Blacks.

2. The probability of approval is consistently higher for Whites than for Blacks, but this gap is substantially moderated by education. The substantial effect of race is reduced at higher levels of education, falling from a 63% difference at the lowest level of education to a 3% gap at the highest level. In turn, the negative effect of education on the probability of approval is strongly moderated by race, with a steeper slope for Whites than for Blacks.

The first statement is an accurate description of how the probability of approval varies by race and education, while the second is an inaccurate description of how education moderates the effect of race on the probability of approval. Although the first statement is technically accurate, I think many readers would fill in the blanks and implicitly translate it into a version of the second statement. One way to appropriately describe the moderation is from the plots in Panel B of Figure 5.3, in which the prediction curves from the main effects model are superimposed onto the plot of predicted values from the interaction effects model:

3. The probability of approval is consistently higher for Whites than for Blacks, but this gap is moderated by education, falling from a 63% difference at the

lowest level of education to a 3% gap at the highest level. Much of this large decrease is due to the nonlinear relationship of the probability of approval to the predictors, which is apparent from the comparison of the *gray* and *black* *curves*. Relative to the main effects model, the Black–White gap at lower levels of education (<10 years) is even larger than the main effects model would predict, rising to 26% larger at the minimum value of education. But the Black–White gap becomes increasingly small relative to the main effects prediction as education increases, falling to half the main effect's gap at the highest level of education. Education has a larger negative effect on the probability of approval for Whites than for Blacks. Compared with the main effects model, the slope of the education effect is somewhat steeper for Whites and somewhat shallower for Blacks (Figure 5.3).

Compared with Statement 2, Statement 3 provides a more nuanced interpretation of the relationship and a more accurate portrayal of the extent to which education moderates the effect of race, and vice versa.

Readers who are convinced by the examples about the problem of confounding and not interested in the mathematical foundation and technical details of this problem should skip ahead to the section titled "Revisiting Options for Display and Reference Values."

MATHEMATICALLY DEFINING THE CONFOUNDED SOURCES OF NONLINEARITY

I develop and discuss formulas to show the confounded sources of nonlinearity for the case of a model with a single interaction between F and M (the same steps could be followed to demonstrate this for other interaction forms). Suppose you want to construct and interpret a scatterplot of the predicted outcome ($\hat{\mu}$) against the focal variable F for the selected display values of the moderator M (d different values) and a constant set of reference values (Z_R) for the other predictors in the model.[3] That is, you plot separate curves of the outcome ($\hat{\mu}_d$) on the y-axis against F on the x-axis, one curve for each of the d display values of M. Visually, you would look at two things to understand the pattern of the relationship between $\hat{\mu}$, F, and M: (1) the vertical gaps between the curves and how they change as F increases and (2) the slope of each curve and how they vary across the display values of M. In the first case, we are comparing the outcome's predicted values, and in the second case, we are comparing slopes.

Confounding in Comparing Predicted Values

We can put together the two components of a GLM (Equations 2.1 and 2.2) to define the predicted value of the observed outcome ($\hat{\mu}$) as

$$\hat{\mu} = g^{-1}\left(a + b_1 F + b_2 M + b_3 F \times M + b_z Z\right) \tag{5.1}$$

Substituting M's display values (d) and Z's reference values (Z_R) into this equation provides the formula for calculating the predicted values of the outcome:

$$\hat{\mu}_d = g^{-1}\left(a + b_1 F + b_2 d + b_3 F \times d + b_z Z_R\right)$$

Rearranging the terms and defining $\tilde{a} = a + b_z Z_R$, we get

$$\hat{\mu}_d = g^{-1}\left(\tilde{a} + b_2 d + \{b_1 + b_3 d\} F\right) \tag{5.2}$$

The vertical gaps between the curves are mathematically defined by differences in $\hat{\mu}$ at a given value of F. Let's consider the vertical gap between the curves for two pairs of the moderator's display values, between $d = 1$ and $d = 0$ and between $d = 0$ and $d = -1$: that is, $\hat{\mu}_1 - \hat{\mu}_0$ and $\hat{\mu}_0 - \hat{\mu}_{-1}$. Substituting these values of d into Equation 5.2 and subtracting the predicted values for each paired comparison, we get

$$\begin{aligned}
\hat{\mu}_1 - \hat{\mu}_0 &= g^{-1}\left(\tilde{a} + b_2 + \{b_1 + b_3\} F\right) - g^{-1}\left(\tilde{a} + \{b_1\} F\right) \\
\hat{\mu}_0 - \hat{\mu}_{-1} &= g^{-1}\left(\tilde{a} + \{b_1\} F\right) - g^{-1}\left(\tilde{a} - b_2 + \{b_1 - b_3\} F\right)
\end{aligned} \tag{5.3}$$

The magnitude of these differences at a given value of F depends not only on the size of the coefficients for F, M, and their product term (b_1, b_2, and b_3) but also on the function $g^{-1}(\ )$, which transforms the linear function of the predictors into the predicted value of the observed outcome. This simply makes the obvious point that the vertical gaps we see between the curves simultaneously reflect two things:

1. The difference between the index function values, which conceptually represents the structural effect of the interaction specification—that is, for the two pairs of curves

$$\begin{aligned}
\left(\tilde{a} + b_2 + \{b_1 + b_3\} F\right) \text{ compared with } \left(\tilde{a} + \{b_1\} F\right) &= b_2 + b_3 F \\
\left(\tilde{a} + \{b_1\} F\right) \text{ compared with } \left(\tilde{a} - b_2 + \{b_1 - b_3\} F\right) &= b_2 + b_3 F
\end{aligned} \tag{5.4}$$

This specifies through $g^{-1}(\ )$ how the magnitude of the vertical gap varies as F increases or decreases.

2. How the link function $g^{-1}(\ )$ transforms index function values into predicted values. Because $g^{-1}(\ )$ is nonlinear, the point on the $g^{-1}(\ )$ curve from which differences in the index function value are evaluated is consequential because the slope of the $g^{-1}(\ )$ function is not constant (i.e., it is nonlinear). Changing the evaluation point will necessarily affect the amount of difference between the two predicted value curves and the extent of change along the predicted value curve.

Applying Point 2 to the vertical gap between the pair of curves at a given value of F, the difference between $\hat{\mu}_1$ and $\hat{\mu}_0$ is calculated starting at a point on the $g^{-1}(\ )$ function corresponding to the index function value for $\hat{\mu}_0$. But the difference between $\hat{\mu}_0$ and $\hat{\mu}_{-1}$ is calculated beginning at a point corresponding to the index function value for $\hat{\mu}_{-1}$. Thus, the vertical gaps between the two pairs of curves at a given value of F are different in general,[4] even though the index function differences are equal in magnitude. Moreover, as F changes, the extent of change in the gaps between the curves similarly reflects both sources of differences—the modeled interaction and the nonlinearity of the link function.

What is less obvious is that plotting comparable curves from a main effects model will often also visually appear like an interaction effect due to the nonlinearity inherent in the link function, as the four panels in Figure 5.1 showed. We can derive a prediction equation for a main effects model from the equation for the interaction model by setting the interaction coefficient b_3 equal to 0 and using * to denote coefficients and estimates from the main effects analysis:

$$\hat{\mu}_d^* = g^{-1}\left(\tilde{a}^* + b_2^* d + b_1^* F\right) \tag{5.5}$$

Now the vertical gaps between the two pairs of curves are defined by

$$\hat{\mu}_1^* - \hat{\mu}_0^* = g^{-1}\left(\tilde{a}^* + b_2^* + b_1^* F\right) - g^{-1}\left(\tilde{a}^* + b_1^* F\right)$$
$$\hat{\mu}_0^* - \hat{\mu}_{-1}^* = g^{-1}\left(\tilde{a}^* + b_1^* F\right) - g^{-1}\left(\tilde{a}^* - b_2^* + b_1^* F\right) \tag{5.6}$$

and the differences in the index function for each pair of curves is equal to b_2^*. As was the case for the interaction model, these gaps will be unequal because the differences are calculated at dissimilar points along the $g^{-1}(\ )$ function. Furthermore, because the predicted value formula includes $b_1^* F$, the size of each gap will vary as F changes, which moves the evaluation point to different locations along the $g^{-1}(\ )$ function.

To make these points more concrete, let's apply them to the logistic regression analysis of the probability of approval from a model without the interaction between education and race. This produced the plotted curves in Panel B of Figures 5.1 and 5.3, a curve for the predicted probability by education for Whites and a second curve for Blacks using education as the focal variable (F). Specifying the link function for logistic regression yields (Fox, 2008, p. 380):

$$\text{Predicted probability of approval} = \hat{P}_d^* = g^{-1}\left(\hat{\eta}^*\right) = \frac{1}{1 + e^{-\left(\tilde{a}^* + b_2^* d + b_1^* F\right)}} \tag{5.7}$$

Substituting in the coefficients for education and race (−0.21547 and 2.18301, respectively) and the value for $\tilde{a}^*$ (−0.79703, the constant plus the sum of the other predictors times their means),

$$\hat{P}_d^* = \frac{1}{1 + e^{-(-0.79703 + 2.18301 \times d - 0.21547 \times Education)}} \tag{5.8}$$

The expressions for the plotted curves from the main effects specification for Whites and Blacks in Figures 5.1 and 5.3 are defined by substituting the display values for Whites ($d = 1$) and Blacks ($d = 0$) into Equation 5.8 and simplifying the expressions for the index function:

$$\hat{P}_{White}^* = \frac{1}{1 + e^{-\left(-0.79703 + 2.18301 \times 1 - 0.21547 \times Education\right)}} = \frac{1}{1 + e^{-\left(1.38597 - 0.21547 \times Education\right)}}$$
$$\hat{P}_{Black}^* = \frac{1}{1 + e^{-\left(-0.79703 + 2.18301 \times 0 - 0.21547 \times Education\right)}} = \frac{1}{1 + e^{-\left(-0.79703 - 0.21547 \times Education\right)}} \tag{5.9}$$

Subtracting $\hat{P}_{Black}^*$ from $\hat{P}_{White}^*$ defines the vertical gap between the Black and White curves. What may not be readily apparent is that the size of this gap will vary as education levels change, even though there is a constant structural effect of education (−0.21547).

Figure 5.4 illustrates how and why this happens. It shows the standard logistic curve and the points along the curve corresponding to the predicted probabilities from the

FIGURE 5.4   ●   EXAMPLE POINTS OF EVALUATION ON STANDARD LOGISTIC CURVE, MAIN EFFECTS MODEL

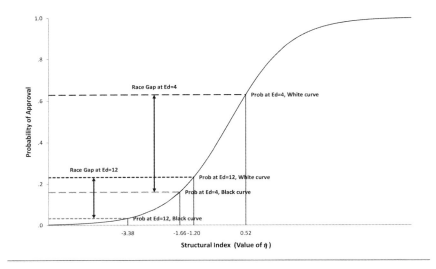

FIGURE 5.4   ●   EXAMPLE POINTS OF EVALUATION ON STANDARD LOGISTIC CURVE, MAIN EFFECTS MODEL

Black and the White formulas for two values of education, 4 and 12 years. How do we find the race gap at each education value? Start with the *small dashed gray line* for the predicted probability for *Ed* = 12 for Blacks ending at $\hat{\eta} = -3.38$. The corresponding predicted probability for Whites is found by moving along the *x*-axis by 2.18301 units (the magnitude of the structural coefficient for race) to $\hat{\eta} = -1.20$, the *small dashed black line*. Notice that we are starting from a point on the logistic curve where the curve's slope increases slowly and the change in the predicted probability is 0.1987 (the difference between the *small dashed black* and *gray lines*). In contrast, the starting point for evaluating the race gap for *Ed* = 4 years is the *larger dashed gray line*, which is at a point $\hat{\eta} = -1.66$ (where the curve rises much more sharply for the same change along the *x*-axis of 2.18301 units to $\hat{\eta} = 0.52$). Consequently, the race gap in predicted probabilities (the difference in the *large dashed gray* and *black lines*) is substantially larger at 0.4682 *in the absence of any interaction effects.*

For the model with an interaction between education and race, we could follow an analogous set of steps to get the following expressions for the Black and White predicted probability curves:

$$\hat{P}_{White} = \frac{1}{1 + e^{-(1.77413 - 0.24886\,Education)}}$$

$$\hat{P}_{Black} = \frac{1}{1 + e^{-(-2.08519 - 0.08582\,Education)}} \tag{5.10}$$

Figure 5.5 shows the results from using these to find the predicted probability of approval for Blacks and Whites for education equals 4 and 12 years. As in Figure 5.4, the starting point for evaluating the size of the vertical gap differs for the two values of education. Again, the predicted probability for Blacks is at a place with a shallower slope for *Ed* = 12 than for *Ed* = 4, setting the stage for a larger gap when *Ed* = 12. Unlike Figure 5.4, in which we found the predicted probability for Whites by moving

FIGURE 5.5 ● EXAMPLE POINTS OF EVALUATION ON STANDARD LOGISTIC CURVE, INTERACTION EFFECTS MODEL

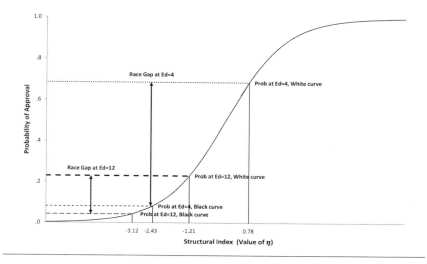

Example Points of Evaluation on Standard Logistic Curve, Interaction Effect Model.

an equal distance along the x-axis from each starting point, we move a larger distance for Whites when $Ed = 4$ (3.21 units from $\hat{\eta} = -2.43$ to 0.78) than when $Ed = 12$ (1.91 units from $\hat{\eta} = -3.12$ to -1.21), due to the modeled interaction of race and education. (Of course, if the interaction coefficient had been positive, the movement along the x-axis would have been smaller when $Ed = 4$.)

This illustrates how the vertical gaps between plots of the predicted values of the outcome depend on both the structural modeling of an interaction effect and the nonlinearity of the link function; that is, such visual displays combine (confound) these two sources of nonlinearity. And in the absence of an interaction specification, the nonlinearity of the link function can give rise to the appearance of an interaction effect. The next section demonstrates the comparable problems in visual displays when comparing slopes.

Confounding in Comparing Slopes

We can derive expressions for the slopes of the curves for the interaction effects and main effects models to show that the slopes analogously depend on how the effect of F is specified in the structural model and on the point on the $g^{-1}(\ )$ function at which the slope is evaluated. The slopes of the curves for the interaction effects model and the main effects model are the partial derivatives of $\hat{\mu}_d$ and $\hat{\mu}_d^*$ with respect to F from Equations 5.2 and 5.5, respectively. Applying the chain rule for derivatives and keeping in mind that η and η^* are equal to the index functions for each model, the partial derivatives are

$$\frac{\partial \hat{\mu}_d}{\partial F} = \left(b_1 + b_3 d\right) \times \frac{\partial}{\partial \eta} g^{-1}\left(\tilde{a} + b_2 d + \{b_1 + b_3 d\} F\right) \tag{5.11}$$

$$\frac{\partial \hat{\mu}_d^*}{\partial F} = b_1^* \times \frac{\partial}{\partial \eta^*} g^{-1}\left(\tilde{a}^* + b_2^* d + b_1^* F\right) \tag{5.12}$$

The structure of these expressions makes clear that the slope depends on both factors. The coefficients specifying the effect of F in each structural model—the terms inside the dashed line boxes—are multiplied by a function of the link function $g^{-1}(\ )$. For the main effects model, the boxed expression is just the unconditional effect of F (b_1^*), but the slope of each curve defined by $M = d$ will still be different at a given value of F and will change with F due to the nonlinear link function. In contrast, the slopes from the interaction effects model will vary for both reasons.

Let's illustrate this with the same logistic regression example and compare the slopes for Whites and Blacks of the predicted probability of approval $\left(\hat{P}_d \text{ or } \hat{P}_d^* \right)$ by education (Ed). See the "Derivations and Calculations" section at the end of the chapter for the derivation of these calculating formulas for the slopes for the main effects model:

$$\frac{\partial P_{White}^*}{\partial Ed} = -0.21547 \times P_{White}^* \times \left(1 - P_{White}^*\right) \text{ where } P_{White}^* = \frac{1}{1 + e^{-(1.38597 - 0.21547 \times Ed)}}$$

$$\frac{\partial P_{Black}^*}{\partial Ed} = -0.21547 \times P_{Black}^* \times \left(1 - P_{Black}^*\right) \text{ where } P_{Black}^* = \frac{1}{1 + e^{-(-0.79703 - 0.21547 \times Ed)}} \quad (5.13)$$

This specifies that the education slope for the two curves will vary for Blacks and Whites at a given education level despite the absence of a structural interaction effect, because the predicted probabilities for Whites and Blacks differ, and that the magnitude of the differences in the slopes will change across education levels. Look back at Figure 5.4 at the places on the standard logistic curve where we would evaluate the Black and White education slopes. When *Education* = 4, the logistic curve is changing more slowly where we would calculate the education slope for Blacks $\left(\hat{P}_{Black}^* = 0.1599, \hat{\eta} = -1.66 \right)$ than for Whites $\left(\hat{P}_{White}^* = 0.6281, \hat{\eta} = 0.52 \right)$. Consequently, the White slope (−0.0503) is 1.7 times the size of the Black slope (−0.0290) at that education level. And while both slopes are smaller when *Education* = 12—because each is evaluated at a point with a lesser slope along the logistic curve—the White slope (−0.0383) is 5.6 times the size of the Black slope (−0.0068).

The calculating equations for the education slopes from the interaction effects model (see the "Derivations and Calculations" section at the end of the chapter) show the second source of nonlinearity in the education slopes in the scatterplots, the conditional-on-race structural effect of education:

$$\frac{\partial P_{White}}{\partial Ed} = -0.24886 \times P_{White} \times \left(1 - P_{White}\right) \text{ where } P_{White} = \frac{1}{1 + e^{-(1.77413 - 0.24886 \times Ed)}}$$

$$\frac{\partial P_{Black}}{\partial Ed} = -0.08582 \times P_{Black} \times \left(1 - P_{Black}\right) \text{ where } P_{Black} \frac{1}{1 + e^{-(-2.08519 - 0.08582 \times Ed)}} \quad (5.14)$$

Thus, when we compare the two predicted probability curves, the magnitudes of their slopes differ at any value of education because the conditional effects diverge (−0.2489 vs. −0.0858) *and* because each conditional effect is multiplied by a different factor. The multiplicands vary with the value of the conditional effects of education as well as the effect and display values of its moderator (race). In Figure 5.5, note that the evaluation points for the White predicted probabilities are at a steeper part of the standard logistic curve than the corresponding evaluation points for the Black predicted probabilities, just as they were for the main effects model. Thus, the factors in Equation 5.14 by which education's conditional effects are multiplied to calculate the slope for each racial group are quite different. At 12 years of education, for example, the White conditional education effect (−0.2489) is multiplied by 0.1779, yielding a slope of −0.0443; the Black conditional education effect (−0.0858) is multiplied by 0.0318, resulting in a slope of −0.0027. At 4 years of education, the multiplicand is 0.2336 for Whites and 0.1343 for Blacks, giving slopes of −0.0581 and −0.0115, respectively.

Thus, when you examine the pattern of the slopes, the difference in patterns is generated in part by the differences in the structural conditional effects of the focal variable and in part by the nonlinearity of the link function. Table 5.1 shows the calculating formulas for the slopes of the focal variable for the common link functions identified

TABLE 5.1 ● MULTIPLICANDS FOR CALCULATING THE SLOPE OF THE RELATIONSHIP OF THE PREDICTED OUTCOME[a] WITH THE FOCAL VARIABLE FOR SOME COMMON LINK FUNCTIONS

Link Function	Structural Conditional Effect of F Is Multiplied by[b] $\frac{\partial}{\partial \eta} g^{-1}(\eta) =$	Second Derivative (Defines Domain for Convex or Concave Function)
Identity	1	0
Logit	$\dfrac{e^{-\eta}}{\left(1+e^{-\eta}\right)^2}$	$\dfrac{-e^{\eta}\left(e^{\eta}-1\right)}{\left(1+e^{\eta}\right)^3}$ $\begin{cases} \text{concave for } \eta > 0 \\ \text{convex for } \eta < 0 \end{cases}$
Probit	$\dfrac{1}{\sqrt{2\pi}}e^{-\frac{1}{2}\eta^2}$	$\dfrac{-\eta}{\sqrt{2\pi}}e^{-\frac{1}{2}\eta^2}$ $\begin{cases} \text{convex for } \eta > 0 \\ \text{concave for } \eta < 0 \end{cases}$
Log	e^{η}	ηe^{η} $\begin{cases} \text{concave for } \eta > 0 \\ \text{convex for } \eta < 0 \end{cases}$
Inverse	$-\dfrac{1}{\eta^2}$	$\dfrac{2}{\eta^3}$ $\begin{cases} \text{concave for } \eta > 0 \\ \text{convex for } \eta < 0 \end{cases}$
Inverse square	$-\dfrac{2}{\eta^3}$	$\dfrac{6}{\eta^4}$ concave
Square root	$-\dfrac{1}{2}\times\dfrac{1}{\eta^{\frac{3}{2}}}$ for $\eta > 0$	$\dfrac{0.75}{\eta^{\frac{5}{2}}}$ concave since $\eta > 0$
Log-log	$e^{-\eta}e^{-e^{-\eta}}$	$e^{-\eta}e^{-e^{-\eta}}\left(e^{-\eta}-1\right)$ $\begin{cases} \text{convex for } \eta > 0 \\ \text{concave for } \eta < 0 \end{cases}$
Complementary log-log	$-e^{-\eta}e^{-e^{-\eta}}$	$e^{-\eta}e^{-e^{-\eta}}\left(1-e^{-\eta}\right)$ $\begin{cases} \text{concave for } \eta > 0 \\ \text{convex for } \eta < 0 \end{cases}$

a. Not included are models with multiple outcomes/categories (e.g., multinomial logistic or ordinal regression models).

b. $\eta = a + b_z Z_R + b_1 F + b_2 M_1 + b_3 M_1 F +$ additional terms for other moderators of F and product terms with F evaluated at the display values of the moderator.

by Fox (2008, p. 380). For each, the structural conditional effect of F (as determined by the GFI method, presented in Chapter 2) would be multiplied by the listed function. These formulas are general and not specific to a particular specification of the interaction effect; that is, they apply to single-moderator models, multiple-moderator models, three-way interaction models, and so on.

REVISITING OPTIONS FOR DISPLAY AND REFERENCE VALUES

In terms of the display values of the focal and moderating variables used to create displays of the predicted outcome ($\hat{\mu}$), I would apply the same guidelines I suggested in Chapter 4 for linear link models. To me, the influence of the linear perception bias and confounding on correctly evaluating the degree of moderation of the effect of F reinforces the importance of considering a wide range of display values. And it suggests the need for alternatives to predicted value displays using the observed outcome variable's metric, as I discuss in the next section.

But first, it is important to consider how the choice of reference values for the other predictors (Z) in a nonlinear link model is more consequential than it is in a linear link model. In a linear link model, the result of using different reference values is to add or subtract the same value when calculating any predicted value. This would just shift an entire visual display by the same amount, but it would not affect the difference between any pair of predicted values. And it would not influence the calculation of the slope. In contrast, changing the reference values in a nonlinear link model alters the point along the curve generated by the $g^{-1}(\ )$ link function at which predicted values or slopes are calculated. As the prior section demonstrated, this shifts predicted values and slopes to a varying degree for different combinations of values of the focal and moderating variables because it alters the value of $\tilde{a}$ used in the underlying prediction formula in Equation 5.1. Consider the logistic example in Figure 5.5. Changing $\tilde{a}$ is equivalent to shifting each of the evaluation points along the x-axis by the same amount. But the corresponding changes on the y-axis would not be equal because the $g^{-1}(\ )$ function by definition translates equal changes in x into unequal changes in y. As a result, different segments of a visual display will shift by different amounts, and the gap between pairs of predicted values will shift differentially. The slopes calculated at each evaluation point will similarly change nonlinearly.

How does this affect choosing among and applying the three rubrics (discussed in Chapter 4) for picking reference values for Z—central tendency values, representative (substantively interesting) values, or "as-observed" values? One issue to consider when choosing among these rubrics is the degree to which the visual display of the predicted outcome of the interaction variables is influenced by the confounding of the two sources of nonlinearity: (1) the interaction effect in the structural component and (2) the nonlinearity of the link function. The representative values rubric creates a set of visual displays of the relationship of the observed outcome and the interacting variables, one for each combination of representative values chosen by the researcher. On the one hand, this highlights the reality that in nonlinear models there is no single effect of any predictor on the observed outcome independent of the values of the other predictors. On the other hand, making comparisons among the sets of displays to interpret the interaction effect is both cumbersome and very much subject to confounding of the nonlinearity of the link function with the (nonlinear) interaction effects. The advantage of using the central tendencies or as-observed rubrics is that these choices provide a single visual display of the relationship to interpret that can be considered typical of the sample, albeit typical in different ways. But the disadvantage is that such a single display does not consequently make clear how the display of the relationship is contingent on the reference values chosen.

A related issue is that the central tendency rubric uses a single, set reference value to which each combination of values of the focal and moderating variables (multiplied by their respective coefficients) is added to get the predicted value of the modeled outcome $(\hat{\eta})$. This intermediate predicted value is then transformed into the predicted value of the observed outcome by using the link function, $\hat{\mu} = g^{-1}(\hat{\eta})$. The process the as-observed rubric uses to calculate $\hat{\mu}$ for each combination of the focal and moderating variables' values should in general introduce more variability into the estimate of $\hat{\mu}$ because it calculates the predicted outcomes at up to N reference points—which are evaluated at different points along the $g^{-1}(\ )$ curve—and averages them together.

To see why this is so, consider an alternative calculation of the predicted value for the central tendency rubric that is parallel to the process the as-observed rubric uses. Figure 5.6 illustrates this with step-by-step parallel calculations. For simplicity and

---

FIGURE 5.6 ● PARALLEL CALCULATION OF THE PREDICTED VALUE OF THE OBSERVED OUTCOME USING THE AS-OBSERVED $(\hat{\mu}_{AO})$ AND CENTRAL TENDENCY $(\hat{\mu}_{CT})$ RUBRICS

Let $FM_{f,m} = b_1f + b_2m + b_3f \times m, b_zZ_i = \sum\limits_{\substack{j,Z_j \neq \\ F,M,F^*M}} b_jZ_{i,j}$ and $b_zZ_R = \sum\limits_{\substack{j,Z_j \neq \\ F,M,F^*M}} b_j\bar{Z}_j$

As Observed	Central Tendency
For $F = f$ and $M = m$ calculate for each case i	For $F = f$ and $M = m$ calculate for each case i
$\hat{\eta}_{i,AO} = a + FM_{f,m} + b_zZ_i$ $\hat{\mu}_{i,AO} = g^{-1}(\hat{\eta}_{i,AO})$ $\quad = g^{-1}(a + FM_{f,m} + b_zZ_i)$	$\hat{\eta}_{i,CT} = a + FM_{f,m} + b_zZ_R$ $\hat{\mu}_{i,CT} = g^{-1}(\hat{\eta}_{i,CT})$ $\quad = g^{-1}(a + FM_{f,m} + b_zZ_R)$
Predicted value for each case varies with b_zZ_i.	Predicted value for each case is the same.
Average the predicted values across all cases	Average the predicted values across all cases
$\hat{\mu}_{AO} = \dfrac{1}{N}\sum\limits_{i=1}^{N}\hat{\mu}_{i,AO}$ $\quad = \dfrac{1}{N}\sum\limits_{i=1}^{N}g^{-1}(a + FM_{f,m} + b_zZ_i)$	$\hat{\mu}_{CT} = \dfrac{1}{N}\sum\limits_{i=1}^{N}\hat{\mu}_{i,CT}$ $\quad = \dfrac{1}{N}\sum\limits_{i=1}^{N}g^{-1}(a + FM_{f,m} + b_zZ_R)$ $\quad = \dfrac{1}{N}N \times g^{-1}(a + FM_{f,m} + b_zZ_R)$ $\quad = g^{-1}(a + FM_{f,m} + b_zZ_R)$

concreteness, the example uses the interaction of F with a single moderator M for logistic regression. Both rubrics use the specified values of the focal and moderating variables (f and d, respectively) in the calculations. For the other predictors (Z), the as-observed rubric uses each case's own values to calculate the casewise predicted value. In contrast, the central tendency rubric assigns to each case the mean of the other predictors (Z)—or, equivalently, the value from another measure of central tendency—to calculate the casewise predicted value. The overall estimated predicted value is then calculated by averaging together the casewise predicted values. The as-observed rubric averages the N different casewise values, while the central tendency rubric averages the same value across cases. Hence, there is the potential for more variability—and sensitivity to nonlinear confounding—when using the as-observed rubric.

Another implication of the differences between these two calculating rubrics is the possibility that the resulting pattern of the visual display for the as-observed predicted values would be influenced to a greater degree by the nonlinearity of the link function than if using the central tendency rubric. This would be the case if the as-observed predicted values are systematically calculated at points along the link function's curve that are steeper than where the central tendency predicted values are calculated. How likely this is is determined by both the link function $g^{-1}(\ )$ and the characteristics of the sample data. The basis for this is an application of Jensen's inequality (Florescu & Tudor, 2014, p. 436) to the expected value of a function of a random variable compared with that function evaluated at the random variable's expected value. For a convex function $C[\ ]$ of a random variable X with finite expectation,[5]

$$\mathcal{E}\big(C[X]\big) \ge C\big(\mathcal{E}[X]\big)$$

That is, the expected value of a function of a random variable is greater than or equal to the function calculated at the random variable's expected value. Applying this to the rubrics, the as-observed rubric estimates the expected value of the function $g^{-1}(\ )$ of the random variable η_i:

$$\mathcal{E}\big[\hat{\mu}_{AO}\big] = \mathcal{E}\big[g^{-1}(\eta_i)\big]$$

The central tendency rubric estimates the function $g^{-1}(\ )$ at the expected value of the random variable η_i:

$$\mathcal{E}\big[\hat{\mu}_{CT}\big] = g^{-1}\big[\mathcal{E}(\eta_i)\big]$$

Applying the inequality principle above,

$$\begin{aligned}
\mathcal{E}\big[\hat{\mu}_{AO}\big] &\ge \mathcal{E}\big[\hat{\mu}_{CT}\big] \text{ if } g^{-1}(\ ) \text{ is a convex function} \\
\mathcal{E}\big[\hat{\mu}_{AO}\big] &\le \mathcal{E}\big[\hat{\mu}_{CT}\big] \text{ if } g^{-1}(\ ) \text{ is a concave function}
\end{aligned} \qquad (5.15)$$

Several of the link functions in Table 5.1 are convex functions (increasing at a nondecreasing rate), most notably the log link that is used for count models. It is likely then that the estimated $\hat{\mu}_{AO} \ge \hat{\mu}_{CT}$ because this holds for their expected values. Consequently, $\hat{\mu}_{AO}$ will usually be evaluated at a steeper section of the inverse link function curve. Thus, the as-observed visual display will quite likely be more influenced by the nonlinearity of the link function than will the central tendency visual display for log link models.

The remaining link functions are convex for one part of their domain but concave for another part; the dividing line is $\eta = 0$. In these cases, $\hat{\mu}_{AO}$ is not necessarily greater than $\hat{\mu}_{CT}$. But, depending on the actual distribution of $\hat{\eta}_{i,AO}$, it may be the case that $\hat{\mu}_{AO}$ is at a steeper point on the inverse link function curve than $\hat{\mu}_{CT}$. For example, consider a logistic regression analysis and the case where the as-observed predicted probability $\hat{\mu}_{AO}$ is less than 0.5—especially the farther from 0.5 it is. In this case, the average of $\hat{\eta}_{i,AO}$ will be less than 0 because the predicted probability is 0.5 when the average of $\hat{\eta}_{i,AO}$ equals 0. Consequently, the actual distribution of $\hat{\eta}_{i,AO}$ is likely to be predominantly on the convex section of the logistic curve, and hence it is likely that $\hat{\mu}_{AO} \geq \hat{\mu}_{CT}$, which puts $\hat{\mu}_{AO}$ on the steeper section of the logistic curve. Conversely, when $\hat{\mu}_{AO} > 0.5$, it is likely that the $\hat{\eta}_{i,AO}$ values are primarily on the concave section of the curve. Thus, it is likely that $\hat{\mu}_{AO} \leq \hat{\mu}_{CT}$, which again puts $\hat{\mu}_{AO}$ on the steeper section of the logistic curve.

Because the representative values rubric uses reference points chosen by the researcher, there is no way to draw a global conclusion concerning whether its visual displays are more or less affected by the confounding from the nonlinearity of the link function. I would observe that one way in which representative values are sometimes chosen is to highlight contrasts—that is, to pair high values of predictors that have positive effects with low values of predictors that have negative effects, and vice versa. In this case, such choices may exacerbate the confounding unless the predicted values are at comparably steep locations on the inverse link function.

I consider the choice of a rubric for selecting reference values for the other predictors in the model an individual preference but would note that the central tendencies and the as-observed rubrics are the most commonly used. In the end, there is no perfect choice among the options for rubrics, and it is up to each practitioner to make his or her own decision. I hope this discussion provides helpful background for making an informed choice among the options. I turn next to discussing three approaches to deal with the confounding in order to provide an accurate representation of the interacting variables' relationship to the outcome, specifically the extent to which the predictors moderate one another's effects.

SOLUTIONS

In this section, I propose the use of three alternatives that avoid the confounding due to the nonlinearity of the link function when interpreting interaction effects using visual displays of the predicted values. I illustrate these with three examples that differ in terms of the type of interaction effect, the level of measurement of the predictors and outcomes, and hence the type of visual display, and the type of GLM analysis. Rather than describe and apply the options one at a time to the examples, I will describe all three options first and then apply all of them to the examples in turn. This will help demonstrate that there is no single option that is always the best (in my opinion) but should also let you decide which one you prefer. My bottom-line recommendation is to try all three in any given analysis to see which one helps you tell the story most effectively.

   1. *Effect displays:* The first option is to use an effect display, as discussed in Chapter 3, to describe and interpret the structure of the interaction effect—how the

effect of *F* on the outcome is contingent on one or more moderators. This can be used either in place of a visual display of the predicted outcome or as a supplement to help discuss a predicted outcome display. In the former case, you would avoid the problem of confounding altogether by not presenting a predicted value visual display. But many researchers and readers do not find an effect display as intuitive and accessible as a predicted value visual display. The supplementary use of an effect display is to rely on it to interpret the degree of moderation to inform your discussion of the visual display of the predicted outcome. This avoids the confounding problem because it separates the interpretation of the overall relationship between the outcome and the interacting predictors (using the display of predicted values) from an assessment of the extent of moderation (using the effect display).

2. *Dual-axis predicted value displays:* A second possibility is to construct the display using the metric of the transformed outcome (η) to define the predicted outcome axis. Such a visual display is not subject to the linear perception bias because the predicted value in the η-metric is a linear function of the predictors. As such, we have a visual reference point for determining how different the slopes of the predicted value curves are, and changing the reference values of the noninteracting predictors uniformly shifts the values of the visual display up or down. The drawback is that the η-metric is typically not an intuitive one. This problem can be largely overcome by presenting the visual display with dual axes: a primary axis in the η-metric used to plot the predicted values and a secondary axis labeled with a set of μ-metric values whose position on the secondary axis corresponds to their η-metric values. The observed-metric values provide a context for understanding the changes in the model-metric predicted values.

For some GLMs, you can use an estimated standard deviation of the outcome in the model metric to give meaning to the model-metric expected values in standard deviation units. Binomial logistic regression, probit analysis, and ordinal regression models all can be derived as a model-metric outcome, which is an underlying latent interval variable whose variance is estimable, and the multinomial logistic's model metrics can be developed from a utilities perspective as a standardized latent measure for each pair of outcome categories (Long & Freese, 2014, chap. 8). Last, the η-metric for a count model is the log of the count, and its variance can be estimated with the delta method.

3. *Superimposed displays:* The final option is to begin with a visual display (in the μ-metric) of how the predicted values vary with the interacting predictors. Then you superimpose a display of the predicted values from the corresponding main effects model, as I did in Figure 5.3. The display from the model with interaction effects provides a picture of the overall relationship. Comparing how the interaction effects model's predicted outcome values diverge from the main effects model's predictions permits an accurate assessment of the degree of moderation among the interacting variables free of the confounding due to the nonlinear link function.

Example 1: Two-Way Nominal-by-Nominal Interaction

This example is a binomial logistic regression analysis of the approval of an interracial marriage ban predicted by the interaction between region of upbringing (South, non-South) and race (Black, White), as well as other predictors. The model also specifies that race moderates the effect of education, which is considered in Example 2. Consequently, I do not discuss the race effect here because race's effect is moderated by both region and education.

Effect Display Option. Figure 5.7 presents the effect display using region as the focal variable and race as the moderator. The effect of region is positive (0.021) but not significant for Blacks because the confidence interval includes 0, while the region effect (1.388) is positive and significant for Whites. Because this example is from a logistic regression, these effects can be directly interpreted as effects on the log odds or, equivalently, on a latent interval–level outcome (Long & Freese, 2014, p. 235). Thus, the effect of region for Whites is a difference in the log odds of approval of 1.39 units between Whites raised in the South versus the non-South, while for Blacks the (nonsignificant) effect of region is a difference of only 0.02 units.

While Figure 5.7 visually portrays the magnitude of this difference, we lack a meaningful metric for interpreting what these differences mean. How large is a difference in the log odds of 1.39? As noted above, logistic regression effects can be *y*-standardized in the η-metric and then interpreted as an effect in standard deviation units. This provides a metric for judging the size of the effects. The latent outcome's standard deviation is 2.392, so the effect of region for Whites is a difference of more than 0.5 standard deviation (1.3878/2.3922 = 0.58). That is, approval of a ban is 0.58 standard deviation higher for Whites raised in the South versus outside the South. The effect of region for Blacks is not only insignificant but also small in magnitude, with a difference in approval of less than 0.01 standard deviation between Blacks raised in the South versus outside the South. Figure 5.8 presents the effect display rescaled in the standardized η-metric. This effect display clearly conveys the degree to which race moderates the effect of region.

Dual-Axis Display Option. For a nominal-by-nominal interaction, the usual choice for a predicted value display is a bar chart. For the dual-axis option, the height of the bars is scaled in the η-units of the transformed outcome, and the chart is labeled with two *y*-axes, as shown in Figure 5.9. The axis on the left reports the η-scale used to graph the bars—in this case, the log odds of approval. The axis on the right shows the values in the observed outcome's metric—in this case, the predicted probability of

---

FIGURE 5.7 ● ERROR BAR (CONFIDENCE INTERVAL) CHART, ONE MODERATOR

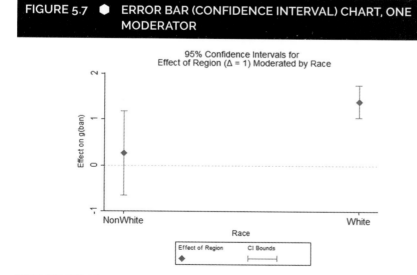

95% Confidence Intervals for
Effect of Region (Δ = 1) Moderated by Race

ICALC command: effdisp , plot(type(errbar)) ndig(0) pltopts(ylab(-1(1)2)). The plot( ) option specifies an error bar chart, and the ndig( ) and pltopts( ) options customize the *y*-axis labels.

FIGURE 5.8 ● LATENT STANDARDIZED METRIC ERROR BAR CHART, ONE MODERATOR

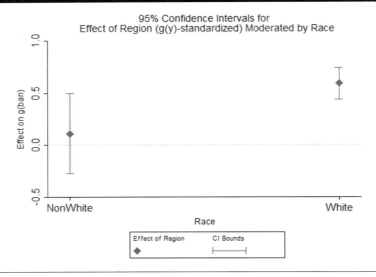

ICALC command: effdisp , effect(b(sdy)) plot(type(errbar)) ndig(1) pltopts(ylab(-.5(.5)1)). The effect( ) option sets the *y*-metric to latent standard deviation. The plot( ) option gets an error bar chart, and the ndig( ) and pltopts( ) options customize the *y*-axis labels.

approval—located vertically according to their values transformed into the η-metric. The probability axis value labels range from a probability of 0.05 to 0.50, in unequally spaced increments in value and in vertical location because their corresponding log odds values are not equidistant. The pair of bars on the left reports the predicted approval in the two regions for Blacks, while the bars on the right show regional differences for Whites.

Each bar in Figure 5.9 is labeled with its predicted log odds. In looking at this, keep in mind that small probabilities correspond to large log odds in the negative direction and are represented by long bars. The bar chart clearly shows the absence of a regional effect for Blacks but a notable regional effect for Whites. For Blacks, the difference in the height of the bars is visually negligible in the log odds metric, and the difference in the corresponding predicted probabilities is only 0.001. In contrast, there is a substantial visual disparity between the two regions for Whites, with those raised in the South (the *gray bar*) having a higher log odds of approval—lower negative number and shorter bar—than those raised outside the South. For Whites, the regional difference in predicted log odds translates into a difference in probabilities of .254, which is substantial in comparison with the observed mean probability of approval of .209.

Although Figure 5.9 makes clear that race moderates the regional effect to a great extent, you might be wondering why this is any better than a simpler graphic display with a single *y*-axis scaled in terms of the predicted probabilities. My answer is that for bar charts there may not be an apparent advantage because we judge differences in the focal variable's effects across the values of the moderator categorically. But we will see in later examples that this option can be very useful for scatterplots when the *x*-axis variable is interval level. And the next section demonstrates an option for bar charts of the predicted outcome that does have an advantage for interpreting a nominal-by-nominal interaction.

FIGURE 5.9 ● BAR CHART OF PREDICTED OUTCOME WITH DUAL AXES, ONE MODERATOR

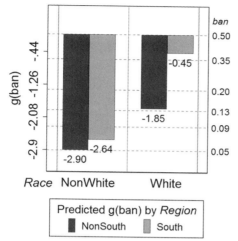

g(ban) by the Interaction of Region and Race, Dual Outcome Metric Axes.

ICALC command: outdisp, plot(type(bar)) out(metric(model) dual) ndig(2) blabopts(size(medsmall) orient(horizontal)) /// pltopts(ytit(, ma(r+4)) ti(,size(*.9))). The plot( ) option specifies a bar chart, the out( ) option calculates predicted values in the model metric and creates dual y-axis labeling, ndig( ) formats the bar labels with 2 decimals, the blabopts( ) option formats the bar labels, and the pltopts( ) option formats titles.

Superimposed Display Option. In this option, a secondary display from a main effects only model is added to the primary visual display of the predicted observed outcome derived from the interaction effects model. In Figure 5.10, the primary display from the interaction effects model is a pair of bar charts, one for Blacks and one for Whites, showing the predicted probability of approval for each region within race. Within the panel for each race, a second pair of bars is added that reports the predicted probability of approval for each region for that race derived from the main effects only model. The interaction model's bars are outlined with *solid lines*, while the main effects model's bars are outlined with *dashed lines*.

This display effectively portrays the pattern of the relationship between approval, region, and race from the interaction specification that we previously saw in the other options. Examining the leftmost pair of bars within each panel shows the following:

- For Blacks, the pair of bars for region are small in magnitude (probability ≈ .05) and barely distinguishable from each other. This indicates no difference in approval between the regions for Blacks.

- For Whites, in contrast, the pair of bars are each larger in magnitude (probabilities of .14 and .39) but also show a substantial difference in the probability of approval between the regions (.25).

This figure also clearly demonstrates that much of what initially appears to be moderation of the region effect by race is in fact a result of the nonlinearity of the link function and not due to the interaction specification. This becomes most apparent by visually comparing the predicted probabilities in the panel for Whites from the interaction specification (left-hand pair of bars) with those from the main effects

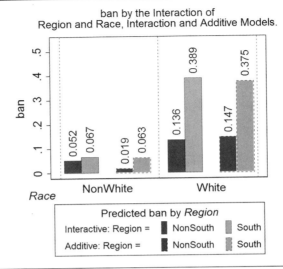

FIGURE 5.10 ● BAR CHART OF PREDICTED OUTCOME, ADDED MAIN EFFECTS PREDICTIONS, ONE MODERATOR

ban by the Interaction of
Region and Race, Interaction and Additive Models.

ICALC command: outdisp, plot(type(bar)) out(metric(obs) main(main) atopts((means)_all)) ndig(3) pltopts(ylab(0(.1).5)). The plot( ) option specifies a bar chart, the out( ) option calculates predicted values in the observed metric with other predictors' reference values set to their means and superimposes main effects model predictions, and the ndig( ) and pltopts( ) options format the *y*-axis and the bar labels.

specification (right-hand pair of bars). Note that the bars for each region for Whites are relatively similar in height, whether from the main effects or interaction effects model. The result of the interaction specification for Whites is to increase the gap between the regions from 0.231 in the main effects model to 0.254 in the interaction specification. In contrast, comparing the predicted probabilities from the two specifications in the panel for Blacks shows that the difference between the regions is reduced from .031 in the main effects results to .001 in the interaction effect specification. This indicates that the story is at least as much about a diminished regional effect for Blacks as it is about an increased regional effect for Whites.

In sum, race moderates the effect of region, resulting in a nonsignificant, negligible regional difference among Blacks but a significant and somewhat magnified regional difference among Whites (about 10% larger). While the degree to which race moderates region is more modest than what might be naively concluded from an examination of just the interaction effect results, we should not lose sight of the totality of the approval-by-region-by-race relationship. The bottom line is that there are negligible regional differences in the probability of approval among Blacks (0.001) but substantial regional differences in approval among Whites (0.254).

Example 2: Two-Way Interval-by-Nominal Interaction

This example is from the same logistic regression results used in Example 1 but considers how race (Black, White) moderates the effect of education (in years) on approval of an interracial marriage ban. The effects display option shows the effects scaled in standard deviation units of latent approval. For the dual-axis option and the superimposed option, the visual displays are scatterplots of the predicted outcome on the *y*-axis (scaled as log odds and predicted probability, respectively) against education on the *x*-axis, with separate lines/curves for Whites and Blacks.

Effect Display Option. Figure 5.11 shows how the effect of education on approval is negative but changes across the categories of race. For Blacks, as education increases by 1 year, approval is predicted to decrease by 0.04 standard deviation, but this effect is not significantly different from 0 (the 95% confidence bounds include an effect value of 0). For Whites, the effect is significant and suggests that 1-year increases in education are associated with a decline in approval of 1/10 of a standard deviation. To put it another way, the difference between someone who finishes high school and someone who finishes an advanced degree (an 8-year difference) is predicted to be 0.8 standard deviation.

Dual-Axis Option. Figure 5.12 plots the log odds of approval against education separately for Blacks (*solid line*) and Whites (*dashed line*). The left-hand *y*-axis shows the log odds scale used for plotting, and the right-hand *y*-axis shows the probability scale, which provides a sense of magnitude in the more meaningful probability metric. Visually, it is apparent that approval declines more steeply with education for Whites than for Blacks. In log odds terms, the change per year of education for Blacks is 0.10, and for Whites it is 0.24; that is, the rate of decline with education is twice as fast for Whites as it is for Blacks.

How big is the magnitude of the decline; that is, how big is a decline of 0.24 in the log odds of approval for Whites? To give some sense of this, consider the difference between 8 and 12 years of education. For Whites, the probability of approval changes by .212 (you can approximate this by looking at the probability reference lines—the difference between a little less than .56 and a little more than .29). Given that the mean probability of approval is .21, this suggests a substantial decline. In contrast, for Blacks, the decline in approval is only .024. Note that how much the

---

FIGURE 5.11 ● ERROR BAR PLOT IN STANDARD DEVIATION UNITS, ONE MODERATOR

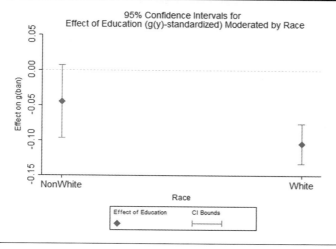

ICALC command: effdisp , plot(type(errbar)) effect(b(sdy)) pltopts(ylab(-.15(.05).05)) ndig(2). The plot( ) option specifies error bar chart, the effect( ) option sets the *y*-metric to latent standard deviation, and the pltopts( ) option customizes the *y*-axis labels.

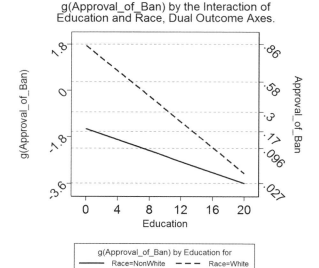

FIGURE 5.12   ◆   PREDICTED MODEL-METRIC SCATTERPLOT WITH DUAL AXES

g(Approval_of_Ban) by the Interaction of Education and Race, Dual Outcome Axes.

ICALC command: outdisp, plot(type(scat)) out(metric(model) dual atopts((means) _all)). The plot( ) option specifies a scatterplot, and the out( ) option calculates predicted values in the model metric using other predictors' reference values set to their means and creates dual *y*-axis labeling.

probability changes for 4 years' difference in education will vary depending on at which value of education you start the comparison. For example, if you go 4 years higher, the change in probability of approval between 12 and 16 years is .130 for Whites and .016 for Blacks. This is reflected in the unequal spacing along the *y*-axis of the predicted probability reference lines.

All in all, this dual-axis scatterplot provides a clear indication of the degree to which race moderates the effect of education on approval. We know from other results that the effect of education is not significant for Blacks, and this scatterplot shows that approval decreases with education at twice the rate for Whites as it does for Blacks. Showing the predicted probability reference lines permits a more intuitive understanding of the size of the education effect.

Superimposed Predictions Option. This option also is an effective choice for this example, as the predicted probability curves in Figure 5.13 demonstrate. The *black curves* for the interaction model show the predicted probability of approval by education for Whites (*dashed line*) and Blacks (*solid lines*). Visually, we can see a much shallower slope (education effect) for Blacks than for Whites. Across the range of education, the predicted probability for Whites decreases by .807 but only by .132 for Blacks. Comparing these curves with the *light gray curves* for the main effects model, we see that much of the gap between the curves for Whites and Blacks is due to the main effects of education and race, not to the interaction specification. The plot also makes clear how race moderates the education effect—it flattens the curve for Blacks while steepening it for Whites. As did the dual-axis plot in Figure 5.12, this plot provides an accurate portrayal of how the probability of approval depends on race and education, without distorting the role of the interaction specification.

FIGURE 5.13 ● **PREDICTED OBSERVED-METRIC SCATTERPLOT WITH ADDED MAIN EFFECTS MODEL PREDICTIONS**

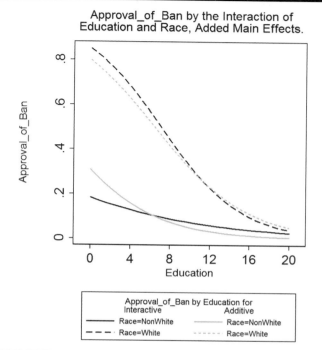

ICALC command: outdisp, plot(type(scat)) out(metric(obs) main(main) atopts((means) _all)) pltopts(ylab(0(.2).8)). The plot( ) option specifies a scatterplot, the out( ) option calculates predicted values in the observed metric with other predictors' reference values set to their means and superimposes main effects model predictions, and the pltopts( ) option formats the *y*-axis labels.

Example 3: Two-Way Interval-by-Interval Interaction

This example uses the results from a negative binomial regression model predicting the number of poor mental health days reported during the past month, including an interaction between work–family conflict and occupational SEI. Work–family conflict is on a scale ranging from 1 (*never*) to 4 (*often*), with a mean of 2.19 and a standard deviation of 0.78. SEI is a typical occupational status scale ranging from 17.6 to 97.2, with a mean of 50.9 and a standard deviation of 19.2. I focus on how SEI moderates work–family conflict, but some of the figures also provide information about how work–family conflict moderates SEI. Because these are interval-level measures, I use scatterplots for the dual-axis and superimposed plot options to portray the relationship between poor mental health days and work–family conflict, with lines/curves for this relationship plotted separately for selected values of SEI spanning nearly the full range of SEI values (20, 45, 70, 95).

Effect Display Option. Figure 5.14 shows that the effect of work–family conflict on log number of poor mental health days is positive across the range of sample SEI values but is not statistically significant for $SEI < 29.4$. The effect of conflict on the log number of poor mental health days increases by a factor of 5.6 as SEI rises, from a 0.15 increase per unit of conflict to a 0.85 increase in the log number of days. To give some context for judging the size of the effect of conflict, the smallest effect (0.15) represents an increase of 0.03 standard deviation in the log number of poor mental health days per unit increase in conflict. And the largest effect (0.85) is an increase of nearly one fifth of a standard deviation (0.18) per unit of conflict. Clearly, conflict has a relatively substantial effect among workers with a high SEI and a much more modest effect among workers with low occupational status. An equivalent effect display plot for the effect of SEI moderated by conflict would show that SEI has a significant negative effect on the log number of days but only when *conflict* < 2.82, which includes a little more than one quarter of the sample.

Dual-Axis Option. The plot of the log number of poor mental health days by conflict similarly shows a sizable increase in the effect of conflict as SEI changes from near its minimum to near its maximum in Figure 5.15. The shallowest slope occurs when *SEI* is 20 (*solid line*), showing an increase of about 0.5 log days across the range of work–family conflict, while the steepest slope is when *SEI* is 95 (*small dashed line*), representing a change of 4.4 log days. Thus, the effect of conflict is nine times greater for workers in the highest status jobs than for those in the lowest status jobs. We can use the scale on the right-hand *y*-axis in number of days to give a more intuitive feel for the size of these changes. This indicates that in low-SEI jobs (*solid line*), number of poor mental health days is almost 2 days higher for workers with the most work–family conflict than for those with the least. And the magnitude of the change in the number of days of poor mental health between least and most frequent work–family conflict increases greatly as SEI rises. For *SEI* = 45 (*medium dashed line*), the difference is a little less than 5 days; for *SEI* = 70 (*large dashed line*), the disparity is about

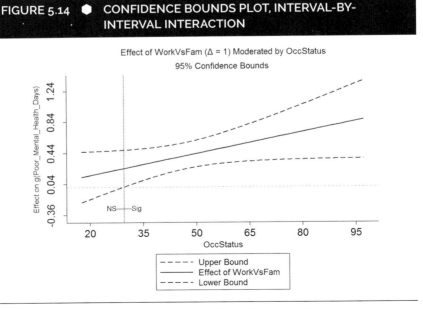

FIGURE 5.14 ● CONFIDENCE BOUNDS PLOT, INTERVAL-BY-INTERVAL INTERACTION

ICALC command: effdisp, plot(type(cbound)) ndig(2). The plot( ) option specifies a confidence bounds plot, and ndig( ) formats the number of decimals for *y*-axis labels.

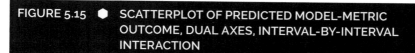

FIGURE 5.15 ● SCATTERPLOT OF PREDICTED MODEL-METRIC OUTCOME, DUAL AXES, INTERVAL-BY-INTERVAL INTERACTION

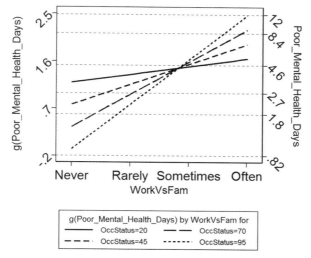

ICALC command: outdisp, plot(type(scat)) out(metric(model) atopt((means) _all) dual) pltopts(ylab(-.2(.9)2.5)). The plot( ) option specifies a scatterplot, the out( ) option calculates predicted values in the model metric using other predictors' reference values set to their means and creates dual y-axis labeling, and the pltopts( ) option formats the y-axis.

7.5 days; and for *SEI* = 95 (*small dashed line*), the increase is very close to 11 days. (To get the values used in this interpretation, I used OUTDISP to calculate tables of predicted values in both the model and the observed metrics.)

As was the case for the previous example, the dual-axis display is an effective device for GLMs with a nonlinear link function. We use the η-metric axis to judge how much the effect of an interval variable is moderated by another predictor and the μ-metric axis to provide an understanding of the magnitude of the focal variable's effects in a more accessible metric.

This plot also reveals how work–family conflict moderates the effect of SEI on poor mental health days, although it takes more work to glean that information from Figure 5.15 than it would to produce a plot reversing the roles of the focal and moderator variables. Nevertheless, SEI's effect is represented by the vertical gap between the lines at each value of conflict, and the changing magnitude/sign of the gaps as conflict increases from left to right shows how SEI is moderated by conflict. At the lowest levels of work–family conflict (1 = *never*), the lines for lower SEI are above the lines for higher SEI, indicating a negative effect of SEI—that is, more poor mental health days for lower SEI jobs than for higher SEI jobs. For example, workers in the lowest SEI jobs are expected to report about 2.25 more days of poor mental health than workers in the highest SEI jobs. But the vertical distances between the lines decrease as conflict increases until the order reverses when conflict reaches a scale value a little less than 3 (*sometimes*). From that point on, SEI has a positive effect on poor mental health days, which increases in magnitude as conflict increases. At the highest level of work–family conflict (4 = *often*), workers in the highest SEI jobs are predicted to have almost 7 more days of poor mental health than workers in the lowest SEI jobs.

Superimposed Effects Option. The results from the interaction effects model in Figure 5.16 are plotted with *black lines* and show the same clear pattern of moderation portrayed in the dual-axis plots. The effect of work–family conflict on poor mental health days is the least for workers in low-SEI jobs (*solid line*) and much larger in the highest SEI jobs (*small dashes*). I would argue that this visual pattern is actually an underestimate of the degree to which SEI moderates the effect of conflict. Compare the relative slopes and locations of the main effects curves—the *lighter gray lines* in the background—with each other, and consider how that pattern differs from the pattern for the interaction effects curves. This is perhaps easier to see in Figure 5.17, which separately plots each pair of interaction model and main effects model prediction lines for a display value of SEI. Look first at the upper-left plot in Figure 5.17 for the lowest SEI prediction lines. Compared with the other plots in Figure 5.17, the lowest SEI prediction line for the main effects model (*dashed line*) has the steepest slope for conflict's effect. But the prediction line for the interaction model (*solid line*) has the shallowest curve. Looking across the subplots, the higher the SEI, the smaller is

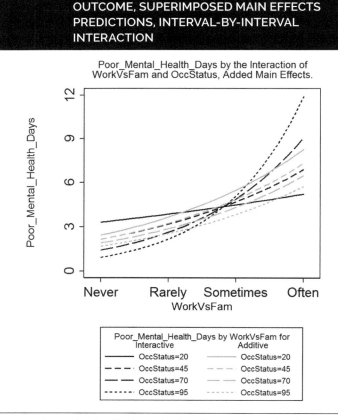

FIGURE 5.16   ⬢   SCATTERPLOT OF PREDICTED MODEL-METRIC OUTCOME, SUPERIMPOSED MAIN EFFECTS PREDICTIONS, INTERVAL-BY-INTERVAL INTERACTION

ICALC command: outdisp, plot(type(scat)) out(metric(obs) atopt((means) _all) main(main)) pltopts(ylab(0(3)12)). The plot( ) option specifies a scatterplot, the out( ) option calculates predicted values in the model metric using other predictors' reference values set to their means and creates dual *y*-axis labeling, and the pltopts( ) option formats the *y*-axis.

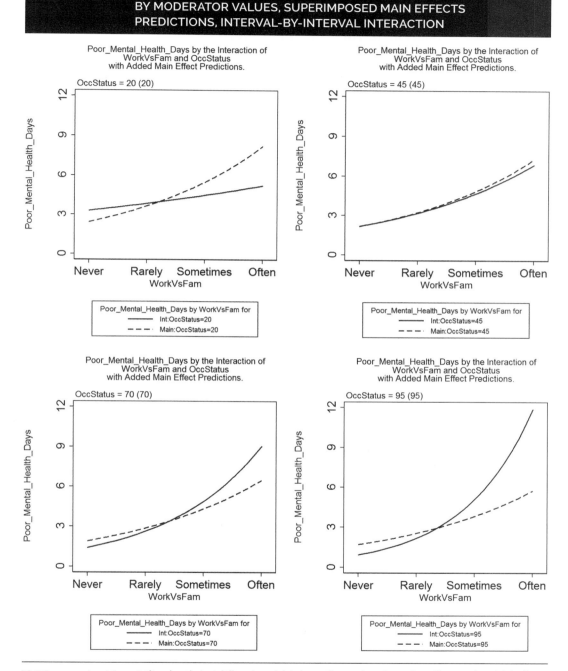

FIGURE 5.17 ● SCATTERPLOT OF PREDICTED MODEL-METRIC OUTCOME SEPARATELY BY MODERATOR VALUES, SUPERIMPOSED MAIN EFFECTS PREDICTIONS, INTERVAL-BY-INTERVAL INTERACTION

ICALC command: outdisp, plot(type(scat) single(1)) out(metric(obs) atopt((means) _all) main(main)) pltopts(ylab(0(3)12)). The plot( ) option specifies a scatterplot for each conflict value, the out( ) option calculates predicted values in the observed metric with other predictors' reference values set to their means and superimposes main effects model predictions, and the pltopts( ) option formats the *y*-axis.

the slope of the conflict main effects. Among the interaction effect prediction lines, the opposite is true—the higher the SEI, the greater the slope of the conflict effect. Thus, the moderation of conflict's effect by SEI reversed the pattern created by the nonlinearity of the link function, which is why I argue that the interaction effects display by itself underestimates how much SEI moderates work–family conflict.

Example 4: Three-Way Interval-by-Interval-by-Nominal Interaction

This final example is a negative binomial regression predicting the number of voluntary association memberships from the interaction of age by education by sex, perceived social class, and population size of the place where a respondent lives. Age and education are both interval-level indicators measured in years (with mean = 43.9 and 12.6, standard deviation = 17.1 and 3.1, respectively). Sex is a dichotomous indicator (1 = *female*, 0 = *male*), with 57.5% of the cases coded as female. To illustrate the interpretation of this three-way interaction, I treat age as the focal variable, education as the first moderator, and sex as the second moderator. Specifically, this means that for each option, I create two visual displays of the effect of age moderated by education, one for women and one for men.

Effect Display Option. Figure 5.18 makes it obvious that education moderates the effect of age quite differently for women than for men. For women, the effect of age is positive and significant at lower years of education and declines in size as education increases—older women are expected to have more memberships than younger women, but this difference is smaller for more highly educated women. The age effect stays positive and significant until about 12 years of education and changes to an increasingly negative effect around 15 years of education—older women have fewer memberships than younger women. But this negative effect is never significant. In contrast, the effect of age for men is always positive and increases in magnitude as education rises—older men are predicted to have more memberships than younger

FIGURE 5.18 ● EFFECT DISPLAY, THREE-WAY INTERACTION

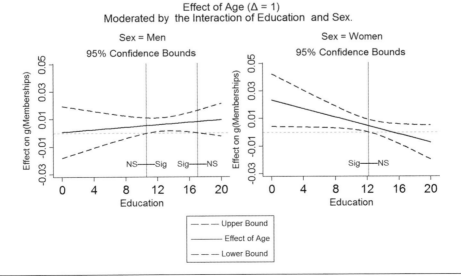

ICALC command: effdisp, plot(type(cbound)) ndig(2) pltopts(ylab(-.03(.02).05, labsize(*.8))). The plot( ) option specifies a confidence bounds plot, and the ndig( ) and pltopts( ) options format the *y*-axis labels.

men, and the gap grows as education increases. But the age effect is significant for men only when education is between 11 and 17 years.

In sum, education moderates the effect of age in opposite directions for men and women—a positive effect that slowly increases with education for men versus an initial positive effect that more rapidly decreases with education for women and then becomes a negative effect. To anticipate what the other options show, when the pattern of moderation is as dramatically different as this, all of the options should readily portray this. In addition, a standard graphic in the μ-metric should also accurately reveal the basic patterns of the relationship between the outcome and the interacting predictors when the differences are so apparent.

Dual-Axis Option. I suspect that most readers find the predicted values plot in Figure 5.19 much easier to use to understand the age–membership relationship and how it is moderated by education and sex. For both men and women, the age–ln(membership) prediction lines are vertically ordered by education—the line for the lowest education is at the bottom, and the line for the highest education is at the top. But the pattern of these lines is quite distinct for women and men:

- For women, the vertical distance between the age–ln(membership) lines is greatest at the youngest age and diminishes as age increases. The slope of the age–ln(membership) lines changes from positive and relatively steep for women with the fewest years of education (0.0213) to negative and much less steep for those with the most years of education (−0.0077). Thus, older women are predicted to have more memberships than younger women when their education levels are low. But this difference diminishes and then reverses direction as education increases: Among the most highly educated, younger women are expected to belong to more organizations than older women.

- For men, the gap between the age–ln(membership) lines is smallest at the youngest age and slowly widens as age increases, consistently indicating that older men are expected to have more memberships than younger men. The predicted age–ln(membership) lines change from a shallow positive slope for the least educated men (0.0026) to a relatively steeper positive slope for the most educated men (0.0082). In contrast to women, older men are always expected to have more memberships than younger men, and this disparity increases as education rises.

For those GLMs for which you can estimate the standard deviation of the modeled outcome (e.g., this negative binomial regression), you can get a better sense of the magnitude of these differences by translating them into standard deviation changes in the modeled outcome across the range of the focal variable. The labels on the left-hand y-axis are scaled in standard deviations of log number of memberships, from 2.3 standard deviations below the mean to 1 standard deviation above. For women with the least education (*solid line*), log membership increases by 1.5 standard deviations from age 18 to 88 years, and for those with the most education (*long-dash-dot line*), log membership decreases by 0.5 standard deviation. For men, the age effects are increases in log membership by 0.2 standard deviation for the least educated and by 0.6 for the most educated.

Alternatively, you can assess how big the differences in the effect of age are by translating them into the number of memberships metric on the right-hand y-axis. For the least educated women (*solid line*), the predicted change from age 18 to 88 years

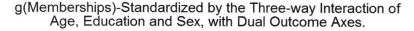

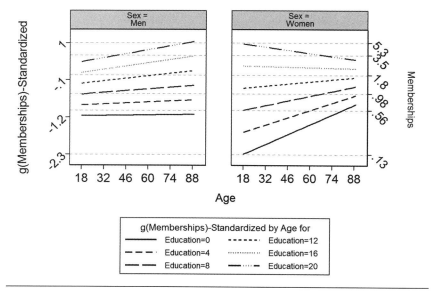

FIGURE 5.19 ● PREDICTED MODELED OUTCOME WITH DUAL AXES, THREE-WAY INTERACTION

ICALC command: outdisp , plot(type(scat)) out(metric(model) sdy atopt((means) _all) dual) tab(def). The plot( ) option specifies a scatterplot, the out( ) option calculates predicted values in the model metric using other predictors' reference values set to their means and creates dual *y*-axis labeling, and the tab( ) option creates a default table of predicted values in the model metric.

is an increase of 0.58 membership, while for the most educated women (*long-dash-dot line*), it is a decrease of 2.42 memberships. For men, the predicted change from age 18 to 88 years is an increase of 0.10 membership for the least educated (*solid line*) and an increase of 2.56 memberships for the most highly educated (*long-dash-dot line*). Notice that for women—but not for men—the relative magnitude of the changes in the number of memberships metric (largest for the most educated and smallest for the least educated) is the opposite of the relative size of the changes in the ln(membership) metric. This reversal for women—but not for men—illustrates the confounding of the two sources of nonlinearity and how it can problematize the interpretation of the relationships in the μ-metric. (*Aside:* I did not use the value labels on Figure 5.19 for the numbers reported for the change in memberships. Rather, they come from the corresponding modeled or observed outcome predicted values table.)

We can also extract information about the moderation of the education effect by age from these plots. The plotted lines are visually sorted such that the line for those with the least education is at the bottom of the scatterplot and each successive line is for the next higher level of education. This ordering indicates that, at any age, the log number of memberships is expected to be higher for those with more education. How the size of the gap between the lines changes as age increases reveals how age moderates the effect of education. For women, these plots show a positive effect of education on the log number of memberships. This effect diminishes at older ages, as shown by the narrowing of the gaps between the lines as age increases. For men, we similarly see a positive effect of education from the vertical ordering of the lines. But the opposite pattern of moderation by age is seen: The effect of education increases at older ages.

Superimposed Main Effects Option. Figure 5.20 shows the relationship of age and the predicted number of memberships from the interaction specification with *black lines* and the relationship based on the main effects model with *light gray lines* in the background. For men, the scatterplot makes apparent the same pattern of results for this relationship that we saw from the other two options. The age–membership lines are vertically ordered by education, with the line for the least educated (*solid line*) at the bottom and moving progressively upward with the education level to the line at the top for the most educated (*long dash, two shorter dashes*). The slopes of the age–membership lines are similarly ordered by education from shallowest to steepest as education increases. Thus, at older ages, the predicted number of memberships is higher than at younger ages, and this difference is larger at higher levels of education. But also notice that much of this reflects the nonlinearity of the negative binomial link function—the same general pattern is also evident for the main effects model prediction lines. Relative to the main effects model prediction lines, the interaction model's prediction lines become progressively steeper as education levels rise. Thus, for men, education moderates the age effect by enhancing the main effects' pattern.

In contrast, for women, it is hard to fully reconcile the results shown in Figure 5.20 with those portrayed by the dual-axis option (Figure 5.19). Figure 5.20 does indicate

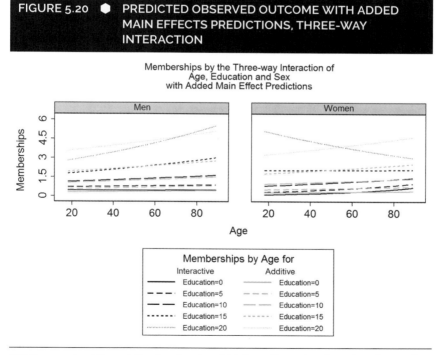

FIGURE 5.20 ● PREDICTED OBSERVED OUTCOME WITH ADDED MAIN EFFECTS PREDICTIONS, THREE-WAY INTERACTION

ICALC command: outdisp , plot(type(scat)) out(metric(obs) atopt((means) _all) main(main)) pltopts(ylab (0(1.5)6)). The plot( ) option specifies a scatterplot, the out( ) option calculates predicted values in the observed metric with other predictors' reference values set to their means and superimposes main effects model predictions, and the pltopts( ) option customizes the *y*-axis labels.

that the age effect changes from a positive effect on membership at lower levels of education to a negative effect at the upper end of the education distribution. But the relative magnitudes of the age effect across years of education appear quite different in the two figures. In Figure 5.20, the steepest slope is for women with 20 years of education, and this visually dominates the magnitude of the age–membership relationship at all the lower levels of education. Moreover, the size of the age effect portrayed by the other predicted outcome lines appears relatively similar. This is the opposite of what Figure 5.19 shows, in which the steepest slope is for the least educated women. And the next two steepest slopes are for women with 20 and 5 years of education, respectively, which are relatively similar in the magnitude of the age effect.

Thus, while the overall pattern of the relationship between membership and age as moderated by education for women is clear in both figures, they provide very different impressions of the extent of the moderation. Comparing the main effects and the interaction models' predictions (*gray* and *black lines*, respectively) in Figure 5.20 provides some insight into the reason for this. The main effects model's predictions indicate that when education level is high the age–membership prediction line is calculated in a region of the negative binomial prediction function where a given change in the index function predicts larger outcome changes than the same index function change would predict when education is low. This is shown by the increasing gaps between the gray prediction lines as education level increases. The result is to minimize the differences between the age–membership prediction lines in the observed metric from the interaction model at low education and to heighten them when education is high. Consequently, the plot in the μ-metric suggests modest moderation of the age effect when education is at lower levels and much greater moderation at the highest levels of education. In contrast, the plot in the η-metric reveals greater moderation at both ends of the education distribution.

With so much of the focus in this chapter on the concerns with the confounding of the nonlinearity of interactions with GLM link functions and on the nuances of interpreting the moderated effects properly, it is important not to lose sight of what is often of at least equal interest—describing the bigger picture of the expressed relationship between the interacting predictors and the observed outcome. Presenting and discussing either a plot or a table of the predicted outcome varying across the predictors' display values—like Figure 5.20 (without the superimposed main effects predictions) or Table 5.2—is usually the most effective method for this purpose. And you can draw on your analyses of the patterns of moderated effects to help you explain the overall story. For this example, I would draw the audience's attention to three points:

1. The model predicts that women have fewer memberships than men across most combinations of age and education. This difference is greatest when education and age are both in the lower or the upper parts of their distributions and diminishes when both are in the middle range. And it reverses to more predicted memberships for women than for men when age and education are mismatched at the opposite ends of their distributions.

2. Similarly, for most permutations of education and sex, increasing age predicts a higher number of memberships. For men, the age effect remains positive and uniformly rises across education levels. But for women, the age effect increases slightly with education before declining. It turns increasingly negative at 15 years of education and beyond.

TABLE 5.2 ● PREDICTED OUTCOME VALUES, OBSERVED METRIC, THREE-WAY INTERACTION

Memnum by the Three-Way Interaction of Age, Education, and Sex

Sex	Age	Education					
		0	4	8	12	16	20
Men	18	0.471	0.674	0.963	1.376	1.967	2.812
	32	0.476	0.697	1.021	1.496	2.192	3.212
	46	0.481	0.722	1.084	1.627	2.443	3.668
	60	0.485	0.747	1.150	1.769	2.722	4.189
	74	0.490	0.773	1.219	1.923	3.033	4.784
	88	0.495	0.800	1.294	2.091	3.380	5.463
Women	18	0.133	0.275	0.571	1.184	2.456	5.097
	32	0.184	0.350	0.666	1.268	2.415	4.598
	46	0.255	0.445	0.778	1.359	2.374	4.148
	60	0.353	0.566	0.908	1.455	2.334	3.742
	74	0.489	0.720	1.059	1.559	2.294	3.375
	88	0.678	0.916	1.237	1.670	2.255	3.045

Note: Memnum = number of memberships.

ICALC command: outdisp, tab(save(Pred_Memnum.xlsx)) out(metric(obs) atopt((means) _all)) ndig(3). The tab( ) option saves a formatted predicted values table to an Excel file, the out( ) option specifies the calculation of predicted values in the observed metric with other predictors' reference values set to their means, and ndig( ) sets the cell format to 3 decimals.

3. Unlike sex or age, education always has a positive relationship with the number of memberships. But the predicted increase in membership grows larger with age for men but grows smaller with age for women.

SUMMARY AND RECOMMENDATIONS

In this chapter, I argued that care is needed to interpret visual displays of the predicted outcome as it varies with a set of interacting variables when the results are from GLMs with nonlinear link functions. The usual approach of interpreting bar charts and scatterplots scaled in the outcome's observed (μ) metric runs the risk of misinterpretation of the interaction effect for two reasons: (1) the nonlinearity of the link function can create the appearance of an interaction effect when there is none and (2) we lack a standard against which to visually judge the existence and degree of the interaction effect. In models with linear link functions, we have an implicit visual standard against which to assess the extent of interaction; for example, parallel straight lines in a scatterplot would indicate no interaction. Without such a standard, it is easy to misinterpret the degree to which the focal variable is moderated by the other predictors with which it interacts. I stressed the important caveat that visual displays in the μ-metric accurately portray the pattern of the relationship between the outcome and the interacting variables but can be prone to misrepresentation of the degree of modeled moderation.

I suggested and illustrated with examples three options for helping avoid such mis-interpretations: (1) effect displays in the η-metric, discussed in Chapter 3; (2) bar charts and scatterplots with dual y-axes, one in the η-metric used to plot results along the vertical dimension and one showing μ-metric labels with between-label distances scaled in the η-metric; and (3) bar charts and scatterplots scaled in the observed μ-metric with a predicted outcome plot from a main effects model superimposed in the background. As the four examples demonstrated, none of these options is always the best choice, but I do have recommendations concerning which ones may work best in different circumstances:

1. I suggest that you always start by creating an effect display for your own initial use in understanding the pattern of the interaction specification. Laying that foundation of understanding is useful for helping you understand and interpret the results from the other options. Although I find this option very useful for my own understanding, many researchers (and readers) are conditioned to expect visual displays to show outcome values, not effects, and find effect displays harder to comprehend.

2. For a nominal-by-nominal interaction, the dual-axis option is often not useful because the x-axis is categorical, especially when there are more than two categories on the x-axis. Bar charts are somewhat less susceptible to the linear perception bias because we interpret the effects as differences between qualitative categories rather than as differences per unit of the x-axis predictor. Depending on the link function relating the μ- and η-metrics, dual-axis bar charts can be somewhat confusing, as they were in Example 1. This example had a log link function, and the values of the predicted outcome in the μ-metric were all less than 0. Consequently, scaling the predicted outcome values in the log metric created a bar chart in which the longer bars represented smaller outcome values.

3. There is one circumstance in which a dual-axis display will usually be a clearly superior alternative to a superimposed effects display—namely, when the pattern created by the confounding induced by a nonlinear link function is counter to the pattern produced from the modeled interaction specification. What I mean by this is that the visual display produced from the interaction specification is not simply reinforcing the pattern produced by a main effects only model but also changing the nature of the pattern altogether. In this situation, I would strongly recommend using a dual-axis display.

    Example 4 provides a demonstration of what I mean by both situations. Men and women had the same main effects pattern—a positive slope for the membership–age relationship that was steeper at higher levels of education. For men, the modeled interaction specification exaggerated this pattern because the results specified a larger age slope as education increased. Consequently, the pattern of moderation portrayed in the dual-axis and the superimposed effects plots were similar overall. For women, the interaction specification resulted in a pattern that was qualitatively as well as numerically different from the main effects pattern—the slope for the membership–age relationship was smaller at higher levels of education and changed sign from positive to negative. As a result, the pattern of moderation appeared different in the dual-axis versus the superimposed effects plots. If I were choosing a visual display only for men, then either option would be viable. But for women, I would not use the superimposed effects display given the nuanced differences (described above) in what it suggests about the moderation of the age effect by education.

But my bottom-line recommendation is to try all three options in any given analysis to see which one works most effectively to help you tell a story about the interaction effect.

The next chapter begins Part II, turning from principles to applications. Chapter 6 is a reference guide for the ICALC Stata toolkit commands. In it, I define and describe each command and its options, present brief examples of the output produced by each command with annotations, and offer guidance on how to customize the output (especially graphs). The subsequent chapters provide detailed application examples for a number of the most commonly used GLM models: OLS regression, binomial logistic regression and probit analysis, MNLR, ordinal regression models, and Poisson and negative binomial regressions (including zero-inflated models). The concluding chapter briefly considers how these tools can be applied to other models (e.g., Tobit models and survival analyses) and to the interaction between a linear predictor and a predictor specified as a nonlinear functional form (e.g., a quadratic).

DERIVATIONS AND CALCULATIONS

Equation 5.13 for Slope of Logistic Prediction Function, Main Effects Model

Applying Equation 5.12 to the main effects model prediction function for logistic regression yields a simple formula for the slope of $P^* = Prob\,(Y = 1|x$ from main effects model) with respect to F (Long & Freese, 2014, p. 240):

$$\frac{\partial P^*}{\partial F} = b_1^* \times P^* \times \left(1 - P^*\right)$$

where for the main effects model

$$P^* = \frac{1}{1 + e^{-\left(\tilde{a}^* + b_2^* d + b_1^* F\right)}}$$

For the empirical example, $\tilde{a}^* = -0.79703$, $b_1^* = -0.21547$, and $b_2^* = 2.18301$. d indexes the display values for race (0 = *Black* and 1 = *White*), Ed is the focal variable

F, and P^* is the probability of approval of a legislative ban. Consequently, the calculating formulas for the slope of the probability of approval with respect to education for Whites and Blacks are

$$\frac{\partial P^*_{White}}{\partial Ed} = -0.21547 \times P^*_{White} \times \left(1 - P^*_{White}\right)$$

where

$$P^*_{White} = \frac{1}{1 + e^{-\left(-0.79703 + 2.18301 \times 1 - .21547 \times Ed\right)}} = \frac{1}{1 + e^{-\left(1.38597 - 0.21547 \times Ed\right)}}$$

and

$$\frac{\partial P^*_{Black}}{\partial Ed} = -0.21547 \times P^*_{Black} \times \left(1 - P^*_{Black}\right)$$

where

$$P^*_{Black} = \frac{1}{1 + e^{-\left(-0.79703 + 2.18301 \times 0 - 0.21547 \times Ed\right)}} = \frac{1}{1 + e^{-\left(-0.79703 - 0.21547 \times Ed\right)}}$$

Equation 5.14 for Slope of Logistic Prediction Function, Interaction Model

For the interaction effects model, starting from Equation 5.11, we now use the slope of $P = Prob(Y = 1 \mid x$ from interaction effects model$)$ and thus,

$$\frac{\partial P}{\partial F} = \left(b_1 + b_3 d\right) \times P \times \left(1 - P\right)$$

and for the interaction effects model

$$P = \frac{1}{1 + e^{-\left(\tilde{a} + b_2 d + \{b_1 + b_3 d\}F\right)}}$$

For the empirical example, $\tilde{a} = 2.08519$, $b_1 = -0.08582$, $b_2 = 3.85933$, and $b_3 = -0.16304$. Now the calculating formulas for the slope of the probability of approval with respect to education for Whites and Blacks are

$$\frac{\partial P_{White}}{\partial Ed} = -0.24886 \times P_{White} \times \left(1 - P_{White}\right)$$

where

$$P_{White} = \frac{1}{1 + e^{-\left(-2.08519 + 3.85933 \times 1 + \left(-0.08582 - 0.16304 \times 1\right) \times Ed\right)}} = \frac{1}{1 + e^{-\left(1.77413 - 0.24886 \times Ed\right)}}$$

and

$$\frac{\partial P_{Black}}{\partial Ed} = -0.08582 \times P_{Black} \times \left(1 - P_{Black}\right)$$

where

$$P_{Black} = \frac{1}{1 + e^{-\left(-2.08519 + 3.85933 \times 0 + \left(-0.08582 - 0.16304 \times 0\right) \times Ed\right)}} = \frac{1}{1 + e^{-\left(-2.08519 - 0.08582 \times Ed\right)}}$$

CHAPTER 5 NOTES

1. By the term *parallel curves*, I mean that each curve is vertically offset from the other curves by a constant amount.
2. You might be wondering if the semblance of an interaction is due to misspecifying the main effects model by excluding a needed interaction effect. But the results in Panel D are from an analysis in which there is no significant interaction between sex and political views and yet there is still the suggestion of an interaction effect.
3. A comparable logic can be applied to show the confounding when using the "as-observed" option to specify the reference values for the other predictors.
4. There can be instances in which the gaps at a specific value of F will be equal. For example, in a binomial logistic analysis, if there is a value of F at which $\hat{\mu}_0 = 0.5$, the gaps will be equal.
5. A function $f(x)$ is convex if its second derivative is always positive across the domain of x and concave if its second derivative is always negative across the domain of x.

APPLICATIONS

ICALC **TOOLKIT**

Syntax, Options, and Examples

OVERVIEW

The ICALC toolkit for Stata consists of a setup tool plus four separate tools for understanding the focal variable's effect on the outcome as moderated by its interacting variables. Instructions for downloading and examples are available at www .icalcrlk.com. Almost all the calculations, tables, and visual displays in Chapters 1 to 5 were created using these tools. After overviewing these tools, I describe the syntax and options for using ICALC in Stata, provide examples of specifying the commands, and briefly discuss the output they produce. The tools are listed in the order in which they would typically be applied in an analysis. Options are available to save into an Excel file the data represented in tables or used to produce plots so that you can use other software to produce or edit tables and graphs. Subsequent chapters cover in-depth application examples for different estimation techniques.

INTSPEC **Tool: Interaction Specification**

This setup tool must be run before you can use the other tools. You must specify the elements of the interaction model, identifying the main effect variables, the two-way effect terms, and the three-way effect terms (if any). You may choose a display name and/or set display values for each of the interacting variables with custom options.

GFI **Tool: Gather, Factor, and Inspect**

This tool defines and works with the formula describing the structure of the focal variable's interaction effect on the "modeled" outcome.[1] As discussed earlier, this formula is the foundation for understanding interaction effects. The GFI tool does the following:

1. Produce the algebraic expression for the effect of the focal variable on the modeled outcome and how it changes with the moderators. If the focal variable is a nominal measure consisting of C categories, it defines $C - 1$ formulas, one for each of the dummy variables representing the focal

variable. It optionally reports the algebraic expression as a factor change when applicable (see Long & Freese, 2014, p. 179).

2. Construct a table describing if and when the focal variable's effect changes sign (positive to negative or negative to positive) as F's effect varies with the values of the moderators.

3. Draw a path diagram–like graphic showing the structure of the interaction effects of the focal and moderating variables—that is, a visual representation of the algebraic expressions.

SIGREG **Tool: Significance Regions**

This produces information about the significance region for the focal variable's effect. You can explore whether the effect of the focal variable on the modeled outcome remains significant across the range of values of each moderator or combinations of values across all the moderators. By default, it analyzes the effects on the modeled outcome using the estimated coefficients. For empirically derived significance regions, it can alternatively analyze the effects on the observed outcome, factor changes (if applicable), or any of the SPOST13 marginal effects calculated by its *mchange* command (see Long & Freese, 2014, pp. 166–171). Specifically, the SIGREG tool will do the following:

1. Find, if possible, a mathematical expression of the conditions under which the significance of the focal variable's effect changes or not. This is defined by boundary values in the distribution of the moderator(s)—the exact points in the range of a moderator's values at which the focal variable's effect changes from not significant to significant or vice versa. It reports boundary values in tables and/or graphs.

2. An empirically derived definition of the significance region can be presented in tabular form. The table's columns and/or rows correspond to values of the moderator(s). The cells of the table report three pieces of information about the focal variable's effect on the modeled outcome conditioned on the moderator's values: its numeric magnitude, sign, and statistical significance. You may instead request significance regions for effects on the observed outcome as factor changes (when applicable) or as any of the SPOST13 marginal effects. The regions of significance versus nonsignificance and sign changes are visually highlighted by varying fonts and added symbols. There is an option to save the formatted table to an Excel file with additional formatted highlighting.

EFFDISP **Tool: Graphic Displays of the Moderated Effect**

These graphics are the visual counterparts to the significance region tables created by SIGREG, providing information about both the varying magnitude and the significance of the focal variable's moderated effect. Multiple plots are produced if there is more than a single moderator or if the focal variable is categorical with more than two categories. You can plot the effects on either the observed or modeled outcome. You can construct the following:

1. *Line plots of the effect with or without confidence bounds for interval moderators:* The *y*-axis represents the size of the effect and the *x*-axis indexes the moderator. It plots the moderated effect and can add lines to show its

upper and lower confidence bounds. The plot shows how the focal variable's effect changes with the moderator and the optional bounds show if the conditional effect is significant at any given value of the moderator.

2. *Drop-line or error bar plots for a categorical moderator:* The categories of the moderator are shown on the *x*-axis with a marker above the axis indicating the value of the moderated effect on the *y*-axis. An error bar chart also draws lines up and down from the marker to show the confidence interval for the focal variable's effect.

3. A contour plot can be drawn if the interaction specification includes two or more interval-level moderators. The contour plot is a 2D representation of the 3D relationship of the focal variable's effect as it varies with the two moderators. The 2D plane indexed by the moderators is divided into areas defined by ranges of values of the moderated effect. Each area (range) has a different shaded color. Darker shades represent larger (more positive) effect values and lighter shades represent smaller (more negative) effect values. Area significance markers can also be added.

OUTDISP **Tool: Display of a Predicted Outcome by the Interacting Variables**

This tool creates tables and graphs that display predictions of either the observed or the modeled outcome. The modeled metric can be scaled in its natural metric or in standard deviation units (when applicable). The visual displays can be bar charts, scatterplots, or contour plots. As discussed in Chapter 5, such visual displays for nonlinear link models confound the nonlinearity of the link function with the nonlinearity of the interaction effect. This tool provides two options for dealing with that issue:

1. If you predict the observed outcome, you have the option to also display predictions from a main effects model. Such main effects model predictions typically appear nonlinear due to the link function. Comparing the pattern of the main effects model's predictions with those of the interaction effects model can help isolate the interaction effect from the effects of the nonlinearity of the model. This option is available for bar charts and scatterplots but not for contour plots.

2. If you use the modeled outcome, the visual display shows only the nonlinearity of the interaction model. But the modeled outcome's metric is often not intuitive. For many common GLMs, you can scale the modeled outcome in standard deviation units to provide a more meaningful metric. You can also show two outcome axes, one labeled according to the modeled outcome metric and one according to the observed outcome metric to provide some context. The option to label a second axis in the observed metric is not available (or possible) for multinomial logit models.

INTSPEC: SYNTAX AND OPTIONS

intspec , **focal**(*varname*) **main**((*varlist1, suboptions*) (*varlist2, suboptions*) ...)
  int2vars(*varlist*) [**int3vars**(*varlist*) dv**name**(*string*) **eqname**(*string*)
  ndigits(#) **nrange**(#) **abbrevn**(#) **sumwgt**(*string*)]

INTSPEC **Syntax and Options**	**Description**
focal(varname)	Name of the focal variable
main((varlist, suboptions) repeat)	Set info for each focal and moderator main effect variable
varlist	varname, if interval variable
	fvname, if factor variable
	varlist, if a set of dummy variables
	Relative order determines moderator numbering; moderator#1 is first moderator listed; moderator#2 is second moderator listed; focal set can be in any order
suboptions, name(string)	Display name for the variable; default is first varname in varlist, abbreviated to 12 characters
range(numlist)	Set display values to label/define tables and graphs for interval variable with numlist. Max no. of digits in numlist entry defines no. of digits for display values
	Or
	minmax, min to max in increments = (max-min)/nrange
	meanpm1, mean ± 1 *SD*
	meanpm2, mean ± 1 *SD*, ± 2 *SD*
	meanpm1mm, min, mean ± 1 *SD*, max
	meanpm2mm, min, mean ± 1 *SD* and ± 2 *SD*, max
	Default is range(minmax) with default nrange = 5
	For a single dummy variable can specify range(0/1)
	Not needed for i.varname; ignored if included
int2vars(varlist)	List the two-way interaction terms; must be ordered as focal-by-moderator#1 focal-by-moderator#2; if a three-way interaction also moderator#1-by-moderator#2
int3vars(varlist)	Lists the three-way interaction terms if any; must be ordered as focal-by-moderator#1-by-moderator#2
dvname(string)	Display name for dependent variable. Default is varname of dependent variable
eqname(string)	Use for multiequation models to specify which equation's coefficients are analyzed
ndigits(#)	Integer no. of digits used in default display value labels if range( ) not specified. Default ndig(2)
nrange(#)	nrange+1 = default no. of increments in range(minmax)
abbrevn(#)	Specifies the character length used to abbreviate names; must be an integer; default is abbrevn(12)
sumwgt(string)	**No** weights specified on estimation command not used to calculate summary statistics

One-Moderator Example

Consider the negative binomial regression analysis given by

nbreg pmhdays wfconflict sei wfcbysei age educ childs female if wrkstat < = 4

This predicts the number of poor mental health days reported by a worker in the last month (*pmhdays*) by the effects of job status (*sei*), work–family conflict (*wfconflict*), and their interaction (*wfcbysei = wfconflict × sei*), as well as age, education, number of children, and sex.

Suppose the focus is on how work–family conflict affects poor mental health days and how conflict's effect varies with job status. The focal variable is *wfconflict*, so I specify

    focal(wfconflict)

followed by information about both main effects—*wfconflict* and *sei*—using the main option:

main((sei , name(Job_Status) range(17(20)97))
        (wfconflict, name(W_F_Conflict) range(1/4)))

The name( ) specification may not include spaces, so note the use of an underscore (_) instead to define display names for the variables. The range( ) suboption defines display values (e.g., axis labels) for each variable. *sei* will have labels of 17, 37, 57, 77, and 97, while *wfconflict*'s display values are 1, 2, 3, and 4.

Last, I identify the two-way interaction term in the *nbreg* analysis:

    int2(wfcbysei)

You can also set a display name for your dependent variable, the abbreviation length for alphanumeric labels, and the number of digits in value labels. For instance, to set the outcome variable's display name, 13-character alphanumeric labels, and whole number values (zero digits), the following is used:

    dvname(Poor_M_H_Days) abbrevn(13) ndig(0)

Putting this all together, the syntax for this example would be

intspec focal(wfconflict)   ///
    main((sei , name(Job_Status) range(17(20)97)) ///
            (wfconflict, name(W_F_Conflict) range(1/4))) ///
    int2(wfcbysei) dvname(Poor_M_H_Days) abbrevn(13) ndig(0)

ICALC echoes back how it interprets your specification of the components of the interaction:

```
Interaction Effects on Poor M H Days Specified as

    Main effect terms: wfconflict  sei
    Two-way interaction terms: wfcbysei
```

```
These will be treated as: Focal variable = wfconflict ("W_F_Conflict")
   moderated by interaction(s) with
      sei ("Job_Status")
```

Typically, you would also probe how the effect of job status is moderated by work–family conflict. The only change needed to do so is to change focal(wfconflict) to focal(sei).

Suppose I instead used factor-variable notation to specify the interaction:

nbreg pmhdays c.wfconflict##c.sei age educ childs female if wrkstat < =4

Then, the intspec syntax would use *c.varname* in place of *varname* and use *c.focalvar#c .moderator* for the interaction variable (see Stata help for factor variable for details on specifying factor variables and the advantages they offer for properly treating categorical predictors and interaction effects in postestimation calculations.):

intspec , focal(c.wfconflict)   ///
     main((c.sei , name(Job_Status) range(17(20)97)]   ///
     (c.wfconflict, name(W_F_Conflict) range(1/4))) ///
int2(c.wfconflict#c.sei) dvname(Poor_M_H_Days) abbrevn(13) ndig(0)

Do not specify the two-way interaction in the int2( ) option with two # symbols—c.wfconflict##c.sei. This would include extra terms because it expands to

c.wfconflict c.sei c.wfconflict#c.sei

But you *do* need to use ## in the estimation model syntax as above; otherwise, it will neither include nor estimate the main effects coefficients, but only a coefficient for the interaction term.

The only required elements in the INTSPEC syntax are the variable names for the focal and moderator variables and their interactions. So you can simplify what you have to specify by using the default specifications for any other options or suboptions. For example,

intspec , focal(c.wfconflict) main((c.sei) (c.wfconflict)) int2(c.wfconflict#c.sei)

Two-Moderator Example

Consider an OLS regression of number of children (*childs*) in which family income (*faminc10k*) interacts with education (*educ*) and birth cohort (*cohort*), but there is neither a three-way interaction of income, education, and cohort nor a two-way interaction of cohort and education:

reg childs c.faminc10k##i.cohort c.faminc10k##c.educ sibs i.race religintens
   if age>39

As discussed in Chapter 4, you could separately explore how family income (scaled in $10K) is moderated by education from how family income is moderated by cohort following the syntax used in Example 1, once for each moderating variable. For instance, for education moderating income,

```
intspec , focal(c.faminc10k)  ///
        main( (c.educ,  name(Education) range(0(2)20))  ///
               (c.faminc10k,  name(Family_Inc) range(0(2)24)))  ///
        int2(c.faminc10k#c.educ) ndig(0)
```

More typically, you would explore the concurrent moderation of family income by education and cohort. To do so, add the highlighted specifications for cohort as moderator#2 in the main( ) and in the int2( ) options:

```
intspec , focal(c.faminc10k)  ///
     main( (c.educ,  name(Education) range(0(2)20))  ///
            (c.faminc10k,  name(Family_Inc) range(0(2)24))  ///
            (i.cohort,  name(Cohort)) )  ///
        int2(c.faminc10k#c.educ c.faminc10k#i.cohort)
```

ICALC interprets this as follows:

```
Interaction Effects on childs Specified as

       Main effect terms: faminc10k educ 2.cohort 3.cohort 4.cohort
       Two-way interaction terms: c.faminc10k#c.educ  i.cohort#c.
faminc10k

    These will be treated as: Focal variable = faminc10k ("Family_Inc")
       moderated by interaction(s) with
           educ ("Education")
           2.cohort 3.cohort 4.cohort ("Cohort")
```

Notice the ordering of the names in the main( ) and int2( ) options. Because c.educ is listed before i.cohort in main( ), c.educ is the first moderator and i.cohort is the second moderator. For int2( ), you must list focalvar-by-moderator1 first, followed by focalvar-by-moderator2. That is, the interaction of family income and education is listed first, and the interaction of family income and cohort is listed second. Note that because I did not specify the dvname( ) option, the variable name for the outcome, childs, is used in the header.

Three-Way Interaction Example

A three-way interaction between education (*educ*), age (*age*), and sex (*female*), predicting the number of voluntary associations to which a respondent belongs, is estimated from a negative binomial regression:

```
nbreg memnum c.age##c.ed##i.female size class
```

I need to specify the focal variable, include all three predictors in the main( ) option, include all three two-way interaction variables in the int2( ) option, and add the int3( ) option:

```
intspec focal(i.female) ///
        main( (c.age , name(Age) range(18(10)88))     ///
              (c.ed , name(Education) range(0(4)20))     ///
              (i.female , name(Sex) range(0/1)))     ///
          int2( c.age#i.female c.ed#i.female c.age#c.ed ) ///
          int3(i.female#c.age#c.ed) dvname(Memberships) ndig(0)
```

ICALC interprets this as follows:

```
Interaction Effects on Memberships Specified as

    Main effect terms: 1.female  age ed
    Two-way interaction terms: i.female#c.age i.female#c.ed
c.age#c.ed
    Three-way interaction terms: i.female#c.age#c.ed

  These will be treated as: Focal variable = 1.female ("Sex")
    moderated by interaction(s) with
        age ("Age")
        ed ("Education")
    and with "Age" x "Education"
```

For an initial run, you could simplify this by relying on defaults for the main( ) option and just specify the names of the focal and moderator variables and their interactions:

intspec, focal(i.female) main((c.age) (c.ed) (i.female)) ///
int2(c.age#i.female c.ed#i.female c.age#c.ed) int3(i.female#c.age#c.ed) nd(0)

Several Important Details in Specifying the Interaction

1. The order of variables is important when you specify the variables in int2vars( ) or in int3vars( ). It has to be *focalvars_by_moderator1*, then *focalvars_by_moderator2*, and so on.

    a. Moderators are numbered according to the relative order in which you list them in the main( ) option: The first listed is moderator#1, the second is moderator#2, and so on.

    b. It doesn't matter where you list the focal variable inside main( ) because ICALC knows which one is the focal variable.

2. If using factor variables,

    a. when you specify the variables for int2vars( ) or int3vars( ), use only a single #, not a double ##. For instance, if the model you ran was *regress outcome focvar##modvar* . . . , then you would specify *int2vars(focvar#modvar)*. [If you specify ## it expands to include the main effects term, which is not what you want!]

    b. for a nominal factor variable, if you specify a base category other than the first, such as *ib3.myvar* setting the base value to 3, you must include this every place you list *myvar*:

        focal(ib3.myvar) main((ib3.myvar, ...)) int2(ib3.myvar#c.mymod)
            int3(ib3.myvar #c.mymod#i.myothermod)

3. If you have a three-way interaction, don't forget to specify the *modvar1_by_modvar2* interaction in int2vars( ). (This is one I sometimes forget too.)

4. The ordering issue applies to only what you specify for the ICALC syntax, *not* what you specify in the model you estimated, with one exception. You can switch the roles of the focal and moderator variables without reestimating your model. Suppose you have a two-way interaction of

var1 × *var2*. You can first treat *var2* as the focal variable and then treat *var1* as the focal variable in different applications of a tool:

```
logit outcome var1##var2
intspec focal(var2) main( (var2, ...) (var1 , ...) ) int2vars(var2#var1)
        gfi , ...
intspec focal(var1) main( (var2, ...) (var1 , ...) ) int2vars(var1#var2)
        gfi , ...
```

The exception is if you use the spost( ) suboption for the SIGREG tool, then you must list variables in the same order in the estimation model, in main( ), in int2( ), and in int3( ).

GFI TOOL: SYNTAX AND OPTIONS

<u>gfi</u>, [<u>factorchg</u> <u>ndigits</u>(#) <u>path</u>(type , <u>title</u>(*string*) <u>name</u>(*string*) <u>boxwidth</u>(#) <u>ygap</u>(#) <u>xgap</u>(#) <u>ndigits</u>(#))]

GFI **Tool Options**	Description
factorchg	Algebraic expression for the moderated effect of focal variable also shown as a factor change if this keyword is specified
ndigits(#)	Number of digits after the decimal for effects and coefficients in the algebraic expression and sign change table. Default = **4**
path(type, suboptions)	Create path diagram of structure of interaction effect
type	focal or all. focal shows only coefficient values for variables involving the focal variable. all shows all coefficient values. Default is focal
suboptions title(string)	Title for path diagram
name(string)	Stores as memory graph with name given by string
ndigits(#)	No. of digits used to report coefficient values in diagram
boxwidth(#) , ygap(#) , xgap(#)	Used to fine-tune graph. boxwidth sets width of boxes; ygap sets vertical distance between boxes; xgap sets horizontal distance between boxes. Defaults are boxwidth(1.25), ygap(.625), xgap(1.25)
How moderator range( ) specification from *intspec* is used	Defines points at which moderated effect is calculated in sign change analysis. For categorical variables, all categories define calculation points.

In addition to the display in the results window, gfi returns the sign change table information in three matrices. r(SC) contains the moderated effect of the focal variable at the display values of the moderator(s); r(SCcol) contains the value of Moderator 1 at which the focal variable's effect changes sign (for given values of the other moderators, if any); if applicable, r(SCrow) contains the value of Moderator 2 at which the focal variable's effect changes sign (for given values of the other moderators).

One-Moderator Example

For a single-moderator interaction model, I usually don't find the path diagram useful given the simplicity of the interaction model. But it can be informative when either the focal or the moderating variable is categorical with more than two categories. The only GFI option I would add is ndigits() to change the default coefficients formatting given what I know about the scale of the coefficients. For this example, I use the intspec syntax specified above, but I use job status as the focal variable, not work–family conflict. And I change the default coefficient format with the option ndig(5). This produces the following results from the *intspec* and *gfi* commands to which I have added annotations and explanations.

```
. intspec focal(sei) ///
>       main( (sei ,  name(Job_Status) range(17(20)97))  ///
>       (wfconflict, name(W_F_Conflict) range(1/4))) ///
> int2(wfcbysei) dvname(Poor_M_H_Days) abbrevn(13) ndig(0)

Interaction Effects on Poor_M_H_Days Specified as

     Main effect terms: sei  wfconflict
     Two-way interaction terms: wfcbysei

  These will be treated as: Focal variable = sei ("Job_Status")
     moderated by interaction(s) with
         wfconflict ("W_F_Conflict")
```

Always check if ICALC's above interpretation of your interaction specification matches what you want. The GFI output begins with the algebraic expression for the moderated effect of the focal variable below. This shows that the main effect of job status is negative (−0.02619), but it becomes less negative as job status increases (changing by 0.00929 per unit of *wfconflict*).

```
. gfi, ndig(5)

GFI Information from Interaction Specification of
Effect of Job_Status on g(Poor_M_H_Days) from Negative Binomial Regression
-------------------------------------------------------------------

Effect of sei =
     -0.02619 + 0.00929*wfconflict
```

If you also specify the factorchg option, you get a second algebraic expression written as a factor change for those estimation techniques for which factor changes can be calculated. This includes count models such as the negative binomial regression used in this example:

```
. gfi,  factor ndig(5)

GFI Information from Interaction Specification of
Effect of Job_Status on g(Poor_M_H_Days) from Negative Binomial
Regression
-------------------------------------------------------------------
```

```
Effect of sei =
   -0.02619 + 0.00929*wfconflict


   Factor Change Effect (1 unit change in) sei =
      e^-0.02619 * e^( 0.00929*wfconflict) =

   0.97415 * 1.00933^wfconflict
```

The factor change expression indicates that when work–family conflict is at its lowest level (a value of 1), poor mental health days decreases by a factor of 0.98324 ($= e^{-0.02619} e^{0.00929 \times 1}$) for each unit increase in job status. And this factor becomes less negative/more positive for higher levels of work–family conflict.

As discussed in Chapter 2, you evaluate the moderated effect of job status across the range of possible (or actual sample) values for work–family conflict to determine if the job status effect changes sign. In the intspec syntax, I set display values for work–family conflict to range from its minimum (1) to its maximum (4). The sign change analysis table below reports the effect of job status across this range of conflict values. This table tells us that the job status effect changes from negative at low levels of conflict to positive at higher levels. The "Sign Changes" row reports the exact value of conflict at which the job status effect changes sign (2.818); it would report "Never" if the sign did not change. The "% Positive" line reports that 26% of the sample has a positive moderated effect of job status, calculated using the GFI expression above.

```
Sign Change Analysis of Effect of Job_Status
on g(Poor_M_H_Days), Moderated by W_F_Conflict (MV)
------------------------------------
             |      Job_Status
When         |    --------------------
W_F_Conflict=|
-------------+------------------------
           1 | Neg  b =   -0.01690
           2 | Neg  b =   -0.00760
           3 | Pos  b =    0.00169
           4 | Pos  b =    0.01099
-------------+------------------------
Sign Changes |  when MV= 2.817953
-------------+------------------------
% Positive   |         26.5
------------------------------------
```

Two-Moderator Example

To illustrate the GFI tool for two moderators, I start with the intspec syntax developed above for family income moderated by education and by cohort and then apply the GFI tool requesting a path diagram. The path *type* is set to "all" to show all coefficient values in the interaction. Coefficients in the path diagram are formatted with three places after the decimal [ndig(3)], and a title is specified.

```
. intspec  focal(c.faminc)  ///
>        main((c.educ , name(Education) range(0(2)20)) ///
>               (c.faminc , name(Family_Inc) range(0(2)24)) ///
>               (i.cohort , name(Cohort) range(1/4))) ///
>        int2(c.faminc#c.educ c.faminc#i.cohort) ndig(0)
```

```
Interaction Effects on childs Specified as

    Main effect terms: faminc  educ 2.cohort 3.cohort 4.cohort
    Two-way interaction terms: c.faminc#c.educ i.cohort#c.faminc

  These will be treated as: Focal variable = faminc ("Family_Inc")
     moderated by interaction(s) with
        educ ("Education")
        2.cohort 3.cohort 4.cohort ("Cohort")
```

A quick reminder: You need to order the main effects variables in main( ) and two-way interaction variables in int2( ) to correspond to one another. Because *c.educ* is listed in main( ) before *i.cohort*, its interaction with *c.faminc10k* must be listed first in int2( ). You can verify this by the order in which ICALC lists the main effect and two-way interaction terms above. The multiple factor variables for the *i.cohort* main and interaction effects are automatically listed in the correct order. But if you use your own dummy variables, you need to order them correctly.

```
. gfi,  ndig(4) path(all, ndig(3))   ///
>        title(Interaction Effects of Family Income by Education and
>              Family Income by Cohort))


  GFI Information from Interaction Specification of
  Effect of Family_Inc on g(childs) from Linear Regression
  ---------------------------------------------------------------------------------

  Effect of Faminc =
       -0.0487 + 0.0065*educ - 0.0257*BabyBoom - 0.0282*WWII
       -0.0645*DeprEra
```

The GFI expression indicates a negative main effect of family income on the number of children (−0.0487), which moves toward a positive value as education increases (0.0065 change per year of education). But the cohort moderators all have negative coefficients, meaning that the income effect is larger (more negative) for each of the three older cohorts compared with the most recent (the reference category). Note that the dummy indicators are labeled with the value labels for cohort. These would be *2.cohort*, *3.cohort*, and *4.cohort*, if the cohort does not have value labels.

| | | When Cohort = | | | | Sign Changes |
When Education=		PostBoom	BabyBoom	WWII	DeprEra		given MV1
0		Neg b= −0.0487	Neg b= −0.0744	Neg b= −0.0769	Neg b= −0.1132		Never
2		Neg b= −0.0356	Neg b= −0.0614	Neg b= −0.0639	Neg b= −0.1002		Never
4		Neg b= −0.0226	Neg b= −0.0483	Neg b= −0.0508	Neg b= −0.0872		Never

Effect of Faminc

6	Neg b= -0.0096	Neg b= -0.0353	Neg b= -0.0378	Neg b= -0.0742	Never
8	Pos b= 0.0034	Neg b= -0.0223	Neg b= -0.0248	Neg b= -0.0611	Sometimes
10	Pos b= 0.0164	Neg b= -0.0093	Neg b= -0.0118	Neg b= -0.0481	Sometimes
12	Pos b= 0.0294	Pos b= 0.0037	Pos b= 0.0012	Neg b= -0.0351	Sometimes
14	Pos b= 0.0424	Pos b= 0.0167	Pos b= 0.0142	Neg b= -0.0221	Sometimes
16	Pos b= 0.0555	Pos b= 0.0297	Pos b= 0.0272	Neg b= -0.0091	Sometimes
18	Pos b= 0.0685	Pos b= 0.0428	Pos b= 0.0403	Pos b= 0.0039	Never
20	Pos b= 0.0815	Pos b= 0.0558	Pos b= 0.0533	Pos b= 0.0169	Never

| Sign Changes
given MV2 | when M1=
7.477 | when M1=
11.429 | when M1=
11.814 | when M1=
17.396 | |

Percent of in-sample cases with positive moderated effect of Faminc = 74.0

The sign change analysis table makes it easy to see how the moderated effect of family income changes with education and cohort. Reading down a column indicates that for each cohort the family income effect is negative at lower levels of education—higher income is related to fewer children—but turns positive at higher levels of education—higher income is related to a greater number of children. The bottom row reports the education value at which the income effect changes from negative to positive for each cohort. For the Post-Boom cohort, the change occurs at about 7.5 years. For the next two older cohorts, the sign changes around 11 to 12 years and for the oldest cohort, the income effect is positive only for the most highly educated (>17 years).

Comparing across any row, income has the smallest negative (and largest positive) effect for the Post-Boom cohort, it has the largest negative (and smallest positive) effect for the Depression Era cohort, and the next two cohorts have very similar effects of an in-between magnitude. The rightmost column indicates whether or not the sign of the income effect changes across the four cohorts at any given level of education. Between 8 and 16 years of education, the direction of the income effect changes across cohorts, but otherwise each cohort has the same signed effect of family income. Overall, about three quarters of the sample (74.0%) have a positive effect of family income, as reported just below the table.

What is not so readily apparent from this table is that the change in the income effect with education for each cohort is exactly the same (an increase of 0.013 per 2 years of education) because there is no three-way interaction of income by education by cohort. And similarly, the income effect changes with cohort by exactly the same amount at any level of education. When I apply the EFFDISP and/or OUTDISP tool to this analysis, this will become clear.

The path diagram in Figure 6.1 has boxes for the focal and moderator variables in the leftmost column, for the two-way interaction terms in the second column, and for the outcome in the rightmost column. If you have a three-way interaction, it is shown in the third column before the outcome box. The boxes report the variables and their coefficients that constitute the interaction model. The interaction terms are expressed as focal × moderator variable using display or label names to make the conceptual and mathematical relationships clearer. The lines are drawn to connect main effects terms to the relevant two-way interaction terms (and to connect two-way terms to the relevant three-way terms if present). Intersecting lines—where one line stops with its arrowhead on the other line—indicate that the effects of the corresponding variables interact. This provides a visual and conceptual map of the interaction specification. You know that family income and education interact because their lines intersect and lead to their corresponding two-way interaction term. Similarly, family income and cohort interact. But education and cohort do not interact because their lines do not intersect. *Note that this is entirely dependent on what you have specified; if you do not specify an interaction on the intspec command, it does not show on the path diagram even if it was estimated by your model.*

You can also read off the algebraic expression for the effect of the focal variable from the path diagram. Start with the focal variable's effect shown in its box—*drop the focal variable name*—and then trace all arrows from left to right. Add/subtract all

---

FIGURE 6.1 ● PATH DIAGRAM FROM THE GFI TOOL, TWO-MODERATOR EXAMPLE

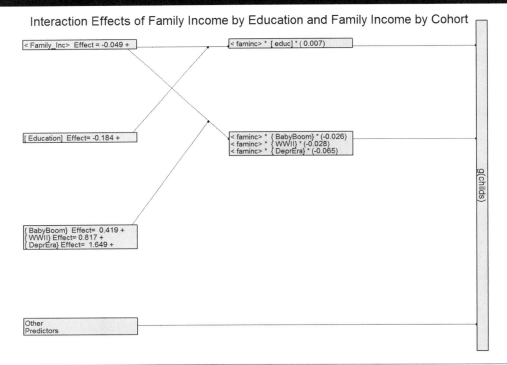

Interaction Effects of Family Income by Education and Family Income by Cohort

moderator × coefficient expressions from subsequent boxes connected to the focal variable (exclude duplicates if you come to the same box twice)—*again dropping the focal variable name*. For example, two arrows originate in the <Family_Inc> box (upper left), and you follow each one, adding terms as you go. The income effect is as follows:

$$Income\ effect = -0.049$$

$$+0.007 \times Educ \quad \text{from horizontal arrow to second column}$$
$$-0.026 \times BabyBoom \quad \text{from diagonal arrow to second column, Line 1}$$
$$-0.028 \times WWII \quad \quad \text{Line 2}$$
$$-0.065 \times DeprEra \quad \quad \text{Line 3}$$

In this example, you follow the same process for each of the moderators because I set path *type* to "all." For education, start with its box (middle, Column 1) and trace its one upward diagonal arrow to where it intersects the family income arrow and then goes to a two-way box (top, Column 2):

$$Education\ effect = -0.184$$

$$+0.007 \times faminc10k \quad \text{from diagonal arrow to second column}$$

For cohort, start with its box (third from top, Column 1) and trace its one upward diagonal arrow to where it intersects the family income arrow and then goes to a two-way box (middle, Column 2). Because cohort has multiple categories, you repeat the process of reading and constructing the moderated effect for each category of cohort, matching the corresponding line (entry) from each box connected by arrows:

$$BabyBoom\ effect = 0.419$$

$$-0.026 \times faminc10k \quad \text{First line in each connected box}$$
$$WWII\ effect = 0.817$$

$$-0.028 \times faminc10k \quad \text{Second line in each connected box}$$
$$DeprEra\ effect = 1.649$$

$$-0.065 \times faminc10k \quad \text{Third line in each connected box}$$

SIGREG TOOL: SYNTAX AND OPTIONS

sigreg, [<u>sigl</u>ev(#, adjtype) <u>effect</u>(*type(suboptions)*) save(filepath
    <u>tabl</u>e <u>matr</u>ix) <u>ndigi</u>ts(#) concise nobva <u>plotj</u>n(graphname, skip#)
    pltopts(string)]

SIGREG **Tool-Specific Options**	Description
siglev(#, adjtype)	# is nominal alpha level for significance testing adjtype = bonferroni, sidak or potthoff adjustment for multiple tests, can specify potthoff in combination with either bonferroni or sidak. Default is siglev(.05)

(Continued)

(Continued)

SIGREG **Tool-Specific Options**	Description
effect(type(suboptions))	Type of coefficient value reported in significance region table for delta unit change in focal variable
type	b estimated coefficient
	factor factor change
	spost marginal/discrete change calculated by the SPOST13 *mchange* command
	default effect(b)
suboptions for type = b, factor b(delta) or factor(delta)	delta
	1 1 unit change default
	sd 1 standard deviation change
	# any nonzero real number change
	sdy estimated effect (*b*) scaled in standard deviation units of g(y). Only for logit, probit, mlogit, ologit, oprobit, poisson, nbreg, zip, zinb, stcox
	sdyx sdy effect calculated for a 1 standard deviation change in the focal variable
suboptions for type = spost spost(amtopt( ) atopt( ))	
amtopt( )	Specify amount-of-change options for SPOST13 *mchange* command. Must include am( ) with a single entry (see Long & Freese, 2014, p. 168).
	default amtopt(am(one))
atopt( )	Specify content for the at( ) option of the *margins* command for predictors other than focal or moderators. Default[a] atopt((asobs) _all)
save(filepath table matrix)	
table	Save the formatted significance region table to Excel file in the name and location given by filepath
matrix	Save results used to construct the table (moderated coefficient, its variance and *p* level) to filepath
ndigits(#)	# is an integer for number of digits in tables. Default ndig(2)
concise	Limit details in boundary value analysis report; numeric boundary values only if within moderators' sample range; no derivative values
nobva	No boundary value analysis if keyword specified
plotjn(graphname, skip#)	Request plot of JN boundary values; only available if you have two moderators
graphname	Store as memory graph named graphname
skip#	Specify an integer skip# to improve plot readability by skipping points between markers; default is 10

SIGREG **Tool-Specific Options**	**Description**
pltopts(string)	Specify two-way graph options to customize appearance (e.g., line colors). These do not always work as expected. Use the graph editor if not.
How moderator range( ) specification from *intspec* is used	Defines points at which the moderated effect is calculated for empirical significance region table. For nominal variables, all categories define calculation points. For analysis with 2+ moderators, it defines points at which boundary values are found for the other moderator.

a. The default is the same as for the *margins* command since ICALC and SPOST13 rely heavily on *margins*. I prefer to evaluate at the means of the other predictors. Hence, most examples specify atopts((means) _all).

One-Moderator Example

I use the same *intspec* syntax as for the GFI analysis, so it is shown as "intspec …" in the output from the SIGREG tool below. The *sigreg* command—shown here and in the output—and what the options mean are as follows:

sigreg, sig(.05) save(c:\temp\sigtable_ch6_ex1.xlsx tab mat) ndig(5)

> sig(.05) sets the significance level for testing to 0.05 with no adjustment for multiple tests.

> save( ) creates an Excel file in the specified path filename containing the significance region table and intermediate result matrices because I specified the suboptions tab and mat.

> ndig(5) sets the number of decimal places in the numeric results to five.

The first table in the output shows the boundary values at which the significance of job status changes with work–family conflict. You read the table from left to right, with a pair of result columns for each boundary value. In theory, there can be two boundary values, but a boundary value could be outside the moderator's sample range and thus not applicable, or there could be no real-value solution because the significance never changes. The first boundary value is at *Conflict* ≥ 1.728, where the effect of job status changes to not significant. The second boundary value is −26.23, which is not applicable because the value is less than the minimum value of job status, marked (<min) in the table. The boundary values indicate that job status has a significant effect for workers who experience low levels of work–family conflict (1–1.728) but is not significant for workers with greater work–family conflict.

The second output table—an empirically derived significance region table—is more useful and accessible, especially when you have more than a single moderator. The cells in the table show the magnitude, direction, and significance of the effect of job status as these change with work–family conflict. The cells are formatted to visually highlight sign changes and significance changes, as indicated by the key below the table. This table makes it quite apparent that the effect of job status changes sign from negative to positive with increasing work–family conflict and that the effect of job status is only significant for workers in employment situations with the lowest levels of conflict. Moreover, this tool functions essentially as a calculator to get the effect of the focal variable for any desired set of values of the moderator(s).

The final two lines of output indicate that the significance region table and matrices containing intermediate calculations (moderated coefficient values, their variance, and their significance level) were written into sheets labeled *tab_1* and *mat_1* of an Excel file. The next example, a three-way interaction, includes an illustration of the contents of the Excel file.

```
. intspec …

Interaction Effects on Poor_M_H_Days Specified as

     Main effect terms: sei  wfconflict
     Two-way interaction terms: wfcbysei

  These will be treated as: Focal variable = sei ("Job_Status")
    moderated by interaction(s) with
        wfconflict ("W_F_Conflict")
.
. sigreg,  sig(.05 ) save(c:\temp\sigtable_ex1.xlsx tab mat ) ndig(5)
concise
        …
Boundary Values for Significance of Effect of Job_Status on Poor_M_H_Days
Moderated by W_F_Conflict
   Critical value Chi_sq = 3.841 set with p = 0.0500


+-------------------------------------------------------------------------------------------+
|               | When            | Sig Changes          | When            | Sig Changes      |
|               | W_F_Conflict >= |                      | W_F_Conflict >= |                  |
|---------------+-----------------+----------------------+-----------------+------------------|
| Effect of sei |    1.72815      | to Not Sig [-3.33430]| -26.23393 (< min)| to Sig [0.00469] |
+-------------------------------------------------------------------------------------------+

      Note: Derivatives of Boundary Values in [ ]

    Significance Region for the Effect of Job_Status
    on Poor_M_H_Days at Selected Values of W_F_Conflict
------------------------------------------------------------
                            At W_F_Conflict=
Effect of      |     1         2         3         4
---------------+--------------------------------------------
        Sei    | -0.01690*  -0.00760   0.00169   0.01099
------------------------------------------------------------
      Key: Plain font  = Pos, Not Sig    Bold font*    = Pos, Sig
           Italic font = Neg, Not Sig    Italic font*  = Neg, Sig

Formatted Significance Region table written to sheet tab_1 of
c:\temp\sigtable_ch6_ex1.xlsx

Significance Region matrices written to sheet mat_1 of
c:\temp\sigtable_ch6_ex1.xlsx
```

Three-Way Interaction Example

For this example, I use the same SIGREG options used in the prior example but add the plot( ) option to plot the boundary values. This option is only available for interaction models with two or more moderators.

```
sigreg, sig(.05 ) save(Output\Figure_6_3_A.xlsx tab mat ) ndig(3) ///
    plot(Bound_Values, 6) pltopts(ytit( , ma(r +2)) tit( , size(*.9)))
```

The content of plot( ) indicates that the boundary value plot will be stored in a memory graph named "Bound_Values" and that six markers are skipped between each marker that is shown. Trial and error led to the choice to set a skip value of six and to report three digits. pltopts( ) customizes the *y*-axis and plot titles.

The output below begins with the *intspec* syntax and output, followed by the *sigreg* syntax and output. SIGREG lists the options that you specify—or their defaults if you don't specify them—for the boundary value analysis, and then for the significance region table so you can verify that these are what you want. Two boundary values tables report how the effect of sex changes significance across combinations of age and education values. The first shows at what age the sex effect changes significance for a given value of education, while the second describes at what education level the sex effect changes significance for a given value of age. The boundary values plot opens in a separate graph window. The final piece of the output is the formatted significance region table.

```
. intspec …

Interaction Effects on Memnum Specified as

        Main effect terms: 1.female  age ed
        Two-way interaction terms: i.female#c.age i.female#c.ed c.age#c.ed
        Three-way interaction terms: i.female#c.age#c.ed

    These will be treated as: Focal variable = 1.female ("Sex")
        moderated by interaction(s) with
            age ("Age")
            ed ("Education")
        and with "Age" x "Education"

. sigreg, sig(.05) save(c:\temp\sigtable_ch6_ex1.xlsx tab mat) ndig(3)
>    plot(Bound_Values, 6) pltopts(ytit( , ma(r +2)) tit( , size(*.9)))

Boundary Value Analysis Options Specified or Default

        Skip BVA  =  no
        Details
      BV Plot saved as  Bound_Values

Significance Region Table Options Specified or Default

        Critical value Chi_sq = 3.841 set with p = 0.0500
        Effect type = b   (1 unit difference)
        Decimals reported in tables = 3
        Sig Region results saved:  c:\temp\sigtable_ch6_ex1.xlsx tab mat
```

Sex Effect Significance, Boundary Values for Age on g(Memberships) Given Education
   Critical value Chi_sq = 3.841 set with p = 0.0500

Effect of Sex	When Age >=	Sig Changes	When Age >=	Sig Changes
At Education =				
0	52.601	to Not Sig [-0.401]	-66.320 (< min)	to Sig [0.009]
2	53.655	to Not Sig [-0.403]	-53.338 (< min)	to Sig [0.012]
4	55.134	to Not Sig [-0.405]	-38.536 (< min)	to Sig [0.017]
6	57.325	to Not Sig [-0.402]	-21.889 (< min)	to Sig [0.025]
8	60.737	to Not Sig [-0.383]	-3.852 (< min)	to Sig [0.045]
10	66.006	to Not Sig [-0.309]	14.106 (< min)	to Sig [0.100]
12	71.630	to Not Sig [-0.162]	29.907	to Sig [0.262]
14	NA	Never	NA	Never
16	NA	Never	NA	Never
18	25.881	to Not Sig [-0.072]	-99.519 (< min)	to Sig [0.002]
20	34.436	to Not Sig [-0.192]	-382.862 (< min)	to Sig [0.000]

   Note: Derivatives of Boundary Values in [ ]

Sex Effect Significance, Boundary Values for Education on g(Memberships) Given Age
   Critical value Chi_sq = 3.841 set with p = 0.0500

Effect of Sex	When Education >=	Sig Changes	When Education >=	Sig Changes
At Age =				
18	10.463	to Not Sig [-1.031]	17.474	to Sig [0.689]
28	11.742	to Not Sig [-1.730]	18.273	to Sig [0.627]
38	13.056	to Not Sig [-2.923]	22.422 (> max)	to Sig [0.258]
48	13.855	to Not Sig [-2.793]	-43.323 (< min)	to Sig [0.009]
58	13.964	to Not Sig [-1.622]	6.475	to Sig [0.599]
68	13.473	to Not Sig [-0.727]	10.633	to Sig [0.848]
78	NA	Never	NA	Never
88	NA	Never	NA	Never

   Note: Derivatives of Boundary Values in [ ]

The first row of the first boundary value analysis table indicates that at *Education* = 0, the sex effect changes to not significant when *Age* > 52.601, while the second boundary value marks a change to a significant sex effect when *Age* > −66.320 (well below the minimum possible age value). The same general pattern applies to education values 0 to 12. The sex effect at the specified education level changes to significant at a younger age (second set of boundary values) and then turns not significant at an older age (first set). At *Education* = 14 to 16, the significance of the effect of sex does not change with age within the sample's age range. Last, for those with *Education* = 18 to 20, the sex effect turns not significant at an age between 25 and 35 years and remains not significant within the remaining sample range of age.

ASIDE: *CONCISE* OPTION

This option suppresses the reporting of the derivatives, reports a calculated boundary value outside the sample range as "NA" and marks its significance change as "NA," and reports "Never" if the effect does not change significance within the sample range of the moderator.

The second boundary values table shows that at the three youngest age values the boundary values of education at which the sex effect changes significance increase with age. The sex effect becomes not significant when education values are about average (10–13), but it changes back to significant at the upper end of the education distribution (17.4, 18.3, and above education's maximum [22.4]). At the middle three ages (48, 58, and 68), the sex effect becomes significant—look at the second boundary value—at education values toward the bottom of the education distribution (below education's minimum [−43.3], 6.5, and 10.6). But the boundary value for the change to not significant does not vary much, staying around 14 years of education in each case. And at the oldest ages, the significance of the sex effect never changes.

It is often difficult, as in this case, to see and provide a clear overall summary of the significance region from the boundary values tables. A boundary values plot (Figure 6.2) provides the same information in a potentially more accessible fashion. The vertical dimension is education, and the solid line and dashed line show how the significance of the sex effect varies with education at different ages. In this case, the region above the solid line but below the dashed line demarcates nonsignificant sex effects, while the region above the dashed line or below the solid line represents significant effects.

The horizontal dimension is age and the points with a • symbol or a × symbol mark where the significance of the sex effect changes with age for a given education level. The sex effect is significant in the region to the right of the • symbol but to the left of the × symbol—the region roughly bounded by education 0 to 12 and age 18 through 48–68 years. It is also significant in the area to the left of the × symbol when the × symbol is to the left of the • symbol.

Putting the two sets of criteria together, the sex effect is significant for highly educated adults only if they are in the youngest age brackets. And the sex effect is significant for all but the highest education levels when age is less than a varying upper limit—71.6 for *Education* = 12 dropping to 52.6 years when *Education* = 0. Note that the lowest and highest values of the boundary value lines are truncated to make the plot more readable.

I suspect that most readers would agree that although the boundary value plot is easier to read than the boundary values tables, you have to work hard to extract and explain the information from the plots. The last piece of SIGREG output—an empirically derived significance region table—is realistically the most useful and accessible. The cells in the table below show the effect of sex for combinations of education and age values. They are formatted to visually highlight both sign changes and significance changes as indicated by the key below the table. This table makes it quite easy to see that the effect of sex changes sign and is significantly positive only for the most highly educated in the youngest age brackets. Up through age 48, the effect of

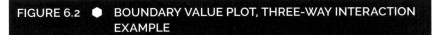

FIGURE 6.2 ● BOUNDARY VALUE PLOT, THREE-WAY INTERACTION EXAMPLE

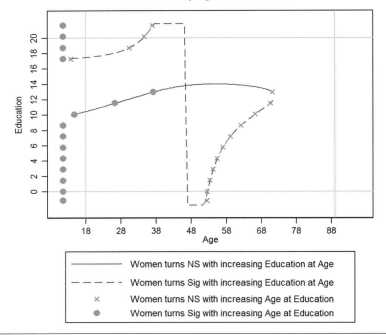

Boundary Value Plot for Significance of Women on g(Memberships)
Moderated by Age and Education

sex is negative and significant for respondents with below-average education. For the next two highest age brackets, the sex effect is significant only for those in the middle of the education distribution. For the oldest adults, there is no significant sex gap.

```
Significance Region for Effect of Female (1 unit difference)
n g(Memberships) at Selected Values of the Interaction of Age and Education
-------------------------------------------------------------------------------
                                        At Age=
At Education = |    18       28       38       48       58       68       78       88
--------------+----------------------------------------------------------------
           0  | -1.268*  -1.042*  -0.816*  -0.590*  -0.364   -0.138    0.089    0.315
           2  | -1.082*  -0.895*  -0.709*  -0.522*  -0.335   -0.149    0.038    0.225
           4  | -0.896*  -0.749*  -0.601*  -0.454*  -0.307   -0.160   -0.012    0.135
           6  | -0.709*  -0.602*  -0.494*  -0.386*  -0.278   -0.171   -0.063    0.045
           8  | -0.523*  -0.455*  -0.387*  -0.318*  -0.250*  -0.182   -0.113   -0.045
          10  | -0.337*  -0.308*  -0.279*  -0.250*  -0.221*  -0.193   -0.164   -0.135
          12  | -0.151   -0.161   -0.172*  -0.182*  -0.193*  -0.204*  -0.214   -0.225
          14  |  0.036   -0.014   -0.064   -0.115   -0.165   -0.215   -0.265   -0.315
          16  |  0.222    0.132    0.043   -0.047   -0.136   -0.226   -0.315   -0.405
          18  |  0.408*   0.279    0.150    0.021   -0.108   -0.237   -0.366   -0.495
          20  |  0.595*   0.426*   0.258    0.089   -0.079   -0.248   -0.416   -0.585
```

```
-----------------------------------------------------------------------------------
     Key: Plain font  = Pos, Not Sig     Bold font*   = Pos, Sig
          Italic font = Neg, Not Sig     Italic font* = Neg, Sig
```

Formatted Significance Region table written to sheet tab_1 of Output\Figure_6_3_A.xlsx

Significance Region matrices written to sheet mat_1 of Output\Figure_6_3_A.xlsx

While this table is easy to read in the results window, the font effects such as bold and italic are not preserved if you try to cut and paste the table. While you can save a picture of the table, you cannot edit the table's contents or format. Thus, SIGREG has an option to write the formatted table to an Excel file and further enhance the visual highlighting by using different fill patterns for positive and negative significant effects. This table is shown in Figure 6.3, along with a table of the saved matrices of intermediate results from the Excel file. The three matrices contain the moderated coefficients (the effect of sex for combinations of age and education values), the variance of the moderated coefficients, and the p level for the moderated coefficient tested against a hypothesized value of 0 (no effect).

Advanced Options: Factor Change Coefficients, Coefficients Scaled in Standard Deviations of $g(y)$, and SPOST13 Marginal Effects

This section is written for readers well-versed in the concepts of factor change and marginal effects. The effect( ) option provides considerable flexibility for the content of the effect reported in the significance region table. The examples above used the default, which reports the estimated moderated effect for a one unit change in the focal variable. You can alternatively calculate the focal variable's effect for other amounts of change in the focal variable (e.g., a standard deviation), as a factor change effect or rescaled in standard deviations of $g(y)$ (for some GLMs), or you can calculate any of the marginal effects options for the SPOST13 *mchange* command. I will illustrate the use and interpretation of these briefly for this example and in more detail in later chapters. For readers unfamiliar with the concepts of marginal effects or factor changes, I highly recommend Long and Freese (2014, chap. 4) as well as specific application chapters relevant to your analyses. For details on the *mchange* command, read the Stata help entry for *mchange* after installing the SPOST13 package. The presentation in this section presumes a working knowledge of this material.

Factor Changes

To get factor changes instead of estimated coefficients, use the effect (factor (*delta*)) option. *delta* specifies the amount of change in the independent variable for which the factor change is calculated. While the most common value for *delta* is 1, any real number may be specified. And "sd" requests calculations using a 1 standard

Effect of Women (1 unit difference) Moderated by the Interaction of Age and Education on g(Memberships), Formatted to Highlight Sign and Significance

Education	Age							
	18	28	38	48	58	68	78	88
0	*−1.268**	*−1.042**	*−.816**	*−.590**	*−.364*	*−.138*	.089	.315
2	*−1.082**	*−.895**	*−.709**	*−.522**	*−.335*	*−.149*	.038	.225
4	*−.896**	*−.749**	*−.601**	*−.454**	*−.307*	*−.160*	*−.012*	.135
6	*−.709**	*−.602**	*−.494**	*−.386**	*−.278*	*−.171*	*−.063*	.045
8	*−.523**	*−.455**	*−.387**	*−.318**	*−.250**	*−.182*	*−.113*	*−.045*
10	*−.337**	*−.308**	*−.279**	*−.250**	*−.221**	*−.193*	*−.164*	*−.135*
12	*−.151*	*−.161*	*−.172**	*−.182**	*−.193**	*−.204**	*−.214*	*−.225*
14	.036	*−.014*	*−.064*	*−.115*	*−.165*	*−.215*	*−.265*	*−.315*
16	.222	.132	.043	*−.047*	*−.136*	*−.226*	*−.315*	*−.405*
18	**.408***	.279	.150	.021	*−.108*	*−.237*	*−.366*	*−.495*
20	**.595***	**.426***	.258	.089	*−.079*	*−.248*	*−.416*	*−.585*

Key	
Plain font, no fill	Pos, Not Sig
Bold *, filled	Pos, Sig
Italic, no fill	Neg, Not Sig
Bold Italic *, filled	Neg, Sig

Coefficient, Var(Coef) and Significance Level Matrices for Effect of Women at Selected Values of the Interaction of Age (Col) and Education (Row)

Coefficient		18	28	38	48	58	68	78	88
	0	−1.26832	−1.04217	−0.81602	−0.58987	−0.36372	−0.13757	0.088582	0.314732
	2	−1.08203	−0.89534	−0.70865	−0.52196	−0.33527	−0.14858	0.038111	0.224801
	4	−0.89575	−0.74852	−0.60128	−0.45405	−0.30682	−0.15959	−0.01236	0.134871
	6	−0.70946	−0.60169	−0.49392	−0.38615	−0.27837	−0.1706	−0.06283	0.04494
	8	−0.52317	−0.45486	−0.38655	−0.31824	−0.24993	−0.18161	−0.1133	−0.04499
	10	−0.33689	−0.30803	−0.27918	−0.25033	−0.22148	−0.19263	−0.16377	−0.13492
	12	−0.1506	−0.16121	−0.17181	−0.18242	−0.19303	−0.20364	−0.21424	−0.22485
	14	0.035689	−0.01438	−0.06445	−0.11451	−0.16458	−0.21465	−0.26472	−0.31478
	16	0.221976	0.132449	0.042922	−0.04661	−0.13613	−0.22566	−0.31519	−0.40471
	18	0.408263	0.279277	0.15029	0.021303	−0.10768	−0.23667	−0.36566	−0.49464
	20	0.59455	0.426104	0.257658	0.089211	−0.07924	−0.24768	−0.41613	−0.58457

Var(Coef)		18	28	38	48	58	68	78	88
	0	0.261556	0.157713	0.091274	0.062238	0.070606	0.116378	0.199553	0.320132
	2	0.192181	0.116306	0.067431	0.045555	0.050678	0.082801	0.141924	0.228046
	4	0.133996	0.081462	0.047347	0.03165	0.034372	0.055512	0.095071	0.153049
	6	0.086999	0.05318	0.031022	0.020524	0.021687	0.03451	0.058994	0.095139
	8	0.05119	0.03146	0.018456	0.012177	0.012623	0.019796	0.033694	0.054318
	10	0.026571	0.016303	0.009648	0.006608	0.007181	0.011368	0.019169	0.030584
	12	0.013141	0.007708	0.0046	0.003817	0.00536	0.009228	0.015421	0.023939
	14	0.0109	0.005675	0.003311	0.003806	0.00716	0.013375	0.022448	0.034382
	16	0.019847	0.010205	0.00578	0.006573	0.012582	0.023809	0.040252	0.061913
	18	0.039983	0.021297	0.012009	0.012118	0.021625	0.04053	0.068832	0.106532
	20	0.071309	0.038952	0.021997	0.020443	0.03429	0.063538	0.108188	0.168239
p-level		18	28	38	48	58	68	78	88
	0	0.013139	0.008684	0.006913	0.018057	0.171057	0.686757	0.842813	0.578035
	2	0.013578	0.008656	0.006353	0.014464	0.136406	0.605612	0.919422	0.637822
	4	0.014404	0.008727	0.005721	0.010704	0.097935	0.498183	0.968024	0.730283
	6	0.016159	0.009077	0.005043	0.007031	0.058719	0.358433	0.795879	0.88416
	8	0.020759	0.010333	0.004436	0.003927	0.026118	0.196768	0.537067	0.846926
	10	0.038763	0.015844	0.00448	0.002073	0.00896	0.070821	0.236857	0.440415
	12	0.188938	0.066334	0.011302	0.003152	0.008375	0.034018	0.08448	0.146151
	14	0.732468	0.848637	0.262706	0.063422	0.051781	0.063449	0.077263	0.089575
	16	0.115108	0.189823	0.572384	0.565391	0.224894	0.143614	0.116186	0.103841
	18	0.041177	0.05566	0.17024	0.846554	0.464006	0.239758	0.163399	0.129649
	20	0.025983	0.030851	0.082341	0.532659	0.668728	0.325805	0.205823	0.154099

deviation change in the focal variable, as in these results for the one-moderator example:

```
sigreg, sig(.05 ) ndig(3)  effect(factor(sd))
…
Significance Region for Factor Change Effect of Job_Status (1 s.d.
  difference) On Poor_M_H_Days at Selected Values of W_F_Conflict

--------------------------------------------------------------
                               At W_F_Conflict=
Effect of      |       1        2          3          4
---------------+----------------------------------------------
        Sei    |     0.723*    0.864      1.033      1.234
--------------------------------------------------------------
Key: Plain font  = Pos, Not Sig    Bold font*    = Pos, Sig
     Italic font = Neg, Not Sig    Italic font*  = Neg, Sig
```

A factor change measures the proportionate increase or decrease in the predicted outcome corresponding to a specified amount of increase in the focal variable. A factor change greater than 1 shows an increase in the predicted outcome, while a factor change less than 1 represents a decrease in the predicted outcome. Specifically, the factor change effect of 0.723 for job status when *Work–family conflict* = 1 means that for a one standard deviation increase in job status, the predicted number of poor mental health days decreases by a factor of 0.723. That is, the number of days for someone with a higher job status is only about three quarters of the number of days for those with a 1 standard deviation lower job status. Note that this factor change is in italics in the table, denoting a "negative" effect, although the number is a positive value; that is, the outcome decreases as the predictor increases.

When *Work–family conflict* = 2, the factor change effect for job status is 0.864, indicating a smaller effect of job status; poor mental health days decreases by only a factor of 0.864. And as work–family conflict continues to rise, the factor change effect of job status becomes greater than 1.0, meaning that poor mental health days increase rather than decrease with job status. At the highest level of work–family conflict, the number of days increases by a factor of 1.234—the number of days is 23% higher for someone with a higher job status. Factor change interpretations are frequently used for logistic regression (binomial, multinomial, and ordered), for count models (Poisson, negative binomial, and their zero-inflated extensions), and for survival analyses (e.g., Cox proportional hazards model).

Scaled by *g*(*y*)'s Standard Deviation

To rescale the estimated coefficients to indicate how much *g*(*y*) changes in standard deviations for a one-unit change in the focal variable, use the effect (b(sdy)) option:

```
sigreg, sig(.05 ) ndig(4) effect(b(sdy))

  ...

  Significance Region for Effect of Job_Status (1 unit difference)
  on g(Poor_M_H_Days)-standardized at Selected Values of W_F_Conflict
  --------------------------------------------------------------
                                    At W_F_Conflict=
                                   -----------------------------
  Effect of      |      1        2         3          4
  ---------------+----------------------------------------------
        Sei      |  -0.0088*   -0.0040    0.0009    0.0057
  --------------------------------------------------------------
      Key: Plain font  = Pos, Not Sig    Bold font*   = Pos, Sig
           Italic font = Neg, Not Sig    Italic font* = Neg, Sig
```

Because these results are from a negative binomial regression, the rescaled effects can be interpreted as changes in standard deviations of the natural log of the count outcome. For example, at the lowest level of work–family conflict, a one-unit increase in SEI predicts that the log number of poor mental health days will decrease by 0.0088 standard deviation. The related specification—effect (b(sdyx))—recalculates these effects to represent standard deviation changes in *g*(*y*) for a 1 standard deviation increase in the focal variable. In the example, this would give an effect value of −0.169, indicating that a one standard deviation increase in SEI decreases the log number of poor mental health days by 0.17 standard deviation. This option is available for GLMs for which the metric of $g(y)$ has some intrinsic meaning. This includes count models (scaled as log counts) as well as logistic, probit, MNLR, and ordinal regression models (scaled as a latent interval outcome with an estimable standard deviation).

SPOST13 Marginal Effects

If you have installed SPOST13, you can use ICALC to calculate SPOST13 marginal effects for interactions. But you must (a) estimate your model using factor-variable notation and (b) list your variables in the same order on the *intspec* command—in main( ), int2( ), and int3( )—as on the estimation model command. For a generic example with two moderators,

```
estimation_command outcome focvar##modvar1 focvar##modvar2
    othervars
intspec focal(focvar) main( (focvar , ...) (modvar1, ...) (modvar2, ...) ) ///
    int2( focvar#modvar1 focal#modvar2 )
```

Then, to access the SPOST13 suite of marginal effects through ICALC, you specify the effect type as "spost" and use the suboptions amtopt( ) and atopt( ) to specify information for the SPOST13 *mchange* command. amtopt( ) should contain the amount-of-change options you want for *mchange*. The am( ) suboption controls the type of marginal effect calculated. *You can specify only one type* from the list below reproduced from the *mchange* help file in Stata:

Option	Amount of change
binary	Change from 0 to 1.
marginal	Marginal rate of change.
one	Change by 1.
range	Change over range or trimmed range; see trim().
sd	Change by the regressor's standard deviation or by the amount specified by delta(). Delta is a synonym for sd.

A one-unit discrete change—am(one)—is a common choice because it applies to both interval and nominal predictors. It measures the change in the predicted outcome corresponding to a one-unit change in the focal variable. You use atopt( ) to specify the values of the predictors other than the focal and moderator variables at which to evaluate the specified marginal effect of the focal variable. As discussed in Chapters 4 and 5, the most common choices are to evaluate at the means of the other predictors—atopt((means) _all)—or to use the average across the observed values in the sample—atopt((asobs) _all). For example, to get a discrete change measure of job status's effect moderated by work–family conflict, evaluated at the means of the other predictors, we use the following:

```
. nbreg pmhdays c.sei##c.wfconflict age educ childs female  if wrkstat <=4
    ...
. intspec focal(c.sei) main((c.sei, name(Job_Status) ///
> range(17(20)97)) (c.wfconflict, name(W_F_Conflict) range(1/4))) ///
> int2(c.sei#c.wfconflict) dvname(Poor_M_H_Days) abbrevn(13) ndig(0)
```

In the *intspec* command, note that main( ) lists *c.sei* first and then *c.wfconflict*, and, similarly that int2( ) lists the interaction term with *c.sei* first and *c.wfconflict* second (*c.sei#c.wfconflict*), the same order as in the *nbreg* estimation command.

```
. sigreg,  sig(.05 ) ndig(5) ///
>          effect(spost(amtopt(am(one)) atopt((means) _all)))

    ...

    Significance Region for SPOST Change Effect of Job_Status
      on Poor_M_H_Days at Selected Values of W_F_Conflict
    _____

                                       At W_F_Conflict=
    Effect of       |      1       2        3          4
    ----------------+------------------------------------------
            Sei     |  -0.03290*  -0.02307   0.00800    0.08099
    _____
        Key: Plain font  = Pos, Not Sig   Bold font*  = Pos, Sig
            Italic font = Neg, Not Sig    Italic font* = Neg, Sig


    Spost Effect for  sei specified as amount =  +1  calculated with
       at((means) _all )
```

The entries in the significance region table are discrete change measures, indicating the predicted change in the number of poor mental health days for one additional unit of Job Status. At the lowest level of work–family conflict, the number of poor mental health days decreases by 0.033 for workers with otherwise average levels on the other predictors. When *Conflict* = 2, the decrease with an increase in job status is smaller—0.023 days—which is not significantly different from zero. At higher levels of conflict, the discrete change becomes positive but remains nonsignificant. In this case, creating a table with discrete changes in the outcome corresponding to 1 standard deviation increases in job status provides a better sense of the magnitude of the job status effect. I get this by changing the suboption in amtopt( ) from am(one) to am(sd):

```
. sigreg,  sig(.05 ) ndig(5) ///
>          effect(spost(amtopt(am(sd)) atopt((means) _all)))

    ...

    Significance Region for SPOST Change Effect of Job_Status
      on Poor_M_H_Days at Selected Values of W_F_Conflict
    _____
                                       At W_F_Conflict=
    Effect of       |      1       2        3          4
    ----------------+------------------------------------------
            Sei     |  -0.54319*  -0.41304   0.15578    1.71848
    _____
        Key: Plain font  = Pos, Not Sig   Bold font*  = Pos, Sig
            Italic font = Neg, Not Sig    Italic font* = Neg, Sig
```

This tells us that, among workers with low levels of work–family conflict and average values on the other predictors, those who have a 1 standard deviation higher job status experience symptoms of poor mental health about half a day less (−0.543). As work–family conflict increases, the apparent protective effect of job status decreases in magnitude to four tenths of a day and then turns positive, indicating a possibly deleterious effect of higher job status in workers with higher levels of work–family conflict. At the highest level of conflict, workers with higher job status experience poor mental health symptoms 1.7 days more frequently than workers with job status 1 standard deviation lower. But this difference is not statistically significant. This section has provided a taste of the possibilities for using the SIGREG tool to calculate the numeric value and statistical significance of different types of marginal effects. The subsequent application chapters will discuss and illustrate other options for interpreting interaction effects in linear models and GLMs.

EFFDISP TOOL: SYNTAX AND OPTIONS

effdisp, [<u>plot</u>(<u>type</u>(*plottype*) <u>name</u>(*graphname*) <u>keep</u> <u>save</u>(*filepath*) <u>freq</u>(*base*)) effect(*type*(*suboptions*)) <u>ci</u>lev(#, adjtype) <u>ndig</u>its(#) pltopts(*string*) <u>sigmark</u> ccuts(*numlist*) <u>heatmap</u>]

EFFDISP **Tool-Specific Options**	**Description**
plot(type(plottype) suboptions)	Plot focal variable's effect varying with moderator(s)
type(plottype)	plottype is keyword for cbound for a confidence bounds plot, default if interval first moderator errbar for an error bar plot, default if categorical first moderator line for a connected-line plot, interval first moderator drop for a drop-line plot, usually categorical first moderator contour for a contour plot, only if interval first and second moderators
Suboptions name(graphname) keep save(filepath)	Store plot as memory graph with name graphname Store any intermediate graphs used to create final graph Save the plotting data and frequency distribution to Excel file with name and location given by filepath

(Continued)

(Continued)

EFFDISP **Tool-Specific Options**	Description

freq(base)	Add relative frequency distribution of first moderating variable or first by second moderator to the plot.
	base can be
	
	    tot for distribution of first moderator
	    sub for distribution of the first moderator within levels of the second
	    subtot for joint distribution of the first and second moderators relative to total sample size

effect(type(suboptions))	Type of coefficient value plotted for change of delta units in focal variable. Use these keywords:
type	b estimated coefficient factor factor change spost marginal/discrete change calculated by the SPOST13 *mchange* command default effect(b)
suboptions for type = b, factor b(delta) or factor(delta)	Delta 1 1 unit change default sd 1 standard deviation change # any nonzero real number change sdy estimated effect (b) scaled in standard deviation units of g(y). Only for logit, probit, mlogit, ologit, oprobit, poisson, nbreq, zip, zinb sdyx sdy effect calculated for a 1 standard deviation change in the focal variable
suboptions for type = spost spost(amtopt( ) atopt( )) amtopt( ) atopt( )	Specify amount-of-change options for SPOST13 *mchange* command. Must include am( ) with a single entry (see Long & Freese, 2014, p. 168) default amtopt(am(one)) Specify content for the at( ) option of the *margins* command for predictors other than focal or moderators default[a] atopt((asobs) _all)

cilev(#, adjtype)	# is confidence interval level (0.95 for a 95% CI)
	adjtype = bonferroni, sidak, or potthoff adjustment for multiple tests, can specify potthoff in combination with either bonferroni or sidak
	default ci(.95)

ndigits(#)	# is an integer for number of digits in tables. Default ndig(2)

EFFDISP **Tool-Specific Options**	**Description**
sigmark	If sigmark keyword present, add visual displays to denote significant and nonsignificant effects on line, drop, or contour plots
pltopts(string)	Specify two-way graph options to customize appearance of plots (e.g., change line colors). These will often work but not always or as expected. Use the graph editor if pltopts ( ) is unsuccessful
Contour plot only options ccuts(numlist)	numlist defines contour cut points. Default = 6 equal steps from min to max of moderated effect
heatmap	Stata twoway contour option for how similar height areas are portrayed
How moderator range( ) specification from *intspec* is used	Defines display values/labels on axis plotting first moderator (contour plots: also second moderator axis). For additional moderators, defines calculation points at which the plot is repeated.

a. The default is the same as for the *margins* command since ICALC and SPOST13 rely heavily on *margins*. I prefer to evaluate at the means of the other predictors. Hence, most examples specify atopts((means) _all).

One-Moderator Example

For the interaction of job status by work–family conflict, a confidence bounds plot is the preferred choice because the moderator (work–family conflict) is an interval variable. It plots the focal variable's moderated effect as a smooth curve across the range of the moderator (divided into 50 equally spaced points), as well as curves for the upper bound and lower bound of the confidence interval evaluated at each point. After setting up the interaction with INTSPEC, you can get a confidence bounds plot—a line plot with 95% confidence bounds—by specifying just effdisp with no options because confidence bounds is the default plot for an interval-level moderator:

```
intspec focal(sei) main( (sei , name(Job_Status) range(17(20)97)) ///
        (wfconflict, name(W_F_Conflict) range(1(.5)4))) int2(wfcbysei) ndig(0)
effdisp
```

Keep in mind that this produces a plot of the values of the *effect* of job status on poor mental health days, not a plot of the values of the *outcome*. Look first at the solid black line for the effect of job status in Figure 6.4. This tells us that the effect of job status is negative until work–family conflict rises above 2.82—this is the point at which the solid black line crosses the horizontal light gray line marking a zero effect—and then turns positive. That is, at lower levels of conflict, increasing job status is related to fewer poor mental health days; but at higher levels of conflict, job status is related to more poor mental health days. The plot also indicates that the effect of job status is only significant when work–family conflict is less than 1.73. You can see this from examining either where the confidence bounds (*dashed lines*) first bracket the light gray line for a zero effect or where the *vertical light gray* line marks the change from significant (Sig) to not significant (NS). In this discussion, I used the results from the prior GFI and SIGREG tools to report the value at which the job status effect's sign changes and the value at which the effect's significance changes. If you want a simpler plot of the moderated effect of job status without confidence bounds and change in significance markers, you can get a connected-line plot by specifying the plot type as "line":

```
effdisp, plot(type(line))
```

There are two things you can do to improve this plot. One is to change the format for the *y*-axis labels to display only two digits by adding the option ndig(2) to the *effdisp* command. The other is to add information about the distribution of sample cases across the values of work–family conflict. The distributional information will indicate what proportion of the sample experiences a significant effect of job status. To do this, add freq(tot) within the plot( ) option.

I also like to specify a graph name so that the graph does not disappear when the next graph command is executed and to save the plotting data to Excel. With these options, the syntax is

```
effdisp, ndig(2) plot(freq(tot) name(StatusWFC) save(Output\StatusWFC.xlsx) )
```

The spike plot at the bottom of the confidence bounds plot (Figure 6.5) shows that around 25% of the sample (the leftmost spike) have low enough work–family conflict such that their job status significantly affects their number of poor mental health days.

Figure 6.6 shows the first few rows of the plotting data saved to the Excel file. It indicates the names of the focal and moderator variables, the name of the memory graph

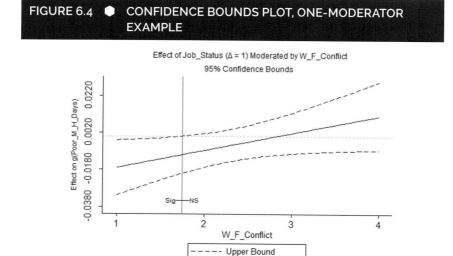

FIGURE 6.4 ● CONFIDENCE BOUNDS PLOT, ONE-MODERATOR EXAMPLE

FIGURE 6.5 ● REVISED CONFIDENCE BOUNDS PLOT, ONE-MODERATOR EXAMPLE

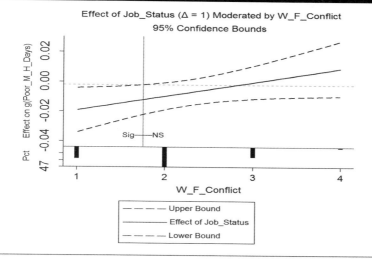

FIGURE 6.6 ● EXCERPT FROM PLOTTING DATA SAVED IN EXCEL FILE, ONE-MODERATOR EXAMPLE

Effect (Δ = 1) Plot	Focal= Job_Status	upb	bmod	lowb	M1Value	M1Label
	Plot Name = StatusWFC	−0.00199	−0.0169	−0.03181	1	1.00
	Mod1= W_F_Conflict	−0.00189	−0.01634	−0.03079	1.06	
	Mod2=	−0.00178	−0.01578	−0.02978	1.12	
	Mod3=	−0.00166	−0.01522	−0.02878	1.18	
		−0.00154	−0.01467	−0.02779	1.24	
		−0.0014	−0.01411	−0.02682	1.3	
		−0.00125	−0.01355	−0.02585	1.36	
		−0.00109	−0.01299	−0.0249	1.42	
		−0.00091	−0.01243	−0.02396	1.48	
		−0.00072	−0.01188	−0.02303	1.54	

(Plot Name), and the plotting data in the subsequent columns—the upper confidence bound (*upb*), the moderated effect (*bmod*), the lower confidence bound (*lowb*), the values of the first moderator that define the *x*-axis (*M1Value*), and labels corresponding to the values of *M1Value* that match any predefined labels of the moderator.

For this example, an error bar plot would be a reasonable alternative given the construction of the work–family conflict scale as seven values ranging from 1 to 4 in half-unit increments. The error bar plot places a diamond marker at the point y = effect of the focal variable for each value/category of the moderator and x = the moderator category/value. The marker is connected by vertical lines to horizontal bars, indicating the upper and lower bounds of the confidence interval. This can be done as follows:

- In the *intspec* command, change the range option for wfconflict to range(1(.5)4)

- In the *effdisp* command, add type(errbar) within the plot( ) option

As did the confidence bounds plot, the error bar plot (Figure 6.7) clearly shows that the effect of job status changes from negative to positive when conflict is greater than 2.5. And the effect is only significant at the two lowest levels of conflict—where it is negative—because the upper and lower bounds do not contain the zero-effect line.

Other than the plot(s), EFFDISP produces minimal output, as shown below. It echoes back its reading of the options and reports what if any information has been saved to an Excel file as shown for the error bar plot output:

```
. effdisp, ndig(2) plot( type(errbar) name(StatusWFC_EB)  freq(tot)  ///
>           save(Output\StatusWFC_Errbar.xlsx) )


Effect type Specified

   Effect type = b   (1 unit difference)

Plot Options Specified

   type =  errbar
   name =  StatusWFC_EB
   save =  Output\StatusWFC_Errbar.xlsx
   freq =  tot
   sigmark =  Yes

95% Confidence intervals calculated with critical value Chi_sq =    3.841

Plot data written to sheet plotdata_1 of c:\temp\StatusWFC_Errbar.xlsx

Frequency distribution data written to sheet frqdistdata of
>           Output\StatusWFC_Errbar.xlsx
```

Two-Moderator Example

When you have multiple moderators, you need to decide which one to treat as the first moderator to define the x-axis for the plot of the focal variable's moderated effect. The display values of the other moderators—specified by the range( ) suboption—define repeated *focal-by-moderator#1* plots, one for each combination of values of the other moderators. For a confidence bounds plot, you would typically use an interval moderator as the first moderator. But for an error bars plot, you would choose a categorical moderator as the first moderator. In this example, family income

FIGURE 6.7 ● ERROR BAR PLOT, ONE-MODERATOR EXAMPLE

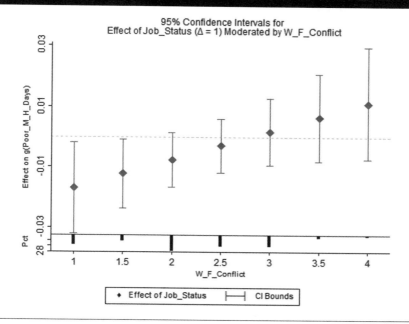

is moderated by education (an interval variable) and birth cohort (a categorical variable). Thus, you could choose either a confidence bounds plot against education or an error bar plot against birth cohort.

In such a case, I think you get the most detail by using the confidence bounds plot that shows the changes in the focal variable's effect across the full range of the interval predictor with the plot repeated for each category of the categorical moderator, whereas an error bar plot would incorporate the full range of the categorical predictor but the repeated plots would be for selected values of the interval moderator. For repeated plots, *effdisp* automatically creates panels containing two to three plots. If the number of plots is a multiple of 3, each panel contains three plots. Otherwise, for an even number of plots, each panel contains two plots, while for an odd number of plots, the last panel will have three plots and the other panels will have two plots.

For this example, I start with confidence bounds plots of the family income effect against education, with separate plots for the four categories of birth cohort. This uses the same options as in the prior example. I show the *intspec* and *effdisp* syntax in boldface in the output to make clear how to modify them later to use cohort as the first moderator.

```
. intspec  focal(c.faminc10k)  ///
>          main((c.faminc10k , name(Family_Inc) range(0(2)24)) ///
>              (c.educ , name(Education) range(0(2)20)) ///
>              (i.cohort , name(Cohort) range(1/4)))  ///
>          int2(c.faminc10k#c.educ c.faminc10k#i.cohort) ndig(0)
```

```
Interaction Effects on Childs Specified as

    Main effect terms: faminc10k  educ 2.cohort 3.cohort 4.cohort
    Two-way interaction terms: c.faminc10k#c.educ i.cohort#c.faminc10k

 These will be treated as: Focal variable = faminc ("Family_Inc")
    moderated by interaction(s) with
        educ ("Education")
        2.cohort 3.cohort 4.cohort ("Cohort")
 .
. effdisp,  ci(.95) plot(type(cbound) name(IncEdCohort) freq(tot)
>         save(Output\IncEdCohort.xlsx))

Effect type Specified

     Effect type = b    (Δ = 1)
Plot Options Specified

  type =  cbound
  name =  IncEdCohort
  save =  Output\IncEdCohort.xlsx
  freq =  tot
  sigmark = Yes

95% Confidence intervals calculated with critical value F =   3.850 .

Plot data written to sheet plotdata_1 of Output\IncEdCohort.xlsx

Frequency distribution data written to sheet frqdistdata of
>        Output\IncEdCohort.xlsx
```

This produces four plots organized into two panels with two plots each (Figure 6.8). Notice that the title also includes a panel label. Similarly, the graph window in Stata identifies each panel using the graph name you specify, to which it adds the suffix Pan_# to identify the panel number.

Looking across the four plots, focus first on where the *solid black line* for the family income effect crosses the *light gray horizontal line* of no effect (0). This indicates that the family income effect changes from negative to positive at education values ranging from 7.5 to 17.4 years, depending on the cohort. Second, observe that the effect lines for family income have the same slope for each cohort because there is no three-way interaction. Last, determine when the upper and lower bound lines do or do not contain the zero effect line or, equivalently, where there is a vertical line marking a change in significance of the family income effect. You can see that for the two youngest cohorts (Post-Boom and Baby Boom) family income only has a significant effect in the upper part of the education distribution, and that effect is positive. For the WWII cohort, the family income effect is never significant, while for the oldest cohort (Depression Era), the family income effect is only significant in the bottom of the education distribution and that effect is negative.

These plots focus your attention more on how education moderates the family income effect, but you can also see the moderating effect of cohort. Notice that the moderated effect line is highest on the *y*-axis for the Post-Boom cohort, lower on the *y*-axis and similar for the Baby Boom and WWII cohorts, and much

FIGURE 6.8 ● CONFIDENCE BOUNDS PLOT, TWO-MODERATOR EXAMPLE

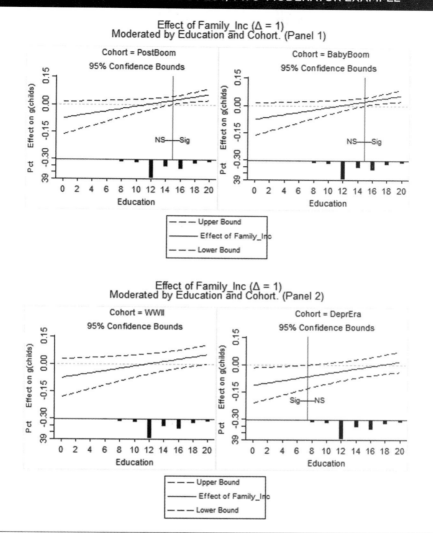

lower for the Depression Era cohort. This is much easier to see if I change the order of the moderators and create error bar plots of the family income effect against cohort, with separate plots for selected years of education. To accomplish this, I need to make changes to the syntax for *intspec* and for *effdisp*. In *intspec*, I need to switch the order of the terms for each moderator in the main ( ) option and in the int2 ( ) option so that the cohort terms come first. It also makes sense to change the range suboption for education, which would produce 11 repeated plots. Setting range(0(4)20) creates six repeated plots, a more manageable number to interpret.

```
intspec  focal(c.faminc10k) ///
        main((c.faminc10k , name(Family_Inc) range(0(2)24)) ///
            (i.cohort , name(Cohort) range(1/4))        ///
            (c.educ , name(Education) range(0(4)20))) ///
        int2(c.faminc10k#i.cohort c.faminc10k#c.educ)
```

And in *effdisp* change the plot type to "errbar" because cohort is categorical:

```
effdisp, ci(.95) plot(type(errbar) name(IncCohortEd) req(tot) save(Output\ ///
    IncCohortEd.xlsx)) ndig(2) pltopts(ylab( -.3(.15).15 , labsize(*.8)))
```

Look at any of the resulting six error bar plots shown in the two panels in Figure 6.9, which now focuses our attention on how cohort moderates the family income effect. The relative vertical position of the diamond markers for the family income effect for each cohort makes it immediately obvious that the family income effect is more positive (and less negative) for the Post-Boom cohort than it is for the Baby Boom and WWII cohorts, who are similar and have a more positive (and less negative) family income effect than the Depression Era cohort. Also note that the vertical distance between the markers is the same in every plot because there is not a three-way interaction between income, cohort, and education.

In practice, I would present one set of plots of the focal variable's effect against Moderator A repeated for Moderator B. However, I would produce the second set of plots switching the roles of the two moderators for myself to understand how the focal variable's effect changes with Moderator B. As a result, I should be able to clearly discuss the single set of plots to explain how both A and B moderate the focal variable's effect.

Three-Way Interaction Example

The prior example illustrated the use of repeated confidence bounds or error bar plots to explore how the focal variable's effect is moderated by two predictors. The same approach can be used for a three-way interaction, but a contour plot is sometimes a good alternative if both moderators are interval level. I illustrate this with the three-way interaction example of sex by age by education as they affect the number of memberships, treating sex as the focal variable with age and education as the moderators.

For a contour plot, the distinction between the first and second moderators determines which one defines the *x*-axis and which the *y*-axis, which as you will see does not really matter. The syntax for the *effdisp* options specifies the plot type as contour, and the sigmark option adds markers to the plot to denote areas of significant or nonsignificant effects of the focal variable. I begin with *intspec* to set up the interaction, again paying attention that age as the first moderator is listed first in main( ) and its interaction with *i.female* is listed first in int2( ). There is only one term in int3( ), so the ordering issue is moot. Because the initial contour plot had very busy labels, I respecified the range( ) suboption on the *intspec* command for age and education from what was used for running the *gfi* and *sigreg* commands. Alternatively, I could

FIGURE 6.9 ● ERROR BAR PLOTS, TWO-MODERATOR EXAMPLE

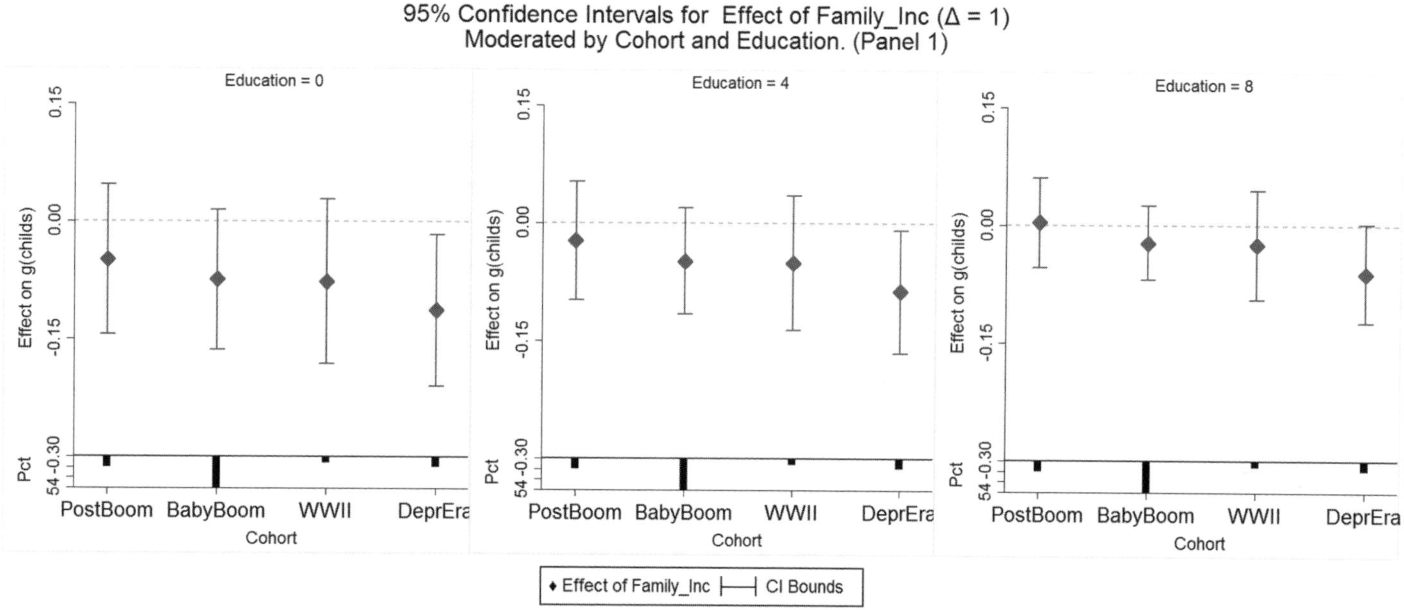
95% Confidence Intervals for Effect of Family_Inc (Δ = 1)
Moderated by Cohort and Education. (Panel 1)
Education = 0
Education = 4
Education = 8
Effect on g(childs)
0.15
0.00
-0.15
-0.30
Pct
54
PostBoom
BabyBoom
WWII
DeprEra
Cohort
Effect of Family_Inc
CI Bounds

(Continued)

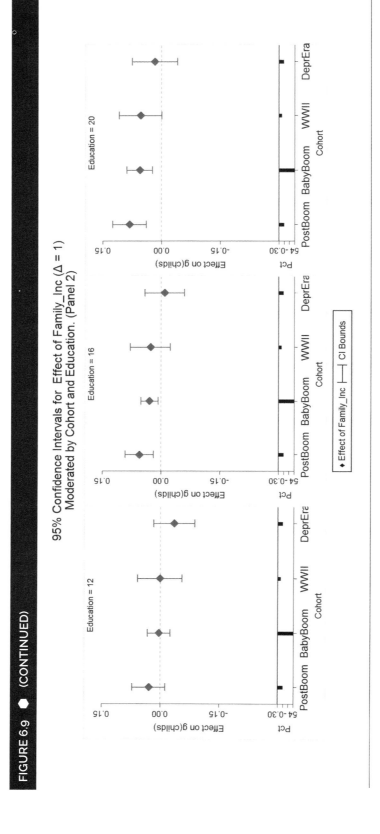

FIGURE 6.9 ⬡ (CONTINUED)

95% Confidence Intervals for Effect of Family_Inc (Δ = 1)
Moderated by Cohort and Education. (Panel 2)

Chapter 6 ■ ICALC Toolkit

have added pltopts( ) to the *effdisp* command below, but this would not affect the labeling on the frequency distribution contour plot:

```
. intspec focal(i.female) main( (c.age , name(Age) range(18(15)88)) ///
> (c.ed, name(Education) range(0(4)20)) (i.female, name(Sex) range(0/1))) ///
> int2(i.female#c.age i.female#c.ed c.age#c.ed ) int3(c.age#c.ed#i.female) ndig(0)
```

Interaction Effects on Memberships Specified as

.
.
.

```
. effdisp, ci(.95) plot( type(contour) name(AgeEdSex) )  sigmark
```

Effect type Specified

  Effect type = b   (Δ = 1)

Plot Options Specified

  type = contour
  name = AgeEdSex
  sigmark = Yes

95% Confidence intervals calculated with critical value Chi_sq =   3.841.

The different colors (shades) on the plot in Figure 6.10 indicate the magnitude of the effect of sex on the log number of memberships as it varies with age and education. *Darker shades* show more positive effects, and *lighter shades* indicate

FIGURE 6.10 ⬡   EFFDISP CONTOUR PLOT, THREE-WAY INTERACTION EXAMPLE

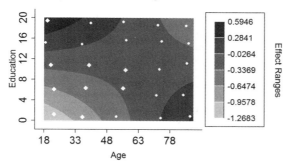

Effect of Sex (Δ = 1) on g(Memberships)
Moderated by the Interaction of Age and Education

larger-in-magnitude negative effects. You can see how age moderates the effect of sex by picking a given level of education and noting the pattern of change in the contours from left to right as age increases, as well as how that pattern changes across levels of education. Starting at the bottom left, for 0 to 4 years of education, the area changes from the lightest shade on the left to the next to darkest shade on the right and includes five or six contours. That is, there is considerable change in the effect of sex with age from negative to positive at the lowest levels of education. In contrast, for 8 to 12 years of education, the path contains only two or three contours in medium shades, primarily representing the negative effects of sex, which become smaller in magnitude with increasing age. And at the top of the education distribution (16–20 years), the left to right path again encompasses more contours (four or five) but changes from the darkest shade at the left to lighter shades at the right. This indicates that the effect of sex changes from positive to negative with increasing age at the highest levels of education. Also note that the left to right width of the contours is smallest at the extremes of education and widest in the middle, a further indicator of the increased pace of change in the effect of sex at the top and bottom of the education distribution.

There is a somewhat similar pattern for how education moderates the effect of sex, which is revealed by tracing a path from the bottom to the top of the contour plot for a given level of age. At the younger ages (18–33 years), the path includes six or seven contours, denoting substantial change in the effect of sex with education. Specifically, the sex effect is most negative—the lightest shade—for those with little education and most positive—the darkest shade—for those with the most education in this age-group. In the region of ages 48 to 63 years, you see very little change in the effect of sex, with the area dominated by the medium shade (effects from -0.399 to -0.142). In the vicinity of age 78 and up, the bottom to top path contains three or four contours showing an increasing rate of change in the sex effect with education, with a change from positive effects for those with little education to negative effects for those with higher levels of education.

The *large white diamond* markers denote areas in which the effect of sex is significantly different from zero, and the *small white circle* markers show where the effect of sex is not significant. The significant effects of sex are concentrated in the lower-left quadrant—the bottom halves of the age and education ranges—where they are all negative effects. The only other area with a significant effect of sex is for the youngest adults with the highest education (top left).

You can customize the cut points for defining the contours by adding the ccuts(numlist) option to the *effdisp* command. When the moderated effect ranges from negative to positive, I would recommend setting zero as one of the cut points to help visualize where the moderated effect changes sign. For this analysis, I would suggest starting the cut points at -1.2 (just above the minimum shown in Figure 6.11) and incrementing by 0.3 and ending at 0.6 (slightly more than the maximum moderated effect) (plot not shown):

```
effdisp, ci(.95) plot( type(contour) name(AgeEdSex) ) sigmark ccuts(-1.2(.3).6)
```

There is also an option to create a smaller-in-size contour plot of the joint distribution of the moderators, shown in Figure 6.11. It is placed on the graph to the right of the primary contour plot. Specify freq(subtot) in the plot( ) option of *effdisp*. This tells us that there are relatively few sample cases in the region where the effect of sex is significant.

FIGURE 6.11 ● FREQUENCY PLOT FOR CONTOUR PLOT, THREE-WAY INTERACTION EXAMPLE

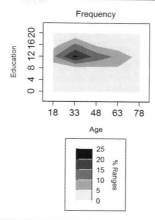

Although the contour plot can take some degree of effort to work through, it has the advantage of portraying the moderating effect of both education and age with a single, relatively concise plot. The alternative would be a confidence bounds plot of the effect of sex against age, repeated for selected values of education, which would show the moderating effect of age easily. You would most likely need to create a second set of plots of the effect of sex against education for selected values of age in order to more easily understand the moderating effect of education.

OUTDISP TOOL: SYNTAX AND OPTIONS

outdisp, [outcome(metric(*predtype*) atopt(*marginspec*) sdy dualaxis
   mainest(*estname*)) plot(type(*plottype*) name(*graphname*) single(*detail*)
   save(*filepath*) freq(*base*) keepfrq table(rowvar(*type*) save(*filepath*)
   freq(*base*) abs) ndigits(#) ccuts(numlist))pltopts(*string*)]

outdisp (no options specified) produces a default table of predicted values in the observed metric, setting reference values for predictors not part of the interaction using the margins specification at((asobs) _all).

OUTDISP **Options**	**Description**
outcome(suboptions)	Define how outcome is predicted and labeled
suboptions metric(predtype)	obs use observed outcome's metric (default)
	model use model-link transformed outcome metric
atopt(marginspec)	Specify content for the at( ) option of the *margins* command for predictors other than focal or moderators. Default[a] atopt((asobs) _all)

(Continued)

(Continued)

OUTDISP **Options**	Description
sdy	Is valid only with metric(model). Labels the primary *y*-axis in model metric standard deviation units (if applicable)
dualaxis	Is valid only with metric(model). Adds a second *y*-axis labeled in the observed outcome metric. Not valid for mlogit
mainest(estname)	Is valid only with metric(obs). Adds main effects predictions to interaction effects display. Must have used estimates store estname to store main effects model estimates
plot(suboptions)	Create graph of outcome by focal and moderating variables
suboptions type(plottype)	Keywords for plottype scat for scatterplot, default if focal is interval bar for bar chart, default if focal is categorical contour for contour plot, only interval focal and first moderator
name(graphname)	Store as memory graph named graphname
single(detail)	Define detail shown on single graph. Default is outcome–focal plot with results for all display values of moderator#1 in the same plot, and to combine two to three focal–outcome plots for moderator#2 values on the same graph detail = 1 shows separate outcome–focal plots for each display value of moderator#1 detail = 2 shows all moderator#1 results on same outcome–focal plot but a single graph for each display value of moderator#2 detail = all shows separate outcome–focal plots for each combination of moderator#1 – moderator#2 display values
save(filepath)	Save plotting data and frequency distribution to Excel file with name and location given by filepath
freq(base)	Add a plot of relative frequency distribution of focal or focal by moderator#1 base can be tot for distribution of only focal sub for distribution of focal within levels of moderator#1 subtot for joint distribution of focal and moderator#1 relative to total sample size
keepfrq	Store the separate frequency distribution graphs used to create the final frequency graph
table(suboptions)	Make table of predicted outcome by focal and moderators
suboptions rowvar(type)	*type = focal* or *mod* to define table rows. Default <u>rowvar</u>(focal)
save(filepath) freq(base) abs	Same as for plot suboptions Same as for plot suboptions Font size scaled by absolute value of table entry

OUTDISP **Options**	**Description**
ndigits(#)	# is integer number of digits in tables. Default is ndig(4)
ccuts(numlist)	Only for contour plot. numlist defines the contour cut points. Default = 6 steps from min to max of predicted outcome
pltopts(string)	Specify two-way graph options to customize appearance of plots (e.g., line colors). These will often work but not always or as expected. The graph editor can be used if this is unsuccessful
How focal and moderator range( ) specifications from *intspec* are used	For plots, defines display values/labels on the axis plotting the focal variable (contour plots: also for first moderator axis). For tables, sets display values/labels of row and column. For both plots and tables, moderators' range( ) defines the calculation points at which the plot or table calculations are repeated

a. I set the default to be the same as for the *margins* command since ICALC relies heavily on the *margins* command. I prefer to evaluate at the means of the other predictors. Hence most examples specify atopts((means) _all).

Limitation: If you are not using factor-variable notation, there is a limit of 70 on the number of combinations of display values (*DV*) for the focal and moderator variables when using OUTDISP. For an interaction with only one moderator: $\#focalDV \times \#moderator1DV \le 70$. For two moderators (with or without a three-way interaction): $\#focalDV \times \#moderator1DV \times \#moderator2DV \le 70$. To illustrate, in the three-way interaction example, the focal variable (sex) has two display values, and the moderators (age and education) have five and six display values, respectively. The number of combinations is $2 \times 5 \times 6 = 60$. If the model were estimated without using factor-variable notation, this meets the limit. But if there was a sixth display value for age, this would exceed the limit of 70: $2 \times 6 \times 6 = 72$.

One-Moderator Example

I use the OUTDISP tool to create both a table and a plot of the number of poor mental health days predicted by the interaction of job status by work–family conflict. The tab(row (mod)) option specifies a table of predicted values with rows defined by the moderator (work–family conflict) and columns by the focal variable (job status). To draw a graph, I use the plot( ) option with the suboptions type(scat) to indicate a scatterplot and name(StatusWFC) to store the plot as a memory graph. The outcome( ) option is not specified because I want the defaults for this initial application. In the output, *outdisp* lists the options that you specify and the defaults for those that you do not specify. This shows that predicted values are in the outcome variable's observed metric and are calculated with the "as observed" option in the Stata *margins* command to set values for predictors not part of the interaction specification.

```
. intspec …
┇
. outdisp,   plot(type(scat) name(StatusWFC) )   tab(row(mod))

Outcome Options Specified or Default

   metric =  obs
   atopt =  (asobs) _all
```

```
        Table Options Specified or Default

           row =   mod

        Plot Options Specified or Default

           type =   scat
           name =   StatusWFC
```

The plot type is specified as "scat," which draws a scatterplot. The *y*-axis represents the predicted number of days and the *x*-axis indexes job status (focal variable), and separate curves are drawn for predicted days by job status for each display value of work–family conflict (moderator). The table of predicted values is the next output in the result window. With the default number format of four decimal places, the table has more detail than necessary—two decimal places would be sufficient, so specify ndig(2) in the *outdisp* command.

The story that the table appears to tell is fairly straightforward. Comparing across rows in the lower half of the conflict range of values, we can see that the number of mental health days steadily decreases with job status at a somewhat higher rate for workers with the lowest conflict levels. In contrast, in the upper half of the conflict range, the number of days increases with job status much more quickly for those at the highest conflict levels. You can also examine the pattern of change across columns. This shows that the predicted number of days increases with higher levels of conflict, and the rate of change is much steeper for higher levels of job status. The upper scatterplot in Figure 6.12 clearly shows the same story about how the mental health days–job status relationship changes with work–family conflict, as would a scatterplot that graphed mental health days against conflict for different levels of job status.

```
Predicted Value of Poor_M_H_Days by the Interaction of Job_Status with
>     W_F_Conflict.

----------------------------------------------------------------------------
            |                          Job_Status
 W_F_Conflict |      17        37        57        77        97
--------------------+-------------------------------------------------------------
            1 |   3.6963    2.6364    1.8804    1.3412    0.9566
            2 |   4.1832    3.5932    3.0864    2.6510    2.2771
            3 |   4.7343    4.8973    5.0658    5.2402    5.4205
            4 |   5.3580    6.6746    8.3148   10.3579   12.9032
----------------------------------------------------------------------------
```

You may have noticed that I said this output appears to tell a fairly straightforward story. I use this qualifier because it is not clear to what extent the apparent degree of moderation of the effect of job status by conflict is due to their interaction and to what extent it is due to the nonlinearity of the count model. If you instead examine the predicted log number of days—the modeled metric of the outcome—you can see the nature of the interaction effects without the confounding of the model's nonlinearity. To do this, use the outcome(metric(model) dual) option. This specifies that predictions are in the modeled (transformed) outcome metric, and the suboption dual adds a *y*-axis on the right side of the plot labeled with the outcome's observed metric:

outdisp, out(metric(model) dual) plot(type(scat) name(StatusWFCdual))

FIGURE 6.12 ◆ OUTCOME PLOTS IN OBSERVED AND MODEL
METRICS, ONE-MODERATOR EXAMPLE

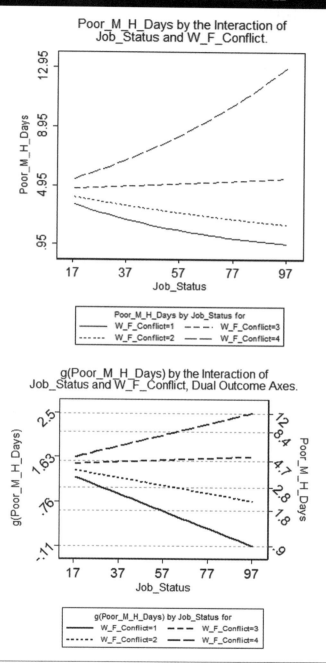

This is the lower scatterplot in Figure 6.12. While the general outline of the story is
the same, this plot shows that the absolute magnitude of the slope of the predicted
days–job status line is in fact much larger (1.75 times the size) at the lowest level
of work–family conflict (−0.513) than at the highest level of work–family conflict

(0.292). The plot in the observed metric (top) exaggerates the positive slope of the curve for the highest conflict because the curve moves in the direction of steeper slopes along the observed metric prediction curve. Conversely, the negative slope for the lowest conflict level is diminished because the curve moves in the direction of flatter slopes along the observed metric prediction curve; the overall change in the observed outcome metric along the low-conflict curve (−2.740) is only about one third the size of that along the high-conflict curve (7.545).

I want to reiterate the point I made in Chapter 5 that predictions in the observed metric are valid representations of the relationship between the outcome and the interacting predictors. But if your interest is in the underlying structure of the interaction effects and the degree of moderation, then working with predictions made in the observed metric can be misleading.

There is a second option to help disentangle the two sources of nonlinearity; you can create a plot produced using the predicted outcome from a main-effects-only model. Examining the degree to which the resulting curves do not appear parallel will illustrate the nonlinearity created by the estimation technique. And comparing the interaction effect predicted curves with the main effects predicted curves provides some context for interpreting the interaction effect patterns. To do this, you must first run and store the results from a main-effects-only model and then specify the suboption main (*estname*) in the outcome( ) option. *Be careful to ensure that the last model estimated before you run the intspec command is the interaction model, not the main effects model.* The following syntax shows how to do this: Run the main effects model, and store its results in "*holdmn*"; run the interaction model and store its results in "*intmod*"; then run *intspec* and other ICALC commands. (You can specify different names for the stored estimates than what I use in the example.)

```
*** Estimate main effects model and store with name holdmn

nbreg pmhdays  sei wfconflict age educ childs female if wrkstat <=4
est store holdmn

*** Estimate interaction model and store with name intmod

nbreg pmhdays c.sei ##c.wfconflict age educ childs female  if wrkstat <=4
est store intmod

intspec focal(c.sei) main( (c.sei , name(Job_Status) range(17(20)97)) ///
       (c.wfconflict, name(W_F_Conflict) range(1/4)) int2(c.sei#c.wfconflict) ///
       dvname(Poor_M_H_Days) abbrevn(13) ndig(0)

outdisp,  outcome(main(holdmn)) plot(type(scat) name(StatusWFC_mn) )
```

This produces the display in Figure 6.13 of the interaction predictions (*black lines*) with the main effects predictions (*light gray lines*) superimposed. The most obvious difference is that the main effects curves show a negative job status effect at all levels of conflict, while the interaction plot shows a negative effect at low levels of conflict but a positive one at high levels. But take note how the curves in the main effects plot become steeper corresponding to higher levels of work–family conflict *despite the absence of an interaction effect.* This reinforces what I concluded from the modeled metric plot: The moderating effect of work–family conflict is exaggerated at higher levels of conflict—particularly because this effect is positive—and correspondingly the moderation is muted at lower levels of work–family conflict—especially because the effect is negative—when using predictions in the observed outcome metric.

It can be easier to compare the main and interactive effect predictions if each pair of interaction predicted curves and main effects predicted curves are shown together on the same plot. You can request this with the single(1) suboption for plot( ), where the value 1 specifies that separate plots be created for each display value of the first moderator (in this case, the only moderator). The separate plots are named using the graphname you specify with a suffix of _1 for the first plot, _2 for the second plot, and so on, and each is labeled below the title as "Moderator Name = value label (value)." The individual plots in Figure 6.14 make it relatively easy to see that for the two lowest levels of work–family conflict the interaction predicted curve is steeper than the main effects predicted curve (upper plots) while for the next level of conflict the magnitude of the slopes are relatively similar but opposite in sign. But for the highest level of conflict, the magnitude of the interaction predicted curve is again somewhat steeper. While these comparisons can sensitize you to the type and degree of nonlinearity introduced by your estimation technique, I find the plots of predictions in the modeled metric easier and more useful for interpreting and understanding the nature of the interaction effect and how much each variable moderates the effect of the other.

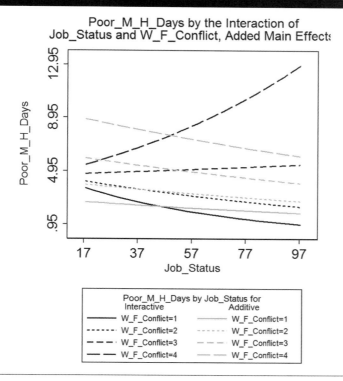

FIGURE 6.13 ● OUTCOME PLOTS FROM INTERACTIVE AND ADDITIVE MODELS, ONE-MODERATOR EXAMPLE

FIGURE 6.14   ●   INTERACTIVE AND ADDITIVE MODEL PLOTS WITH SINGLE(1) OPTION, ONE-MODERATOR EXAMPLE

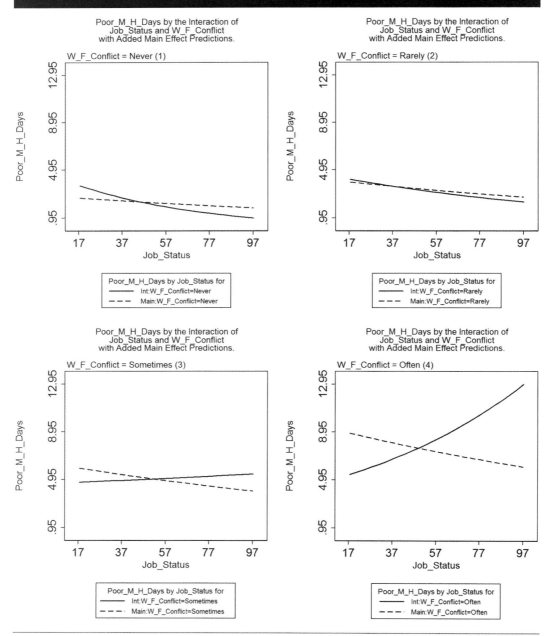

Two-Moderator Example

This example is an interval focal variable (family income) moderated by another interval variable (education) and by a categorical variable (birth cohort). For the predicted outcome plot, it makes the most sense to use birth cohort as the second moderator because this produces separate plots for each birth cohort. Each plot shows the predicted number of children as it varies with the interaction of family income and education. You can choose either a scatterplot or a contour plot to show the relationship between the predicted number of children in the interaction between family income and education. I start with a contour plot to demonstrate how to specify and interpret it. I then produce a scatterplot, which I think most readers will agree is much easier to interpret. The first step is to respecify the display values for family income and education in order to put fewer labels on the axes, which requires rerunning the *intspec* command and changing the range ( ) suboption for income and education. The *outdisp* syntax is then straightforward, simply specifying the type of plot and a graph name.

```
intspec focal(c.faminc10k) ///
    main((c.educ , name(Education) range(0(4)20)) ///
        (c.faminc10k , name(Family_Inc) range(0(6)24)) ///
        (i.cohort , name(Cohort) range(1/4))) ///
    int2(c.faminc10k#c.educ c.faminc10k#i.cohort) ndig(0)
outdisp, plot(type(contour) name(IncEdCohort) )
```

This creates two panels, each with plots of the predicted number of children by family income and education for two different cohorts (Figure 6.15). The *darker shades* represent a higher predicted number of children, and the *lighter shades* show a lower predicted number. To interpret the income effect, look at a particular value (or region) of education on the horizontal axis, and observe how the shades change from the top to the bottom as income changes from low to high. Starting with the Post-Boom cohort in Panel A, there are primarily vertical bands of the same shade, indicating relatively little change in the number of children as income increases at most levels of education.

For example, at 8 years of education, tracing a vertical path from the bottom to the top shows no change at all in the color, denoting a relatively constant predicted number of children as income increases. The same pattern is evident for education between 0 and 10 years. At the very bottom of the education distribution, the predicted number of children decreases somewhat with increasing income as shown by the change from the darkest shade to the next lighter shade. And at the top part of the education distribution (12–20 years), there is similarly some change in the number of children as income rises but here the number of children increases as income increases—the shades get darker. The same pattern holds for the Baby Boom and WWII cohorts and to a lesser degree for the Depression Era cohort. For the oldest cohort (Depression Era), there is more of an income effect, as shown by a larger area in which a vertical path representing the income effect crosses more shade boundaries.

To interpret the effect of education, pick an income level on the vertical axis, and follow a horizontal path from left to right to determine the effect of education. The horizontal paths usually encompass three or four different shades, except at the very highest income levels, where the education path contains only two shades. In all cases, the predicted number of children decreases as education increases, but there is a smaller effect of education at higher levels of income—fewer shade changes (two)—than at lower levels of income (four).

FIGURE 6.15 ● CONTOUR OUTCOME PLOT, TWO-MODERATOR EXAMPLE

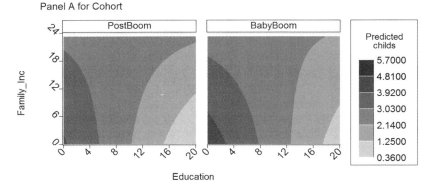

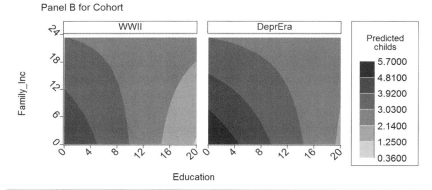

You can customize the cut points for defining the contours by adding the *ccuts(numlist)* option to the *outdisp* command. In this example, I would respecify the contour values to span the range from 0 to 6 in increments of 1 child by running

```
outdisp, plot(type(contour) name(IncEdCohort)) ccuts(0(1)6)
```

I think you can more easily see and understand the same patterns by looking at a scatterplot, which you can produce just by changing the type(contour) suboption to type(scat). This also produces two panels each with two scatterplots (Figure 6.16). The scatterplot for each birth cohort consists of lines showing how the number of children changes with family income, with each line representing the prediction for a selected display value of education (in this case every 4 years from 0 to 20). What is immediately apparent from the scatterplots is that the slope of the child–income lines changes from negative to positive. Specifically, the income effect changes from a negative effect at lower levels of education (the top two to five lines in each scatterplot) to a positive effect at higher levels of education. You can also see the education effect and how it changes with income from the scatterplots by examining the

FIGURE 6.16 ● OUTCOME SCATTERPLOT, TWO-MODERATOR EXAMPLE

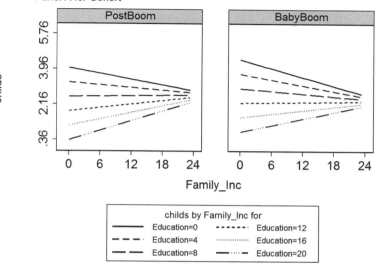

childs by the Two-way Interactions of
Family_Inc with Education and Family_Inc with Cohort.

Panel A for Cohort

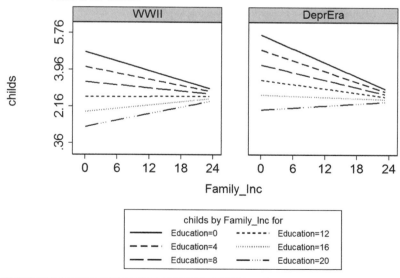

childs by the Two-way Interactions of
Family_Inc with Education and Family_Inc with Cohort.

Panel B for Cohort

vertical distance between a pair of lines as income increases from left to right. This reveals that the effect of education is always negative (the line for a lower level of education is always above the lines for higher levels of education). Furthermore, the education effect is largest at low levels of family income (larger vertical gaps) but becomes progressively smaller as income increases.

A table of predicted values is sometimes a good option, especially if formatted with the larger values shown in larger font sizes. For a linear link function, it is often as easy to interpret as a scatterplot. For a nonlinear link function, I would recommend using a predicted values table in the modeled metric, not in the observed metric, to avoid the problem of confounding. For this example, I get the predicted values in the observed metric since OLS has a linear link function. The syntax for *outdisp* includes the table( ) option with its save( ) suboption to write the formatted table in an Excel file, and ndig(2) to format the cells with two decimal places:

```
outdisp, tab(save(Output\Table_6_1.xlsx)) ndig(2)
```

To interpret the effect of family income as moderated by education in Table 6.1, pick one of the cohorts and compare the pattern in a column of how the number of children changes with income for a given level of education with the child–income relationship for other levels of education (columns). It is readily apparent that the income effect is negative for lower levels of education, but declines in magnitude until it turns increasingly positive at higher levels of education. For example, the first three columns for the Baby-Boom cohort have relatively similar predicted numbers of children for the maximum income row ($240K). But in the minimum-income row (<$5K), the predicted number of children is much larger—showing a negative effect of income—while the predicted number of children declines as education rises. Conversely, in the last three columns, the highest income row has the highest predicted number of children, which changes only slightly with education, while the minimum income row has the smallest predicted number of children, which declines notably as education rises.

The same general pattern characterizes the income effect in each of the cohorts, but the cohorts differ in both the magnitude of the income effect and the level of education at which the income effect switches from negative to positive. A careful examination of the income effect for each cohort shows that the effect of income on number of children is always the steepest negative (shallowest positive) for the Depression Era cohort but the steepest positive (shallowest negative) for the Post-Boom cohort. For the other two cohorts, the income effect is similar in magnitude and lies between these two extremes.

Three-Way Interaction Example

This last example illustrates the creation of bar charts by focusing on how the effect of being female on the number of voluntary association memberships varies with age and education, specified as a three-way interaction. The bar charts are organized as a pair of bars—predicted number of memberships for women and for men—for each display value of age, with that chart repeated for each display value of education.

Let's start with a bar chart with predictions in the observed outcome metric, number of memberships. I respecified the ranges for the moderators on the *intspec* command to define the values of age and education at which the predicted outcome values are calculated for women and for men. After some experimentation, I decided to set the display values of age as every 20 years from age 18 to 78 and for education as every 6 years from 2 to 20. On the *outdisp* command, I include type(bar) within the plot( ) option because this is not the default plot type for these measures.[2] I also use the atopts( ) suboption in the outcome( ) option to specify that the predicted values are calculated at the means of predictors which are not part of the interaction specification:

TABLE 6.1 ⬢ CHILDS BY THE TWO-WAY INTERACTIONS OF FAMILY_INC WITH EDUCATION AND WITH COHORT

Cohort	Family_Inc	Education					
		0	4	8	12	16	20
PostBoom	0	4.03	3.29	2.56	1.82	1.09	0.35
	6	3.74	3.16	2.58	2.00	1.42	0.84
	12	3.44	3.02	2.60	2.18	1.75	1.33
	18	3.15	2.89	2.62	2.35	2.09	1.82
	24	2.86	2.75	2.64	2.53	2.42	2.31
BabyBoom	0	4.45	3.71	2.98	2.24	1.51	0.77
	6	4.00	3.42	2.84	2.27	1.69	1.11
	12	3.55	3.13	2.71	2.29	1.87	1.44
	18	3.11	2.84	2.58	2.31	2.04	1.78
	24	2.66	2.55	2.44	2.33	2.22	2.11
WWII	0	4.84	4.11	3.38	2.64	1.91	1.17
	6	4.38	3.81	3.23	2.65	2.07	1.49
	12	3.92	3.50	3.08	2.66	2.23	1.81
	18	3.46	3.20	2.93	2.66	2.40	2.13
	24	3.00	2.89	2.78	2.67	2.56	2.45
DeprEra	0	5.68	4.94	4.21	3.47	2.74	2.00
	6	5.00	4.42	3.84	3.26	2.68	2.11
	12	4.32	3.90	3.47	3.05	2.63	2.21
	18	3.64	3.37	3.11	2.84	2.57	2.31
	24	2.96	2.85	2.74	2.63	2.52	2.41

```
intspec focal(i.female) main( (c.age , name(Age) range(18(20)78)) ///
    (c.ed , name(Education) range(2(6)20)) (i.female , name(Sex) range(0/1))) ///
    int2(i.female#c.age i.female#c.ed  c.age#c.ed ) int3(c.age#c.ed#i.female ) ///
    dvname(Memberships) abbrevn(13) ndig(0)

outdisp, plot(type(bar) name(AgeEdSex)) outcome(atopts((means) _all)) ndig(2) ///
    pltopts(plotreg(ma(t+2)))
```

This produces the four sets of bar charts shown in Figure 6.17. Each set shows the predicted membership for women (*gray bar*) and men (*black bar*) and how those predictions vary with age at a given level of education. Several patterns are obvious at first sight for the predicted number of memberships:

- It increases with education—the height of every bar rises across the four panels of education.

- It always increases with age for men—within each set, the height of the black bars rises.

- It increases with age for women who have 14 years or less of education—the heights of the gray bars rise. But decreases with age for more highly educated women—the heights of the gray bars decline.

Consequently, a close scrutiny of the bar chart shows that the effect of being female changes with age differently across the values of education:

- At low education levels, all but the oldest women (>76 years) have fewer predicted memberships than men. But the difference declines with age and reverses at older ages.

- At *Education* = 8, women have fewer predicted memberships at all ages, with the gap decreasing as age increases.

- At *Education* = 14, the youngest women (<33 years) have more predicted memberships, but this reverses and men have increasingly more predicted memberships at older ages.

- At the highest education level, young and middle-aged women (<54 years) have more predicted memberships than men, but older women have increasingly fewer predicted memberships than older men.

These bar charts clearly show that the effect of being female changes considerably with age and education, with the differences most apparent at the top and bottom of the education distribution.

I suspect that most readers would focus on the pattern of sex differences that are most obvious at the highest education levels and unconsciously pay less heed to those at the lowest education levels. But the magnitude of the difference in the bars is less at low education values, in large part due to the nonlinear link. Thus, adding a comparative plot of the main effects predictions can help provide some context for interpretation. To do this, add the suboption main(holdmn) to the outcome( ) option specification on *outdisp*.

```
outdisp, plot(type(bar)) outcome(main(holdmn) atopts((means) _all)) ///
    pltopts(plotreg(ma(t+2)))
```

This adds a pair of bars for women's and for men's number of memberships predicted from the main effects model next to each pair of bars of predicted memberships from the interaction model (Figure 6.18). Looking at the set of bar charts for *Education* = 2, you can judge the size of the sex gap in the interaction model relative to the size of the

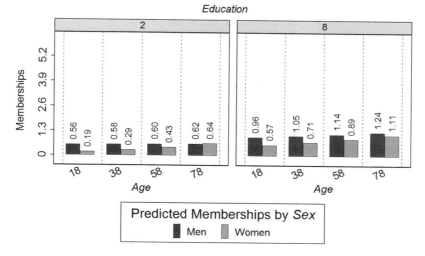

FIGURE 6.17 ● OUTCOME BAR CHARTS, THREE-WAY INTERACTION EXAMPLE

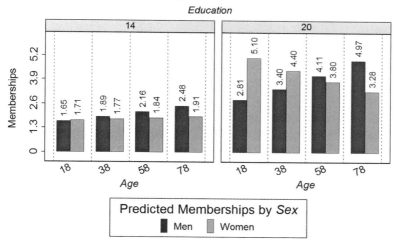

bars in the main effects model. Because there is virtually no difference between the women and men in the main effects model, the interaction model predictions suggest a considerable advantage for men of young and middle ages. For the youngest women, the gap is about 82% of the size of the main effects bars, and about 60% for the next age-group. In proportionate terms, this is larger than the female advantage for these ages at the highest education levels (73% and 29%).

FIGURE 6.18 ● OUTCOME BAR CHARTS FOR INTERACTIVE AND ADDITIVE MODELS, THREE-WAY INTERACTION EXAMPLE

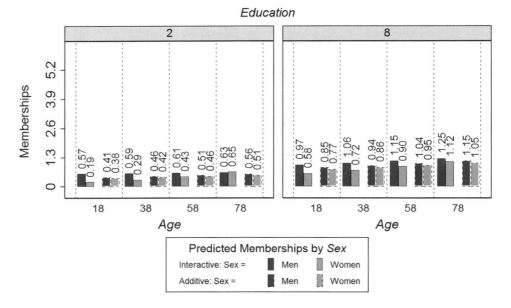

Memberships by the Three-way Interaction of
Sex, Age and Education, Interaction and Additive Models.

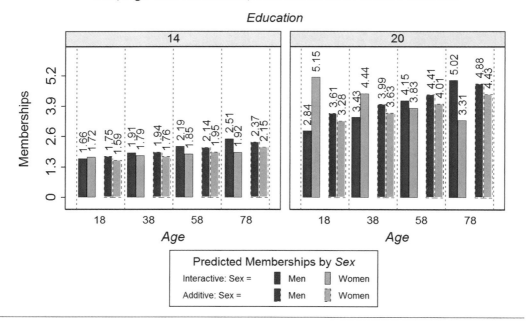

Memberships by the Three-way Interaction of
Sex, Age and Education, Interaction and Additive Models.

However, as I have been arguing, creating a plot in the modeled metric rather than the observed metric shows the pattern and magnitude of moderation more clearly. To get this from ICALC, respecify the suboptions in outcome( ) to set the metric to "model" in standard deviation units and request the dual labeling of the axes:

```
outdisp, plot(type(bar) name(AgeEdSexMod)) ///
    outcome(metric(model) sdy dual atopts( (means) _all))
```

When looking at the bar charts in Figure 6.19, keep in mind that the outcome metric is the log of number of memberships in standard deviations, which means that a larger negative value represents a smaller number of memberships. For instance, in the first set of bar charts for *Education* = 8, women aged 18 years belong to fewer organizations (about −1.0 standard deviation of logged memberships, or about 0.20 memberships) than men aged 18 years (about −0.5 standard deviation of logged memberships, or about 0.50 membership). You can now quite readily see that the moderation of the sex gap by age is greatest at the two extremes of education level. Moreover, it is larger for those with the least education than for those with the most. In both cases, there is an age at which there is virtually no sex gap, and the two largest sex gaps are in the bar charts for *Education* = 2 (ages 18 and 38).

You could also glean from these bar charts the same set of patterns described in the bullet points above, albeit some of the patterns take more work to process mentally given the log metric. Consider the very first bullet point that the number of memberships for both men and women of any age increases with education. Compare the bar for men aged 18 years with 2 years of education with those for men with 8 and 14 years of education, and the first impulse is to think that this shows a decline because the size of the bar is progressively smaller for those with 8 and 14 years of education. But this actually represents an increase in logged memberships from −0.99 to −0.51 to −0.04 standard deviation, followed by a more intuitive increase to a positive number of logged memberships (0.44 standard deviation). I leave it as an exercise for readers to see how the other patterns manifest themselves in Figure 6.19.

NEXT STEPS

The data and the syntax commands (Stata do-files) for the examples in this chapter are available for download at www.icalcrlk.com. As a first step in learning how to apply the ICALC tools, I would encourage readers to download and play with these examples. Make sure you can reproduce the results in this chapter before you try applying them to your own analyses. The remaining chapters cover detailed applications of the use of the ICALC tools for a variety of GLMs and types of interaction specifications. These will provide a more holistic approach to the interpretation of interaction effects by applying each tool in sequence to the interaction effects to tell the story of the relationship between the outcome and the interacting variables. They will also treat each variable in the interaction in turn as the focal variable to avoid a one-sided emphasis on only one part of the interaction specification. I begin with OLS regression in the next chapter as the simplest case because the tabular and visual results of the predicted outcome are not confounded by nonlinearity in the link function.

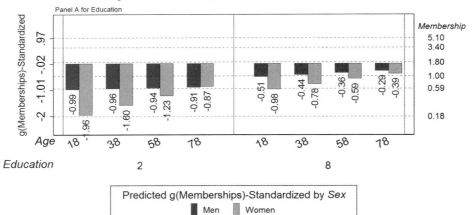

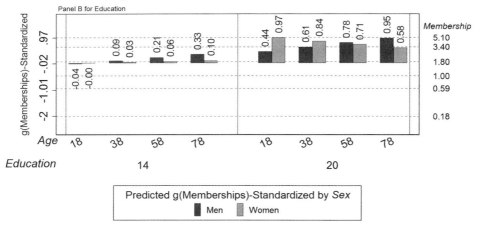

CHAPTER 6 NOTES

1. In linear link models, the modeled outcome is the observed outcome, while in nonlinear link models, the modeled outcome is a transformation of the observed outcome (e.g., in logistic regression the modeled outcome is the $\ln odds = \ln \frac{probability(y=1)}{1- probability(y=1)}$).

2. A bar chart is the default for a nominal focal variable and a nominal moderator, and this example has a nominal focal and an interval moderator.

LINEAR REGRESSION MODEL APPLICATIONS

Because this is the first application chapter, let me preview the basic structure and content that each application chapter has. I introduce the estimation technique by describing the situations and types of data for which the technique is appropriate and commonly used. I next summarize the properties that identify it as a GLM and discuss the meaning of the modeled outcome and its metric if they are different from the observed outcome. I then identify key diagnostic procedures, and I end the introduction with a note concerning the source of the data for the application examples. The remainder of the chapter is typically two start-to-finish application examples: (1) a single-moderator example and (2) a multiple-moderator and/or three-way interaction example. Each application example begins with a description of the dependent and independent variables and an overview of the results of diagnostic procedures and tests of the statistical significance of the interaction effects.

To briefly review, each GLM is built around a linear function of a set of predictors; that is, a sum of coefficients multiplied by predictors. The predictors may include interval variables, polynomials or other functions of an interval variable, categorical variables expressed as a set of binary indicators, and/or sets of predictors multiplied together (i.e., interaction terms). A specific GLM is distinguished by two key properties. The *link function* mathematically defines the transformation of the linear function of the predictors into the expected value of the outcome variable Y. Equivalently, we use the inverse of the link function, which describes the modeled outcome as a mathematical function of the observed outcome. The second distinguishing property is the *conditional distribution function*, which characterizes the probability distribution of the outcome conditional on the predictors.

OVERVIEW

Properties and Use of Linear Regression Model

Data and Circumstances When Commonly Used

Linear regression models (LRMs) are typically appropriate choices for analyzing interval or ratio outcome measures for which the mean of the outcome conditional on the predictors can be described by a linear function of the predictors. Note that the predictors can include categorical variables or a nonlinear function of a predictor, such as a polynomial or logarithmic expression, as well as interaction terms represented by the product of two or more predictors. But there are interval or ratio outcome measures for which an LRM can be a problematic choice. These include truncated or censored outcomes, badly skewed outcomes, mathematically bounded outcomes (e.g., a proportion), and count outcomes. A solution sometimes employed is to find a transformation of the outcome that makes an LRM appropriate (see Fox, 2008, chap. 4; Kaufman, 2013, chap. 3). The other, more common solution is to use a different member of the GLM family appropriate to the data situation.

Published Examples.

- Auspurg, Hinz, and Sauer (2017) used generalized least squares regression to analyze the fairness of baseline pay as portrayed in a series of vignettes to test theories of justice evaluation. Among other interactions, they tested whether the effect of the gendered pay ratio of respondents' job is moderated by whether respondents evaluate a man's base pay or a woman's base pay. They interpreted the interaction by describing how the focal variable's effect changed.

- In a two-stage least squares analysis of countries' tax revenues in the early modern era, Karaman and Pamuk (2013) estimated a three-way interaction between war pressure, urbanization, and regime representativeness. They created and interpreted a table of predicted tax revenues at selected values of urbanization and of real and hypothetical regime type.

- Cortes and Lincove (2016) studied the college admissions fit of student applicants (probability of a mismatch) using OLS regression and found an interaction between the applicants' high school class rank and their race/ethnicity. They discussed the interaction coefficients to interpret the interaction effect.

- Using OLS regression, Rambotti (2015) reanalyzed prior country-level research of the effect of income inequality and poverty on life expectancy to add a two-way interaction between income inequality and poverty. Rambotti interpreted the interaction by discussing predicted life expectancy plotted against inequality for low- and for high-poverty countries.

GLM Properties

The LRM is a GLM with an identity link function—Y equals the linear function of the predictors without a transformation of the prediction function—and a normal (Gaussian) conditional distribution function. In this instance, the modeled outcome is identical to the observed outcome. The LRM is a class of models with these characteristics, which are further differentiated by the assumptions they make about the error terms. OLS assumes that the error terms are normally distributed with equal variance (homoscedasticity) and zero covariance/uncorrelated errors (Greene, 2008, p. 111). Other models in this class relax the assumption of equal variance and/or zero covariance.[1] Common examples include WLS, which allows for a nonconstant error variance; time-series models, which usually specify autocorrelated errors; panel models; and least squares with robust standard errors, which may permit both correlated errors and unequal error variances. The empirical applications in this chapter use OLS in the two-moderator example and WLS in the one-moderator example.

Diagnostic Tests and Procedures

The diagnostic tests that I reviewed in Chapter 1 can all be used for LRMs. Plotting the residuals (typically studentized residuals) against the predicted outcome and against individual predictors provides an evaluation of overall model fit and possible model misspecification. These are essential for limiting the possible influence of a misspecified model on testing for and estimation of interaction effects. Similarly, such residual plots, leverage analysis, and influence statistics can help identify potential influential outliers to examine. There are also numerous tests of the assumptions of homoscedasticity and uncorrelated errors specific to OLS regression, panel models, and time-series analyses (for details, see Greene, 2008, chaps. 8, 9, and 19, respectively). Last, checking for collinearity among the predictors—apart from the necessary functional collinearity among interaction terms—is also standard.

Data Source for Examples

Both application examples use data from the 2010 GSS; the Stata data file (GSS_2010. dta) can be downloaded at www.icalcrlk.com, as can the Stata do-files used for the examples.

SINGLE-MODERATOR EXAMPLE

Data and Testing

The dependent variable is the frequency of sexual intercourse per month. The predictors consist of four interval measures (age in years, SES, number of children, and frequency of attending religious services) and two nominal measures (female-identified and never married status), each represented by a dummy variable. Both the frequency of sex and attendance at religious services were recoded from ordinal categories to a monthly frequency by annualizing the response category and dividing by 12.[2] The sample was restricted to respondents aged over 24; an additional 431 cases were excluded due to missing information on one or more of the variables.[3] Diagnostic testing indicated no outliers or influential cases but did suggest the presence of heteroscedasticity. A WLS regression was successful in correcting for heteroscedasticity and is used throughout the example. A sensitivity analysis showed no appreciable differences in the results reported for this example from using WLS versus OLS versus OLS with robust standard errors versus an OLS analysis of a variance-stabilizing transformation of the outcome (square root). Moreover, I treat the outcome measure of sexual intimacy as continuous to provide an example of a WLS analysis with heteroscedasticity, even though it has a limited number of observed values from the underlying continuous measure. Although doing so is somewhat contrived, the results are robust to alternative estimation choices and operationalizations of the outcome (negative binomial regression or ordinal logit).

The model includes an interaction between age and SES, conceptually justified by the expectation that the well-established negative effect of age on sexual activity would be diminished by the knowledge and resources available to those with higher SES. The WLS regression results report a t statistic for the interaction coefficient of 3.75 ($p < .001$). This indicates that the coefficient for the age-by-SES product term is statistically significant. Although the main effect coefficients for age and SES are each significant, this is not useful information in this case because 0 is not a valid value for either predictor. It is not meaningful to know that the effect of SES is significant for respondents who are 0 years old or that the effect of age is significant for respondents with *SES* = 0.

In the following sections, I cover the interpretation of the effect of age as moderated by SES and then the effect of SES as moderated by age, applying in turn the ICALC tools. The ICALC command lines are bolded in the Stata output for ease of identification.

The Effect of Age Moderated by SES

Setup With INTSPEC **Tool**

The first step is to use *intspec* to define the interaction details for the other tools:

```
. intspec focal(c.age) main( (c.age, name(Age) range(25(10)85)) ///
>          (c.ses, name(SES) range(17(10)97))) int2(c.age#c.ses) sumwgt(no)

Interaction Effects on Sexfrqmonth Specified as
-------------------------------------------------------

    Main effect terms: age  ses
    Two-way interaction terms: c.age#c.ses

  These will be treated as: Focal variable = age ("Age")
    moderated by interaction(s) with
       ses ("Ses")
```

Always check the output from *intspec* to verify that the interaction information is correct before proceeding to apply the other tools. The options/information specified are as follows:

- focal( ) declares *age* as the focal variable.

- main( ) contains the variable names of the main effect variables (*c.age* and *c.ses*) and their display names for the output, and range( ) lists the values used for display and/or calculation. The range specification for age—25(10)85—specifies that age's display values range from 25 to 85 in increments of 10. The range for SES is 10 to 97 in steps of 10.

- int2( ) specifies the variable name of the two-way interaction term *c.age #c.ses*.

- sumwgt(no) tells ICALC not to use the weights specified on the estimation command when calculating summary statistics. *This is crucial when your weights are analytic weights because the Stata margins command automatically uses the weights even when not listed in the margins command syntax unless you explicitly tell it not to do so.*

GFI **Analysis**

We start with the GFI tool to identify and extract the expression defining the effect of age and then to determine when, if at all, the sign of the effect of age changes. The option ndig(4) sets the display format for the coefficients to four digits after the decimal.

```
. gfi , ndig(4)

GFI Information from Interaction Specification of
Effect of Age on g(Sexfrqmonth) from Linear Regression
---------------------------------------------------------------------
Effect of Age =
   -0.2014 + 0.0013*Ses


Sign Change Analysis of Effect of Age
on g(Sexfrqmonth), Moderated by Ses (MV)


     ---------------------------------------
                   |              Age
     When          |      -------------------
     Ses=          |
     ------------+-------------------------
            17   | Neg b =     -0.1786
            27   | Neg b =     -0.1651
            37   | Neg b =     -0.1517
            47   | Neg b =     -0.1383
            57   | Neg b =     -0.1249
            67   | Neg b =     -0.1114
            77   | Neg b =     -0.0980
            87   | Neg b =     -0.0846
            97   | Neg b =     -0.0711
     ------------+-------------------------
     Sign Changes |         Never
     ------------+-------------------------
     % Positive   |          0.0
     ---------------------------------------
```

The GFI expression tells us that the base effect of age is negative (−0.2014) and that *it becomes smaller in magnitude as SES increases*. Note that the italicized phrase would be incorrect if SES could have negative values. The sign change analysis table shows that within the range of sample values of SES, the effect of age is negative and never changes. The values of the moderated age effect change considerably. The largest magnitude effect (−0.1786) is more than 2.5 times the size of the smallest (−0.0711).

Significance Region Analyses:
SIGREG **and** EFFDISP **Tools**

The SIGREG tool lets us explore whether the age effect changes in statistical significance as SES varies and the pattern and size of the changes in the age effect. I initially specify only one option: ndig(3) sets the number of digits for reporting the age effect to three.

```
Boundary Values for Significance of Effect of Age on g(sexfrqmonth) Moderated by SES
   Critical value F = 3.848 set with p = 0.0500
```

	When SES >=	Sig Changes	When SES >=	Sig Changes
Effect of age	116.745 (> max)	to Not Sig [-0.346]	254.014 (> max)	to Sig [0.035]

```
   Note: Derivatives of Boundary Values in [ ]
```

```
Significance Region for Effect of Age (1 unit difference)
on g(sexfrqmonth) at Selected Values of SES
```

					At SES=				
Effect of	17	27	37	47	57	67	77	87	97
age	-0.179*	-0.165*	-0.152*	-0.138*	-0.125*	-0.111*	-0.098*	-0.085*	-0.071*

```
Key: Plain font  = Pos, Not Sig    Bold font*   = Pos, Sig
     Italic font = Neg, Not Sig    Italic font* = Neg, Sig
```

Briefly, the boundary values report shows that the effect of age does not change significance within the sample range of SES values (17.1–19.2). There are changes in the age effect's significance that occur well beyond the maximum possible SES value of 97.2, when $SES = 116.7$ and $SES = 254.0$. The significance region table shows this invariant significance as well as the actual changes in the values of the age effect quite clearly. The effect of age on the frequency of intimacy is always negative. At the minimum SES (17), a 1-year increase in age would predict a 0.18 decline in the monthly frequency of intercourse. At the other end of the SES index, a 1-year increase in age would predict a 0.07 decrease in the frequency.

I think that the age effect's magnitude is better conveyed by reporting the age effect for a 10-year difference. And, given that the age effect is always negative and significant, limiting the display values for SES to every 20 units instead of every 10 units—the range( ) suboption for SES on *intspec*—will produce a more succinct and readable table. Such results can be produced using the effect( ) option as a calculator to get the focal variable's effect for any amount of change in the focal variable reported for each of the display values of the moderator. The syntax below accomplishes this recalculation by making the highlighted changes in the syntax for *intspec* and *sigreg*.

```
intspec focal(c.age) main( (c.age, name(Age) range(25(10)85)) ///
       (c.ses, name(SES) range(17(20)97))) int2(c.age#c.ses) ndig(0) ///
           sumwgt(n0)
sigreg, ndig(3) effect(b(10))
```

The results for the effect of age on frequency of intimacy for 10-year differences in age are as follows:

```
Significance Region for Effect of Age (10 unit difference)
on g(sexfrqmonth) at Selected Values of SES
```

			At SES=		
Effect of	17	37	57	77	97
age	-1.786*	-1.517*	-1.249*	-0.980*	-0.711*

```
Key: Plain font  = Pos, Not Sig    Bold font*   = Pos, Sig
     Italic font = Neg, Not Sig    Italic font* = Neg, Sig
```

A 10-year difference in age predicts a decline of close to 2 times a month in the frequency of having sex for those with very low SES (17). For someone just above the mean of SES[4] (57), the decline is much less at 1.2 times fewer per month. And at high SES (97), a 10-year difference in age reduces the frequency of intercourse by less than once a month (0.7). Keeping in mind that the mean frequency of intimacy is 4 times per month, the reduction in the 10-year age effects with increasing SES appears substantial, as does the magnitude of the 10-year age effect at any level of SES. For example, the age effect at $SES = 97$ is a reduction equal to 18% of the mean, while the age effect when $SES = 17$ is 44% of the mean.

An alternative way to visualize how SES moderates the size, sign, and significance of the age effect is with a confidence bounds plot, the default plot type for an interval-by-interval interaction created with the EFFDISP tool. I add the freq(tot) option to display the relative frequency distribution of SES below the plot and use the name( ) suboption within the plot( ) option to store the plot as a memory graph with the specified name. This produces a report of the plot options in the results window and the graph shown in Figure 7.1.

```
. effdisp , effect(b(10)) plot(name(Age_by_SES_by_frq) freq(tot)) ndig(1)


Plot Options Specified

   name =  Age_by_SES_b_frq
   freq =  tot
   type =  cbound by default
```

This plot is a visual counterpart to the significance region table for 10-year age effects. The *solid black line* shows that the effect of age decreases in magnitude from about −1.8 to about −0.8 as SES rises from 17 to 97. The spike plot at the bottom of the graph shows a somewhat even distribution of cases between $SES = 27$ and $SES = 77$ with somewhat larger concentrations in the neighborhood of 37 and 67, places at

FIGURE 7.1 ● EFFDISP **PLOT FOR ONE MODERATOR**

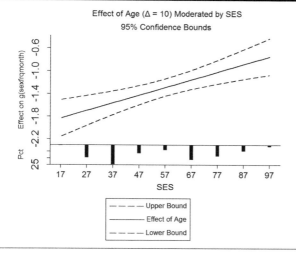

which the age effect is still fairly substantial. Because the confidence bounds (*dotted lines*) never contain the value 0, the age effect is always significant. Given that neither the sign of the age effect nor its statistical significance changes with SES, I find the effect display unnecessary. I would skip the effect display plot and instead present the significance region table or possibly an outcome display, as discussed next.

Outcome Displays: OUTDISP **Tool**

The OUTDISP tool produces tables or plots of the predicted outcome as it changes with the focal and moderating variables. Specifying the display values is especially consequential for creating tables because the predicted outcome is shown in the table only for the display values of the focal and moderating variables. For scatterplots and bar graphs, the moderators' display values similarly define and limit the calculation points (except that all the categories of a nominal variable are used to define calculation points). The focal variable's display values define axis labels, but the predicted outcome is calculated and shown across the full range of the focal variable's values. In general, pick the number and spacing of display values to capture important changes in the pattern of the relationship of the outcome with the interacting predictors.

As we just saw, the moderated age effect does not change sign or significance across the sample values of SES. Thus, using four display values across close to the full range of SES sample values (20–95 in increments of 25) works well to portray the relationship. This requires respecifying the range( ) suboption for SES on the *intspec* command and then running the *outdisp* command:

```
intspec focal(c.age) main( (c.age, name(Age) range(25(10)85)) ///
    (c.ses, name(SES) range(20(25)95))) int2(c.age#c.ses) ndig(0)
outdisp, outcome(atopt((means) _all)) plot(name(Age_by_SEI_Sex)) ///
    table(default)
```

The *outdisp* options specify the following:

- outcome(atopt((means) _all)) sets how the values of the predictors not part of the interaction are treated in calculating the predicted values of the monthly frequency of intercourse; that is, they are set to their means.[5]

- plot(name(SexFrq_by_Age_by_SES)) requests the creation of a plot that is stored during the duration of the Stata session with the specified name. Because the type( ) suboption is not listed, ICALC creates the default plot type, which is a scatterplot for an interval-by-interval interaction.

- table(default) specifies creation of a table of predicted values with the rows defined by the focal variable by default.

Reading down a column of the predicted values table indicates how the monthly frequency of having sex is predicted to change with age at the given level of SES. The pattern is easy to discern: The frequency of intercourse declines with age, and the amount of decline is less at higher levels of SES. A simple way to see this for a two-variable interaction—and explain it to an audience—is to compare for each column the change in the predicted outcome for the youngest age (25) with that for the oldest age (85). The reduction in the predicted number of times of having sex with age changes sharply with SES, as shown in Table 7.1:

- 10.473 for *SES* = 20

- 8.459 for *SES* = 45

- 6.444 for *SES* = 70

- 4.430 for *SES* = 95

The scatterplot in Figure 7.2 also portrays the pattern quite well. The lines in the scatterplot trace the predicted frequency of intercourse by age for each level of SES and make it quite obvious that all the slopes are negative. Similarly, how SES moderates the age effect is clear. The *solid line* for *SES* = 20 has the steepest slope. The slope is somewhat shallower when *SES* = 45 (*medium dashed line*), even shallower when *SES* = 70 (*large dashed line*), and shallower yet when *SES* = 95 (*small dashed line*).

Recap

The GFI and SIGREG results provided useful initial information about the moderation of the age effect by SES—namely, that the age effect is always negative and statistically significant across the range of SES values. The significance region table clearly summarized the relationship between age and frequency of intercourse and how that was moderated by SES. The confidence bounds plot produced by the EFFDISP tool also showed these patterns in a straightforward way. The application of the OUTDISP tool created both a table and a scatterplot of the predicted frequency of intimacy as it varies with age and how that relationship changes with SES. Both showed the pattern of the effect of age and its moderation by SES in a way that is easy to see and understand and, consequently, easy to present to an audience.

I suspect that many readers who are accustomed to seeing interaction effects interpreted using predicted values found the significance region table and confidence bounds plot less useful and harder to interpret, in part, because they are unfamiliar. If for nothing else, they are useful for identifying what the changes are in the pattern of the focal variable's effect and where they occur. This informs the choice of

TABLE 7.1 ● PREDICTED VALUES TABLE FROM STATA OUTPUT

Predicted Value of sexfrqmonth by the Interaction of Age with SES.

Age	SES			
	20	45	70	95
25	8.6077	7.4937	6.3797	5.2658
35	6.8622	6.0840	5.3057	4.5275
45	5.1167	4.6742	4.2317	3.7892
55	3.3712	3.2645	3.1577	3.0509
65	1.6257	1.8547	2.0837	2.3127
75	-0.1198	0.4449	1.0097	1.5744
85	-1.8653	-0.9648	-0.0643	0.8361

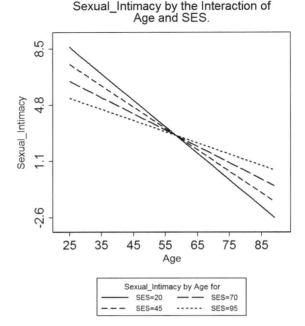

FIGURE 7.2 ● OUTDISP SCATTERPLOT FOR AGE MODERATED BY SES

display values for the moderator to show all the changes in the pattern. Hopefully, their utility will become more apparent when applied to more complex interactions or in situations where the moderated effect changes sign, as it does when we turn to interpreting the effect of SES moderated by age.

The Effect of SES Moderated by Age

One of the common errors in interpreting interaction effects that I discussed in Chapter 1 is to focus on only one of the predictors in the interaction and how it is moderated by the other. This provides an incomplete picture of the relationship between the outcome and the interacting predictors. As a case in point, we know how age affects the frequency of intercourse and how that is contingent on SES. But we know very little about how SES affects the outcome and how that is moderated by age. Thus, I reverse the roles of age and SES to interpret how SES as the focal variable is moderated by age.

Applying the ICALC **Tools**

The ICALC syntax that we use is nearly identical to what we just used for age as the focal variable. So I will walk through the example's Stata output and only note the changes in the command lines without a detailed explanation. Initially, I declare SES as the focal variable on the *intspec* command by changing the focal(c.age) option to focal(c.ses) and then use the same specifications for the *gfi* and *sigreg* commands.

```
intspec focal(c.ses) main( (c.age, name(Age) range(25(10)85)) ///
>         (c.ses, name(SES) range(17(10)97))) int2(c.age#c.ses) ndig(0) sumwgt(no)
```

```
Interaction Effects on Sexfrqmonth Specified as

      Main effect terms: ses  age
      Two-way interaction terms: c.age#c.ses

   These will be treated as: Focal variable = ses ("Ses")
     moderated by interaction(s) with
        age ("Age")
```

. **gfi** , **ndig(4)**

```
GFI Information from Interaction Specification of
Effect of Ses on g(Sexfrqmonth) from Linear Regression
-----------------------------------------------------------------------


Effect of Ses =
   -0.0781 + 0.0013*Age

Sign Change Analysis of Effect of Ses
on g(Sexfrqmonth), Moderated by Age (MV)


         ---------------------------------------
                   |            Ses
   When            |    --------------------
   Age=            |
         ------------+-----------------------
         25   | Neg b =    -0.0446
         35   | Neg b =    -0.0311
         45   | Neg b =    -0.0177
         55   | Neg b =    -0.0043
         65   | Pos b =     0.0092
         75   | Pos b =     0.0226
         85   | Pos b =     0.0360
         ------------+-----------------------
   Sign Changes |  when MV= 58.17982
         ------------+-----------------------
   % Positive   |        29.6
         ---------------------------------------
```

The algebraic expression for SES's effect tells us that its base effect on the frequency of having sex is negative (−0.0781) but becomes less negative (more positive) as age increases (0.0013). The sign change analysis indicates that the SES effect is initially negative but becomes smaller in magnitude as age increases and eventually turns positive. The row labeled "Sign Changes" reports that the sign change occurs when *Age* = 58.2; that is, for those 58 years and younger, there is a negative relationship between SES and frequency of intercourse; but for those older than 58, the relationship is positive. The last row in the table indicates that about one third of respondents have a positive effect of SES and two thirds have a negative effect.

Because the effect of SES changes sign, we want to know for what values of age the effect is significant. It could be that a range of both positive and negative effects is significant, that only positive effects are significant, or that only negative effects are significant. Knowing this is obviously consequential for how we understand the nature of the relationship between SES and the frequency of intimacy. Running the *sigreg* command produces the information we need.

Boundary Values for Significance of Effect of SES on **g(sexfrqmonth)** Moderated by Age
Critical value F = **3.848** set with p = **0.0500**

	When Age >=	Sig Changes	When Age >=	Sig Changes
Effect of ses	48.875	to Not Sig [-0.655]	69.251	to Sig [0.463]

Note: Derivatives of Boundary Values in [ ]

Significance Region for Effect of SES (1 unit difference)
on g(sexfrqmonth) at Selected Values of Age

				At Age=			
Effect of	25	35	45	55	65	75	85
ses	*-0.045**	*-0.031**	*-0.018**	*-0.004*	0.009	**0.023***	**0.036***

Key: Plain font = Pos, Not Sig   **Bold font*** = Pos, Sig
     Italic font = Neg, Not Sig   *Italic font** = Neg, Sig

The boundary values analysis provides exact information on the change in significance of the effect of SES:

- When *Age* ≥ 48.875, the SES effect changes from significant to nonsignificant.

- When *Age* ≥ 69.251, the SES effect becomes significant again.

For this example, the significance region table works well to present how age moderates the SES effect because you can readily see the changing sign, magnitude, and statistical significance of the moderated effect. You can supplement discussion of the table with the exact age values at which the sign and significance of the effect change. But rather than show the SES effect as one-unit differences, I would present 1 standard deviation differences. And I would save the formatted table to Excel for presentation. You add two options to *sigreg* to do this:

```
sigreg , ndig(3) effect(b(sd)) save(Output\Table_7_2.xlsx table)
```

The effect( ) option specifies calculating effects as 1 standard deviation changes in SES. In the save( ) option, the keyword "table" saves the formatted table in the specified file and location.

Table 7.2 presents the formatted significance region table. This shows that the effect of SES on the frequency of having sex changes from negative and significant to positive and significant across the age range. For the youngest respondents (age 25), the frequency of intercourse is almost one time a month fewer for someone with 1 standard deviation higher SES. At age 45, the SES effect is a decline in frequency of about one third of a time per month, and the SES effect becomes not significant at age 48.875. The effect turns positive around age 58 but does not become significant until age is slightly more than 69 years. At age 75, a 1 standard deviation higher SES predicts someone having sex almost one-half time more per month, increasing to 0.7 times more for someone aged 85.

Applying the EFFDISP tool creates a confidence bounds plot by default that is also a very effective presentation choice in this context. Inside the plot( ) option, the name( ) suboption stores the plot as a named memory graph for the duration of your Stata session, and the freq( ) option adds a spike plot of the frequency distribution of age to the plot shown in Figure 7.3. The *solid line* shows how the moderated effect of SES

| TABLE 7.2 ● EFFECT OF SES MODERATED BY AGE, FORMATTED TO HIGHLIGHT SIGN AND SIGNIFICANCE |

Effect of SES (1 *SD* difference) on *g*(sexfrqmonth)							
Effect of SES	Age (years)						
	25	**35**	**45**	**55**	**65**	**75**	**85**
	*−0.884**	*−0.618**	*−0.351**	*−0.085*	0.182	**0.448***	**0.715***
Key							
Plain font, no fill	Pos, not Sig						
Bold*, filled	Pos, Sig						
Italic, no fill	Neg, not Sig						
Bold italic*, filled	Neg, Sig						

Note: SES = socioeconomic status; SD = standard deviation.

changes with age, while the two *dotted lines* represent the upper and lower confidence bounds for the SES effect. At a given age, when the *dotted lines* bracket the horizontal zero reference line, this indicates a nonsignificant effect.

```
effdisp, plot(name(SexFrq_SES_by_Age) freq(tot)) ndig(2)
```

The areas of significance and nonsignificance on the plot are shown by the two *vertical reference lines*. For ages below the first vertical line around age 49, SES has a negative and significant relationship with the frequency of having sex. For those between

| FIGURE 7.3 ● CONFIDENCE BOUNDS PLOT FROM EFFDISP |

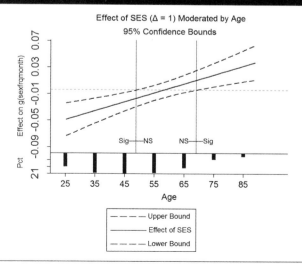

the two vertical lines—ages 49 through 69—SES does not have a significant effect on the frequency of having sex. And for those above age 69, the effect of SES turns positive and significant. Note from the spike plot that a plurality of cases have negative and significant effects of SES. The next largest group—ages 49 to 68—have no significant effect of SES, and the smallest group—over age 72—experience a positive relationship between SES and frequency of intercourse.

The OUTDISP tool creates tables and/or plots of the predicted outcome as it varies with the focal and moderating variables, which are the most commonly used displays for interpreting interaction effects. For this example, we want to pick display values to highlight how the SES effect varies across the range of ages from a positive to a negative effect of similar magnitude, and where the SES effect experiences notable transitions. Setting the age display to range from 26 through 81 in steps of 11 years will achieve this. The lines in a scatterplot (rows in the predicted values table) showing the relationship between SES and the frequency of sex contingent on age will be at or near where the SES effect

- has its minimum and maximum values (25, 85),

- changes from significant to nonsignificant (49),

- changes sign (58), and

- changes from nonsignificant back to significant (69).

To accomplish this, we change the range( ) suboption for age in the *intspec* command to range(26(11)81) and then run the *outdisp* command with a plot( ) option to store the scatterplot to a memory graph, a table( ) option to create a table of predicted values, and the outcome( ) option to set the reference values for the other predictors, as shown in this excerpt from the Stata output.

```
. intspec focal(c.ses) main( (c.age, name(Age) range(26(11)81)) ///
    …

. outdisp, outcome(atopt((means) _all)) plot(name(SexFrq_by_SES_by_Age)) >
>       table(save(Output\Table_7_3.xlsx)
    …
```

The scatterplot in Figure 7.4 shows the progression of the SES effect on the frequency of intercourse from its most negative effect at age 26 (*solid line*) through its most positive effect at age 81 (*large-dash-and-dot line*). The third, fourth, and fifth lines from the top mark the transition from negative and barely significant (age 48), to barely positive and nonsignificant (age 58), to positive and barely significant (age 70). This scatterplot makes it easy to see the effect of SES and the frequency of having sex contingent on age. But while it is possible with some effort to tease out the effect of age contingent on SES from this plot, it is not easy to do so. I think this point is even truer for pulling out the effect of SES moderated by age from the scatterplot designed to highlight the effect of age moderated by SES in Figure 7.2. Presenting and interpreting both scatterplots would solve this problem.

A good alternative, at least for two-way interaction effects, is to report and interpret a table of predicted values from which you can equally well see and interpret both sides of the interaction effect. Look at Table 7.3, which presents the predicted values for frequency of having sex by age and SES. The rows are defined by SES and the

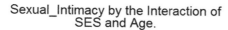

FIGURE 7.4 ● OUTDISP SCATTERPLOT OF SES MODERATED BY AGE

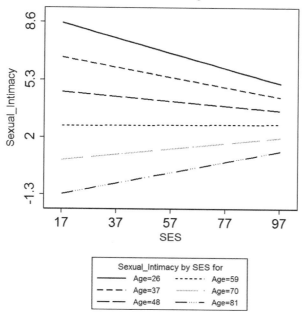

Note: SES = socioeconomic status.

TABLE 7.3 ● PREDICTED VALUES TABLE FROM OUTDISP, ONE MODERATOR

	sexfrqmonth by the Interaction of SES With Age					
	Age					
SES	**26**	**37**	**48**	**59**	**70**	**81**
17	8.5628	6.5985	4.6341	2.6697	0.7054	−1.2590
37	7.6985	6.0296	4.3607	2.6917	1.0228	−0.6461
57	6.8342	5.4607	4.0872	2.7138	1.3403	−0.0331
77	5.9698	4.8918	3.8138	2.7358	1.6578	0.5798
97	5.1055	4.3229	3.5404	2.7578	1.9753	1.1927

columns by age, and the cell entries report the predicted frequency of intercourse for the combination of SES and age values. This table that ICALC saved to an Excel file is formatted with font sizes proportional to the cell values. Reading down a column shows the SES effect for the given value of age. In each of the first three columns, you can see that the predicted frequency of intercourse declines with SES as you read down the column—note that the font sizes are largest in the first row and then shrink as you read down the column.

You can also see that the rate of decline with SES diminishes as age increases across the columns. At age 26, the change in sexual activity from lowest to highest SES is a reduction of about 3.5 times per month ($8.56 - 5.11 = 3.45$), while at age 48, the reduction is only about once per month ($4.63 - 3.54 = 1.09$). Examining the next three columns, the SES effect has become positive, increasing across the columns. The increase in sexual activity with SES at age 59 is hardly different from 0 ($2.76 - 2.67 = 0.09$), but at age 81, the increase is about two thirds as large ($1.19 - [-1.26] = 2.45$) as the reduction in sexual activity with SES at age 25.

Examining the rows reveals the effect of age contingent on SES, which is even easier to discern because the frequency of having sex at every value of SES decreases from left to right—as do the font sizes—in each row as age increases. And the magnitude of the predicted drop in sexual activity declines from 9.82 at *SES* = 17 to about one third as large (3.91) at *SES* = 97. You can also see this by observing that the highest frequency of sexual activity in a row is for age 26 and that a row's maximum value steadily declines across the range of SES values. At the same time, the smallest frequency of sexual activity is always for age 81, and a row's minimum steadily increases across the range of SES values. Hence, the difference between the falling maximum and the rising minimum (i.e., the effect of age) steadily declines with SES.

Recap

The essence of how age moderates the relationship between SES and frequency of having sex is well documented by the GFI and SIGREG results: SES has a significant negative relationship with sexual activity for adults under age 49 but a positive relationship for those older than 69. Each of the summary displays of this relationship—the significance region table, the confidence bounds plot, the predicted values scatterplot, and the table of predicted values—made this overall pattern clear.

Summary and Recommendations

I explored a variety of techniques for probing this interval-by-interval interaction effect. The pattern of the relationship between monthly frequency of sexual activity and the interaction of age and SES is not complex. So you have many good choices to help you understand and then to explain to an audience the nature of the relationship. I think that presenting and interpreting any of the following could be effective:

1. The two confidence bounds plots for the effect of age on sexual activity contingent on SES and for the effect of SES on sexual activity contingent on age. This is my least preferred option in this example. These plots take more work to explain to most audiences, in part because they are unfamiliar and, likely, many readers will be confused at first because they expect to see predicted values rather than moderated effects. Thus, it may not be worth the effort to use them when other options show the pattern well and are easier to explain.

2. The two formatted significance region tables for the effect of age on sexual activity moderated by SES and for the effect of SES on sexual activity moderated by age. I would suggest reporting effects for 1 standard deviation

changes in the focal variable in both tables. In addition to creating a consistent presentation and discussion of effects, this also permits a comparison of the magnitude of the SES effect versus the age effect. I would also recommend removing the fill pattern highlighting. I think it is more distracting than helpful in this straightforward example. We will see its utility in later examples with multiple moderators or a three-way interaction.

3. The pair of scatterplots, one with plotted lines showing the predicted frequency of sex varying with age for selected values of SES and the other with plotted lines for the predicted frequency of intercourse changing with SES at selected values of age. These are visually appealing and easy to comprehend and interpret for an audience.

4. The table of predicted sexual activity as it changes simultaneously with age and SES. While the table has somewhat less immediate visual impact than the other options, it has the distinct advantage that you only need to present and discuss a single display of the relationship. Other than explaining that the font sizes are proportional to the predicted values, it requires little in the way of explanation or instruction about how to read the table and what it shows.

In the end, the choice is a matter of two factors—first, who your audience is and how accessible you think they would find the different modes of reporting the interactive relationship, and second, and equally important, what works best for you, what method you find the most intuitive and comprehensible. The better you can understand and relate to the technique for reporting the interaction effect, the better you will be able to use it to tell a story that others will understand.

TWO-MODERATOR EXAMPLE

Data and Testing

This analysis regresses the respondent's number of children on the interaction of family income and birth cohort, the interaction of family income and education, and also predictors of the number of siblings, religious intensity, and race. Children, income (in $10K), education, and siblings are interval measures. Birth cohort is a four-category nominal variable (Depression Era, WWII, Baby Boom, and Post-Boom[6]) represented by three dummy variables using Post-Boom as the reference category. Race is a three-category nominal variable (White, Black, and Other) represented by two dummy variables using White as the reference.

The sample is restricted to respondents aged 40 and over, to make it more likely that childbearing has been completed. An additional 218 cases were excluded due to missing information on one or more of the variables.[7] Diagnostic tests identified a small cluster of 15 outliers, but they were neither unusual (unrealistic) in their characteristics nor influential in affecting the estimation results and conclusions.[8] Statistical tests for heteroscedasticity indicated its possible presence, but diagnostic plots suggested a small degree of variation and were more consistent with nonlinearity in the effects of income and education, such as an interaction. A sensitivity analysis comparing the results for OLS versus OLS with robust standard errors demonstrated inconsequential differences, both overall and in terms of testing for the presence of the interaction effects.

This analysis demonstrates how to interpret interactions when there are two moderators of the same focal variable, family income by cohort and by education. Reversing the roles of the focal and moderator variables, it illustrates how to interpret an interaction between a multicategory focal variable and an interval moderator (cohort by family income), as well as a second example of an interval-by-interval interaction (education by family income). Conceptually, the interaction between economic conditions and birth cohort to predict fertility behavior has a long history of study, going back to the Easterlin (1961) hypothesis. The expectation is that the income effect should be larger (more negative) for more recent birth cohorts. An income-by-education interaction on fertility behaviors has also long been studied (e.g., Halli, 1990; Simon, 1975).

Including the income-by-education interaction in the model is supported by the t test of its coefficient ($t = 2.47$, $p = .14$) in the full model, which also includes the income-by-cohort interaction. The statistical grounds for including the income-by-cohort interaction terms in the model is to some degree a judgment call, depending on how you test the set of three interaction parameters (see Chapter 1). A global test of the simultaneous removal of the three parameters from the full model is not significant ($p = .166$), but the global test cannot take into account that we would expect all the three interaction parameters to be negative and hence that their sum should be less than 0.

You can do a one-tailed test of this under a null hypothesis that the sum of the three parameters is greater than or equal to 0. This results in a significant t statistic rejecting the null hypothesis ($t = -1.82$, $p = .035$). Alternatively, we could test the three parameters individually with a one-tailed test, adjusting the significance level for multiple testing, and conclude that the interaction term should be included if any of the three are negative and significant. This procedure also supports including the income-by-cohort interaction in the model because the Depression-Era-by-income coefficient is negative and significant ($t = -2.26$, $p = .012$, Sidak-adjusted $\alpha = .017$). Since the directional tests of the cohort-by-family-income interaction are significant, I include the family-income-by-cohort interaction in the model.

Strategy for Interpreting Two-Moderator Interaction Models

When you have multiple moderators of the same focal variable, you need to decide whether to interpret it one focal variable–moderator pair at a time or all of them simultaneously. In this instance, the first option is to interpret how the family income effect differs by cohort, with education set to reference values, and then how the income effect varies by education, with cohort limited to reference values. The second option is to interpret the effect of income as it changes across combinations of education's display values and cohort's display values. I prefer the second choice because it is a more holistic view. For LRMs and other linear link models, this choice is inconsequential because you will see exactly the same pattern for the income-by-education interaction at any value of cohort and analogously for income-by-cohort interaction.

Keep in mind that you should reverse the roles of the focal and moderator variables, which defines two focal-by-single-moderator interactions to explore—that is, (1) birth cohort's effect moderated by income and (2) education's effect moderated by income. I recommend interpreting these first. And then I would interpret the double-moderator component: how the family income effect is moderated by cohort and education. This will make an outcome display of the predicted number of children varying with income, cohort, and education simultaneously—which we will consider at the end of the section on interpreting family income moderated by cohort and education—easier to interpret because we will have an understanding of all the underlying patterns in hand.

The Effect of Birth Cohort Moderated by Family Income

INTSPEC **Setup and** GFI **Analysis**

These commands are shown in boldface at the top of the Stata output for the GFI analysis. I define birth cohort as the focal variable with the focal(i.cohort) option. The main( ) option includes birth cohort and its moderator family income (*c.faminc10k*). Because *i.cohort* is a nominal factor variable, ICALC will automatically include all its categories as display/calculation values, so the range( ) suboption would be ignored if it were listed. I set the display values for family income as $10K to $190K in $30K increments with the suboption range(1(3)19) because income is coded in units of $10K.

The *gfi* command has only the ndig(3) option to set the format for coefficients in the tables to three digits after the decimal. Given the scale of the coefficients for birth cohort and its interaction with family income, this provides three to four digits of information, which is what I prefer.

```
. intspec  focal(i.cohort)  ///
>         main((c.faminc10k , name(Family_Inc) range(1(3)19)) ///
>              (i.cohort , name(Cohort))) ///
>         int2(c.faminc10k#i.cohort ) ndig(0)
     ...
.
. gfi ,  ndig(3)

GFI Information from Interaction Specification of
Effect of Cohort on g(Childs) from Linear Regression
-------------------------------------------------------------------

Effect of BabyBoom =
    0.419 - 0.026*faminc10k

Effect of WWII =
    0.817 - 0.028*faminc10k

Effect of DeprEra =
    1.649 - 0.065*faminc10k
```

When	Cohort					
Family_Inc=	BabyBoom		WWII		DeprEra	
1	Pos b =	0.393	Pos b =	0.789	Pos b =	1.585
4	Pos b =	0.316	Pos b =	0.705	Pos b =	1.391
7	Pos b =	0.239	Pos b =	0.620	Pos b =	1.197
10	Pos b =	0.162	Pos b =	0.535	Pos b =	1.004
13	Pos b =	0.085	Pos b =	0.451	Pos b =	0.810
16	Pos b =	0.008	Pos b =	0.366	Pos b =	0.616
19	Neg b =	-0.069	Pos b =	0.281	Pos b =	0.423
Sign Changes	when MV= 16.298		Never		Never	
% Positive	92.0		100.0		100.0	

Birth cohort has three effects in the regression analysis corresponding to the three included cohort indicators. Each effect represents the predicted difference in the number of children between an included cohort and the reference cohort (Post-Boom). Thus, the GFI and sign change analyses report information about each effect as it varies with family income. The GFI results indicate that each of the included cohorts has a positive main effect coefficient but a negative interaction effect coefficient. This means that each of these cohorts has a larger predicted number of children than the Post-Boom cohort when *Income* = 0 but the difference diminishes as income increases. The rate of decline is similar for the Baby Boom and WWII cohorts and less than half the rate for the Depression Era cohort. The sign change analysis tells us that only the Baby Boom effect turns negative (fewer predicted children than the Post-Boom cohort) and that this occurs at a fairly high income level ($162,980).

Significance Region Analyses: SIGREG and EFFDISP Tools

Because the differences among cohorts in the predicted number of children declines with family income, this raises the question of whether the differences remain significant, especially since they change at different rates. I start exploring this using the *sigreg* command with minimal options (shown in bold at the top of the output), specifying the significance level to use (.05) and the number of digits to report in tables. An important point to keep in mind is that the details of these results are contingent on the choice of the reference category for birth cohort. For instance, if we used the Depression Era cohort for the reference, almost all of the moderated effects for the three included cohorts would be negative rather than positive. But the results and hence the stories we tell would be consistent in the meaning of the overall cohort effect on the number of children.

```
Boundary Values for Significance of Effect of Cohort on g(childs) Moderated by Family Inc
    Critical value F = 3.850 set with p = 0.0500
```

Effect of Cohort	When Family_Inc >=	Sig Changes	When Family_Inc >=	Sig Changes
BabyBoom	5.919	to Not Sig [-0.457]	-2.338 (< min)	to Sig [0.142]
WWII	11.575	to Not Sig [-1.070]	-11.095 (< min)	to Sig [0.202]
DeprEra	15.213	to Not Sig [-1.356]	376.357 (> max)	to Sig [0.001]

```
    Note: Derivatives of Boundary Values in [ ]
```

```
    Significance Region for Effect of Cohort (1 unit difference)
    on g(childs) at Selected Values of Family_Inc
```

Effect of	At Family Inc=						
	1	4	7	10	13	16	19
BabyBoom	0.393*	0.316*	0.239	0.162	0.085	0.008	*-0.069*
WWII	0.789*	0.705*	0.620*	0.535*	0.451	0.366	0.281
DeprEra	1.585*	1.391*	1.197*	1.004*	0.810*	0.616	0.423

```
    Key: Plain font  = Pos, Not Sig     Bold font*   = Pos, Sig
         Italic font = Neg, Not Sig     Italic font* = Neg, Sig
```

The boundary values analysis pinpoints the income level at which each of the cohort effects changes from significant to not significant. This occurs at $59,190 (5.919 × $10,000) for the Baby Boom cohort, about $116K for the WWII cohort, and about $152K for the Depression Era cohort. You can see this visually in the formatted significance region table—note where the cell entries change from having a * symbol to not—which also shows the Baby Boom effect turning negative but remaining nonsignificant (italicized font).

The effect values show the differences in the predicted number of children between each cohort compared with the Post-Boom cohort; positive values indicate that the

Post-Boom cohort has a smaller predicted number than the comparison cohort. The Post-Boom cohort has the lowest predicted number of children, with the exception of the Baby Boom cohort at high income levels, but this disparity is not significant. The other two cohorts always have a predicted number of children greater than the Baby Boom and Post-Boom cohorts. And the Depression Era cohort has a higher predicted number of children than the WWII cohort except at very high incomes.

Another good presentation option is a confidence bounds plot for each of the included cohort effects. Because the default plot is needed, I only specify on the *effdisp* command the option ndig(1) to format the *y*-axis labels to one decimal place:

effdisp, ndig(1)

Figure 7.5 presents the three confidence bounds plots. The *horizontal thin reference line* separates the negative from the positive effects and reaffirms that only the Baby Boom cohort's effect turns negative. The *vertical reference lines* mark the change from significant to not significant for each cohort, as well as documenting the different income levels at which this occurs for each cohort.

Outcome Displays: OUTDISP **Tool**

For a multicategory focal variable, the results in outcome displays are not contingent on the choice of the base reference category because the outcome's predicted values are displayed for all the categories. Thus, they are typically a better option to convey the nature of the interaction effect in this situation than significance region tables or effect displays. I request the default plot from the *outdisp* command by specifying the options with plot(def) and use the moderator to define rows in the predicted values table with tab(row(mod)). The default plot for a categorical focal variable such as birth cohort is a bar chart for each of the display values of the moderator. I use the same *intspec* command as before, and the *outdisp* command produces the following output and bar charts.

FIGURE 7.5 ● EFFDISP FOR MULTIPLE-CATEGORY FOCAL VARIABLE

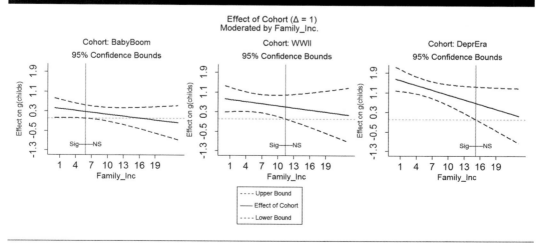

```
. intspec  focal(i.cohort)   ///
>          main((c.faminc10k , name(Family_Inc) range(1(3)19)) ///
>                 (i.cohort , name(Cohort) range(1/4))   ) ///
>          int2(c.faminc10k#i.cohort ) ndig(0)

    ...

.

. outdisp, out(atopt((means) _all)) plot(def) table(row(mod)) ndig(2)
```

Outcome Options Specified or Default

```
    metric =  obs
    atopt =  (asobs) _all
```

Table Options Specified or Default

```
    row =  mod
```

Plot Options Specified or Default

```
    type =  bar
```

Predicted Value of Childs by the Interaction of Cohort with Family_Inc.

```
-------------------------------------------------------------------
                     |                    Cohort
        Family_Inc |  PostBoom   BabyBoom    WWII     DeprEra
-------------------+-----------------------------------------------
                 1 |    1.57       1.96      2.36       3.15
                 4 |    1.69       2.00      2.39       3.08
                 7 |    1.81       2.05      2.43       3.00
                10 |    1.93       2.09      2.46       2.93
                13 |    2.05       2.13      2.50       2.86
                16 |    2.17       2.17      2.53       2.78
                19 |    2.29       2.22      2.57       2.71
-------------------------------------------------------------------
```

Comparing the rows in the predicted values table reveals how the predicted differences among cohorts in number of children change across family income levels. The first five rows show the predicted number of children increasing from the youngest cohort (Post-Boom) to the oldest cohort (Depression Era). Doing some calculations in your head (or otherwise) shows diminishing differences among the cohorts at higher income levels. Moreover, the pattern of cohort differences alters in the last two rows while generally continuing to equalize. You can also contrast the patterns in the columns to see how the effect of family income is contingent on birth cohort. The Post-Boom cohort exhibits a steady increase in the predicted number of children with rising family income, the Depression Era cohort shows a smaller steady decline with income, and the other two cohorts show increases that are less in magnitude.

The bar charts in Figure 7.6 reveal these patterns without you having to do calculations in your head. The height of the bars represents the predicted number of children in a birth cohort, and the bars are labeled with the predicted value. When family income is $130K or less (the first five bar charts), there is a stair-step pattern in which the number of children increases from the most recent cohort to the oldest cohort. But the size of the steps diminishes as income rises. Beyond $130K, the cohort differences continue to level off, and the relative order of the number of children across cohorts changes.

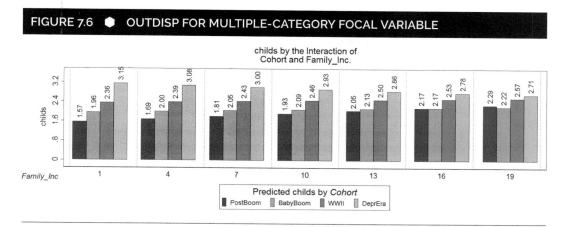

FIGURE 7.6 ● OUTDISP FOR MULTIPLE-CATEGORY FOCAL VARIABLE

Next, compare the height of the bar for each cohort individually across income levels. The Post-Boom cohort exhibits a steady increase in the predicted number of children with rising family income, the Depression Era cohort shows a steady decline with income, and the other two cohorts show increases that are much smaller in magnitude. This provides a first take on how family income is differentially related to the number of children. But the income effect is also moderated by education, so these values represent the income effect at education's reference value (its mean). In the next section, I probe how education's effect on the number of children is moderated by family income, and in the process, we learn more about the children–income relationship.

The Effect of Education Moderated by Family Income

INTSPEC **Setup and** GFI **Analysis**

The *intspec* command has the same structure as for the cohort-by-income interaction just analyzed, with the substitution of education (*c.ed*) and its information as the focal variable, as shown in the Stata output. The *gfi* command now reports four digits for the coefficient values in the tables.

```
. intspec  focal(c.ed)  ///
>          main((c.faminc10k , name(Family_Inc) range(1(3)19)) ///
>               (c.educ , name(Education) range(0(4)20)) ) ///
>          int2( c.faminc10k#c.educ ) ndig(0)
    ...
. gfi , ndig(4)

GFI Information from Interaction Specification of
Effect of Education on g(Childs) from Linear Regression
-----------------------------------------------------------------

Effect of Educ =
   -0.1836 + 0.0065*Faminc10k
 Sign Change Analysis of Effect of Education
 on g(Childs), Moderated by Family_Inc (MV)
```

```
---------------------------------------
              |      Education
When          |   --------------------
Family_Inc=   |
--------------+------------------------
           1  | Neg b =    -0.1771
           4  | Neg b =    -0.1576
           7  | Neg b =    -0.1381
          10  | Neg b =    -0.1186
          13  | Neg b =    -0.0990
          16  | Neg b =    -0.0795
          19  | Neg b =    -0.0600
--------------+------------------------
Sign Changes  |         Never
--------------+------------------------
% Positive    |          0.0
---------------------------------------
```

The GFI expression tells us that the moderated education effect starts as negative when family income is 0 and is predicted to decline by nearly 2/10 of a child for a 1-year difference in education. But the effect becomes less negative as family income rises—the coefficient value increases by 0.0065 with a $10K increase in family income. The sign change analysis shows that the education effect remains negative across the range of income values.

Significance Region Analyses: SIGREG Tool

To determine if the education effect remains significant, I use the *sigreg* command, with the number of digits for reporting set to four. The boundary values analysis indicates that the education effect on number of children is no longer significant once family income is greater than $170,529. The significance region table also shows this, as well as how the magnitude of the education effect is changing. By the income level at which it loses significance, the education effect is about one third of what it is at *Income* = 0. A confidence bounds plot could also be used to show this—*effdisp* with no options—but I prefer the significance region table when the results are simple and straightforward.

Boundary Values for Significance of Effect of Education on **g(childs)** Moderated by Family Inc
Critical value F = **3.850** set with p = **0.0500**

	When Family_Inc >=	Sig Changes	When Family_Inc >=	Sig Changes
Effect of educ	17.0529	to Not Sig [-1.2052]	159.7610 (> max)	to Sig [0.0087]

Note: Derivatives of Boundary Values in [ ]

Significance Region for Effect of Education (1 unit difference)
 on g(childs) at Selected Values of Family_Inc

Effect of				At Family Inc=			
	1	4	7	10	13	16	19
educ	-0.1771*	-0.1576*	-0.1381*	-0.1186*	-0.0990*	-0.0795*	-0.0600

Key: Plain font = Pos, Not Sig   **Bold font***   = Pos, Sig
     Italic font = Neg, Not Sig   *Italic font** = Neg, Sig

Outcome Displays: OUTDISP Tool

I use the *outdisp* command to produce the table below and the plot in Figure 7.7 to portray how the predicted number of children changes with education contingent on family income. As shown at the top of the Stata output, I specify three options:

(1) out( ) to set the reference values for the other predictors in the model to their means, (2) plot( ) to produce the default scatterplot for interval variables stored as a named memory graph, and (3) tab( ) to create a predicted values table with the moderator (income) to define the rows.

```
. intspec  focal(c.ed)  ///
>          main((c.faminc10k , name(Family_Inc) range(1(3)19)) ///
>              (c.educ , name(Education) range(0(4)20)) ) ///
>          int2( c.faminc10k#c.educ ) ndig(0)
    ....

. outdisp, out(atopt((means) _all)) plot(name(Plot1)) tab(row(mod))

Outcome Options Specified or Default

   metric =  obs
   atopt =   (means) _all

Table Options Specified or Default

   row =  mod

Plot Options Specified or Default

   type =  scat
   name =  plot1


Predicted Value of Childs by the Interaction of Education with Family_Inc.
```

Family_Inc	Education 0	4	8	12	16	20
1	4.5489	3.8404	3.1320	2.4235	1.7150	1.0065
4	4.3189	3.6885	3.0581	2.4277	1.7973	1.1669
7	4.0888	3.5365	2.9842	2.4319	1.8796	1.3272
10	3.8587	3.3845	2.9103	2.4361	1.9618	1.4876
13	3.6287	3.2325	2.8364	2.4403	2.0441	1.6480
16	3.3986	3.0806	2.7625	2.4445	2.1264	1.8084
19	3.1686	2.9286	2.6886	2.4487	2.2087	1.9687

Each row in the table shows that the predicted number of children decreases across education levels at a rate that diminishes with income level. For example, between 0 and 20 years of education, the number of children is predicted to drop by 3.54 children when income equals $10K, by 2.37 when income equals $100K, and by 1.20 for an income of $190K. Comparing the columns reveals the relationship between number of children and family income as the relationship varies by education, but keep in mind this is at the reference values for birth cohort. Nonetheless, we see that the income effect changes from a negative one for low levels of education to a positive one at higher levels of education.

Presenting a predicted values plot is often a more convenient choice because it obviates the need for you to write out, as I did, examples of the magnitude of the numeric changes, to include such change calculations in the table, or to leave it to the reader to do head calculations. The upper panel of Figure 7.7 shows the default scatterplot created by the *outdisp* command. The lower panel is the same scatterplot with the

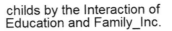

FIGURE 7.7 ● OUTDISP **SCATTERPLOT FOR ONE MODERATOR, DEFAULT AND REVISED**

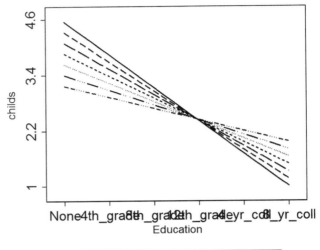

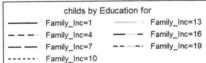

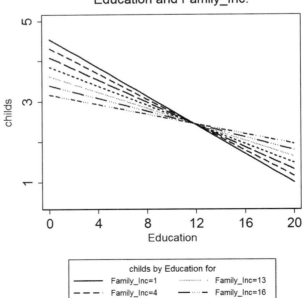

axis labeling cleaned up using the pltopts( ) option, as described in the concluding "Special Topics" section. Each line represents the children–education relationship for a given level of family income. The *solid line* shows that the predicted number of children drops most quickly across education levels when family income is near its minimum. The remaining *dashed lines* show that the children–education relationship becomes less and less steep as income rises. And from the boundary values analysis, we know that the slopes of the lines for the highest two income levels are the shallowest and not significantly different from 0.

In the end, I think that there are three good options for reporting how the effect of education on number of children is moderated by family income (four if you count a confidence bounds plot). The significance region table gives a simple and compact presentation of the information. A predicted values table with an added column of changes in the predicted number of children across the range of education for each level of family income would also be very effective and accessible (see the later discussion of Table 7.5). For a graphical presentation, I would recommend the scatterplot of predicted values both for the visual appeal of the plotted lines and for ease of comprehension and interpretation.

The Effect of Family Income Moderated by Birth Cohort and Education

Analyzing how income moderates the cohort effect and how it moderates the education effect gave insight into how the income effect is moderated by cohort and education separately. I draw on this to discuss how to interpret the simultaneous moderation of family income.

INTSPEC **Setup and** GFI **Analysis**

The *intspec* command is shown in boldface at the top of the Stata output for the GFI analysis below. I first define family income—*c.faminc*—as the focal variable in the focal( ) option. The main( ) option now specifies the three variables constituting the interaction, family income and its moderators birth cohort and education. Because *i.cohort* is a nominal factor variable, ICALC automatically includes all of its categories as display/calculation values, so the range( ) suboption would be ignored if it were listed. When you have two moderators, you should use the one with the fewest display categories as the second moderator for the GIF and SIGREG tools because it will define the columns of 2D tables of results. Given the limited column width in the Stata Results window, you will get easier-to-read displays this way. Note the ordering of the two-way interaction variables in the int2( ) option—*c.faminc#c.educ* is first and then *c.faminc#i.cohort*. Remember that this order must correspond to the relative order in which the moderators are listed in the main( ) option. (The focal variable can be listed anywhere in this ordering.)

In contrast to the prior examples, this is a more complicated interaction specification, with two moderators of family income, one of which is a multicategory nominal variable. So I use the *gfi* command to also produce a path–style diagram of the interaction effects in order to provide a visual display of the algebraic expressions. The suboptions for path( ) request the following:

- Show paths and coefficients for all interaction components in the graph (keyword "all").

- Format the coefficients in the diagram with four decimal places, like the other results.

- Title the diagram as "Interaction of …".

```
. intspec  focal(c.faminc10k)  ///
>          main((c.faminc10k , name(Family_Inc) range(1(3)19)) ///
>                (i.cohort , name(Cohort)) ///
>                (c.educ , name(Education) range(0(5)20)) ) ///
>          int2(c.faminc10k#i.cohort c.faminc10k#c.educ ) ndig(0)
  ...


. gfi ,  ndig(4) path(all, ndig(4) boxw(1.5) ///
>     title("Interaction of Family Income by Cohort, Family Income by Education"))


GFI Information from Interaction Specification of
Effect of Family_Inc on g(Childs) from Linear Regression
----------------------------------------------------------------------

Effect of Faminc =
   -0.0487 - 0.0257*BabyBoom - 0.0282*WWII - 0.0645*DeprEra+ 0.0065*Educ

Sign Change Analysis of Effect of Family_Inc on g(childs)
 Moderated by Cohort (M1) and Education (M2)
```

```
--------------------------------------------------------------------------------
                          Effect of faminc10k
--------------------------------------------------------------------------------
              |        When Education =                            |
When          | -------------------------------------------------- |Sign Changes
Cohort=       |  0        5        10       15       20     |  given M1
------------+--------------------------------------------------------+-----------
              |                                                    |
   PostBoom   |  Neg b=   Neg b=   Pos b=   Pos b=   Pos b= | when M2=
              | -0.0487  -0.0161   0.0164   0.0490   0.0815 |   7.477
              |                                                    |
   BabyBoom   |  Neg b=   Neg b=   Neg b=   Pos b=   Pos b= | when M2=
              | -0.0744  -0.0418  -0.0093   0.0232   0.0558 |   11.429
              |                                                    |
   WWII       |  Neg b=   Neg b=   Neg b=   Pos b=   Pos b= | when M2=
              | -0.0769  -0.0443  -0.0118   0.0207   0.0533 |   11.814
              |                                                    |
   DeprEra    |  Neg b=   Neg b=   Neg b=   Neg b=   Pos b= | when M2=
              | -0.1132  -0.0807  -0.0481  -0.0156   0.0169 |   17.396
              |                                                    |
------------+--------------------------------------------------------+-----------
Sign Changes |   Never    Never  Sometimes Sometimes   Never |
 given M2    |                                                    |
--------------------------------------------------------------------------------

Percent of in-sample cases with positive moderated effect of faminc10k = 74.0
```

Let's start with the GFI's algebraic expression of the moderated effect of family income; remember that income is measured in units of $10K. Because 0 is a valid value for the moderating variables, the main effect coefficient for family income

(−0.0487) has a meaningful interpretation. For someone in the Post-Boom cohort (reference category) who had no formal schooling (*Education* = 0), the number of children is predicted to decline by 0.05 children with a $10K increase in family income. When you have nominal moderators, it is useful to write out the algebraic expression separately for each category. You do this by substituting 0s and 1s into the cohort indicator variables[9] in the GFI expression to calculate the family income effect in each cohort. The Post-Boom cohort is 0 on all three indicators, and the remaining cohorts are coded 1 on their indicator and 0 on the other two. Applying this yields the following:

Post-Boom $\quad -0.0487 - 0.0257 \times 0 - 0.0282 \times 0 - 0.0645 \times 0 + 0.0065 \times \text{Educ}$
$\qquad\qquad = -0.0487 + 0.0065 \times \text{Educ}$

Baby Boom $\quad -0.0487 - 0.0257 \times 1 - 0.0282 \times 0 - 0.0645 \times 0 + 0.0065 \times \text{Educ}$
$\qquad\qquad = -0.0744 + 0.0065 \times \text{Educ}$

WWII $\qquad\quad -0.0487 - 0.0257 \times 0 - 0.0282 \times 1 - 0.0645 \times 0 + 0.0065 \times \text{Educ}$
$\qquad\qquad = -0.0769 + 0.0065 \times \text{Educ}$

DeprEra $\qquad -0.0487 - 0.0257 \times 0 - 0.0282 \times 0 - 0.0645 \times 1 + 0.0065 \times \text{Educ}$
$\qquad\qquad = -0.1132 + 0.0065 \times \text{Educ}$

This calculation makes explicit the meaning of the GFI expression. The baseline negative effect of family income has its smallest magnitude for the Post-Boom cohort but grows larger for the successively older three birth cohorts (negative cohort coefficients), while the family income effect becomes less negative (more positive) as education increases (positive education coefficient). The path diagram in Figure 7.8 shows the structure of the interaction in terms of intersecting arrows leading from the left-hand column boxes to the second-column boxes. Family income and birth cohort interact because their arrows intersect, as do family income and education, but cohort and education do not interact (their arrows do not intersect), nor is there a three-way interaction of income, cohort, and education. The "Special Topics" section shows how to use the path diagram to read off the expressions above for the effect of income in each birth cohort.

The GFI expression raises the question of whether the family income effect changes from negative to positive as education increases and, if so, how that turnover point varies by birth cohort. The sign change analysis, the next part of the output, answers these questions. We can immediately see that the family income effect does in fact become positive at different values of education in each of the birth cohorts. The last column ("Sign Changes Given M1") indicates the sign change point for the birth cohorts. In the Post-Boom cohort, the family income effect becomes positive when *Education* > 7 years. For the Baby Boom and WWII cohorts, this change occurs when *Education* > 11 years, while for the Depression Era cohort, family income has a positive effect only for *Education* ≥ 18 years. The table note indicates that about three quarters of the estimation sample have a positive effect of family income on their number of children.

Significance Region Analyses: SIGREG **and** EFFDISP **Tools**

Especially because the family income effect changes sign, the next step is to explore its significance region to determine whether or not both positive and negative effects

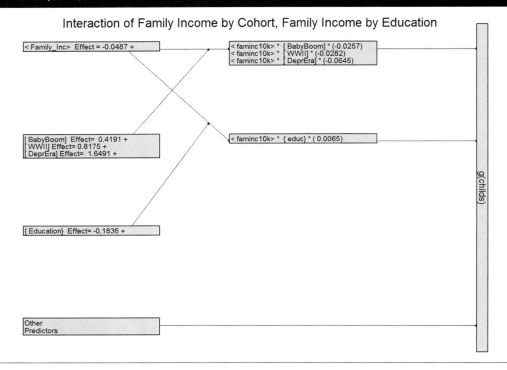

| FIGURE 7.8 ● PATH–STYLE DIAGRAM OF INTERACTION EFFECTS MODEL |

Interaction of Family Income by Cohort, Family Income by Education

are significant. I do this with the *sigreg* command (at the top of the output) with no options because I want the default significance level (.05) and number of decimal places (four) to report the family income effect. A boundary values analysis is only possible for an interval moderator contingent on the other moderator(s). Thus, the output generates a warning message about boundary values for birth cohort contingent on education but reports the boundary values for education dependent on cohort. This analysis shows that for the Post-Boom and Baby Boom cohorts, the effect of family income is not significant unless education is greater than 13 and 14 years, respectively. In contrast, it is only significant for the Depression Era cohort for those with less than 8 years of education and is never significant for the WWII cohort.

Family Inc Effect Significance, Boundary Values for Education on **g(childs)** Given Cohort
Critical value F = **3.850** set with p = **0.0500**

Effect of Family_Inc At Cohort =	When Education >=	Sig Changes	When Education >=	Sig Changes
PostBoom	-43.4555 (< min)	to Not Sig [-0.0231]	13.4457	to Sig [1.6820]
BabyBoom	-13.9665 (< min)	to Not Sig [-0.0471]	14.8670	to Sig [2.5688]
WWII	-21.5109 (< min)	to Not Sig [-0.0396]	20.2875 (> max)	to Sig [0.6130]
DeprEra	7.3759	to Not Sig [-0.3525]	40.9869 (> max)	to Sig [0.0636]

Note: Derivatives of Boundary Values in [ ]

The boundary value results are primarily useful for providing detail when discussing the empirically derived significance region table and other results. This table shows the nature of the changing effect of family income more fully as it reports the sign, magnitude, and significance of the family income effect for each cohort at selected values of education. The formatting of the table makes it easy to see the two regions of significance:

- Positive and significant family income effects for the Post-Boom and Baby Boom cohorts with some college or higher education (more than 13 and 14 years, respectively)

- Negative and significant family income effects for the Depression Era cohort with less than 8 years of education

The significance region table and the content of these two bullet points provide a succinct and easy-to-understand portrayal of how family income predicts number of children differently for birth cohorts and for education levels. If I were to present this table, I would rerun *sigreg* with the save( ) option to save it as the Excel formatted version shown in Table 7.4. With two dimensions in the table and the change in sign, the addition of the fill pattern highlighting is much more effective than the font highlighting in the output table.

```
Significance Region for Effect of faminc10k (1 unit difference)
    on g(childs) at Selected Values of Cohort and Education
-----------------------------------------------------------------
                                    At Cohort=
At Education= |    PostBoom   BabyBoom     WWII       DeprEra
---------------+-------------------------------------------------
          0  |    -0.0487    -0.0744    -0.0769    -0.1132*
          2  |    -0.0356    -0.0614    -0.0639    -0.1002*
          4  |    -0.0226    -0.0483    -0.0508    -0.0872*
          6  |    -0.0096    -0.0353    -0.0378    -0.0742*
          8  |     0.0034    -0.0223    -0.0248    -0.0611
         10  |     0.0164    -0.0093    -0.0118    -0.0481
         12  |     0.0294     0.0037     0.0012    -0.0351
         14  |     0.0424*    0.0167     0.0142    -0.0221
         16  |     0.0555*    0.0297*    0.0272    -0.0091
         18  |     0.0685*    0.0428*    0.0403     0.0039
         20  |     0.0815*    0.0558*    0.0533     0.0169
-----------------------------------------------------------------

Key: Plain font  = Pos, Not Sig    Bold font*  = Pos, Sig
     Italic font = Neg, Not Sig    Italic font* = Neg, Sig
```

Alternatively, we can use the EFFDISP tool to create a set of plots of the family income effect on the *y*-axis against one of the moderators on the *x*-axis, repeated for each display value of the second moderator. A confidence bounds plot requires an interval moderator to define the *x*-axis, while an error bar plot typically uses a categorical moderator to define the *x*-axis. With a mixture of interval and categorical moderators, you get the information from a confidence bounds plot most efficiently. In this instance, we can see how the family income effect varies across all the values of education within a birth cohort, as well as how the income effect varies across all the birth cohort categories. An error bar plot would display the family income effect against all the birth cohorts, with plots for selected values of education; separate plots for the 21 distinct values of education would be a visual overload.

The EFFDISP tool always uses the first moderator to define the *x*-axis, but the *intspec* command used previously lists education as the second moderator. So I respecify *intspec* to list education first. In the *effdisp* command, the plot( ) option generates

TABLE 7.4 ● EXCEL-FORMATTED SIGNIFICANCE REGION TABLE

Effect of faminc10k (One-Unit Difference) Moderated by Cohort and Education on g(childs), Formatted to Highlight Sign and Significance

Education	Cohort			
	Post-Boom	Baby Boom	WWII	DeprEra
0	−0.0487	−0.0744	−0.0769	**−0.1132***
2	−0.0356	−0.0614	−0.0639	**−0.1002***
4	−0.0226	−0.0483	−0.0508	**−0.0872***
6	−0.0096	−0.0353	−0.0378	**−0.0742***
8	0.0034	−0.0223	−0.0248	−0.0611
10	0.0164	−0.0093	−0.0118	−0.0481
12	0.0294	0.0037	0.0012	−0.0351
14	**0.0424***	0.0167	0.0142	−0.0221
16	**0.0555***	**0.0297***	0.0272	−0.0091
18	**0.0685***	**0.0428***	0.0403	0.0039
20	**0.0815***	**0.0558***	0.0533	0.0169
Key				
Plain font, no fill	Pos, Not Sig			
Bold*, filled	Pos, Sig			
Italic, no fill	Neg, Not Sig			
Bold italic*, filled	Neg, Sig			

confidence bounds plots with the specified name for saving them as memory graphs, and pltopts( ) formats the *y*-axis labels:

```
intspec focal(c.faminc10k) ///
    main((c.faminc10k, name(Family_Inc) range(1(3)19)) ///
        (c.educ, name(Education) range(0(5)20)) ///
        (i.cohort, name(Cohort))) ///
    int2( c.faminc10k#c.educ c.faminc10k#i.cohort) ndig(0)

effdisp, plot(type(cbound) name(FamInc)) ndig(1) pltopts(ylab(-.3(.2).3))
```

Figure 7.9 reports for each birth cohort the confidence bounds plots of the effect of family income by education. In each plot, the *solid line* is the moderated effect of family income, and the *dashed lines* show the confidence boundaries for the effect. The *thin gray horizontal reference line* separates negative and positive effects. This shows that the family income effect changes from negative to positive for each birth

FIGURE 7.9 ● EFFDISP CONFIDENCE BOUNDS PLOT FOR TWO MODERATORS

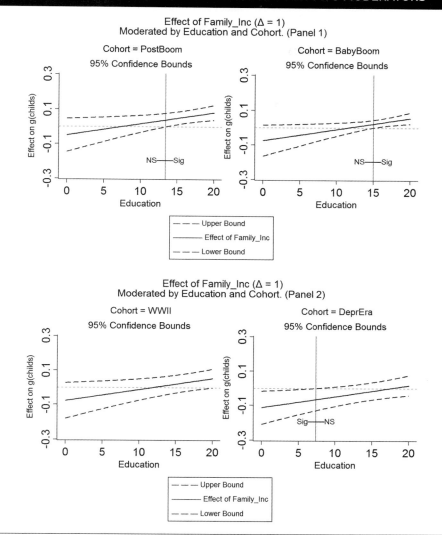

cohort but at different values of education. A vertical reference line if present marks a change in the statistical significance of the family income effect.

For the Post-Boom and Baby Boom cohorts, it is apparent that the effect of family income is significant only at the upper end of the education distribution, where it has a positive effect. The effect of income is never significant for the WWII cohort, and for the Depression Era cohort, the income effect is negative and significant only when *Education* < 8 years. Like the significance region table, the confidence bounds plots provide an accessible and easy-to-interpret portrayal of the moderated effects of family income. In practice, I would also get error bar plots for family income by cohort to see how well they show the patterns of the income effect. In this instance, I think you would find that they also tell the story but it takes more effort to see the patterns.

Outcome Displays: OUTDISP Tool

A display of the predicted values can bring together all the components of the interaction specification. And the prior exploration of changes in the sign, magnitude, and statistical significance of the moderating effects provides a foundation for you to more easily see and interpret the patterns.

The ICALC commands are shown at the top of the output in boldface. I first respecify family income's display values on the *intspec* command (range(1(4.5)19)), with a larger increment between display values (every 4.5 units instead of every 3) to create a more compact predicted values table. The *outdisp* command has five options that create both a table and a scatterplot of the predicted number of children as it varies with family income, birth cohort, and education:

- out(atopt((means) _all)) sets the other predictors' reference values to their means.

- plot(name(Childs_by_FamInc_Cohort_Ed)) names the memory graph and produces the default plot, a scatterplot because the focal variable, family income, is interval.

- ndig(2) sets the format in the predicted values table to report two digits after the decimal.

- tab(def) creates the default table in which the focal variable defines the columns.

- pltopts( ) and its suboptions clean up the look of the initial default scatterplot. ylab and ymtick label the *y*-axis with values and ticks at 0, 2, 4, 6, and with tick marks between these values, respectively. xlab labels the *x*-axis reporting values with $1K scaling. ytit and xtit provide new titles for the *y*-axis and *x*-axis, respectively. See the "Special Topics" section for more details.

(*Note:* For an initial run, you could request results using the defaults very simply with outdisp plot(def) tab(def).)

```
. intspec  focal(c.faminc10k<)  ///
>        main((c.faminc10k , name(Family_Inc) range(1(4.5)19)) ///
>              (c.educ , name(Education) range(0(5)20))  ///
>              (i.cohort , name(Cohort))) ///
>        int2( c.faminc10k#c.educ c.faminc10k#i.cohort) ndig(1)
    ...

.
. outdisp, out(atopt((means) _all)) plot(name(Childs_by_IncCohEd)) ndig(2) ///
>   tab(def) pltopts(ylab(0(2)6) ymtick(1 3 5) ytit("Number of Children") ///
>     xtit("Family Income") xlab(1 "$10K" 7 "$70K" 13 "$130K" 19 "$190K"))
    ...
```

Predicted Value of childs by the Two-way Interactions of Family_Inc with
    Education and with Cohort.

Cohort and Education		Family_Inc				
		1.0	5.5	10.0	14.5	19.0
PostBoom						
	0	3.98	3.76	3.54	3.32	3.10
	5	3.09	3.02	2.95	2.88	2.80
	10	2.21	2.28	2.36	2.43	2.50
	15	1.32	1.54	1.76	1.98	2.20
	20	0.44	0.80	1.17	1.54	1.90
BabyBoom						
	0	4.37	4.04	3.70	3.37	3.03
	5	3.49	3.30	3.11	2.92	2.73
	10	2.60	2.56	2.52	2.48	2.43
	15	1.72	1.82	1.92	2.03	2.13
	20	0.83	1.08	1.33	1.58	1.83
WWII						
	0	4.77	4.42	4.08	3.73	3.38
	5	3.88	3.68	3.48	3.28	3.08
	10	3.00	2.94	2.89	2.84	2.78
	15	2.11	2.20	2.30	2.39	2.48
	20	1.23	1.47	1.70	1.94	2.18
DeprEra						
	0	5.56	5.05	4.54	4.03	3.53
	5	4.68	4.31	3.95	3.59	3.23
	10	3.79	3.58	3.36	3.14	2.93
	15	2.91	2.84	2.77	2.70	2.63
	20	2.02	2.10	2.17	2.25	2.33

While the predicted values table is set up to facilitate the interpretation of the moderated effect of family income by education and birth cohort, it also can be used for interpreting how family income moderates education and how family income moderates cohort. To interpret the effect of education on number of children as moderated by income, we can use the panel of results for any cohort because they will show the identical pattern of magnitude differences in the effect of education. *This would not be true if there were a three-way interaction of income, education, and cohort, or if these results were for a nonlinear link model.* I use the Post-Boom cohort panel for convenience because it is the top one.

Reading down a column shows the effect of education on the predicted number of children for the given level of family income. Each column shows that the predicted number of children declines with education; for instance, at a family income of $55K (5.5), the predicted number of children drops from 3.76 for *Education* = 0 years to 0.80 for *Education* = 20, a difference of 2.96 children. Comparing the columns from left to right, the rate of decline with education decelerates as family income increases: −3.54, −2.96, −2.37, −1.78, and −1.20. And the effect of education on number of children is no longer significant once family income is greater than $170,529, as determined by the boundary value analysis. I would advise readers to verify for themselves that we get the same rate of decline, within rounding error, for the education effect from the other three panels.

Analogously, to interpret the effect of birth cohort, we can use any education row within a panel and compare it across the other panels. Let's use the bottom row in each panel for 20 years of education and start by comparing the Post-Boom cohort with each of the other cohorts. The predicted number of children for the Baby Boom cohort is larger than for the Post-Boom cohort at incomes between $10K and $145K (by 0.39, 0.28, 0.16, and 0.04) but smaller—though not significant—than for the Post-Boom cohort at higher incomes (−0.07). For the other two (older) cohorts, the predicted number of children is always greater than for the Post-Boom cohort, but the differences diminish at higher levels of income and become nonsignificant (at $115,750 for the WWII cohort and at $152,130 for the Depression Era cohort). The overall pattern is a higher predicted number of children for older cohorts compared with younger cohorts, with the set of differences among the cohorts declining with family income and becoming nonsignificant at incomes greater than $160K. The "Special Topics" section at the end of this chapter shows how to estimate the value of the moderator at which the differences in the predicted values among the categories of a nominal focal variable change significance.

Unlike the effect of birth cohort or education, the interpretation of family income's effect changes depending on which panel (cohort) and which row within a panel (level of education) you examine. That is, family income's effect is moderated by both cohort and education. Given the table organization, it is straightforward to discuss how family income's effect varies by education for each cohort. To talk more easily about specifics, I added a column to the predicted values table saved in Excel. The rightmost column in Table 7.5 shows the change in the number of children across the displayed range of income.

Within each cohort, the effect of family income is negative for low levels of education—the predicted number of children declines from left to right—and turns positive at higher levels of education—the predicted number of children rises from left to right. And across the cohorts, the younger the cohort, the smaller the magnitude of the negative effect of income and the larger the size of the positive effect. In the Post-Boom cohort, family income's effect changes to a positive effect when *Education* > 7.48 but does not become significant until *Education* > 13.45. Its initial negative slope is the shallowest, with a change of −0.88 in the number of children, and its final positive slope is the steepest among the cohorts at almost 1.5 children. The changeover points are higher for the effect of family income in the Baby Boom than in the Post-Boom cohort, 11.43 for the sign change and 14.87 for the significance change. And correspondingly, its most negative slope is steeper (−1.34) and its most positive slope shallower (1.00) than for the Post-Boom cohort. The WWII cohort has a very similar sign changeover point in slopes as the Baby Boom cohort (11.81), but the effect of family income does not become significant for any level of education. For the Depression Era cohort, the family income effect stays negative for all but the highest levels of education (>17.40). The negative effect is significant when *Education* < 7.4, but its positive effect is never significant. Not surprisingly, it has the largest magnitude negative slope (−2.04) and the smallest positive slope (0.31).

A good alternative is to use a predicted values plot, for which you do not necessarily need to do side calculations because the magnitudes of the slope for family income are visually apparent. But it can be useful to do so as a check that your visual assessment of steeper and shallower slopes is accurate. Figure 7.10 presents a scatterplot for each cohort, in which the predicted number of children is plotted against family income with separate lines for selected values of education. The *solid line*, which is the highest in each plot, represents the effect of family income for *Education* = 0, and the successive lower lines represent the income effect for progressively higher education levels.

TABLE 7.5 ● OUTDISP PREDICTED VALUES TABLE WITH ADDED CALCULATION

childs by the Two-Way Interactions of Family_Inc With Education and With Cohort							
Cohort	Education	Family Income					Change in Number of Children
Post-Boom	0	3.98	3.76	3.54	3.32	3.10	−0.88
	5	3.09	3.02	2.95	2.88	2.80	−0.29
	10	2.21	2.28	2.36	2.43	2.50	0.30
	15	1.32	1.54	1.76	1.98	2.20	0.88
	20	0.44	0.80	1.17	1.54	1.90	1.47
Baby Boom	0	4.37	4.04	3.70	3.37	3.03	−1.34
	5	3.49	3.30	3.11	2.92	2.73	−0.75
	10	2.60	2.56	2.52	2.48	2.43	−0.17
	15	1.72	1.82	1.92	2.03	2.13	0.42
	20	0.83	1.08	1.33	1.58	1.83	1.00
WWII	0	4.77	4.42	4.08	3.73	3.38	−1.38
	5	3.88	3.68	3.48	3.28	3.08	−0.80
	10	3.00	2.94	2.89	2.84	2.78	−0.21
	15	2.11	2.20	2.30	2.39	2.48	0.37
	20	1.23	1.47	1.70	1.94	2.18	0.96
DeprEra	0	5.56	5.05	4.54	4.03	3.53	−2.04
	5	4.68	4.31	3.95	3.59	3.23	−1.45
	10	3.79	3.58	3.36	3.14	2.93	−0.87
	15	2.91	2.84	2.77	2.70	2.63	−0.28
	20	2.02	2.10	2.17	2.25	2.33	0.31

Note: WWII = World War II; DeprEra = Depression Era.

These plots clearly show the pattern just described above. The family income effect changes from negative to positive as education increases—compare the lines at the top of each plot with those at the bottom—and the steepness of the slopes differs by birth cohort. The Post-Boom cohort has the steepest positive slopes and shallowest negative slopes, and the Depression Era cohort has the shallowest positive slopes and steepest negative slopes, with the other two cohorts similarly in between. Also notice that the angles between the lines in each plot are identical for each cohort, reflecting the fact that the model does not include a three-way interaction of income, education, and cohort.

FIGURE 7.10 ● OUTDISP **SCATTERPLOT FOR TWO MODERATORS**

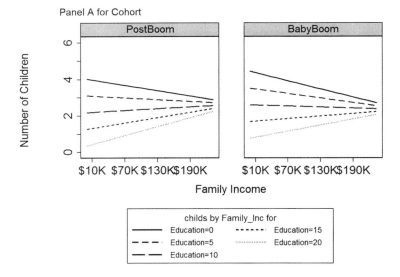

childs by the Two-way Interactions of
Family_Inc with Education and Family_Inc with Cohort.

Panel A for Cohort

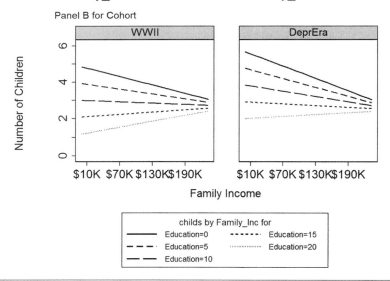

childs by the Two-way Interactions of
Family_Inc with Education and Family_Inc with Cohort.

Panel B for Cohort

The potential drawback of using scatterplots is that they do not lend themselves to describing the effects of education quite as easily or, especially, the effects of cohort. To interpret the education effect (which has the same pattern for each cohort), we can look at the vertical placement of each line as well as how the vertical distance between the plotted lines changes with family income. Because the plotted lines are vertically ordered, with the lines for smaller versus larger values of education having a higher predicted number of children at any value of family income, we know that the effect of education is negative. Furthermore, the gap between the lines diminishes at higher levels of family income, which indicates that the negative effect of education decreases in magnitude as family income increases. From the boundary value analysis, we know that the education effect becomes not significant for incomes greater than $152K. While this provides an accurate description of the education effect, it is harder to explain (and for readers to understand) than using a table or plot designed to highlight the effect of education.

To interpret the effect of birth cohort, use the line for any level of education, and compare it across the four plots. Looking at the left end of the top line for 0 years of education, we can see that the predicted number of children is highest in the Depression Era cohort and smallest in the Post-Boom cohort. But it is not possible to say with any confidence whether the Baby Boom cohort or the WWII cohort has the highest predicted number of children in the scatterplots. We can also deduce that the differences among the cohorts are diminishing. The lines with the steeper decreases in predicted values correspond to the cohorts with the higher initial predicted number of children. Moreover, it appears that the right-hand end of the *solid lines* for the cohorts is less spread out in vertical location than the left-hand ends of the lines. All in all, this is not a very compelling or informative way to describe the moderated effect of birth cohort.

What to Present and Interpret?

Let me start by reemphasizing that you would only present a very limited subset of the results from the analyses we applied to this example. Most are intended to help you, the analyst, better understand the interaction effects so that you can then better explain it to your audience. I outline below what I think are the best options for this example, but these are not necessarily what might work best in other analyses or what might convey the information best for you. Keep in mind that you should discuss how each of the three component variables of the interaction—family income, education, and birth cohort—are related to the outcome. But what works well for interpreting one of these may not work as well for another. For example, the scatterplot in Figure 7.10 is very useful for interpreting the family income effect, while it is difficult to use for the effect of birth cohort. And, as you will see in the next chapter, what works well for a linear link model may not work equally well for a nonlinear link model.

I think that the table of predicted values with the added column of the predicted change in the number of children across the range of income (Table 7.5) is very accessible since it reports predicted values in an intuitive metric: the number of children. And it provides an effective portrayal of how the predicted number of children varies with all three component variables of the interaction. It is effective because the patterns are not complicated and this is a linear link model. It is easy to supplement it with information from the GFI and significance region analyses and can be concisely described. The relationship between number of children and education or between number of children and birth cohort can be described in a short paragraph each. The discussion of the family income effect is longer because it describes how the effect is contingent on two other factors.[10] It also has the advantage that you can use a single display (table) for all three components of the interaction. For added visual impact, you could present the font-size-proportional-to-value-size version of the table, as in Table 7.3.

A second, very good option would be to use three graphic displays of the predicted values, one for each of the variables in the interaction specification, because each graphic would highlight the effects of a different variable on the outcome. That is, I would use scatterplots for family income (Figure 7.10) and for education (Figure 7.7) but a bar chart for birth cohort (Figure 7.6). Although the figures require more space, they are a visually appealing, intuitive, and effective presentation of the relationships for linear link models. And they similarly lend themselves to succinct discussions.

A third good option would be a formatted significance region table for each of the component variables. These highlight changes in the sign and significance of the effects of the component variables and also show the varying magnitude of the effect. For education and family income, I would recommend calculating and reporting the effects for a standard deviation change in the predictor rather than a one-unit change. This would put both predictors on a common scale—standard deviation units—and let you compare the relative magnitude of their effects on the number of children. Depending on your audience, this may take more explanation and be less accessible than relying on the predicted values display, whether tabular or graphical.

Combinations of these three types can also be a quite reasonable option—for example, presenting the predicted values table for the family income effect as well as a bar chart showing the relationship between number of children and birth cohort (Figure 7.6), or presenting formatted significance region tables for the effect of birth cohort and for the effect of education but presenting a predicted values scatterplot for the effect of family income. You may have noticed that I did not include effects displays (confidence bounds plots or error bar plots) among my recommendations. Although I find these visually appealing, I think that many readers are initially confused by them as they expect to see plots of predicted values.

Looking ahead, this chapter concludes with material on several specialized topics or details that would have unduly interrupted the flow of discussion. The next chapter begins our tour through some of the most common GLMs with nonlinear link functions and how to address the challenges of interpretation in those contexts. We begin with binomial logistic regression and probit analysis in Chapter 8.

SPECIAL TOPICS

Customizing Plots With the pltopts( ) Option

Consider the scatterplot in the upper panel of Figure 7.7. Although it is visually easy to interpret, the axis labeling is not ideal. You can create a much cleaner appearance either by using the graph editor in Stata to make changes or by rerunning the *outdisp* command and adding the pltopts( ) option to modify the appearance of the scatterplot. The lower panel was produced with the following command:

```
outdisp, out(atopt( (means) _all)) plot(name(Plot2)) pltopts( xlab(0(4)20) ///
     ylab(1 3 5) ymtick(2 4, tl(**2) ysc(r(.5)) )
```

The content of pltopts( ) consists of Stata two-way graph options that do the following:

- *xlab(0(4)20)* replaces the education value labels on the *x*-axis with their numeric values.

- *ylab(1 3 5)* replaces the noninteger labels for the number of children on the *y*-axis with whole number labels for 1, 3, and 5 children.

- *ymtick(2 4, tl(*2))* adds minor tick marks on the *y*-axis between the new labels at values of 2 and 4, and tl(*2) makes the tick marks twice their usual length.

- *ysc(r(.5))* extends the *y*-axis to a value of 0.5 without a tick mark or a label, to roughly equalize the top- and bottom-margin areas above and below the plotted lines, respectively.

Additional examples of how you can customize plots are shown in the *outdisp* command used to create Figure 7.10, which changes the titles for the *y*- and *x*-axes and adds custom value labels to the *x*-axis. It is beyond the scope of this book to review and explain all of the two-way options and how they might be used to modify plots created with ICALC, including when they won't work. The options are described in the Stata documentation, but I would recommend Mitchell's (2012) book on Stata graphics as an excellent reference on the graph commands and their options, as well as on using the graph editor to customize an already drawn plot. Alternatively, you could specify the save( ) suboption within plot( ) in the *outdisp* command. This will save the data used to create the plot in an Excel file, and then you can construct and customize your own graphics with whatever software platform you prefer.

Aside on Using the Path Diagram for a Multicategory Nominal Moderator

When you have nominal moderators, it is useful to write out the algebraic expression separately for each category. In the discussion of the GFI results for family income moderated by birth cohort and education, I showed how to do this algebraically. You can also read these directly from the path diagram of the interaction model, as I described in Chapter 6, repeating the process for each birth cohort. Start with the family income effect in the top box of the left-hand column, and trace each arrow to the boxes in the second column. That is, follow the horizontal arrow to the top box of the second column, and add the relevant coefficient for the birth cohort, dropping both variable labels. Then go back and follow the diagonal arrow to the other box in the second column, and add its contents, dropping the variable label for family income. For instance, for the WWII birth cohort the family income effect is equal to

-0.0487	from the top left-hand box
-0.0282	from horizontal arrow to 2nd column, top box, Line 2
$+0.0065 \times Educ$	from diagonal arrow to 2nd column, 2nd box

giving $-0.0769 + 0.0065 \times Educ$. For the base birth cohort (Post-Boom), there is no relevant coefficient in the second column's top box, so you add nothing to its expression and proceed to tracing the next arrow.

Testing Differences in the Predicted Outcome Among Categories of a Nominal Variable

I used an iterative computational approach to determine the value of family income at which the differences in the predicted number of children among the birth cohorts were not significantly different from 0. I relied on the *mtable* and *mlincom* commands in SPOST13 for doing the calculations. After describing the step-by-step process, I list a simple program that can be run to do the calculations; it is also available to download from www.icalcrlk.com.

The coded income values ranged from 0.5 to 19.2. So for the first iteration, I generated the predicted number of children for the birth cohorts for the 21 integer values of income from 0 to 20 with the *mtable* command:

mtable, at (faminc10k = (0(1)20) cohort=(1/4)) atmeans stat(pvalue noci) post

This produces 21 sets of estimates for each cohort, for a total of 84 stored estimates. For each set, I tested if the sum of the differences between the estimates for each unique pair of cohorts is equal to 0. With four cohorts, there are six unique pairs of cohorts to compare:

$$\left(\hat{y}_{cohort 2} - \hat{y}_{cohort 1}\right) + \left(\hat{y}_{cohort 3} - \hat{y}_{cohort 1}\right) + \left(\hat{y}_{cohort 4} - \hat{y}_{cohort 1}\right)$$
$$+ \left(\hat{y}_{cohort 3} - \hat{y}_{cohort 2}\right) + \left(\hat{y}_{cohort 4} - \hat{y}_{cohort 2}\right) + \left(\hat{y}_{cohort 4} - \hat{y}_{cohort 3}\right) = 0$$

I performed the test with the *mlincom* command by referring to the position of the stored estimates. For example, the second set of estimates for the birth cohorts (when *faminc* = 1) is stored in positions 5 to 8, and the *mlincom* command is

mlincom 6-5 + 7-5 + 8-5 + 7-6 + 8-6 + 8-7

This produced a test statistic value of 5.149 with a *p* value of .000. I repeated this calculation for all 21 sets of estimates. This showed that the difference was significant at a family income value of 15 ($p = .022$) but not significant at the value of 16 ($p = .051$).

The second iteration increased the precision by one order of magnitude (to the nearest 0.1). I followed the same process to test the differences among cohorts for family income coded values from 15 to 16 in increments of 0.1. This showed that the intercohort differences were first not significant at a family income of 16.0, which represents a value of $160K (16.0 × $10,000). I could have done additional iterations with smaller increments to get additional digits for a more precise income value at which the differences turned nonsignificant.

In reality, I used the program shown in the box to automate each iteration, but it only applies to testing a nominal variable that is moderated by one other variable. After the program is loaded in your Stata session,[11] you access it with the following command:

mcattest mcvar(*nominal variable name*) var2(*moderator name*) vallist
    (*numlist*)

The numlist must be of the form #1/#2 or #1(#2)#3.

The command for the first iteration is

mcattest, mcvar(cohort) var2(faminc10k) vallist(0(1)20)

and for the second iteration,

mcattest, mcvar(cohort) var2(faminc10k) vallist(15(.1)16)

```
program mcattest
syntax, mcvar(varname) var2(varname) vallist(string)
tempname estnm
qui{
est store `estnm'
levelsof `mcvar', loc(nval)
loc ncat : list sizeof nval
mtable , at (`var2' "= (`vallist')" `mcvar'=(`nval')) atmeans stat(pvalue noci) post
mlincom, clear
loc atind = 1-`ncat'
forvalues fi = `vallist' {
      loc atind= `atind' + `ncat'
      loc difftxt ""
      forvalues i=1/`=`ncat'-1' {
      forvalues j=1/`=`ncat'- `i'' {

            if `j' == 1 & `i'== 1 loc difftxt "`difftxt' `=`atind'+`j'' - `atind' "
            if `j' > 1 | `i' > 1 loc difftxt "`difftxt' + `=`atind'+`j'+`i'-1' - `=`atind'+`i'-1' "
      }
      }
      mlincom `difftxt', add rowname("`fi'")
}
}
mlincom
qui est restore `estnm'
end
```

CHAPTER 7 NOTES

1. These models are commonly labeled as GLS because they generalize (relax) the assumptions of equal error variance and zero covariance. I avoid using this label in the text exposition in order to avoid the inevitable confusion about what is meant by labeling a model GLS versus GLM or by making statements such as "a GLS model is one type of GLM."
2. Response categories with a range of values used the midpoint value. The top-coded category for frequency of sex (4+ times per week) was recoded using a value of 4.5 times

per week. Redoing the analyses using a value of 5 times per week produced essentially identical results.

3. In all, 269 cases were missing on frequency of sex, 109 more were missing on attendance at religious services, and 53 more were missing on education.

4. The mean of SES in the sample analyzed is 53.2, which is a bit larger than the mean in the full sample of 52.2.

5. This is specified for didactic reasons. For OLS regression, the *margins* command produces exactly the same results for the "as observed" method of treating the predictors and for the "as means" option. Thus, in practice, this option is unnecessary for OLS analyses.

6. The birth cohorts are defined by respondents' age in 2010: Depression Era (70 years or older), World War II (65–69), Baby Boom (46–64), and Post-Boom (40–45). Translated to year born, the categories are Depression Era (before 1941), World War II (1941–1945), Baby Boom (1946–1964), and Post-Boom (1965–1970).

7. In all, 172 cases were missing on income, an additional 45 were missing on religious intensity, and 1 more was missing on education.

8. These cases all had high leverage values but moderate residuals. They all had the maximum (top coded) value for income, were in the two oldest birth cohorts, and had an atypically small number of children. Removing these cases from the analysis had negligible effects on the estimation results. Hence, the analyses include them.

9. If your nominal variable is contrast coded rather than dummy coded, you would instead substitute the contrast-code values to specify a given cohort.

10. Taking out my commentary from what I wrote above, the description for education was 176 words in five sentences, and for cohort, 117 words in four sentences.

11. mcattest.do can be downloaded from icalcrlk.com from the icalc_spec package. You could either highlight the set of lines for the program and click on the run icon or issue a run command specifying the path and the filename—run *path*/mcattest.do. It is also listed at the end of the downloadable do-file for this example.

LOGISTIC REGRESSION AND PROBIT APPLICATIONS

OVERVIEW

Properties and Use of Logistic Regression and Probit Analysis

This chapter focuses on logistic regression applications, with some brief additional material on Probit analysis. Although these GLMs differ in both their link functions and conditional distributions, they are close to interchangeable in the results they produce, if you take into account differences in scaling (Long, 1997, pp. 47–48, 83). There is a general preference in practice for using logistic regression rather than probit analysis in many but not all situations (Fox, 2008, pp. 340–342; Long, 1997, p. 83; Long & Freese, 2014, p. 209). Hence, my decision to concentrate on logistic regression applications is typical. Note that I will also use the term *logit analysis* interchangeably with the term *logistic regression*. For logistic regression, there are five common ways to interpret the effects (see Long & Freese, 2014, pp. 227–228):

1. As effects on a (standardized) latent outcome

2. As a marginal effect on the probability of the outcome, either a marginal change (instantaneous) or a discrete change in the probability of the outcome

3. Using tabular or visual displays of predicted probabilities

4. As effects on the log odds of the outcome (cannot be used for probit analysis)

5. As factor changes in the odds of the outcome (cannot be used for probit analysis)

My presentation and discussion of how to apply these different modes of interpretation to interaction effects presumes that readers are already familiar with these approaches to interpretation for predictors not part of an interaction effect. If you do not have a basic understanding of these approaches, then I strongly recommend that you read Chapters 4 and 6 of Long and Freese (2014) before you read further here.

Data and Circumstances When Commonly Used

Logistic regression and probit analysis are usually the techniques of choice when analyzing binary (dichotomously coded as 0 or 1) outcome measures for single cases or grouped data[1] for which you have only a single point in time observation of the outcome category for a case. That is, you do not know (or choose not to analyze) the duration of time for which the case has been in an outcome category. If you do have duration data, then some form of event history or survival analysis would be a more appropriate choice (see Chapter 12).

Both logit and probit analyses predict the probability that a case is in the category coded 1, often referred to as the probability of success, as a nonlinear function of the predictors. A plot of the predicted probabilities against the model-specified function of the predictors $\left(g^{-1}(Xb) \right)$ traces an S-shaped curve. Logit and probit are also used in a number of paired-technique analyses, such as selection models to correct for sample selection bias (Greene, 2008, p. 886), a Tobit model for censored data (Greene, 2008, p. 875), or zero-inflated count models (Long, 1997, p. 243). Like the LRM, you can use predictors coded as interval or categorical[2] measures, a nonlinear function of a predictor such as a polynomial or logarithmic expression, and/or interaction terms represented by the product of two or more predictors.

Published Examples.

- To analyze why people support Tea Party local organizations, McVeigh, Beyerlein, Vann, and Trivedi (2014) used logistic regression. They tested whether the effect of respondent's education on their Tea Party support varied with the degree of residential segregation by education in their locale (county). Their interpretation of this significant interaction describes the effect on support of respondent(s)'s education at the county mean of educational segregation and how that effect changes at higher levels of segregation.

- Kim (2017) applied logistic regression to study how productivity and product differentiation within an industry affects a firm's decision whether or not to lobby for trade liberalization legislation. Kim interprets a significant interaction of firm productivity with firm product differentiation with a mention of how the effect of productivity varies with differentiation and a discussion of a plot of predicted probability of lobbying against productivity and differentiation.

- With probit analysis of individual's unemployment status, Ham and Reilly (2002) tested alternative models of labor market operation. Their assessment of the hours-restricted model included a significant age-by-education interaction. Apart from noting its significance, they did not interpret the interaction effect.

- Malmusi, Borrell, and Benach (2010) used logistic regression to estimate the effects of citizenship, immigration recency, social class, and gender on poor health status. They cross-classified citizenship, recency, and class to define categorical groups to represent the three-way interaction of these factors. They interpreted the interaction by describing the effects of groups and discussing bar charts of the predicted probability of poor health.

GLM Properties and Coefficient Interpretation for Logistic Regression

Logistic regression uses a logit link function to define the relationship between the modeled outcome (η_i) and the observed outcome (μ_i):

$$\eta_i = \ln\left(\frac{\mu_i}{1-\mu_i}\right) \text{ or equivalently using } y_i \text{ to represent the outcome } \eta_i = \ln\left(\frac{y_i}{1-y_i}\right)$$

And the distribution of the outcome conditional on the predictors—$y_i|X_i$—is the binomial distribution. The metric of the observed outcome is a probability, specifically the probability that $y_i = 1$. According to the link function, then, the metric of the modeled outcome is the natural log of the odds that $y_i = 1$, commonly referred to as the log odds of y.

Interpretation. Consequently, logistic regression coefficients can be directly interpreted as effects on the log odds of y. A positive (negative) coefficient b means that a one-unit change in the predictor corresponds to an increase (decrease) of b in the log odds of y. An alternative way to think about the modeled outcome is as a latent interval variable underlying the observed dichotomous outcome, which means that the coefficients can also be interpreted as effects on the latent outcome (Long & Freese, 2014, pp. 188–192). More commonly, interpretation of the relationship between a predictor and the outcome is done using the following:

- A factor change (odds ratio) for which the outcome's odds changes by a factor of e^b for a one-unit change in the predictor

- As a marginal effect, often as a discrete change in the probability of success where the outcome's probability changes by a calculated amount for a one-unit change in the predictor; sometimes as a marginal change in the probability of success, where the outcome's probability changes by a calculated amount for an infinitesimal change in the predictor

- Tabular or visual displays of the predicted probability of the outcome as a function of the predictors

GLM Properties and Coefficient Interpretation for Probit Analysis

Probit analysis in contrast uses an inverse normal distribution link function defining the transformation of the observed outcome into the modeled outcome:

$$\eta_i = \phi^{-1}\left(\mu_i\right)$$

And the corresponding conditional distribution for $y_i|X_i$ is the normal distribution. Like logistic regression, the modeled outcome in probit analysis can be developed as an interval latent variable (Long & Freese, 2014, pp. 188–192).

Interpretation. Like logit coefficients, probit coefficients can be interpreted as effects on the latent outcome. But the different link function for probit precludes interpreting the coefficients as effects on the log odds of y or in terms of factor changes. Probit results can also be interpreted using discrete or marginal changes and tabular/visual displays of the predicted probability of success.

Diagnostic Tests and Procedures

As described in Chapter 1, analyses of residuals (standardized Pearson residuals for logistic regression and deviance residuals for probit analysis) should be used to assess possible model misspecification to limit the impact on testing for and estimation of interaction effects. For the same reason, residual analysis can help identify and investigate outliers for their influence on the analytic results. Outliers can also be defined by what Long and Freese (2014, pp. 216–218) describe as least likely observations: those cases with an *Observed outcome* = 0 but a high predicted probability that the *Outcome* = 1, or cases with an *Observed outcome* = 1 but a low predicted probability that the *Outcome* = 1. Once outliers are identified, Pregibon's (1981) influence statistic can be calculated and studied for both logit and probit models; the Stata predict command will calculate this following a logistic regression, and the steps described in Chapter 1 can be used for probit. Finally, checking for collinearity among the predictors—apart from the necessary functional collinearity among interaction terms—is also standard.

Data Source for Examples

Both application examples analyze GSS data. The single-moderator example data are from the 1998 survey (GSS_1998.dta), and the three-way interaction example uses 1987 data (GSS_1987.dta). Both data sets and example do-files can be downloaded at www.icalcrlk.com.

ONE-MODERATOR EXAMPLE (NOMINAL BY NOMINAL)

Data and Testing

This example illustrates the interpretation of a nominal-by-nominal interaction effect, how sex and residential location interact in predicting gun ownership. Gun ownership is a dichotomous measure (coded 1 = *owns a gun*, 0 = *does not own a gun*). Sex is also a dichotomous indicator (1 = *male*, 0 = *female*). Residential location consists of three categories (1 = *central city*, 2 = *suburban/other urban*, 3 = *rural*) and is represented in the model by dummy indicators for suburban and for rural (central city is the reference category). The other predictors are education, age, race, and neighborhood fearfulness. Years of education (0–20) and age (18–89) are interval measures. The two remaining independent variables are dichotomous: race (1 = *Black*, 0 = *White*) and fear of walking in the neighborhood at night (1 = *fearful*, 0 = *not fearful*). The index function to be estimated takes the form

$$g(y_i) = \beta_0 + \beta_1 Sex_i + \beta_2 Suburban_i + \beta_3 Rural_i + \beta_4 Sex_i \times Suburban_i$$
$$+ \beta_5 Sex_i \times Rural_i + \beta_6 Age_i + \beta_7 Age_i^2 + \beta_8 Education_i$$
$$+ \beta_9 Race_i + \beta_{10} Fearful_i$$

The sample of 1,707 excludes 963 respondents not asked the question (950) or missing (13) on gun ownership, with another 124 excluded because they were coded as other races, and an additional 38 cases with missing information on one or more of the other variables in the analysis.

Initial diagnostic tests on the main effects logit model clearly suggested that the effect of age should be modeled as a quadratic functional form, so subsequent analyses included indicators of age and age squared. Analyses of residuals and leverage identified two possibly influential outliers, and 64 more were suggested by the least likely analysis. The outliers were almost all men who do not live in the central city and do not own guns or women who do not live in rural areas who do own guns. A sensitivity analysis excluding these cases produced results identical in the sign and significance of coefficients but generally larger coefficient magnitudes. Diagnostic testing of the probit analysis showed essentially the same results. While the diagnostic results indicate that there likely are predictors missing from the analysis, they also suggest that their inclusion would only strengthen the findings. Consequently, I take a conservative approach and keep these cases in the analysis.

Statistical tests of the sex by residential location interaction terms in the model— $\beta_4 Sex_i \times Suburban_i$ and $\beta_5 Sex_i \times Rural_i$—unambiguously support their inclusion in the model in both the logit and the probit analyses. A global test of the change in the log likelihood if these two terms are excluded is statistically significant ($p < .001$), as are the z tests of each coefficient individually ($p < .001$).

The interpretation of these interaction effects in the logit analysis proceeds in three parts to illustrate how to apply ICALC and discuss the results it produces for each of the five modes of interpretation. Because tabular and visual displays of the predicted values are most effectively used once you understand how each variable moderates the effect of the other, this mode is illustrated last in Part III. The first part explores the moderation of the sex effect by residential location using the other four modes of interpretation, and the second part does the same for the moderation of residential location by sex. Each set of output begins with the ICALC commands used to produce it, highlighted in boldface. Notice that many commands produce the desired and appropriate output with few if any options necessary.

Following the logistic regression results, I briefly demonstrate that the probit analysis produces nearly identical numerical results, adjusting for differences in the scaling of the coefficients, for the applicable modes of interpretation.

Part I: The Effect of Sex Moderated by Residential Location

INTSPEC **Setup and** GFI **Analysis**

The *intspec* command defines sex as the focal variable and location as the moderator. I specified the factor option on the *gfi* command to produce the algebraic expression for the moderated effect of sex, as well as the corresponding expression for factor change effects.

```
.  intspec focal(i.sex) main( (i.sex, name(Sex) range(0/1)) ///
>          (i.location, name(Location) range(1/3))) int2(i.sex#i.location) ndig(3)

Interaction Effects on Gun Specified as

    Main effect terms: 1.sex  2.location 3.location
    Two-way interaction terms: i.sex#i.location
```

```
    These will be treated as: Focal variable = 1.sex ("Sex")
       moderated by interaction(s) with
          2.location 3.location ("Urbanicity")
   .
. gfi , factor

GFI Information from Interaction Specification of
Effect of Sex on g(Gun) from Logistic Regression
----------------------------------------------------------------------

Effect of Male =
    0.7087 + 1.1550*Suburbs + 1.8962*Rural


   Factor Change Effect (1 unit change in) Male =
       e^ 0.7087 * e^( 1.1550*Suburbs) * e^( 1.8962*Rural)


            2.0313 * 3.1740^Suburbs * 6.6605^Rural

Sign Change Analysis of Effect of Sex
on g(Gun), Moderated by Location (MV)


    ------------------------------------
                  |            Sex
    When          |   --------------------
    Location=     |
    ------------+------------------------
    CentralC~y  | Pos b =      0.7087
       Suburbs  | Pos b =      1.8637
         Rural  | Pos b =      2.6049
    ------------+------------------------
    Sign Changes |         Never
    ------------+------------------------
    % Positive   |         100.0
    ------------------------------------
```

The GFI expression for GLMs is always in the metric of the modeled outcome; so for logistic regression, it describes the effect of sex on the log odds of owning a gun. The sign change analysis table indicates that the effect of sex is positive for all possible values of residential location. The GFI expression shows that the baseline effect of sex (male) is a 0.709 higher log odds of gun ownership for men than for women, and the gap increases by 1.155 for suburban residents and by 1.896 for rural residents. Equivalently, we could say that men have a higher log odds of gun ownership in all types of residential locations. Specifically, it is 0.709 higher among central city residents, 1.864 higher among suburban residents (0.709 + 1.155), and 2.605 higher among rural residents (0.709 + 1.896).

To derive these summary statements, I drew on the basics of interpreting the coefficients of dummy variables; the effect of a dummy variable represents the estimated difference between the category coded 1 for the dummy variable and the reference category. I applied this to interpreting the effect of sex as the predicted difference for men versus women. Similarly, this means that the location dummies' coefficients are the predicted difference in the sex effect for a given location versus central city. While this is straightforward to do by hand with experience, the significance region analysis provides these results already calculated for you. I will come back to the interpretation of the factor change effect GFI expression shortly.

Significance Region Analyses: SIGREG **and** EFFDISP **Tools**

Effects on the Log Odds. I apply the *SIGREG* tool multiple times to calculate interaction effect results for different modes of interpretation, starting with the interpretation as effects on the log odds as I just did for the GFI results. For this initial application, I use the *sigreg* and *effdisp* commands without any options.

```
. sigreg

Cannot calculate boundary values when moderator is categorical with > 2 categories
Proceeding to empirically defined significance region analysis


     Significance Region for Effect of Sex (1 unit difference)
   on g(gun) at Selected Values of Location
-------------------------------------------------------
                              At Location=
   Effect of      |   CentralC~y   Suburbs      Rural
   --------------+----------------------------------
         Male     |    0.7087*      1.8637*     2.6049*
-------------------------------------------------------
    Key: Plain font  = Pos, Not Sig    Bold font*   = Pos, Sig
         Italic font = Neg, Not Sig    Italic font* = Neg, Sig

. effdisp

Plot Options Specified

None specified
   type =  errbar by default
```

First, note the warning message that the boundary values analysis could not be done for a moderator with three or more categories. This is not a concern because the significance region table always includes all the categories of the moderator, which provides information equivalent to the boundary value analysis. The significance region table reports the effect of sex for all three residential locations. This readily shows that men have a significantly higher log odds of gun ownership than do women, regardless of location. Moreover, the difference is more than 2.5 times larger in suburban than in central city locations (1.8637/0.7087 = 2.63) and is another 40% larger in rural than in suburban locations (2.6049/1.8637 = 1.40). This shows that the moderating effect of residential location is substantial.

The *effdisp* command produces an error bar chart by default for a categorical moderator, which visually shows these numeric results (see Figure 8.1). None of the confidence intervals contain the value 0, indicating that the effects are all significant. The advantages of this mode of interpretation are that it is simple, it parallels how we interpret effects from OLS regression, the interpretations do not depend on the values of the other predictors, and it does not confound the nonlinearity of the GLM with the nonlinearity of the interaction effect.

Effects on the Latent Outcome. As many have observed, it is difficult to judge the meaning of the magnitude of effects in the log odds metric because it conveys little meaning for most researchers (e.g., Long & Freese, 2014, p. 228; Liao, 1994, pp. 8, 13). One solution that preserves the advantages of directly interpreting coefficients is to use a latent variable framework and rescale the effects as standard deviation changes in the latent outcome, what Long and Freese (2014, p. 181) label as y^*-standardized coefficients. These

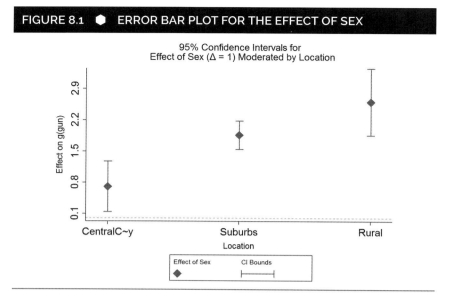

FIGURE 8.1 ● ERROR BAR PLOT FOR THE EFFECT OF SEX

are calculated by dividing effects by the estimated standard deviation of the latent outcome. You can produce such y^*-standardized effects using *sigreg* and the effect(b(sdy)) option. Note that the nobva option skips doing a boundary values analysis.

```
. sigreg , effect(b(sdy)) nobva
...
    Significance Region for Effect of Sex (1 unit difference)
    on g(gun)-standardized at Selected Values of Location
    -----------------------------------------------------
                                   At Location=
    Effect of      |  CentralC~y    Suburbs      Rural
    ---------------+-------------------------------------
           Male    |    0.3372*      0.8868*     1.2395*
    -----------------------------------------------------
      Key: Plain font  = Pos, Not Sig    Bold font*  = Pos, Sig
          Italic font  = Neg, Not Sig    Italic font* = Neg, Sig

    Std. Dev. of latent outcome   =    2.1016
```

These results can be interpreted as effects on the (latent) propensity to own a gun and give a better feel for the size of the sex effect on gun ownership than did the (unstandardized) log odds. Among central city residents, men have a propensity to own guns that is one third of a standard deviation (0.3372) higher than that of women's propensity. For suburban residents, men's propensity is 0.89 standard deviations higher than that of women's; and for rural residents, the disparity in ownership propensity is about 1.25 standard deviations. Again, this suggests large differences in the effect of sex across residential location. Note that the value of the standard deviation of the latent outcome is reported below the significance region table. Personally, I usually find this an informative, intuitive, and easy-to-apply approach to interpretation, which avoids the complexity of interpretations using the probability metric. But not everyone is comfortable with assuming the existence of a latent outcome, especially if the categorical outcome cannot be construed as a choice or a decision.

Factor Change Effects (Odds Ratios). Interpreting effects as factor changes is a different approach to providing a more understandable metric than the log odds. A factor change measures the effect on an outcome from a change in a predictor x by calculating the multiplicative factor by which the outcome changes; that is, it is the ratio of the outcome value after the change in x to the outcome value at the starting value of x. Thus, a factor change value of 1 represents no effect of x because the two outcome values are equal. A factor change greater than 1 represents a positive effect of x because the ending outcome value is greater than its beginning value, and a factor change less than 1 is a negative effect of x because the ending outcome value is smaller than its initial value. It is crucial to recognize that this creates an asymmetry in the size of positive effects versus the size of negative effects. Positive effects have values from greater than 1 up to infinity, while negative effects have values from less than 1 down to 0. This means, for example, that a (positive) factor change of 2.0 is the same magnitude as a (negative) factor change of 0.5. This sometimes makes the interpretation of factor change results, which include both positive and negative effects, a little awkward and potentially confusing.

For logistic regression, a factor change is commonly referred to as an odds ratio because it measures factor changes in the odds of the outcome. It is calculated by exponentiating the logistic coefficient multiplied by a specified amount of change in the predictor x. That is, the factor change equals $e^{b\Delta x}$. To apply this to interaction effects, you exponentiate the focal variable's moderated effect. That is, you exponentiate the GFI algebraic expression. For the current example, this would be

$$e^{0.7087 + 1.1550 \times Suburbs + 1.8962 \times Rural}$$

There is a mathematically equivalent expression, which I think makes more conceptual sense. You separately exponentiate the focal variable's main effect coefficient and each interaction coefficient times moderator term and multiply them together:

$$e^{0.7087} e^{1.1550 \times Suburbs} e^{1.8962 \times Rural}$$

This is the expression for the GFI factor change shown in the earlier output. The first term is the baseline factor change in the odds of gun ownership for men versus women when the moderating variables are all equal to 0. The remaining terms describe the multiplicative factor, which modifies the baseline factor change for values of the moderating variables other than 0.

To interpret this factor change effect GFI expression more concretely, keep in mind that any real number raised to the power of 0 is equal to 1. Thus, when the dummy indicator for *Suburbs* = 0, its term in the GFI expression is equal to 1; that is, $e^{1.550 \times 0} = e^0 = 1$. This means that for a given residential location, the factor change effect is the baseline factor change effect $e^{0.7087}$ multiplied by the factor change effect term for that location.

- The center city factor change effect is $e^{0.7087} = 2.0314$, because *Suburbs* = 0 and *Rural* = 0, so their terms and expressions simplify to 1.

- The suburban factor change effect is $e^{0.7087} \times e^{1.550 \times 1} = 6.4477$ because *Suburbs* = 1, which leaves its term in the expression, but *Rural* = 0, so its term evaluates to 1.

- The rural factor change effect is $e^{0.7087} \times e^{1.8962 \times 1} = 13.5301$ because *Rural* = 1, which leaves its term in the expression, but *Suburbs* = 0, so its term evaluates to 1.

Rather than doing these calculations by hand, the *sigreg* command with the option effect(factor) will do the work for you. The significance region table in the output reports not just the factor change effect of sex for each category of location but also whether that moderated effect is statistically significant. The factor change effect of sex is significant in all residential locations. It is smallest in the central city, where the odds that men own guns are twice as large as the odds for women. In the suburbs, men are 6.5 times more likely than women to own guns, and in rural areas, men are 13.5 times more likely than women to own guns. Like the previous two interpretations, this indicates substantial moderation of the sex effect by location.

```
. sigreg , effect(factor)

   ...

   Significance Region for Factor Change Effect of Sex (1 unit difference)
   at Selected Values of Location
   -------------------------------------------------
                              At Location=
   Effect of     |   CentralC~y   Suburbs     Rural
   --------------+----------------------------------
         Male    |    2.0314*     6.4477*   13.5301*
   -------------------------------------------------
     Key: Plain font  = Pos, Not Sig   Bold font*   = Pos, Sig
          Italic font = Neg, Not Sig   Italic font* = Neg, Sig
```

Discrete or Marginal Changes (Must Have SPOST13 Installed). A marginal change is not valid for categorical variables, but its use will be demonstrated in the second example, a three-way interaction. A discrete change effect is defined as the change in the predicted outcome in its observed metric corresponding to a specified amount of difference in the predictor x, evaluated at the reference values for the other predictors in the model. For logistic regression, discrete change effects refer to changes in the predicted probability of "success"; in our example, the probability of gun ownership. Typically, discrete changes are calculated for a one-unit change in a predictor; for an interval variable, the starting point for the one-unit change is from its reference value, while for a categorical variable the starting point is 0. As I discussed in Chapters 4 and 5, the three common ways to set reference values are using their means, their as-observed values for all sample cases, or researcher-selected representative values. Marginal effects (either marginal changes or discrete changes) calculated using these approaches are sometimes referred to as MEM, AME, and MER, respectively (Long & Freese, 2014; Williams, 2012).

A set of discrete change effects can be calculated to get the moderated effect of the focal variable by setting the moderator(s) to each of their display values in turn and the other predictors to their reference values. In the case of a categorical moderator, you use every category to define its display values. So for this example, you calculate three discrete change effects for sex: one for central city location, one for suburban location, and one for rural location. Let me walk you through doing this with the *sigreg* command with the effect( ) option as follows (readers may want to look back at the description of this option in Chapter 6):

```
sigreg, effect( ///
  spost( ///
      amtopt(am(bin) ) ///
      atopt((means) _all) ))
```

The suboptions listed inside the spost( ) option specify that we want the effect of the focal variable—sex—in the significance region table calculated using one of the SPOST13 calculations of a marginal effect. Which effect to calculate is defined by the two suboptions within spost( ). The amtopt(am(bin)) suboption specifies the amount of change in sex to use in calculations; am(bin)) is an SPOST13 option that indicates that we want the discrete effect of a binary change in sex from 0 to 1. The atopt((means) _all) suboption directs SPOST13 to set the reference values for the other predictors in the model to their means. The amount of change calculated by SPOST13 and the method for setting reference values is shown in a note below the table key. In this case, "amount = Male vs Female" means that the effect is for a binary change from *Sex* = 0 to *Sex* = 1. The reference values are set as the means of the predictors, as shown by the contents within at( ) in the table note.

```
.  sigreg , effect(        ///
>               spost(           ///
>                    amtopt(am(bin) )  ///
>                    atopt((means)  _all)    )) nobva
   ...

   Significance Region for SPOST Change Effect of Sex
   on gun at Selected Values of Location
-----------------------------------------------------
                             At Location=
   Effect of   |   CentralC~y   Suburbs     Rural
-----------------+-----------------------------------
        Male    |    0.1064*     0.3737*    0.5720*
-----------------------------------------------------
     Key: Plain font  = Pos, Not Sig   Bold font*  = Pos, Sig
          Italic font = Neg, Not Sig   Italic font* = Neg, Sig

Spost Effect for  sex specified as amount =  Male vs Female  calculated with
    at(  (means) _all )
```

The significance region table again shows positive and significant effects of sex on gun ownership in each of the residential locations. When you interpret discrete effects, you should indicate how reference values for other predictors were set. For instance, for central city residents who are otherwise average, men's predicted probability of gun ownership is 0.11 higher than for women, while for average suburban and rural residents, men's probability of ownership is even higher than women's by 0.37 and 0.57, respectively. You can also interpret this in percentage terms by referring to the predicted percentage of ownership; for example, for otherwise average rural residents, the predicted percentage of men who own guns is 57 percentage points higher than the percentage of ownership by women.

There are two potential disadvantages to this approach. The first is that the magnitude of the discrete change effects—whether for interaction effects or not—varies with the reference values chosen for the other predictors. Using different reference values could affect the (subjective) judgments we make about the magnitude of effects. This is one of the motivations for using the representative values approach for reference values, which typically uses multiple sets of reference values to explore the varying effects of the predictors. But presenting and discussing multiple sets of effects for a single predictor can be confusing and may not necessarily reveal anything worthwhile. My suggestion would be to explore but not present multiple sets of reference values and to note in the discussion of results if there were any interesting differences.

TABLE 8.1 ●	DISCRETE CHANGE EFFECT OF SEX FROM MODEL WITHOUT INTERACTION EFFECTS		

	Residential Location		
	Central City	Suburban	Rural
Discrete change effect	0.244	0.354	0.411

The more serious concern for interpreting interaction effects is that the discrete change measures are affected by the confounding of the nonlinearity of the interaction effect with the nonlinearity of the GLM link function in the same way that tables or plots of predicted outcomes are, as I discussed in Chapter 5. The point is that we cannot know from the reported discrete effects to what degree the sex effect varies across locations because it is moderated by location and to what degree it varies because the different location values affect the reference point at which the discrete change effect is calculated. And this problem would likely be exacerbated by evaluating the discrete change effects for interaction models at multiple sets of reference values.

To illustrate the problem of confounding, I ran a model without the sex-by-location interaction and calculated the discrete change effect for sex, treating each location category as the reference value in turn. Table 8.1 reports the discrete change effects from this model without an interaction between sex and location. If we were to look at these discrete change effects for sex without knowing that they came from a main effects model, we would almost certainly conclude that there appears to be some moderation of the effect of sex by location. The effects in suburban and rural locations are much larger than the effect in central city locations, 45% and 68% greater, respectively, with a modest increase (16%) in the sex effect between suburban and rural locations.

If you prefer to present discrete changes because of their accessibility and the ease of explaining them to your audience, a reasonable solution is to draw on what you can conclude about the degree of moderation from one of the modes of interpretation that is not affected by this confounding. For example, the y^*-standardized effects on the latent outcome clearly show that the effect of sex changes substantially across residential locations. You could use this information as supplementary evidence to argue that location moderates the effect of sex on gun ownership to a great extent. In Part III, I discuss how outcome displays can be used to interpret interaction effects, along with two different approaches to deal with the issue of confounded nonlinearities.

Part II: The Effect of Residential Location Moderated by Sex

Note: Because I calculate discrete change effects using the sigreg command to interface with SPOST13, I need to rerun the logistic model with the specification of the interaction effects switching the order of the two variables to i.location##i.sex instead of i.sex##i.location (see the discussion in Chapter 6 of the spost( ) suboption). This is shown at the top of the output. Failure to do so would produce nonsensical results.

INTSPEC **Setup and** GFI **Analysis**

To switch the roles of focal and moderator variables, I respecify the *intspec* command to set residential location as the focal variable and sex as the moderator and to make the ordering of these in the int2( ) option the same as in the respecified logistic regression command.

```
. logit gun i.location##i.sex c.age##c.age educ fearnbhd race
    ...
. intspec focal(i.location) main( (i.location, name(Location) range(1/3)) ///
>         (i.sex, name(Sex) range(0/1)) ) int2(i.location#i.sex)
    ...
.
. gfi , factor

GFI Information from Interaction Specification of
Effect of Location on g(Gun) from Logistic Regression
----------------------------------------------------------------------

Effect of Suburbs =
    0.0443 + 1.1550*Male

    Factor Change Effect (1 unit change in) Suburbs =
        e^ 0.0443 * e^( 1.1550*Male)

            1.0453 * 3.1740^Male
Effect of Rural =
    0.4706 + 1.8962*Male

    Factor Change Effect (1 unit change in) Rural =
        e^ 0.4706 * e^( 1.8962*Male)

        1.6010 * 6.6605^Male
Sign Change Analysis of Effect of Location
on g(Gun), Moderated by Sex (MV)


    ------------------------------------------------------------
            |           Location
    When    | ------------------------------------------
    Sex=    |       Suburbs                 Rural
    --------+---------------------------------------------------
    Female  | Pos b =  .04427887    Pos b =  .47059609
      Male  | Pos b =  1.1992764    Pos b =  2.3667902
    --------+---------------------------------------------------
    Sign Changes |     Never                 Never
    --------+---------------------------------------------------
    % Positive   |     100.0                 100.0
    ------------------------------------------------------------
```

In brief, the GFI expression and the sign change analysis provide an initial picture of the pattern of moderation of location by sex, keeping in mind that the effect of the location categories is relative to the log odds of gun ownership in central cities. The effect for suburban location or rural location has the same sign (positive) for both men and women. Moreover, the effect of rural location is greater than the effect of suburban location for both men and women. To interpret this more concretely, I again use the *sigreg* command with different options for the type of effect it calculates, beginning with effects on the log odds.

Significance Region Analyses:
SIGREG **With Varying effect( ) Options**

Effects on the Log Odds. For these effects, I use the *sigreg* command without any options. You also could specify effect(b)—which is the default—to document what the command does. A boundary value analysis can be done for a dichotomous moderator. If the boundary value lies in the interval (0, 1), then it indicates that the significance of the focal variable is different for the category coded 0 than for the category coded 1. Thus, the moderated effect of suburbs changes to significant when *Sex* > 0.2858; that is, it changes from not significant for women to significant for men. Similarly, the rural effect changes from not significant for women to significant for men (*Sex* > 0.0846).

But it is actually easier to ignore the boundary value analysis and instead interpret the results in the significance region table. This shows that for women there is no significant difference in the log odds of ownership either in suburban versus central city areas or in rural versus central city areas. But for men, the log odds of ownership is significantly higher in both suburban and rural areas than in central city locations. The log odds in suburban areas is 1.20 times higher than in central cities, and the log odds in rural areas is 2.37 times higher. While this clearly portrays the varying significance of the locational effect, the magnitude of the significant effects for men are unclear, which is why one of the other options is typically used instead.

```
Boundary Values for Significance of Effect of Location on g(gun) Moderated by Sex
  Critical value Chi_sq = 3.841 set with p = 0.0500
```

Effect of Location	When Sex >=	Sig Changes	When Sex >=	Sig Changes
Suburbs	-0.8907 (< min)	to Not Sig [-4.3148]	0.2858	to Sig [29.8307]
Rural	-1.0732 (< min)	to Not Sig [-4.9488]	0.0846	to Sig [30.4097]

```
   Note: Derivatives of Boundary Values in [ ]
```

```
                Significance Region for Effect of Location (1 unit difference)
                  at Selected Values of Sex
            -------------------------------------------------
                                          At Sex=
                Effect of        |     Female       Male
            --------------+----------------------------
                    Suburbs  |     0.0443       1.1993*
                      Rural  |     0.4706       2.3668*
            -------------------------------------------------
             Key: Plain font   = Pos, Not Sig     Bold font*   = Pos, Sig
                 Italic font = Neg, Not Sig      Italic font* = Neg, Sig
```

Effects on the Latent Outcome. I use the same specification of the *sigreg* command as before to calculate the effects in standard deviation units of the latent propensity for gun ownership. I skip reporting the boundary value analysis in the output as it applies only to effects on the log odds. These results provide a good indication of the magnitude of the location differences in gun ownership for men, which continues to be not significant for women. Among men, the propensity to own a gun is more than half a standard deviation higher in suburban than in central city areas and more than a standard deviation higher in rural versus central city areas. Furthermore, the propensity for rural men to own guns is more than half a standard deviation greater than for suburban men. I determined that this latter difference is significant by rerunning the logistic regression and ICALC commands changing the reference category to suburban (*Location* = 2), which is easy to do with factor variables. I replaced every occurrence of *i.location* with *ib2.location*, where *ib2* signifies that the base category is the category coded 2.[3]

```
. sigreg , effect(b(sdy))
   ...

   Significance Region for Effect of Location (1 unit difference)
   on g(gun)-standardized at Selected Values of Sex
   -----------------------------------------------------------
                                  At Sex=
       Effect of      |     Female        Male
       -------------+---------------------------------
           Suburbs  |      0.0211      0.5706*
             Rural  |      0.2239      1.1262*
   -----------------------------------------------------------
    Key: Plain font  = Pos, Not Sig      Bold font*  = Pos, Sig
         Italic font = Neg, Not Sig      Italic font* = Neg, Sig

   Std. Dev. of latent outcome  =    2.1016
```

Factor Change Effects (Odds Ratios). The *sigreg* command with the effect(factor) option calculates the factor change in the odds of gun ownership shown in the significance region table and the output below. As with the prior two modes of interpretation, the factor change effect of suburban residence and rural residence is significant only for men and indicates a substantial moderation of the location effect by sex. Among men, suburban residents are more than three times as likely (the odds are more than three times higher) to own guns as are central city residents. And rural residents are slightly more than three times as likely to own a gun as are suburban residents (10.6631/3.3177 = 3.21). Like the interpretation using the latent variable approach, these numbers suggest that location is greatly moderated by sex. There is no significant difference among women in the odds of gun ownership but large and significant differences in the odds of ownership among men.

```
. sigreg , effect(factor)
   ...


   Significance Region for Factor Change Effect of Location (1 unit difference)
   on gun at Selected Values of Sex
   -----------------------------------------------------------
                                  At Sex=
       Effect of      |     Female        Male
       -------------+---------------------------------
           Suburbs  |      1.0453      3.3177*
             Rural  |      1.6009     10.6631*
   -----------------------------------------------------------
    Key: Plain font  = Pos, Not Sig      Bold font*  = Pos, Sig
         Italic font = Neg, Not Sig      Italic font* = Neg, Sig
```

Discrete Change Effects. I could and did use exactly the same specification of the *sigreg* command as before because both variables constituting the interaction are categorical. Notice that the table note reports two amounts for the SPOST effect for location because there are two location effects. As the table note shows, one effect compares suburbs with central cities and one effect compares rural areas with central cities. The significance region table for discrete change effects again shows positive and significant effects of location on gun ownership for men but not for women. Specifically, for otherwise average men, the probability of ownership is .27 greater for suburban residents than for central city residents and .53 greater for rural residents than for central city residents. In contrast, for women, the nonsignificant differences in the probability of ownership compared with central city locations are small (.01 for suburbs and .07 for rural areas).

```
. sigreg , effect(          ///
>            spost(                 ///
>                  amtopt(am(bin) )  ///
>                  atopt((means) _all)   ))
      …


   Significance Region for SPOST Change Effect of Location
   On gun at Selected Values of Sex
-----------------------------------------------------
                            At Sex=
       Effect of     |    Female     Male
   --------------+---------------------------
        Suburbs  |    0.0053     0.2726*
        Rural    |    0.0653     0.5309*
-----------------------------------------------------
   Key: Plain font  = Pos, Not Sig    Bold font*  = Pos, Sig
       Italic font = Neg, Not Sig    Italic font* = Neg, Sig

Spost Effect for  location specified as amount =    Suburbs vs CentralCity   Rural vs
CentralCity   calculated with
   at(  (means) _all )
```

This again appears to reveal that sex moderates the location effect to a large extent. But just looking at these results alone, it is unclear how much this might be due to the different reference points defined by sex to evaluate location's discrete effect. Table 8.2 reports the location effects by sex from a model without an interaction, for which a naive interpretation would suggest moderation by sex and larger locational effects for men than for women. While less of a difference in the location effects than for the actual moderated effects, this suggests that part of the apparent moderation of location by sex in the discrete change results is in fact due to the confounded nonlinearities. In Part III on outcome displays, I first produce and interpret a typical table and graph of predicted gun ownership varying with sex and residential location and then apply and discuss the two solutions introduced in Chapter 5.

TABLE 8.2 ● DISCRETE CHANGE EFFECT OF LOCATION BY SEX, MAIN EFFECTS MODEL

	Sex	
Discrete Change Effect of	**Women**	**Men**
Suburban	0.074	0.184
Rural	0.202	0.368

Part III: Outcome Displays With the OUTDISP Tool

I now use the *outdisp* command to create an initial table and graph showing the relationship of the predicted probability of gun ownership to sex and residential location (with the other predictors at their reference values). Because I do not want the default method for reference values (as observed), I specify the contents of the out( ) option to use mean values. The tab(def) and plot(def) options request the default table organization (the focal variable as the rows) and the default plot type (bar chart for a categorical moderator). In the copy of the output, I omit the echoing back of the options specified. I rerun the earlier *intspec* command to return to using sex as the focal variable so that the bar charts are organized as a sex comparison within location.

```
intspec focal(i.sex) main( (i.sex, name(Sex) range(0/1)) ///
>         (i.location, name(Location) range(1/3))) int2(i.sex#i.location) ndig(3)
    …
.outdisp , out(atopt( (means) _all)) plot(def) tab(def) pltopts(ylab(0(.25)1))
    …
```

Predicted Value of gun by the Interaction of Sex with Location.

```
-------------------------------------------------------
                  |            Location
            Sex   | CentralC~    Suburbs     Rural
------------------+------------------------------------
         Female   |    0.14        0.14       0.20
           Male   |    0.24        0.52       0.77
-------------------------------------------------------
```

For a single-moderator model, the patterns in a predicted value table or plot are usually easily understood and explained, especially when both predictors are categorical. Comparing the rows of the table in the output shows that in each location the probability of ownership is lower for otherwise average women than for men: .14 versus .24 in central cities, .14 versus .52 in suburbs, and .20 versus .77 in rural areas. Assessing the column differences reveals a consistent pattern of predicted ownership levels across locations—lowest in central cities, higher in suburbs, and highest in rural areas—but much smaller locational differences among women than among men. For women, the predicted ownership probabilities are .14, .14, and .20, while for men they are .24, .52, and .77. The bar charts in Figure 8.2 effectively portray these patterns visually. The panel for each location contains bars whose height represents the probability of gun ownership for women (*black bars*) and for men (*gray bars*). The bar charts make it obvious that the sex differences in gun ownership—disparity in the heights of the *black* and *gray bars*—increases from central cities to suburbs to rural areas. It is also clear that gun ownership for women does not change much across locations but is considerably different by location for men.

Presenting either the table or the bar chart reveals an overall story that can be easily summarized. Gun ownership is higher among men than among women, with the gap increasing as locations change from more urban to less urban. Moreover, the large variation by location in sex differences appears to be primarily due to men's probability of ownership changing across locations, while the locational differences are small among women. I qualified the last sentence—"appears to be"—because such differences in any analysis could in theory be partly due to how the moderators changing reference values combine with the nonlinearity of the link function. In fact, the discrete change results in Tables 8.1 and 8.2 demonstrated the same two general patterns in the predicted outcome from a model without interaction effects; that is, the probability of ownership

- varied more by location for men than for women and

- showed the smallest sex gap in ownership in central cities and the largest in rural areas.

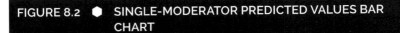

FIGURE 8.2 ● SINGLE-MODERATOR PREDICTED VALUES BAR CHART

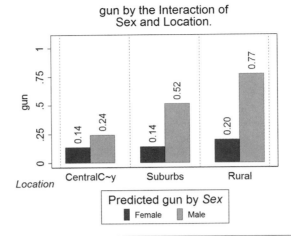

This suggests that the moderating effect of sex on location, and vice versa, partly reinforces the patterns produced by the nonlinear relationship between the probability of ownership and the predictors. Next, let's consider the two alternative approaches to interpretation using predicted value tables and plots that I discussed in Chapter 5.

Adding a Display of Predicted Values From a No Interaction Effects Model

This approach lets you directly compare the patterns from the model with interaction effects to the patterns you would get from an additive model without interaction effects. It creates displays of the plots, which can be arranged side by side to facilitate comparison. To produce these, I quietly ran the additive model and used estimates store to name and store the results followed by restore to make the interaction model results the current estimates. I then included the main( ) suboption in specifying the outcome( ) option on the *outdisp* command. The sequence of commands shown below produced the bar charts in Figure 8.3, but created no output in the Results window other than echoing back what you specified.

```
. quiet logit gun i.sex i.location c.age##c.age educ fearnbhd race
. est store lgtmain
. est restore lgtint
(results lgtint are active now)

. outdisp , plot(type(bar) name(barmain)) ///
> outcome(main(lgtmain) atopt( (means) _all)) ndig(2)
  pltopts(ylab(0(.25)1))
...
```

Each panel in Figure 8.3 shows the predicted probability of gun ownership in one of the residential locations for men (*gray bars*) and for women (*black bars*) from the two models. The additive model results are the pair of bars on the panel's right side and

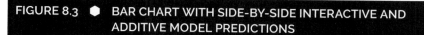

FIGURE 8.3 ● BAR CHART WITH SIDE-BY-SIDE INTERACTIVE AND ADDITIVE MODEL PREDICTIONS

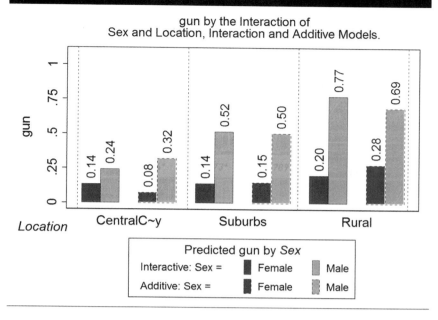

are outlined with *dashes*. Comparing the pair of bars in each location explicitly shows how and how much location moderates the effect of sex from the additive model. For suburban locations, there is virtually no difference between the moderated and the additive effects of sex. In central city locations, the sex effect is diminished by the interaction with almost equal-sized but opposite-signed changes from the additive model results for men's and women's probability of ownership. In contrast, the sex effect in rural areas increases relative to the additive model effects.

We can also use this figure to explore how the location effect is moderated by sex.[4] Start by looking at the additive model's location effect for women. The *black bars* with *dashed* outlines increase in height—predicted probability—from central city locations to suburban locations (by .07) to rural locations (by another .13). The moderated effect of location for women flattens this increase, resulting in no difference in ownership between central city and suburban locations, with rural locations' predicted probability higher by .06. Conversely, the moderated location effect for men (*gray bars solid outline*) expands the additive model's location differences in predicted ownership. The central city–suburban gap increases from .18 to .28, and the suburban–rural difference rises from .19 to .25.

The takeaway message is that the interaction between sex and residential location enhances and modifies the underlying patterns of the additive model. Namely, the sex gap in ownership increases as urbanicity decreases, and the location differences in gun ownership are much larger among men than among women.

Dual-Axis Labeling

This approach avoids confounding the nonlinearity of the interaction effect with the nonlinearity of the GLM link. But there is a potential loss of meaning from its use of the latent propensity of gun ownership as the metric for the predicted outcome. Transforming latent propensity's scale into standard deviations can help add meaning and context to the results, as can labeling a second outcome axis in the observed rather than the modeled metric. To construct such a plot (Figure 8.4), I added three suboptions to the outcome ( ) option. Specifying metric (model) creates the predicted outcome in the modeled metric, the latent propensity, or equivalently the log odds for a logistic regression. The sdy suboption labels the bar chart heights on the left side with the latent propensity of gun ownership in standard deviation units with evenly spaced labels. The dual suboption adds the right-side labels showing probabilities of gun ownership at various strategic points. The pltopts( ) option was added after an initial run to create better labeling of the left-side *y*-axis.

```
. outdisp , plot(type(bar) name(barmodeldual)) tab(def) ///
>       outcome(metric(model) sdy dual atopt( (means) _all) ) ///
   ndig(2)  pltopts(ylab(-.25(.5)1.25) ysc(r(-.5 1.4)))
   …
```

```
Predicted Value of g(gun)-Standardized by the Interaction
of Sex with Location.
```

```
------------------------------------------------------
          |                Location
     Sex | CentralC~    Suburbs      Rural
----------+-------------------------------------------
  Female |     -0.24      -0.22      -0.01
    Male |      0.10       0.67       1.23
------------------------------------------------------
```

The predicted latent propensity for gun ownership by sex and residential location reported in the predicted values table in the output or by the bar charts in Figure 8.4 is informative about the moderation of the sex effect by residential location, and vice versa. It also clearly portrays the overall relationship of these interacting predictors with gun ownership. The sex effect in the bar charts is shown by the difference in the heights of the *black* and *gray bars* in each panel. Men's higher propensity for gun ownership rises as urbanicity decreases, changing from one third of

FIGURE 8.4 ● **BAR CHART WITH DUAL-OUTCOME METRIC AXES**

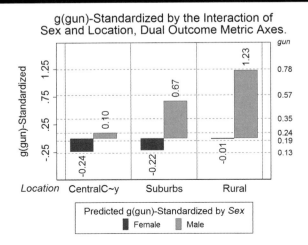

a standard deviation in central city locations ($0.10 - [-0.24] = 0.34$), to nine tenths of a standard deviation in suburban locales ($0.67 - [-0.22] = 0.89$), to 1.25 standard deviations in rural areas ($1.23 - [-0.01] = 1.24$). This definitely shows that location moderates the effect of sex to a large degree.

The location effect is represented by the difference in the heights of bars of the same color. Among women, the propensity to own guns changes very little across residential locations. It is essentially equal to the mean in rural areas and about one fifth of a standard deviation below the mean in central city and suburban locations. Conversely, among men, the propensity to own a gun varies considerably by location, from one tenth of a standard deviation above the mean in central cities, to two thirds of a standard deviation above the mean in suburban areas, to one and one-quarter standard deviations above the mean in rural areas. The contrasting location effects for men and women are an unambiguous indicator that sex greatly moderates the location effect.

Wrap-Up

What to Present to Interpret a One-Moderator Interaction Effect From Logistic Regression

You should begin with a somewhat bare-bones discussion of the GFI algebraic expression and note whether or not the sign of the moderated effect changes, drawing on the sign change analysis. Beyond that, I would recommend picking one device (method) for presenting and interpreting the interaction effect. In making a choice, the most important criterion is that you find the method informative and can use it to communicate well to your audience. Beyond that, the effectiveness of the interpretive devices varies across four criteria, which are specific to your research purposes and specific analysis:

1. How important is it for your purposes to unambiguously portray and discuss the extent of moderation of each component of the interaction by the other component?

2. Are the patterns of the relationship between the outcome and the interacting predictors complex or simple? For example, the patterns in this application are simple because the interacting predictors had three or fewer categories, and the patterns of moderated effects were consistent in the sign of the effect and changed monotonically across categories.

3. How well does the device work for discussing the overall relationship between the outcome and the interacting predictors? And can you effectively use a single table or figure to describe the relationship, or do you need to present a separate table or figure for each component of the interaction?

4. How important is it that the results provided by the device be accessible and easy to explain? That is, how technically sophisticated is your intended audience?

Table 8.3 provides a checklist of the different interpretive techniques and how I would evaluate their use and effectiveness with respect to each criterion, rated as Excellent, Good, or Fair.

TABLE 8.3 ● CRITERIA FOR CHOOSING A METHOD OF INTERPRETATION

Interpretive Device (Method)	Moderation Importance	Complex Patterns	Total Relationship	Accessible
Predicted value table				
Observed metric	Fair	Good	Excellent, 1 table	Excellent
Model metric	Excellent	Good	Excellent, 1 table	Fair, Good if standardized
Predicted value figure				
Plain	Fair	Excellent	Excellent, 1–2 figures	Excellent
Additive model comparison	Good	Excellent	Good, 1–2 figures	Fair
Dual axis in model metric	Excellent	Excellent	Excellent, 1–2 figures	Good
Significance region table				
Discrete change effect	Fair	Good	Good, 2 tables	Excellent
Latent variable standardized	Excellent	Good	Good, 2 tables	Good
Factor change	Good	Good	Good, 2 tables	Fair
Log odds	Fair	Good	Fair, 2 tables	Fair

Comparison of Probit and Logistic Regression Results

I noted earlier that probit and logistic regression almost invariably provide comparable results, which I will demonstrate for this application example. Because the logit and probit coefficients are scaled differently, we cannot directly compare the size of the estimated coefficients but can compare the sign and statistical significance of the coefficients. Taking the scaling into account, an estimated logistic coefficient should be approximately 1.7 times the size of an estimated probit coefficient, and the same applies to their standard errors (Long, 1997, p. 48). Table 8.4 reports the coefficients and p values from the logistic regression and probit analyses, and the last column shows the ratio of the estimated logit to the estimated probit coefficients. The coefficients have the same sign in both analyses. The p values are very similar although they differ somewhat more for those coefficients that are clearly not significant (i.e., close to 0): suburban, fear, and race. Nearly all of the ratios of the size of the coefficients are within the range of 1.60 to 1.85, again with the greatest divergence for the not significant coefficients.

Of greater importance is how the choice of logistic regression versus probit analysis might affect the interpretations and conclusions we draw about the interaction effects. Because the comparison of the logit and probit results indicated a high degree of similarity for all the applicable methods of interpretation,[5] I picked two illustrations. Table 8.5 contrasts significance region table results for the effect of sex moderated by location, and Figure 8.5 compares the bar charts of the predicted probability

TABLE 8.4 ● COMPARISON OF LOGIT AND PROBIT COEFFICIENTS

	Logit		Probit		Logit b / Probit b
	b	**p**	**b**	**p**	
Sex	0.7087	.007	0.3754	.008	1.89
Suburban	0.0443	.429	0.0151	.454	2.93
Rural	0.4706	.089	0.2462	.098	1.91
Sex × Suburban	1.1550	.000	0.7076	.000	1.63
Sex × Rural	1.8962	.000	1.1920	.000	1.59
Age	0.1152	.000	0.0660	.000	1.75
Age squared	−0.0010	.000	−0.0006	.000	1.74
Education	−0.0557	.005	−0.0305	.008	1.83
Fear	−0.0637	.325	−0.0425	.297	1.50
Race	−0.1789	.186	−0.0938	.201	1.91
Intercept	−4.2438		−2.4568		1.73

of gun ownership varying by sex and location. The following syntax produced the results:

```
quiet {
    probit gun i.sex##i.location c.age##c.age educ fearnbhd race
    intspec focal(i.sex) main( (i.sex, name(Sex) range(0/1)) ///
        (i.location, name(Location) range(1/3))) int2(i.sex#i.location)
    noi sigreg , effect(b(sdy))
    noi outdisp , plot(type(bar) name(barprobit)) tab(def) out(atopt ( ///
        (means) _all)) pltopts(ylab(0(.2).8) text( 1 0 Probit" ,size(*1.2) ///
            j(left) place(west))) ///
        ndig(2) plotreg(ma(t +3)))

    logit gun i.sex##i.location c.age##c.age educ fearnbhd race
    intspec focal(i.sex) main( (i.sex, name(Sex) range(0/1)) ///
        (i.location, name(Location) range(1/3))) int2(i.sex#i.location)
    noi sigreg , effect(b(sdy))
    noi outdisp , plot(type(bar) name(barlogit)) tab(def) out(atopt ( ///
        (means) _all)) pltopts(ylab(0(.2).8) text( 1 0 "Logit" ,size(*1.2) ///
            j(left)place(west))) ///
        ndig(2) plotreg(ma(t +3)))
}
```

The significance region analysis reports the moderated effect of sex on the latent propensity to own a gun, with the effects scaled in standard deviation units of the latent

TABLE 8.5 ● SIGNIFICANCE REGION FOR EFFECT OF SEX (1 = *MALE*) ON PROPENSITY TO OWN A GUN, STANDARD DEVIATION UNITS		

	Residential Location		
	Central City	**Suburban**	**Rural**
Probit	0.3194*	0.9215*	1.3336*
Logit	0.3372*	0.8868*	1.2395*

*Significant at $p < .05$.

FIGURE 8.5 ● COMPARISON OF LOGIT AND PROBIT PREDICTED OUTCOME BAR CHARTS

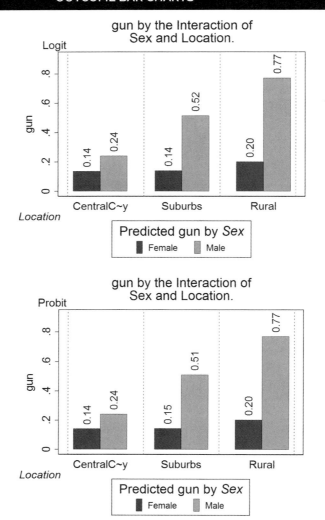

outcome. Because both the logit and probit effects are scaled in the same metric, the moderated effects are directly comparable. The effect of sex on ownership propensity differs between the logit and probit results by less than 8% for any of the residential locations, and the pattern of moderation differs only in small numeric detail. I would summarize each set of results in the same way: The propensity to own a gun among otherwise average men is significantly higher than among women by about one third of a standard deviation in the central city locales, by about nine tenths of a standard deviation in suburban areas, and by 1.25 (1.33) standard deviations in rural locations.

There is even less difference between the probit and logit estimates of the predicted probability of gun ownership by sex and residential location shown in the bar charts. The probabilities differ by less than .0056; that is, the percentage of gun ownership differs by less than one percentage point for any combination of sex and location.

These brief comparisons reinforce the common knowledge that the choice between logistic regression and probit analysis is rarely consequential for the results and the conclusions you draw. If you use ICALC for probit analysis following the syntax and applications of this chapter, the ICALC commands are identical for the logit and probit analyses. The only limitations are that interpretations using the log odds or factor changes are not relevant for probit analysis. Consequently, you cannot get factor change results using the *gfi* or *sigreg* commands; trying to do so should generate an error message.

THREE-WAY INTERACTION EXAMPLE (INTERVAL BY INTERVAL BY NOMINAL)

Data and Testing

This application explores how the interaction of racial contact by education by race affects whether in 1987 people would have supported a law banning interracial marriage (*Ban* = 1) or opposed it (*Ban* = 0). Two of the interacting variables are interval level and one is categorical. Racial contact is an interval index that measures the number of different types of racial contact (0–3) the respondent has. Education is an interval indicator of the number of years of completed education (0–20). And race is a dichotomous indicator (1 = *White*, 0 = *non-White*). The other predictors are education, age, perceived social class rank, and region of birth. Age in years (18–89) and class rank (1 [lowest] to 10 [highest]) are interval measures, while region of birth is a dichotomous variable (1 = *South*, 0 = *non-South*). The index function to estimate is

$$g(y_i) = \beta_0 + \beta_1 Contact_i + \beta_2 Education_i + \beta_3 Race_i + \beta_4 Contact_i \times Education_i$$
$$+ \beta_5 Contact_i \times Race_i + \beta_6 Education_i \times Race_i + \beta_7 Contact_i \times Education_i$$
$$\times Race_i + \beta_8 Age_i + \beta_9 Class_i + \beta_{10} Region_i$$

The sample excludes 322 respondents with missing information on class rank, another 85 with missing information on racial contact, and 38 more cases with missing information on other variables in the analysis.

The diagnostic tests for model fit (Hosmer-Lemeshow statistic and a LOWESS [locally weighted scatterplot smoothing] plot of the predicted probability against the observed) showed no evidence of poor fit for the model. Diagnostic tests initially identified a number of potential outliers because their residuals were well outside the core of the distribution or they had high leverage values. Subsequent probing indicated that three cases were influential; in particular affecting the coefficient size and statistical significance of several lower order interaction terms. A sensitivity analysis of the moderated effects of each component showed that the exclusion of these cases produced significance region tables that were overall similar. But excluding these influential cases generally increased the magnitude of the moderated effect slightly and the size of the significance region. I decided to include these cases so as not to bias the results toward larger effects.

Statistical tests of the contact-by-education-by-race interaction terms—$\beta_4 Contact_i \times Education_i + \beta_5 Contact_i \times Race_i + \beta_6 Education_i \times Race_i + \beta_7 Contact_i \times Education_i \times Race_i$—indicate that they should be included in the logit model. A global test of the change in the log likelihood if these four terms are excluded from the model is statistically significant ($p = .007$) as is the z test of the three-way coefficient β_7 ($p < .001$).

Strategies for Interpreting the Three-Way Interaction

The first example of this chapter provided a foundation of how to apply all of the modes of interpretation for logistic regression interaction effects. For this application example, I instead begin by briefly discussing the GFI expression for each component predictor of the interaction—racial contact, education, and race. I then tell the story of the relationship between support for a legal ban on interracial marriage and these interacting predictors in three ways: (1) using factor changes (odds ratios), (2) using discrete change effects and tables/plots of the predicted probability, and (3) using effects scaled in standard deviation units of the latent variable and tables/plots of the predicted standardized latent variable. The logit command for estimating the interaction model is

```
logit bwlaw c.contact##c.ed##i.racew age region16 class
```

Variations of this command that reorder the three interacting predictors are used to estimate discrete change effects (see the explanation of the effect( ) option and its spost( ) suboption for the *sigreg* command in Chapter 6). Each section begins with the output showing relevant results and the ICALC commands that generated them.

GFI **Results for the Three Predictors**

Note that I do not list in the output the detailed sign change table produced by the *gfi* command because the significance region tables provide an overall picture of the sign changes, albeit a less precise indication of exactly when the sign changes (if it does).

Racial Contact

```
. intspec focal(c.contact) main((c.contact, name(RacialContact) range(0/3)) ///
>   (c.ed, name(Education) range(0(4)20)) (i.racew, name(Race) range(0/1)) ) ///
>   int2( c.contact#c.ed c.contact#i.racew c.ed#i.racew ) ///
>   int3( c.contact#c.ed#i.racew ) ndig(0)
         ...
. gfi,  ndig(3)
```

```
GFI Information from Interaction Specification of
Effect of Racialcont~T on g(Ban) from Logistic Regression
-----------------------------------------------------------------------

Effect of Contact =
   -1.757 + 0.135*ed + 1.448*White - 0.137*ed*White
```

The baseline effect of racial contact (−1.757) is the effect on support for a legal ban for non-Whites (*White* = 0) with no education (*ed* = 0), and indicates that a greater extent of racial contact predicts less support for a legal ban. This negative effect moves toward a more positive value for higher levels of education (0.135) and for Whites relative to non-Whites (1.448), but the combination of higher education and being White (−0.137) diminishes their separate positive moderation of the effect of racial contact. Because race is a dichotomous predictor, I rewrite the GFI expression as two expressions, one for the effect of racial contact for Whites and one for non-Whites, substituting in values for White and regrouping similar terms:

$$
\begin{aligned}
White = 1 \quad & -1.757 + 0.135 \times Educ + 1.448 \times (1) - 0.137 \times Educ \times (1) \\
& = -1.757 + 1.448 + (0.135 - 0.137) \times Educ \\
& = -0.244 - 0.007 \times Educ \\
White = 0 \quad & -1.757 + 0.135 \times Educ + 1.448 \times (0) - 0.137 \times Educ \times (0) \\
& = -1.757 + 0.135 \times Educ
\end{aligned}
$$

Thus, the effect of racial contact for Whites is negative and increases in magnitude as education increases, while the contact effect for non-Whites is initially negative but decreases in magnitude—possibly turning positive—as education increases.

Education

```
. intspec focal(c.ed) main((c.contact, name(RacialContact) range(0/3)) ///
>   (c.ed, name(Education) range(0(4)20)) (i.racew, name(Race) range(0/1)) ) ///
>   int2( c.contact#c.ed c.ed#i.racew  c.contact#i.racew )  ///
>   int3( c.contact#c.ed#i.racew ) ndig(0)
     ...
. gfi,  ndig(3)


    GFI Information from Interaction Specification of
    Effect of Education on g(Ban) from Logistic Regression
    -------------------------------------------------------------------

    Effect of Ed =
  -0.307 + 0.135*Contact + 0.061*White - 0.137*Contact*White
```

The GFI expression for the effect of education parallels what we just saw for racial contact in terms of the signs of the components of the moderated effect. There is an initial negative effect that moves toward the positive direction with increasing contact and for Whites versus non-Whites, but those changes are muted by the combination of greater contact and being White. I can again rewrite the GFI as separate expressions for the effect of education for Whites and non-Whites, which produces

$$
\begin{aligned}
White = 1 \quad & -0.242 - 0.002 \times Contact \\
White = 0 \quad & -0.307 + 0.135 \times Contact
\end{aligned}
$$

In sum, the effect of education for Whites is to decrease support for a legal ban, with a larger negative effect if they have a greater degree of racial contact. The initial effect of education for non-Whites is also negative but lessens in magnitude (possibly turning positive) as racial contact increases. (Some readers might want to go through the step-by-step calculations to reproduce these effects to double-check their understanding.)

Race

```
. intspec focal(i.racew) main((c.contact, name(RacialContact) range(0/3)) ///
> (c.ed, name(Education) range(0(4)20)) (i.racew, name(Race) range(0/1)) ) ///
> int2(c.contact#i.racew c.ed#i.racew c.contact#c.ed ) ///
> int3( c.contact#c.ed#i.racew ) ndig(0)
        …
. gfi,  ndig(3)


GFI Information from Interaction Specification of
Effect of Race on g(Ban) from Logistic Regression
------------------------------------------------------------------

Effect of White =
    1.670 + 1.448*Contact + 0.061*Ed - 0.137*Contact*Ed
```

The positive baseline effect of race on support for a ban (1.670) indicates that support is predicted to be higher for Whites than for non-Whites among those with no racial contact and no formal education. The difference between Whites and non-Whites widens as both contact (1.448) and education (0.061) increase but at a slower rate for the combination of greater racial contact and higher education (−0.137). Unlike the pattern of the moderated effects of contact and education, it is harder to see the overall pattern of the moderation of the race effect because the two-way interactions of race and its moderators are the opposite sign of their three-way interaction. This pattern is made clear by each of the three approaches to telling the story of the relationship between the outcome and the three interacting predictors in the following sections.

Factor Change Interpretation

I use the formatted significance region tables saved in Excel for the moderated effect of each interacting predictor as the basis for discussing how to use factor changes to interpret the relationship between support for a legal ban and the three-way interaction of racial contact, education, and race. Each discussion reports the *sigreg* command followed by the Excel-formatted table in lieu of the Stata output results.

Moderated Effect of Racial Contact

```
. sigreg, effect(factor) save(Output\Table_8_6.xlsx tab) ndig(3)
```

The significance region results in Table 8.6 fill in details elaborating the GFI conclusion that racial contact affects support of a legal ban in a very different way for the two races. For Whites, there is a consistent negative effect of contact on support for a ban—as contact increases by one unit, the odds of support slightly declines to about 95% of its starting size. Keep in mind that factor changes less than 1 represent negative effects on the odds of the outcome. This contact effect is significantly smaller than no effect for education levels representing 89% of the sample of Whites.

	Race	
Education	**Non-White**	**White**
0	*0.173**	*0.735*
2	*0.226**	*0.731*
4	*0.296**	*0.728*
6	*0.388**	*0.725*
8	*0.508**	*0.721*
10	*0.665*	*0.718**
12	*0.870*	*0.715**
14	1.140	*0.711**
16	1.493	*0.708**
18	1.955	*0.705*
20	2.560	*0.702*

TABLE 8.6 ● FACTOR CHANGE EFFECT OF CONTACT (ONE-UNIT DIFFERENCE) MODERATED BY THE INTERACTION OF RACE AND EDUCATION

Key	
Plain font, no fill	Pos, Not Sig
Bold*, filled	Pos, Sig
Italic, no fill	Neg, Not Sig
***Bold italic*, filled**	Neg, Sig

In contrast, among non-Whites, the contact effect changes direction from negative to positive as education levels increase and is statistically significant for education values corresponding to about 10% of the sample of non-Whites. At low levels of education, the contact effect is negative and of greater magnitude than it is for Whites at low levels of education—the odds of support declines to about one third of its initial value. At the highest levels of education, the contact effect for non-Whites turns positive. Although an increase by 1 in contact predicts increased support for a ban by a factor of around 2, the effect is not significant. Overall, this suggests that racial contact is of relatively little consequence for understanding non-Whites' opinions about a legal ban on intermarriage (significant for only 10% of the non-White subsample) but does broadly predict a decrease in Whites' support for a ban for the vast majority of Whites in the sample (89%).

Moderated Effect of Education

```
. sigreg, effect(factor) save(Output\Table_8_7.xlsx tab) ndig(3)
```

The factor change effects of education reported in Table 8.7 demonstrate both the similarities and the differences in the moderated effect of education on support for a

TABLE 8.7 ● FACTOR CHANGE EFFECT OF EDUCATION (ONE-UNIT DIFFERENCE) MODERATED BY THE INTERACTION OF RACE AND CONTACT

Race	Racial Contact			
	0	**1**	**2**	**3**
Non-White	*0.736**	*0.842**	*0.964*	*1.103*
White	*0.782**	*0.780**	*0.778**	*0.776**
Key				
Plain font, no fill	Pos, Not Sig			
Bold*, filled	Pos, Sig			
Italic, no fill	Neg, Not Sig			
Bold italic*, filled	Neg, Sig			

ban for non-Whites and Whites. The education effect is negative (factor change value < 1), decreasing the odds of support for almost all combinations of race and racial contact. But the effect is moderated in different ways by racial contact for Whites and non-Whites. For Whites, the effect of education is always negative and statistically significant and very slightly increases in magnitude—values much less than 1—as racial contact increases. More specifically, a 1-year difference in education predicts a reduction in the odds of support by a factor of 78%.

The education effect for non-Whites is similar in size to the effect for Whites at the two lowest racial contact levels; but only 20% of non-Whites compared with 60% of Whites have such low levels of racial contact and thus experience similar effects of education on their opinions. However, as contact rises further, the education effect for non-Whites turns nonsignificant and then becomes a positive yet still nonsignificant effect at the highest level of contact. In sum, the moderated effects of education and contact evidence a similar moderated relationship with the outcome. Both effects consistently predict a reduction in the odds of supporting a legal ban on intermarriage for Whites but have more limited significant effects on the outcome for non-Whites.

Moderated Effect of Race

```
. sigreg, effect(factor) save(Table_8_8.xlsx tab) ndig(3)
```

The significance region results in Table 8.8 for the moderated effect of race show that, for nearly all combinations of years of education and levels of racial contact, Whites have a much higher odds of supporting a ban than do non-Whites. Among the significant effects of race, the smallest factor change (3.014) indicates that Whites' odds of supporting a ban is predicted to be three times larger than non-Whites' odds of support. And the largest factor change (409.867) represents a nearly 400 times larger odds of support for Whites than for non-Whites. Between these two values, the effect of race follows two parallel patterns of change corresponding to increasing levels of racial contact and of education:

- At the lowest level of racial contact (0), the effect of race increases in magnitude as education rises. This means that the predicted odds of support for Whites relative to non-Whites rises with education (from 14.5 times larger to 89.8 times larger). But at all higher levels of racial contact, the effect of race shows the opposite pattern: The predicted odds of support for Whites compared with non-Whites becomes smaller with increasing education (declining from 39.4 to 14.4 times as large).

- Similarly, at lower levels of education (<11 years[6]), the effect of race grows with increasing racial contact, but the growth in the effect of race is smaller as education levels rise. For instance, when *Education* = 0, the effect of race changes from 5.3 to 409.9, while at *Education* = 10, the change is from 9.7 to 12.3. In contrast, for *Education* ≥ 11, the effect of race diminishes with racial contact but primarily remains a positive effect (factor change > 1), indicating

TABLE 8.8 ● FACTOR CHANGE EFFECT OF RACE (WHITE VS. NON-WHITE) MODERATED BY THE INTERACTION OF EDUCATION AND CONTACT

Education	Racial Contact			
	0	1	2	3
0	5.314	22.621*	96.289*	409.867*
2	5.998	19.407*	62.792*	203.166*
4	6.770*	16.650*	40.948*	100.707*
6	7.641*	14.284*	26.703*	49.919*
8	8.624*	12.254*	17.413*	24.744*
10	9.733*	10.513*	11.356*	12.265*
12	10.986*	9.020*	7.405*	6.080*
14	12.399*	7.738*	4.829*	3.014*
16	13.995*	6.639*	3.149*	1.494
18	15.795*	5.695*	2.054	*0.740*
20	17.828	4.886	1.339	*0.367*
Key				
Plain font, no fill	Pos, Not Sig			
Bold*, filled	Pos, Sig			
Italic, no fill	Neg, Not Sig			
Bold italic*, filled	Neg, Sig			

a higher predicted level of support for Whites than for non-Whites. The effect of race does reverse at the highest level racial contact when *Education* ≤ 18, but these effects are not significant.

Summary

Pulling together the interpretation of the three moderated effects suggests a coherent story of the relationship between support for a legal ban on interracial marriage and the interaction of racial contact, education, and race. Increases in contact predict a reduction in Whites' support of a ban, with the reduction growing larger at higher levels of education. But contact has limited significant effects on the outcome for non-Whites, which changes from reducing to increasing support at higher levels of education. Analogously, rising education predicts a significant decrease in Whites' support of a ban that strengthens slightly with more racial contact. A different pattern holds for non-Whites. The negative effect of education has limited significance, and it grows smaller in magnitude as racial contact increases, turning positive at the highest level of contact.

These different patterns of the effects of education and contact for non-Whites and Whites underlie the more complicated patterns of the effect of race on support for a legal ban. For nearly all combinations of education and racial contact, Whites' support for a ban is predicted to be significantly greater than non-Whites' support. But when racial contact is low, education increases the predicted racial difference in support; while at higher levels of racial contact, education decreases the predicted racial gap. In the same vein, at lower levels of education, racial contact increases the predicted racial disparity in support; but at higher levels of education, racial contact decreases the racial difference. These are consistent with the opposite patterns for non-Whites and Whites of the effect of contact on support when education is low and of the effect of education on support when contact is low.

Standardized Latent Outcome Interpretation

You have two choices for interpreting an interaction effect using the standardized latent outcome approach. You can present and interpret the (Excel-formatted) significance region table for the moderated effect of each component of the interaction, or you can present and interpret tables or plots of the predicted standardized latent outcome. In the latter case, I strongly recommend that you first work through the results in the significance region tables for your own understanding of the relationship. You will then be able to better understand and discuss the outcome displays.

Consequently, I start by discussing each of the moderated effects on the standardized latent outcome, followed by a summary description of the overall relationship, as I just did for the factor change interpretation. I then illustrate the use of visual displays of the predicted standardized latent outcome to interpret the interaction effect, which will draw on what I learned from the significance region table results. Each section begins by listing the ICALC commands that produced the tables or figures discussed but does not otherwise present the Stata output results. Note that the *sigreg* command includes the effect(b(sdy)) option to get the effects scaled in the metric of the standardized latent outcome.

Using Significance Region Tables for Interpretation

Moderated Effect of Racial Contact

```
. intspec focal(c.contact) main((c.contact, name(RacialContact) range(0/3)) ///
> (c.ed, name(Education) range(0(2)20)) (i.racew, name(Race) range(0/1)) ) ///
> int2( c.contact#c.ed  c.contact#i.racew  c.ed#i.racew ) ///
> int3( c.contact#c.ed#i.racew ) ndig(0)
    …
. sigreg, effect(b(sdy)) save(Output\Table_8_9.xlsx tab) ndig(3)
```

Table 8.9 reports that contact has a negative effect on the latent propensity to support a ban for Whites, which slightly increases in size among the significant effects (from -0.136 to -0.144) as education levels rise. The effect is around one seventh of a standard deviation decrease in the propensity of support for each one-unit increase in racial contact, or about 40% of a standard deviation decrease in support between those with minimal racial contact and those with maximal racial contact (3×-0.14). This effect is significant for the 70% of Whites with education between 8 and 16 years. In contrast, the effect of contact for non-Whites changes from a large negative effect at low levels of education—roughly one half of a standard deviation decrease in latent support with a one-unit change in contact—to a comparably large positive effect at the highest levels of education. But the contact effect is significant only for the 9% of the non-White subsample at the bottom half of the education distribution. The overall picture is that contact broadly predicts a reduction in Whites' latent support for a ban with fairly limited effects on latent support for non-Whites.

Moderated Effect of Education

```
. intspec focal(c.ed) main((c.contact, name(RacialContact) range(0/3)) ///
> (c.ed, name(Education) range(0(2)20)) (i.racew, name(Race) range(0/1)) ) ///
> int2( c.contact#c.ed c.ed#i.racew  c.contact#i.racew )  ///
> int3( c.contact#c.ed#i.racew ) ndig(0)
    …
. sigreg, effect(b(sdy)) save(Output\Table_8_10.xlsx tab) ndig(3)
```

The moderated effect of education shown in Table 8.10 follows a pattern analogous to the moderated effect of contact. For Whites, a 1-year increase in education predicts one tenth of a standard deviation decrease in latent support, which is statistically significant but hardly changes in magnitude across the levels of racial contact. The effect of education for non-Whites changes from a negative and statistically significant effect on latent support for a ban to a nonsignificant effect as contact increases from its minimum to its maximum. The significant effects of education at low levels of racial contact are of similar magnitude for Whites and non-Whites. But the effect of education is always significant for Whites and is only significant for the 20% of the non-White subsample with limited interracial contacts.

Thus, like the effect of contact, the effect of education consistently predicts a reduction in latent support for Whites with more limited impact on latent support for non-Whites. Note that the effect of education is greater than the effect of contact on the propensity to support a legal ban for Whites. The predicted change in Whites' latent support from education's minimum to its maximum value is approximately a 2 standard deviation decline ($-0.1 \times [20 - 0]$), whereas the reduction in support across the range of contact levels for Whites is about one third of a standard deviation ($-0.1 \times [3 - 0]$).

TABLE 8.9 ● EFFECT OF CONTACT (ON *G*(*Y*)-STANDARDIZED) MODERATED BY EDUCATION AND RACE

Education	Race	
	Non-White	**White**
0	**−0.731***	−0.128
2	**−0.619***	−0.130
4	**−0.507***	−0.132
6	**−0.395***	−0.134
8	**−0.282***	**−0.136***
10	−0.170	**−0.138***
12	−0.058	**−0.140***
14	0.055	**−0.142***
16	0.167	**−0.144***
18	0.279	−0.146
20	0.391	−0.148
Key		
Plain font, no fill	Pos, Not Sig	
Bold*, filled	Pos, Sig	
Italic, no fill	Neg, Not Sig	
Bold italic*, filled	Neg, Sig	

TABLE 8.10 ● EFFECT OF EDUCATION (ON *G*(*Y*)-STANDARDIZED) MODERATED BY CONTACT AND RACE

Race	Racial Contact			
	0	**1**	**2**	**3**
Non-White	**−0.128***	**−0.072***	−0.015	0.041
White	**−0.103***	**−0.103***	**−0.104***	**−0.1054***
Key				
Plain font, no fill	Pos, Not Sig			
Bold*, filled	Pos, Sig			
Italic, no fill	Neg, Not Sig			
Bold italic*, filled	Neg, Sig			

Moderated Effect of Race

Table 8.11 presents the significance region for the moderated effect of race at different levels of racial contact and education, which is positive and significant except for the combination of the highest levels of education with mainly higher levels of racial contact and the match up of low education and low contact. Given the coding of race, these positive effects mean that the predicted propensity to support a legal ban is higher for Whites than for non-Whites. This difference is significant for all but the 10% of the sample in those combinations of education and contact. The size of the difference in the predicted propensity to support a ban between non-Whites and Whites is greater than one half standard deviation for 84% of the sample, where 22% of the sample has a difference of >1 standard deviation. The effect of race changes across both contact and education in a similar way. At the lowest levels of one moderator, the race gap in predicted support increases as the values of the second moderator increase. But at higher levels of the first moderator, the direction reverses and the race gap diminishes with higher values of the second moderator.

TABLE 8.11 ● EFFECT OF RACE (ON $G(Y)$-STANDARDIZED) MODERATED BY EDUCATION AND CONTACT

Education	Racial Contact			
	0	**1**	**2**	**3**
0	0.696	**1.299***	**1.902***	**2.505***
2	0.746	**1.235***	**1.724***	**2.213***
4	**0.796***	**1.171***	**1.546***	**1.920***
6	**0.847***	**1.107***	**1.368***	**1.628***
8	**0.897***	**1.043***	**1.190***	**1.336***
10	**0.947***	**0.980***	**1.012***	**1.044***
12	**0.998***	**0.916***	**0.834***	**0.752***
14	**1.048***	**0.852***	**0.656***	**0.459***
16	**1.099***	**0.788***	**0.478***	0.167
18	**1.149***	**0.724***	0.300	*−0.125*
20	1.199	0.661	0.122	*−0.417*
Key				
Plain font, no fill	Pos, Not Sig			
Bold*, filled	Pos, Sig			
Italic, no fill	Neg, Not Sig			
***Bold italic*, filled**	Neg, Sig			

Summary of Moderated Effects

Although the numerical details of the preceding discussion of the moderated effects of each predictor have changed compared with the factor change interpretation, both discussions portray the same general picture. Thus, the summary above applies equally well here, so I will not repeat it. Instead, let's turn to the use of tabular or visual displays of the predicted value of the latent outcome, which can be used to explain and interpret the interaction effect.

Using Outcome Displays for Interpretation of the Latent Outcome

You have multiple options for how a tabular or visual display of the predicted latent outcome can be organized, which you control according to the predictor specified as the focal variable and those as the first and second moderators. ICALC then organizes visual displays as follows:

- Table whose cells = predicted values and whose
  - columns are defined by the first moderator,
  - horizontal panels are defined by second moderator, and
  - rows within horizontal panel are defined by the focal variable.

- Scatterplot whose *y*-axis = predicted outcome values and whose
  - *x*-axis is defined by the focal variable,
  - vertical panels are defined by the second moderator, and
  - repeated plotted lines within a vertical panel are defined by the first moderator.

- Bar chart whose *y*-axis = predicted outcome values and whose
  - categorical *x*-axis is defined by the focal variable,
  - vertical panels are defined by the second moderator, and
  - repeated bars within a vertical panel are defined by the first moderator.

In practice, you should try each predictor in turn as the focal variable to decide which organization of the information works best and whether you prefer a tabular display or a graphic display for your analysis. For a three-way interaction, it is typically difficult to use a single table or a plot to easily show and understand the effects of all three interacting predictors. In the following sections, I illustrate the use of a tabular display and then a graphic display for interpreting the three-way interaction. I list the ICALC commands used to produce the tables and graphics I chose as examples at the top of each section. The special topics section at the end of the chapter reports the commands that would produce tables and graphs that use other choices of the focal variable, so that you can try other options for this example to see if they would work better for your understanding.

Tabular Display

```
. intspec focal(c.contact) main((c.contact, name(RacialContact) range(0/3)) ///
>   (c.ed, name(Education) range(0(5)20)) (i.racew, name(Race) range(0/1)) ) ///
>   int2( c.contact#c.ed c.contact#i.racew c.ed#i.racew  )  ///
>   int3( c.contact#c.ed#i.racew ) ndig(0)
      ...

. outdisp, out(metric(model) sdy dual atopt((means) _all)) ///
>        tab(Output\Table_8_12.xlsx)
```

To create the table organization that I decided works best, I specified racial contact as the focal variable, education as the first moderator, and race as the second moderator. The predicted values of the latent propensity to support the legal ban on intermarriage by the three-way interaction of racial contact, education, and race are presented in Table 8.12, based on the ICALC table saved into an Excel file. Education defines the columns—it has separate horizontal panels for non-Whites and Whites—and racial contact defines the rows within each panel. To better see the race effect, I edited the Excel table to add a horizontal panel showing the difference in predicted values for Whites versus non-Whites as it varies with contact and education rather than presenting a second predicted values table. I incorporate information about statistical significance into the following discussion by drawing on the significance region analyses previously discussed. In writing this up for a professional paper, you could add a footnote that the statistical significance results are available from the author on request.

We can determine how education affects the latent propensity for support of the ban by examining the change in predicted values within the rows in the upper two panels. Except for non-Whites with maximal racial contact (*first panel, last row*), the predicted propensity to support a ban declines as education increases. This negative effect diminishes in size with a larger number of racial contacts and changes more so for non-Whites than for Whites. The decline across the range of education is substantial and statistically significant for those who have no racial contact, about a 2 standard deviation decrease in latent support for either non-Whites or Whites. For Whites, this change hardly varies with racial contact. For non-Whites, the size

TABLE 8.12 ● PREDICTED VALUE OF G(BAN)-STANDARDIZED BY THE THREE-WAY INTERACTION OF EDUCATION, RACIAL CONTACT, AND RACE

Race	Racial Contact	Education				
		0	5	10	15	20
Non-White	0	1.0126	0.3739	−0.2647	−0.9033	−1.5420
	1	0.2812	−0.0768	−0.4347	−0.7927	−1.1506
	2	−0.4502	−0.5275	−0.6048	−0.6820	−0.7593
	3	−1.1816	−0.9782	−0.7748	−0.5714	−0.3680
White	0	1.7081	1.1954	0.6828	0.1701	−0.3425
	1	1.5798	1.0623	0.5448	0.0274	−0.4901
	2	1.4515	0.9292	0.4069	−0.1154	−0.6377
	3	1.3232	0.7961	0.2689	−0.2582	−0.7853
White–non-White	0	0.6955	0.8215	0.9475	1.0735	1.1995
	1	1.2986	1.1391	0.9796	0.8200	0.6605
	2	1.9017	1.4567	1.0117	0.5666	0.1216
	3	2.5048	1.7743	1.0437	0.3132	−0.4173

of effect of education on support changes more, and at contact values > 1, it becomes nonsignificant and then turns positive when *Contact* = 3.

To interpret the effect of racial contact, we examine the change within the columns in the upper two panels. For Whites, rising racial contact always predicts falling levels of latent support, which increase slightly in size at higher levels of education; the predicted decline in support across the range of contact grows from 0.38 standard deviations when *Education* = 0 to 0.44 standard deviations at *Education* = 20. The effect of contact for Whites is statistically significant for the 70% of the White subsample with education levels between 8 and 16 years. For non-Whites, the effect of racial contact is negative and significant at low levels of education (≤8 years), turning positive but not significant at higher levels of education (≥14 years). The absolute magnitudes of the significant effects on latent support for Whites are around 0.14 standard deviations and are much larger for non-Whites, ranging from 0.28 to 0.73 standard deviations.

Last, the bottom panel of Table 8.12 reports the White–non-White difference in the predicted latent support of a legal ban and shows how it varies simultaneously with education and racial contact. Except for some combinations of the higher levels of contact with higher levels of education (about 9% of the sample), Whites are predicted to have a significantly greater degree of latent support of a ban than do non-Whites. The significant differences in latent support range from about three quarters of a standard deviation to more than 2 standard deviations. These differences vary with contact and education in analogous ways. When *Contact* = 0, the White–non-White difference in latent support grows larger as education increases. But for higher levels of contact, the difference in support declines as education rises. Similarly, for education levels at the bottom of the distribution (<11 years), the White–non-White difference in latent support increases with racial contact, while the difference decreases with racial contact at higher levels of education.

In sum, there is a complex, but coherent, story of the relationship between support for a legal ban on interracial marriage and the interaction of racial contact, education, and race. Increases in contact predict a reduction in Whites' support of a ban with the reduction growing larger at higher levels of education. But for non-Whites, contact has limited significant effects on the outcome that change from reducing to increasing support as education increases. Analogously, rising education predicts a significant decrease in Whites' support of a ban that strengthens slightly with more racial contact. A different pattern holds for non-Whites. The negative effect of education has limited significance, and it grows smaller in magnitude as racial contact increases, turning positive at the highest level of contact.

These different patterns of the effects of education and contact for non-Whites and Whites underlie the more complicated patterns of the effect of race on support for a legal ban. For nearly all combinations of education and racial contact, Whites' support of a ban is predicted to be significantly greater than non-Whites'. But for no racial contact, education increases the predicted racial difference in support, while at higher levels of racial contact, education decreases the predicted racial gap. In the same vein, at lower levels of education, racial contact increases the predicted racial disparity in support; but at higher levels of education, racial contact decreases the racial difference. These are consistent with the opposite patterns for non-Whites and Whites of the effect of contact on support when education is low and of the effect of education on support when contact is low.

Graphic Display

```
. intspec focal(c.ed) main((i.racew, name(Race) range(0/1)) ///
>   (c.ed, name(Education) range(0(5)20)) ///
>   (c.contact, name(RacialContact) range(0/3))) ///
>   int2( c.ed#i.racew c.contact#c.ed c.contact#i.racew ) ///
>   int3(c.ed#i.racew#c.contact) ndig(0)
    …

. outdisp, out(metric(model) sdy dual atopt((means) _all)) ///
>     pltopts(ylab(-.82 ".05" , custom add axis(2)) ///
>     yline(-.82, lp(shortdash) lc(gs9) lw(*.6))) plot(name(Latent_Educ))
    …


. intspec focal(c.contact) main((i.racew, name(Race) range(0/1)) ///
>   (c.ed, name(Education) range(0(5)20)) ///
>   (c.contact, name(RacialContact) range(0/3))) ///
>   int2( c.contact#i.racew c.contact#c.ed  c.ed#i.racew ) ///
>   int3( c.contact#i.racew#c.ed ) ndig(0)
    …

. outdisp, out(metric(model)sdy dual atopt((means) _all))
>     pltopts(ylab(-.82 ".05, custom add axis(2)) ///
>     yline(-.82, lp(shortdash) lc(gs9) lw(*.6))) plot(name(Latent_Contact))
```

I use two sets of scatterplots to illustrate the interpretation of the three-way interaction effect using visual displays of the standardized latent outcome. The first set (Figure 8.6) plots the predicted latent support for a legal ban on the *y*-axis against education on the *x*-axis, with separate predictions plotted for non-Whites and Whites. This scatterplot is repeated for the four levels of racial contact. These plots are used to interpret the effect of education and the effect of race. Because it is very difficult to tease out the effect of contact from this first set, Figure 8.7 presents a parallel set that plots the predicted latent support against racial contact, separately for non-Whites and Whites, with the scatterplot repeated for selected values of education.

Looking across the four panels in Figure 8.6, it is clear that education systematically reduces latent support for Whites—compare the *dashed lines*—and that the predicted change in support by education is fairly substantial and varies little across the levels of racial contact. The change in latent support across the range of education is about 2 standard deviations in each panel (look at the left-side *y*-axis labels). In the probability metric (right-side *y*-axis labels), these correspond to a decline in the probability of support of about .8. The effect of education for non-Whites, in contrast, changes considerably with the level of racial contact (compare the *solid lines*). At the two lowest levels of racial contact, the effect of education for non-Whites is considerable, a decline in latent support across the range of education of 2.7 and 1.6 standard deviations, respectively. But at *Contact* = 2, the decline with education is much smaller (−0.45 standard deviations) and turns to increased support (0.67 standard deviations higher) when *Contact* = 3.

We can also interpret the race effect (predicted White support minus predicted non-White support) from these scatterplots by looking within each panel at the changing vertical distance between the *solid* and *dashed lines* as education increases, and then comparing that across the panels for different levels of racial contact. First, it is clear that Whites are predicted to have a higher level of latent support than non-Whites except for the highest levels of education and racial contact (note that the *solid* and *dashed lines* cross in the panel for *Contact* = 3 around *Education* = 17). But this racial disparity in support varies with education and racial contact. For those with no racial contact, the difference in support increases as education increases. But at

FIGURE 8.6 ● PREDICTED LATENT SUPPORT BY EDUCATION AND RACE BY CONTACT

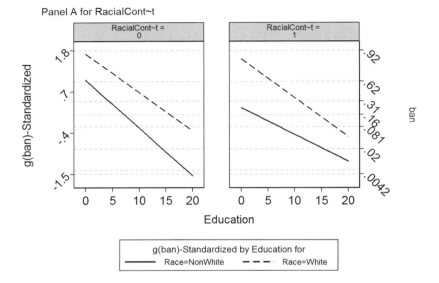

g(ban)-Standardized by the Three-way Interaction of Education, Race and RacialCont~t, with Dual Outcome Axes.

Panel A for RacialCont~t

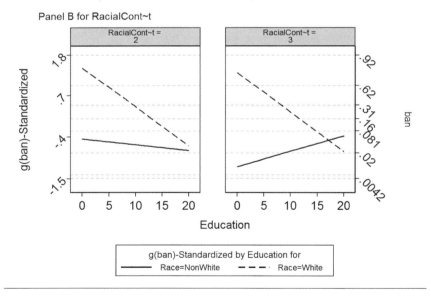

g(ban)-Standardized by the Three-way Interaction of Education, Race and RacialCont~t, with Dual Outcome Axes.

Panel B for RacialCont~t

higher levels of racial contact, the predicted White–non-White difference in latent support decreases as education levels rise.

Turning to the scatterplots in Figure 8.7, it is immediately apparent that the effect of racial contact for Whites changes very little across levels of education (the *dashed line*

FIGURE 8.7 ● PREDICTED LATENT SUPPORT BY CONTACT AND RACE BY EDUCATION

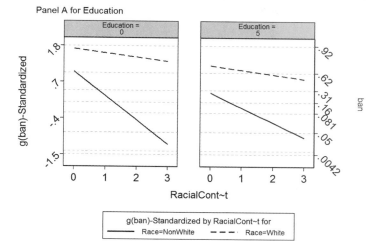

g(ban)-Standardized by the Three-way Interaction of
RacialCont~t, Race and Education, with Dual Outcome Axes.

Panel A for Education

g(ban)-Standardized by RacialCont~t for
Race=NonWhite   ─ ─ ─ Race=White

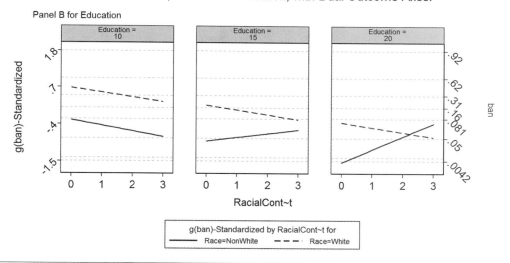

g(ban)-Standardized by the Three-way Interaction of
RacialCont~t, Race and Education, with Dual Outcome Axes.

Panel B for Education

g(ban)-Standardized by RacialCont~t for
Race=NonWhite   ─ ─ ─ Race=White

in each panel). Across the range of contact values, Whites' predicted latent support of a ban declines by between 0.30 and 0.43 standard deviations. The same pattern does not hold true for non-Whites (the *solid lines*). In the first three education panels, non-Whites' predicted latent support decreases with increasing racial contact, by −2.2, −1.4, and −.5 standard deviations, respectively. But at higher levels of education, non-Whites' support increases with rising racial contact by 0.33 and 1.17 standard deviations.

Furthermore, note that how contact moderates the race effect varies with education level in a manner similar to how the moderated race effect by education varied with contact level. At lower levels of education, the White–non-White difference in latent support grows in size with increasing contact; but at higher education levels, the difference in support diminishes with increasing contact. And when *Education* = 20, the plotted lines cross contact indicating that non-Whites' latent support is higher than Whites' for the combination of high education with high contact.

The summary I gave above for the interpretation using a tabular display would work equally well here, so I will not repeat it. I do, however, want to draw readers' attention to the fact that many of the details in the discussion of the scatterplots that fleshed out the results cannot be reliably or precisely read directly from the scatterplots. Rather, I drew these details from tables and analyses not presented in this section; a predicted values table corresponding to each plot is particularly useful. As I suggested earlier, interpretations of outcome displays, particularly graphics, are more effective when you draw on a more detailed understanding of the moderated effects of each predictor. I do the same in the last telling of this story using discrete change effects and predicted probability plots in the next section.

Using Predicted Probabilities for Interpretation

I start this mode of interpretation with a brief look at the significance region tables for the moderated effects calculated as discrete changes to provide background for interpreting tables or graphs of the predicted probability of the outcome. These significance patterns will not necessarily be the same as what we saw for the other ways to calculate effects. These other effects and their significance depend only on the coefficients, their sampling distribution, and the values of the predictors in the interaction. But the discrete change effects estimates and standard errors additionally depend on these same properties for all the other predictors in the model.

After briefly discussing the significance region tables, I turn to telling the story of the three-way interaction effect using scatterplots whose axes are organized in the same way as above for the predicted values in the standardized latent variable metric. After telling the story, I discuss how and why it differs in some respects from the portrayal of the relationship methods that do not use the probability metric. I then discuss how comparisons of the predicted probabilities to a model without interaction effects can help harmonize these differences.

Each set of results lists the commands that generated them. Because ICALC uses SPOST13 to get discrete change effects, the ordering of the interacting predictors is crucial to get correct results. It must be focal variable–first moderator–second moderator in the factor-variable notation on the Stata *logit* command. This same order must be used to list them in the main ( ), int2( ), and int3( ) options on the ICALC *intspec* command. For this reason, I list the *logit* command with each set.

Significance Region Tables
for the Discrete Change Effects

Racial Contact.

```
.logit ban c.contact##i.racew##c.ed  age region16 class
 ...
. intspec focal(c.contact) main((c.contact, name(RacialContact) range(0/3) ) ///
> (i.racew, name(Race) range(0/1)) (c.ed, name(Education) range(0(2)20))) ///
> int2(c.contact#i.racew c.contact#c.ed c.ed#i.racew ) ///
> int3( c.contact#i.racew#c.ed ) ndig(0)
 ...
. sigreg,  effect(spost(amtopt(am(one) center) atopt((means) _all))) ///
>          save(Output\Table_8_13.xlsx tab mat ) ndig(3)
```

The discrete change effect of contact for Whites shown in Table 8.13 is consistently negative—higher contact reduces the probability of supporting a ban—and is significant for the 90% of the White subsample with education between 8 and 17 years. This is similar to the results in Table 8.9 with a key exception. The two sets of results suggest opposite patterns of moderation. The significant discrete change effects of contact decline in magnitude with rising education levels, while the effects on the standardized latent outcome increase in size.

The discrete change effect of contact for non-Whites changes monotonically from negative to positive across education levels. It is significant for the 12% of the non-White subsample with education between 6 and 9 years. The direction of change in the moderated effect for non-Whites is the same as before with a smaller magnitude of change. The moderated effect of contact on latent support is significant for the bottom half of the education distribution. In contrast, the discrete change effect is significant only for a restricted range more toward the middle of the distribution.

A primary reason for the differences in the pattern of moderation for Whites is that the discrete change effect is calculated using a different starting predicted probability for each value of education; and that starting point decreases with education, which moves the starting point further from a probability of .5 and closer to a probability of 0. That is, the discrete change is calculated at a moving point along the logistic curve, where the slope of the curve becomes progressively smaller in magnitude.

TABLE 8.13 ◆ DISCRETE CHANGE EFFECT[a] OF CONTACT MODERATED BY THE INTERACTION OF RACE AND EDUCATION

Education	Race	
	Non-White	White
0	−0.198	−0.038
2	−0.140	−0.053
4	−0.095	−0.068
6	**−0.060***	−0.079
8	**−0.035***	**−0.081***
10	−0.017	**−0.073***
12	−0.005	**−0.059***
14	0.004	**−0.044***
16	0.009	**−0.031***
18	0.012	−0.021
20	0.014	−0.013
Key		
Plain font, no fill	Pos, Not Sig	
Bold*, filled	Pos, Sig	
Italic, no fill	Neg, Not Sig	
Bold italic*, filled	Neg, Sig	

a. One-unit centered change in contact, noninteracting predictors at their means.

Education.

```
. logit ban c.ed##i.racew##c.contact age region16 class

    …

. intspec focal(c.ed) main( (c.ed, name(Education) range(0(4)20)) ///
>(i.racew, name(Race) range(0/1)) (c.contact, name(RacialContact)
range(0/3))) ///
> int2( c.ed#i.racew c.ed#c.contact i.racew#c.contact ) ///
> int3( c.ed#i.racew#c.contact ) ndig(0)

    …

. sigreg, effect(spost(amtopt(am(one) center) atopt((means) _all))) ///
>           save(Output\Table_8_14.xlsx tab mat) ndig(3)
```

The discrete change effect of education in Table 8.14 shows a negative effect of education on support for a ban for both non-Whites and Whites, except for a positive effect at the highest level of racial contact for non-Whites. Note that the pattern of these discrete change results for education differ from the results using the moderated effect on latent support in the same way as do the findings for racial contact. The education effect for Whites diminishes in magnitude as contact rises using discrete change measures, while it increases in size if you use effects on the latent outcome. And for non-Whites, the changes in the effect of education are muted using discrete change effects.

Race.

```
. logit ban i.racew##c.contact##c.ed age region16 class

    …

. intspec focal(i.racew) main((i.racew, name(Race) range(0/1)) ///
>   (c.contact, name(RacialContact) range(0/3)) ///
>   (c.ed, name(Education) range(0(2)20))  ) ///
>   int2(i.racew#c.contact i.racew#c.ed  c.contact#c.ed ) ///
>   int3( i.racew#c.contact#c.ed ) ndig(0)

    …

. sigreg, effect(spost(amtopt(am(one)) atopt((means) _all))) ///
>           save(Output\Table_8_15.xlsx tab mat) ndig(3)
```

TABLE 8.14 ● DISCRETE CHANGE EFFECT OF EDUCATION MODERATED BY THE INTERACTION OF RACE AND CONTACT		
	Race	
Racial Contact	**Non-White**	**White**
0	*−0.011**	*−0.051**
1	*−0.006**	*−0.044**
2	*−0.001*	*−0.036**
3	0.003	*−0.029**
Key		
Plain font, no fill	Pos, Not Sig	
Bold*, filled	Pos, Sig	
Italic, no fill	Neg, Not Sig	
Bold italic*, filled	Neg, Sig	

TABLE 8.15 ● DISCRETE CHANGE EFFECT OF RACE MODERATED BY THE INTERACTION OF CONTACT AND EDUCATION

Education	Racial Contact			
	0	1	2	3
0	0.267	**0.641***	**0.784***	**0.777***
2	0.364	**0.630***	**0.706***	**0.678***
4	**0.438***	**0.584***	**0.604***	**0.559***
6	**0.462***	**0.508***	**0.486***	**0.430***
8	**0.433***	**0.412***	**0.365***	**0.307***
10	**0.366***	**0.311***	**0.255***	**0.203***
12	**0.283***	**0.220***	**0.166***	**0.122***
14	**0.204***	**0.147***	**0.100***	**0.062***
16	**0.140***	**0.094***	**0.055***	0.019
18	**0.092***	**0.059***	0.026	*−0.012*
20	**0.059***	**0.036***	0.008	*−0.037*

Key	
Plain font, no fill	Pos, Not Sig
Bold*, filled	Pos, Sig
Italic, no fill	Neg, Not Sig
***Bold italic*, filled**	Neg, Sig

Note: Binary change in Race, noninteracting predictors at their means.

The discrete change effect on the probability of support for race in Table 8.15 shows greater similarity to the moderated effects on the latent outcome than did contact or education. For nearly all combinations of education and contact, Whites' predicted probability of support is larger than non-Whites'. The pattern of moderation is very similar with a single exception. For the discrete change effects and the effects on the latent outcome, the race effect decreases with education when racial contact is greater than zero. But for *Racial contact* = 0, the race effect on the latent outcome always increases with education, while the discrete change effect increases with education up to 6 years and then the magnitude of the effect diminishes as education rises.

Predicted Probability Plots

```
. ****  Focal = EDUC, MOD1 = RACE, MOD2 = CONTACT
. logit ban c.ed##i.racew##c.contact  age region16 class
   …
. est store lgtint
   …
. ***   OUTDISP
```

```
. intspec focal(c.ed) main((c.ed, name(Education) range(0(5)20)) ///
> (i.racew, name(Race) range(0/1))(c.contact, name(RacialContact) range(0/3))) ///
>  int2( c.ed#i.racew c.ed#c.contact i.racew#c.contact ) ///
>  int3( c.ed#i.racew#c.contact ) ndig(0)
  …
. outdisp,  outcome(atopt((means) _all)) plot(def) pltopts(ylab(0(.3).9))
  …
. ****  Focal = CONTACT, MOD1 = RACE, MOD2 = EDUC
. logit ban c.contact##i.racew##c.ed  age region16 class
  …
. est store lgtint
  …
. intspec focal(c.contact) main((c.contact, name(RacialContact) range(0/3) ) ///
>   (i.racew, name(Race) range(0/1)) (c.ed, name(Education) range(0(5)20)) ) ///
>   int2( c.contact#i.racew c.contact#c.ed i.racew#c.ed ) ///
>   int3( c.contact#i.racew#c.ed )
  …
. outdisp, out(atopt((means) _all)) plot(def) pltopts(ylab(0(.3).9))
```

Figure 8.8 presents scatterplots of the predicted probability of support for non-Whites and Whites as it varies with education, with a separate scatterplot for each level of racial contact. The scatterplot indicates that the predicted probability of Whites supporting a ban (*dashed line*) declines with education and that this drop in support is smaller (but still substantial) as racial contact increases. For non-Whites (*solid line*), the relationship is slightly different. When *Contact* ≤ 2, the predicted probability of support falls as education increases predominantly at a slower rate than for Whites. But when contact is at its maximum (3), the probability of support actually increases with rising education. You can also see that the probability of support is higher for Whites than for non-Whites, except when both education and racial contact are at or near their maximums (lower-right plot). The plots also indicate that the size of the White–non-White difference in support is expected to diminish with increasing education—compare the vertical gap between the two lines as it changes with education—at all levels of racial contact.

It is hard to tease out details of the relationship between support of a legal ban and racial contact from the plots in Figure 8.8, so I created a second set of scatterplots plotting the predicted probability of support against racial contact. Looking across the plots in Figure 8.9, it is obvious that the relationship of support to contact is different for non-Whites and for Whites. For Whites (*dashed line*), the probability of support is always lower when contact is higher, with the size of the drop in support varying with education level, first increasing and then decreasing in magnitude starting at 9 years of education. For non-Whites, the relationship between support and racial contact changes at 11 years of education. At lower years of education, non-Whites' probability of support declines with rising contact levels. Support of a ban then begins to increase with greater amounts of racial contact at a rate that accelerates with rising education levels. Furthermore, the probability of supporting a ban is predicted to be higher for Whites than for non-Whites, except when both education and racial contact are at or near their maximums. The predicted White–non-White difference in support declines with racial contact except at very low levels of education (<3 years).

This telling of the story of the relationship between the probability of support and the interaction of racial contact by education by race is overall similar yet different in a number of details from the interpretations using factor changes or effects on the standardized latent outcome. You may have noticed that I avoided talking about effects for predictors and instead described patterns of change in the probability of

FIGURE 8.8 ● PREDICTED PROBABILITY OF SUPPORT BY EDUCATION BY RACE BY CONTACT

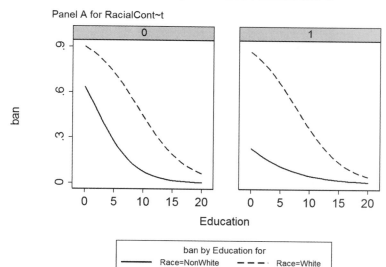

ban by the Three-way Interaction of Education, Race and RacialCont~t.

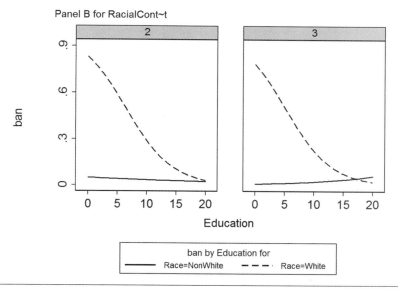

ban by the Three-way Interaction of Education, Race and RacialCont~t.

support. This was done to reinforce my contention that for nonlinear link models, predicted value displays are very good for describing relationships but not always for describing patterns of moderation of effects.

Turning to more substantive differences, the effect of education for Whites relying on the other interpretive methods evidenced a pattern of a larger negative effect as

FIGURE 8.9 ● PREDICTED PROBABILITY OF SUPPORT BY CONTACT BY RACE BY EDUCATION, INTERACTIVE AND ADDITIVE MODELS

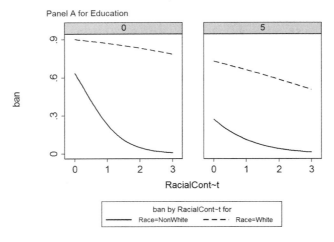

ban by the Three-way Interaction of RacialCont~t, Race and Education.

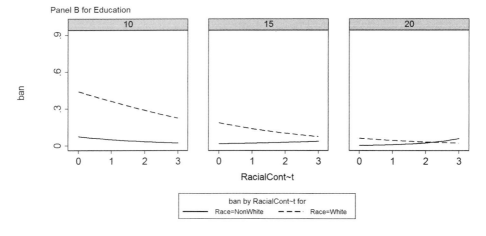

ban by the Three-way Interaction of RacialCont~t, Race and Education.

contact increases, whereas the predicted probability and discrete change descriptions imply a decreasing magnitude of the negative education effect across racial contact. In addition, for non-Whites, the size of the contact effect varied less across education levels in these results than it did earlier. And for Whites, there was a fluctuating magnitude of the contact effect by education levels in contrast to the prior pattern of an increasing effect. Last, when *Racial contact* = 0, the White–non-White difference in the probability of support initially rises with education but then decreases for *Education* ≥ 5. But the other methods of interpretation show that the race effect always increases with education for *Contact* = 0.

The source of these discrepancies is that changes in predicted probabilities incorporate both the nonlinearity of the interaction specification and the nonlinearity of the relationship of the observed outcome to the predictors. One way to partially disentangle these is to compare plots of predicted probabilities from a model without interaction effects—which shows how predicted probabilities change with each component variable solely as a function of the nonlinear link function—with the predicted probability plots from the model with interaction effects. As I discussed in Chapter 5, you can create scatterplots that superimpose the predicted probability from the additive model onto the interactive model predictions to help disentangle the two sources of nonlinearity in the predicted probability plots. These plots—reported in Figure 8.10 highlighting the education effect and in Figure 8.11 highlighting the

FIGURE 8.10 ● PREDICTED SUPPORT BY EDUCATION, INTERACTIVE AND ADDITIVE MODEL PREDICTIONS

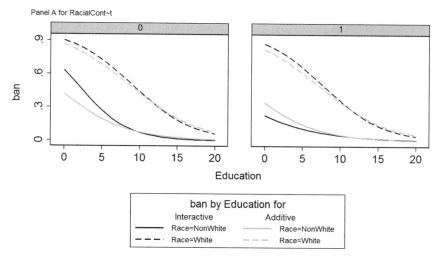

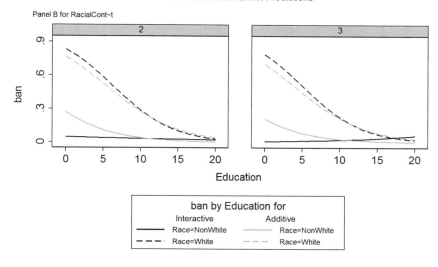

contact effect—were produced by adding the main( ) suboption to the out( ) option to the above *outdisp* commands:

```
outdisp, out(main(lgtmain) atopt((means) _all)) plot(def)  ///
>    pltopts(ylab(0(.3).9)
```

```
outdisp, out(main(lgtmain) atopt((means) _all)) plot(name(def)) ///
>        pltopts(ylab(0(.3).9)
```

Note that you must have previously run a main-effects-only model and stored those estimation results. In this case, I used the name "lgtmain" for the estimates, which I then used in the above commands.

FIGURE 8.11 ● PREDICTED PROBABILITY BY CONTACT, INTERACTIVE AND ADDITIVE MODELS

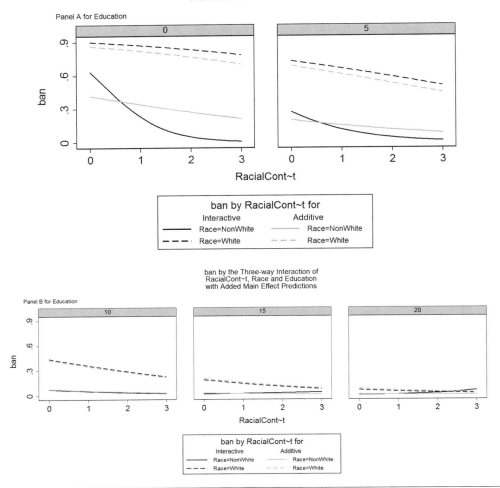

Look first at the plots of the probability of supporting a ban against education on the *x*-axis in Figure 8.10. The *black lines* represent the probability–education curves from the interactive model, and the *gray lines* the predictions from the additive model. The first thing to note is the overall similarity of the two sets of prediction curves. This suggests that much of what appears to be moderation of the education effect by race and contact is actually a reflection of the nonlinear link function. For Whites (*dashed lines*), comparing the prediction curves from the two models shows similar differences in the patterns across the levels of racial contact. The slope of the curves—the education effect—is somewhat steeper in the interactive model at lower levels of education but slightly steeper for the additive model at higher education levels. And these differences slightly increase with racial contact. That is, relative to the additive model predictions, racial contact moderates the education effect for Whites by pushing it in the negative direction, consistent with what the other interpretive approaches showed. For non-Whites, comparing the interactive and additive models' prediction curves suggests that contact moderates this effect more than what is initially apparent. That is, the interactive model predictions diverge even more in the positive direction as racial contact increases.

The superimposed plots of the predicted probability of support against racial contact in Figure 8.11 show a similar result. The interactive model's contact effect for Whites becomes shallower in the absolute, but becomes more negative relative to the additive model predictions as education increases. For non-Whites, the interactive model effect moves even more so in the positive direction compared with the additive model effects than it does in the absolute.

These comparisons of the interactive and additive model comparisons underscore the discrepancies about the pattern and magnitude of moderated effects between the predicted probability method of interpretation versus the factor change or the latent outcome variable approaches described above. One way to think about resolving this paradox is to recognize that without the interaction specification, the predicted probability plots for the effect of contact for Whites, for example, would show a decline in predicted support with contact, with the difference growing at higher levels of education more than it does in the interaction model.

I want to reiterate the important point I've made before that descriptions of the relationship of the outcome with the interacting predictors are a correct and proper representation that is accessible to most audiences. But care is needed in how you write up and discuss the results to avoid discussing patterns of moderation and differences in the size of effects when there are indications, as there are in this example, that the apparent pattern of moderation may reflect the nonlinearity of the estimation technique and not the structure of the interaction effects.

What to Present for a Three-Way Interaction From a Logistic Regression

The criteria and assessment about choosing a method of interpretation presented in Table 8.1 apply equally well in the case of a three-way interaction, although how many tables/figures would be required is different. Specifically, for predicted value figures, you would need two to three figures; and for significance region tables, you would need three tables (one for each predictor). The only caveat is that it is more difficult for a three-way interaction specification to use the additive model plots or discrete change effect significance region tables to understand the nature of the moderated effects. And the same would apply to a two-moderator interaction model.

As before, the crucial question is whether or not it is important to understand the structure of the moderated effects. If so, then the dual-axis plot using a latent standardized variable metric, or a significance region table using either of the standardized latent variable or factor change, are the best choices. If not, then the interpretive methods in the probability metric are also good options, provided you avoid discussing or describing them in terms of how effects are moderated.

SPECIAL TOPICS

Customizing Dual-Axis Scatterplots and Bar Charts

Unlike changing labels on the first *y*-axis, adding or modifying the labels in the probability metric on the second (dual) *y*-axis is not an obvious process. To add a probability metric label at value *prlab*, you first need to calculate its corresponding value in the latent variable metric, *latval*, or in the standardized latent variable metric, *stlatval*:

Logistic: $$latval = \ln\left(\frac{prlab}{1 - prlab}\right)$$

Probit: $$latval = inverse\ normal\ (p), \text{that is}, Prob\ (z \le latval) = prlab$$

For standardized, $$stlatval = \frac{latval - \bar{y}^*}{\hat{\sigma}_{y^*}}$$

where $\hat{\sigma}_{y^*}$ and $\bar{y}^*$ are the standard deviation and mean of the latent outcome, reported beneath the key for the significance region table.

Once you have your pairs of *latval* and *prlab* values or *stlatval* and *prlab* values, you are ready to specify the contents of the pltopts( ) option.

Scatterplot Customization

You use the ylab( ) and yline( ) suboptions to put on the labels and their corresponding grid lines. In ylab( ), you list pairs of *stlatval "prlab"* values—note that there are double quotes around the second value in each pair—and follow the listing of all pairs with the specification

, custom add axis(2)

In yline( ) you list the *stlatval* values followed by the specification

, lp(shortdash) lc(gs9) lw(*.6)

In the three-way interaction example, I initially added two labels to the probability metric axis at values 0.02 and 0.005. The corresponding values in the standardized latent variable metric are

$$-0.82 = \frac{\ln\left(\frac{0.02}{1 - 0.02}\right) - (-1.9026)}{2.4307} \quad \text{and} -1.39 = \frac{\ln\left(\frac{0.05}{1 - 0.05}\right) - (-1.9026)}{2.4307}$$

As described above, I specified pltopts( ) as

```
pltopts(ylab(-.82 ".02" -1.39 ".005", custom add axis(2)) ///
     yline(-.82 -1.39 , lp(shortdash) lc(gs9) lw(*.6)) )
```

The final version of the plot dropped the second label (-1.39 ".005") because it visually overlapped another label.

If your left-side axis is labeled in the unstandardized latent variable metric, follow the same steps but use the *latval* calculated values instead of the *stlatval* calculated values.

Bar Chart Customization

For a bar chart calculated from a one-moderator model, you follow exactly the same steps and specifications just described for the scatterplot. But for a bar chart for a three-way interaction or a two-moderator interaction model, you do not use the ylab( ) suboption to add the labels; instead, you must use the text( ) suboption and continue to use the yline( ) suboption to add labels. The contents of text( ) again starts with a list of the pairs of values but the number 105 is inserted between the values of each pair: *stlatval 105 "prlab"*. This listing of values is followed by

```
, size(*.8) m(I+2) placement(e)
```

Thus, to add the same pairs of values as above pltopts( ) would be specified as

```
pltopts(text(-.82 105 "0.02" -1.39 105 "0.005", ///
     size(*.8) m(I+2) placement(e)) yline(-.82 -1.39 , lp(shortdash) lc(gs9) lw(*.6)) )
```

Alternative Plot Comparing Additive and Interaction Model Predictions

It is not always clear what the difference is between an interaction model prediction line and its matching additive model prediction line in the superimposed plots like those in Figures 8.10 and 8.11. You can get an alternative organization of the prediction lines that will make those differences easier to see if you add the suboption single(1) to the specification of the outcome( ) option on the *outdisp* command. This produces a set of scatterplots—each has only two prediction lines, one from the interaction model and the other from the additive model. This paired comparison scatterplot is repeated for all combinations of the values of the first and second moderators. For example, when education is the focal variable, it creates eight paired comparison scatterplots (two categories of race times four values of racial contact). That is, there are four scatterplots for each race corresponding to the four values of racial contact as shown in Figure 8.12, produced by the syntax

```
. logit ban c.contact i.racew  c.ed age region16 class
  ...
. est store lgtmain

  ...
. logit ban c.contact##i.racew##c.ed  age region16 class
  ...
. est store lgtint

  ...
. intspec focal(c.contact) main((c.contact, name(RacialContact) range(0/3) ) ///
>   (i.racew, name(Race) range(0/1)) (c.ed, name(Education) range(0(5)20)) ) ///
>   int2( c.contact#i.racew c.contact#c.ed i.racew#c.ed ) ///
>   int3( c.contact#i.racew#c.ed ) ndig(0)
  ...
```

```
.outdisp,  outcome(main(lgtmain) atopt((means) _all)))  ///
>  plot(sing(1)) pltopts(ylab(0(.3).9))
```

If you compare these plots with those in Figures 8.10 and 8.11, this alternative organization obviously highlights the difference between the interaction and additive models' predictions quite well. If those differences are fairly substantial, as they are for non-Whites in the top panel, it is still easy to see the pattern of how the interaction model predictions are changing across racial contact. But it is more difficult to see the more subtle pattern of changes for Whites in the bottom panel. It is also

---

FIGURE 8.12 ● ALTERNATIVE COMPARISON PLOT OF ADDITIVE AND INTERACTION MODEL PREDICTIONS

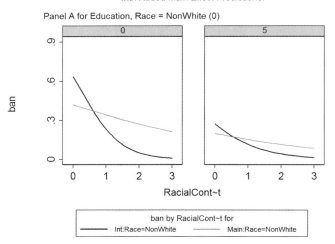

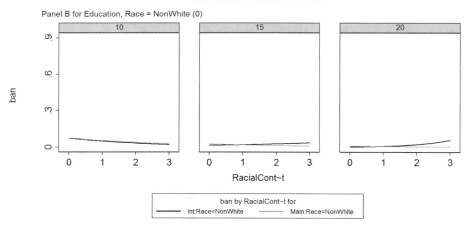

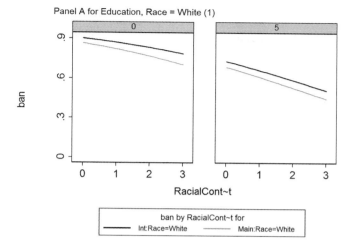

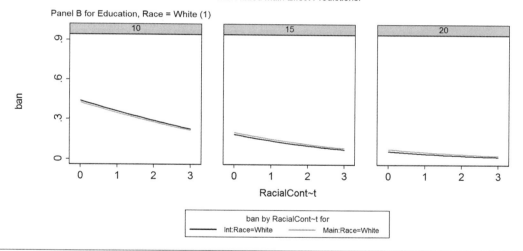

harder to make out the pattern of the difference in the predicted probability of support between non-Whites and Whites. Overall, I find these work best as an unreported supplement for an analyst to look at in formulating his or her interpretation, not as the primary vehicle for interpretation.

CHAPTER 8 NOTES

1. Grouped data refers to situations in which the data are reported as a proportion of the outcome measure aggregated for sets of cases with identical values of the predictors. For example, suppose your data have 22 groups with each group consisting of a unique

combination of predictors' values, the number of cases in the group (group size), and the proportion of successes for that group. Such data can be analyzed by creating a new data set with each of the 22 groups represented by two data points. One data point represents successes coded as 1 on the outcome, a weight defined as proportion of successes × group size, and the predictors reporting the group's (identical) values on the predictors. The second data point represents failures coded as 0 on the outcome, a weight defined as proportion of failures × group size, and predictors with the same values as those reported for the successes. Such group data can be analyzed by specifying a weighted analysis using the weight variable.

2. When all of the predictors are categorical, log-linear analysis is sometimes used, especially in older applications. Log-linear analysis is closely related to logistic regression, and it is possible to specify a log-linear model that is identical to logistic regression, both theoretically and empirically. But the reverse does not hold true; not all log-linear models have a logistic regression equivalent.

3. The commands I used were as follows:

```
logit gun ib2.location##i.sex c.age##c.age educ fearnbhd race
intspec focal(ib2.location) main( (ib2.location, name(Location) ///
    range(1/3)) ///
    (i.sex, name(Sex) range(0/1)) ) int2(ib2.location#i.sex)
sigreg , effect(b(sdy))
```

4. You could reorganize the chart to show location results within sex to see this more easily. Respecify the *intspec* command to switch the roles of focal and moderator variables and then rerun the *outdisp* command.

5. Two of the methods for interpreting logistic regression results—as effects on the log odds or as factor changes—are not applicable to a probit analysis.

6. To verify that the pattern changes at 8 years of education, not at 7 years, I reran the significance region analysis to get single year results. That is, I reset the range for education as 0(1)20 on the *intspec* command.

MULTINOMIAL LOGISTIC REGRESSION APPLICATIONS

OVERVIEW

Properties and Use of Multinomial Logistic Regression

A multinomial logistic regression (MNLR) model for a nominal outcome Y with $M + 1$ categories is a single-estimation model equivalent to a series of M binomial logistic regressions, each contrasting the odds of a category of Y with the reference category (see Long & Freese, 2014, pp. 386–389). Because of this equivalence, all the techniques for interpreting interaction effects for binomial logistic regression can be applied to MNLR results on an equation-by-equation (log odds contrast) or outcome category-by-category basis. While the same interpretive approaches can be applied to MNLR effects, interpreting MNLR results is more complicated due to the multiple comparisons among outcome categories to interpret (Long & Freese, 2014, pp. 386, 411), compounded by the presence of interaction effects.

My presentation and discussion in this chapter of how to apply the modes of interpretation to interaction effects presumes that readers are already familiar with the coverage of these for binomial logistic regression in Chapter 8. I further assume a basic understanding of the MNLR model and approaches to its interpretation for predictors not part of an interaction effect. If you need the latter background, I strongly recommend that you read Chapter 8 of Long and Freese (2014) before you read on.

Data and Circumstances When Commonly Used

Formally, the MNLR model predicts the probability that a case is in each of $M + 1$ outcome categories as a nonlinear function of the predictors, with the coefficients of one of the functions constrained to ensure that the sum of probabilities across the categories for each case is equal to 1. Equivalently, it can be formulated as the log odds of each of M categories versus a reference category as a linear function of the predictors. MNLR's primary use is for the analysis of a nominal-level outcome with three or more (unordered) categories; for a two-category outcome, it is the same as using binomial logistic regression. It is also an alternative to an ordinal regression model of an ordinal outcome when the ordinal regression model results do not meet the parallel regression assumption (see Long & Freese, 2014, p. 386). Like other GLMs, you can use predictors coded as interval or categorical[1] measures; a nonlinear function of a predictor, such as a polynomial or logarithmic expression; and/or interaction terms.

Published Examples.

- Cech, Rubineau, Silbey, and Seron's (2011) analysis of persistence in and switching from engineering majors tested the interactive effect of students' gender and their family plans. While the interaction was not significant, they noted a significant effect of family plans for men but not for women.

- As part of a test of a model of reception–acceptance of political party messages, Dobrzynska and Blais (2008) analyzed how the interaction between respondents' political awareness and campaign stage affected their acceptance of the party message. They interpreted the interaction effect by discussing the coefficients and plots of predicted probabilities of acceptance of message.

- Matsuoka and Maeda (2015) used MNLR to assess how individual and contextual factors affected response/nonresponse to a nationwide survey in Japan. They found two significant interactions: (1) sex of the target respondent by neighborhood crime rate and (2) target building residency size with crime rate. They interpreted the interactions as moderated effects at the mean and the mean plus 1 standard deviation of the moderator.

- In a study of body image dissatisfaction and relative economic deprivation, Esposito and Villaseňor (2017) tested for an interaction between gender and relative deprivation. They interpreted the interaction effect from plots of predicted probabilities of dissatisfaction categories against relative deprivation, separately for men and for women.

GLM Properties and Coefficient Interpretation for MNLR

MNLR analyzes an observed outcome Y_i consisting of $M + 1$ unordered categories numbered (for notational convenience) as $m = 0, 1, 2, \ldots, M$. A generalized logit link function defines the relationship between the observed outcome Y_i and the M modeled outcomes $(\eta_{m,i})$, where $\eta_{m,i}$ is the log odds contrasting the probability $Y_i = m$ with the probability $Y_i = 0$ (the reference category). Let $\pi_{m,i}$ = probability $(Y_i = m)$; then the link function is as follows:

$$\eta_{m,i} = \ln\left(\frac{\pi_{m,i}}{\pi_{0,i}}\right) \text{ for } m = 1, 2, \ldots, M$$

The distribution of the observed outcome conditional on the predictors $(Y_i|X_i)$ is the multinomial distribution. Because the metric of the observed outcome is in probabilities, the link function defines the metric of the mth modeled outcome as the natural log of the odds that $Y_i = m$ versus $Y_i = 0$. Thus, there are M sets of coefficients predicting the log odds of the modeled outcomes from the K predictors:

$$\eta_{m,i} = \ln\left(\frac{\pi_{m,i}}{\pi_{0,i}}\right) = \sum_{k=1}^{K} b_{m,k} x_{k,i} \quad \text{for } m = 1, 2, \ldots, M$$

Interpretation. As in binomial logistic regression, coefficients can be directly inter-preted as effects on the log odds that $Y = m$ versus $Y = 0$, often referred to as the log odds of m versus 0. A positive (negative) coefficient b means that a one-unit change in the predictor corresponds to an increase (decrease) of b in the log odds of m versus 0. Alternatively, you can think about each modeled outcome as a latent interval vari-able (difference in utility) underlying the m versus 0 contrast. The coefficients can be directly interpreted as effects on the corresponding latent outcome, although this is infrequently done. More commonly, interpretation of the relationship between a predictor and the outcome is done using the following:

- A factor change (odds ratio) for which the outcome changes by a factor of $e^{b_{m,k}}$ for a one-unit change in the kth predictor

- A marginal effect, usually a discrete change in the probability of being in category m equal to the calculated change in the probability for a one-unit change in the predictor, or a marginal change in the probability of an infinitesimal change in an interval predictor

- Tabular or visual displays of (1) the predicted probabilities of the $M + 1$ outcome categories against a predictor or (2) the M log odds contrasts to category 0 against a predictor

Using any of these modes of interpretation is complicated by the necessity of making comparisons across multiple outcome categories. When using the probability metric for interpretation, there are $M + 1$ outcome categories across which to interpret the effects of a predictor. When interpreting factor changes or effects on the log odds, there are M unique contrasts to the reference category 0, but there is often interest in contrasts of some categories to a different reference category.

Diagnostic Tests and Procedures

For analyses of residuals, I suggest following Long and Freese's (2014, p. 411) rec-ommendation to apply the standard logistic regression residual, outlier, and influ-ential cases analyses I discussed in Chapter 8 to the series of M binary logit models that correspond to the MNLR. As for any GLM, checking for potential collinearity among the predictors should be done, keeping in mind that there will of necessity be functional collinearity among the interaction terms. There are additional diagnostic tests—the Hausman-McFadden and the Small-Hsaio tests—for the independence of irrelevant alternatives (IIA) assumption inherent in MNLR. The utility of these tests has been questioned, to the extent that Long and Freese (2014, p. 408) advise against using them unless it is to satisfy a journal editor or reviewer. They reiterate the advice of early diagnosticians to do your best to define and analyze categories that are clearly conceptually distinct and cannot be considered substitutes for one another. In that case, the pairwise options can be evaluated independently.

In addition to the usual statistical tests of single coefficients or sets of coefficients for a single prediction equation, there are two additional tests frequently used with MNLR. First, there is often an interest in testing whether the coefficients for a sin-gle predictor across the M prediction equations are all equal to 0—that is, to test whether the predictor has any effect on the outcome as a whole. This is a standard global test that a set of coefficients are all equal to 0, which can be done as either a Wald or an LR test. As with other global tests, you should also examine the signifi-cance of the coefficients for the individual prediction equations in deciding whether or not to keep a predictor in the model. Relatedly, you can also perform a global test of whether or not two outcome categories can be combined by testing the equality of the coefficients for the predicting equations, excluding the constant term.

Data Source for Examples

The data for the single-moderator example are from the 2010 GSS stored in a Stata data set (GSS_2010.dta). The two-moderator interaction example uses data from the 1992 panel of the Survey of Income and Program Participation (SIPP_Occ.dta) analyzed for other purposes by Campbell and Kaufman (2006). Both data sets and example do-files can be downloaded at www.icalcrlk.com.

ONE-MODERATOR EXAMPLE (INTERVAL BY INTERVAL)

Data and Testing

This example interprets an interval-by-interval interaction effect—how education and frequency of attending religious services interact in predicting political ideology. Political ideology for this analysis is a three-category measure (1 = *liberal*, 2 = *moderate*, 3 = *conservative*), and I specify "liberal" as the base category. Education is measured as years of formal education completed (0–20). Religious attendance is operationalized as the monthly number of services a respondent attends,[2] ranging from 0 to 4.33. The other predictors in the model are age, subjective social class, sex, and race. Age is coded in years (18–89), and subjective social class runs from 1 = *lower class* to 4 = *upper class*.[3] Sex is a dichotomous indicator (1 = *female*, 0 = *male*). Race is a three-category indicator represented in the model with dummy variables for Black and for Other Races, with Whites as the reference category. The index functions to be estimated for the contrasts of category *m* = moderate, conservative to the base category (liberal) are

$$g\left(y_{m,i}\right) = \beta_{m,0} + \beta_{m,1}Education_i + \beta_{m,2}Attend_i + \beta_{m,3}Education_i \times Attend_i$$
$$+ \beta_{m,4}Age_i + \beta_{m,5}Class_i + \beta_{m,6}Sex_i + \beta_{m,7}Black_i + \beta_{m,8}Other\ Race_i$$

A supplementary analysis is sometimes estimated specifying "moderate" as the base category. The sample for analysis consists of 1,820 cases. It excludes 146 respondents with missing information on religious attendance, another 62 cases missing on political ideology, and an additional 16 cases with missing information on other variables in the analysis.

Although political ideology appears to be ordinal, the categories are inconsistently ordered across two dimensions as shown by a stereotype logistic regression analysis (see Long & Freese, 2014, pp. 445–454, for a description of this model). And the Brant specification test of the suitability of the data for an ordinal regression model (Long & Freese, 2014, pp. 326–331) also suggests that this outcome should be treated as unordered. Last, the MNLR results for the effect of class illustrate why an ordinal regression model is not appropriate. Class has a negative and significant effect on the log odds of moderate to liberal, a positive and nonsignificant effect for conservative to liberal, and a positive and significant effect on the log odds of conservative to moderate. If an ordinal regression model were appropriate—ordering the categories as liberal to moderate to conservative—each effect would be in the same direction.

Turning to diagnostic tests on the MLNR results, I did tests of the IIA assumption despite the questions about their utility. In this case at least, both tests suggest accepting the null hypothesis that the choice between pairs of categories is independent of other alternatives.[4] Conceptually, the three political ideologies are distinct from one another, and choices between pairs would, I think, be considered independently. For example, when choosing between moderate and conservative, the existence of the liberal option is irrelevant. Analyses of predicted values and residuals from two binary logits indicated no cause for concern in the end. There were some large residuals that were visually distinct as well as some cases with unusually high leverage values, and in both situations, the clearest common factor was high frequency of attendance at religious services. Sensitivity analyses excluding those cases with the largest absolute residuals or with the highest leverage values showed virtually no change in the results. In each sensitivity analysis, one coefficient changed from marginally not significant to significant.

Statistical tests of the education-by-attendance interaction specification in the model—$\beta_{Mod,4} Education_i \times Attend_i$ and $\beta_{Con,4} Education_i \times Attend_i$—support its inclusion in the analysis. A global test of the change in the log likelihood if these two terms are excluded is statistically significant ($p = .001$), as are the z tests of each coefficient individually ($p = .001$). Note, however, that the interaction coefficient for the contrast of the conservative with the moderate category is positive but not significant ($p > .05$). (You can get the coefficients and tests for the contrast of nonbase categories with each other by reestimating the model with "moderate" as the base category. Or if you have SPOST13 installed, you can use the *listcoef* command.)

I illustrate the alternative approaches to interpreting the MNLR interaction effects in five stages, in some cases discussing only the moderated education effect and in some only the moderated attendance effect to minimize the length of the exposition. For background, I begin with a brief discussion of the GFI and sign change analysis results for education as the focal variable and then for attendance as the focal variable. I next use factor changes (odds ratios) to explore the moderated effect of education in choosing among categories of political ideology. I then calculate discrete change effects to discuss the moderated effect of attendance on the probabilities of choosing liberal, moderate, or conservative political ideology. Staying in the probability metric, I demonstrate how to use displays of the predicted probabilities of each category to describe the relationship of political ideology with the interaction of education and attendance. I end the example by discussing the use of displays of the standardized latent outcomes to understand the interaction.

Note that ICALC produces results for either the M contrasts of an outcome category with the base category or for all the outcome categories, depending on what commands you run, as shown in Table 9.1. You can limit the results to a single contrast/outcome category using the eqname( ) option in the *intspec* command. If you want results for contrasts other than with the base category, which is often the case, you must reestimate your model specifying a different base category.

INTSPEC **Setup and** GFI **Analysis**

I ran and stored the estimates from an MNLR using two different specifications of the outcome reference category to make it easy to run supplementary results from the second specification. The primary *mlogit* results use liberal as the outcome reference category, and the supplementary *mlogit* results use moderate. These are shown just above the first *intspec* command, which sets education as the focal variable and attendance as

TABLE 9.1 ● ICALC COMMANDS AND WHETHER CALCULATED FOR LOG ODDS CONTRASTS OR OUTCOME CATEGORIES

ICALC **Command**	**Results for M Categories Contrasted With the Base Category**	**M + 1 Outcome Categories**
GFI	GFI expression for *b* and factor change, sign change analysis for *b*	NA
SIGREG		
Boundary value analysis	Effects on the log odds (*b*)	NA
Significance region table	Factor change effects, effects on the standardized latent outcomes, or effects on the log odds (*b*)	SPOST effects, typically discrete change effects
EFFDISP	Effects on the log odds (*b*)	SPOST effects, typically discrete change effects
OUTDISP	Predicted standardized latent outcomes or predicted log odds	Predicted probabilities

Note: ICALC = Interaction CALCulator; GFI = gather, factor, and inspect; NA = not applicable; SIGREG = significance regions; EFFDISP = effect displays; OUTDISP = predicted outcome displays.

the moderator. This is followed by the *gfi* command to produce the algebraic expression for the moderated effect of education on the log odds of moderate to liberal and of conservative to liberal ideology and the corresponding sign change analyses. These are repeated for the supplementary *mlogit* estimates stored in *mlgt3rd* to get results for conservative to moderate, adding the eqname( ) option to limit the output to the contrast for conservative. I use the second *intspec* command to treat attendance as the focal and education as the moderator variable. In the output, I abbreviate the sign change table results to show only the value of the moderator at which the focal variable's effect changed sign and the percentage of cases with positive effects.

```
. mlogit pol c.educ##c.attendmonth age  class female i.race, base(1)
  ...
. est store mlgtint
  ...
. mlogit pol c.educ##c.attendmonth age  class female i.race, base(2)
  ...
. est store mlgt3rd
  ...
****  FOCAL = educ
. est restore mlgtint
  ...
. intspec focal(c.educ) main((c.educ, name(Education) range(0(4)20)) ///
> (c.attendmonth, name(Attend) range(0(1)4))) int2(c.educ#c.attendmonth) ndig(0)
  ...
. gfi , ndig(4)

GFI Information from Interaction Specification of
Effect of Education on g(Pol[Moderate:Liberal]) from Multinomial Logistic
> Regression
------------------------------------------------------------------------
```

```
Effect of Educ =
   -0.1474 + 0.0454*Attendmonth

GFI Information from Interaction Specification of
Effect of Education on g(Pol[Conservative:Liberal])from Multinomial Logistic
------------------------------------------------------------------------

Effect of Educ =
   -0.1791 + 0.0512*Attendmonth

Sign Change Analysis of Effect of Education
 on g(Pol[Moderate:Liberal]), Moderated by Attend (MV)
    ...
------------+------------------------
 Sign Changes |  when MV= 3.243942
------------+------------------------
 % Positive   |      24.6
------------------------------------


Sign Change Analysis of Effect of Education
 on g(Pol[Conservative:Liberal]), Moderated by Attend (MV)
    ...
------------+------------------------
 Sign Changes |  when MV= 3.495483
------------+------------------------
 % Positive   |      20.1
------------------------------------


. est restore mlgt3rd
    ...
. intspec focal(c.educ) main((c.educ, name(Education) range(0(4)20))  ///
>  (c.attendmonth, name(Attend) range(0(1)4))) int2(c.educ#c.attendmonth) ///
>   ndig(0)  eqname(Conservative)
        ...

. gfi , ndig(4)
    ...
GFI Information from Interaction Specification of
Effect of Education on g(Pol[Conservative:Moderate]) from Multinomial Logistic
>   Regression
------------------------------------------------------------------------

Effect of Educ =
   -0.0318 + 0.0058*Attendmonth
    ...
 Sign Change Analysis of Effect of Education
 on g(Pol[Conservative:Moderate]), Moderated by Attend (MV)
    ...
------------+------------------------
 Sign Changes |      Never
------------+------------------------
 % Positive   |      0.0
------------------------------------
```

The GFI and sign change analyses show that the moderated effect of education on political ideology has a similar structure for the moderate to liberal contrast and for the conservative to liberal contrast but not for the conservative to moderate contrast. For the first two contrasts, the baseline education effect is negative, decreasing the log odds of each contrast, but it declines in magnitude and then turns positive as attendance increases. This is shown clearly in the sign change analysis results; look at the changeover point reported in the "Sign Changes" row.

For those who attend religious services less frequently, education has a negative effect on the log odds of identifying as moderate versus liberal (*Attend* < 3.24 times per month) and conservative versus liberal (*Attend* < 3.50 times per month). But for those who attend more frequently, education has a positive effect. However, the effect on these two outcome contrasts is positive for only 20% to 25% of the sample, as reported in the last row of the tables. For the conservative to moderate contrast, the sign change analysis shows that the education effect is always negative—reducing the log odds of conservative to moderate—and its magnitude diminishes as attendance increases.

Reversing the roles of the focal and moderator variables, the next set of output reports the moderated effect of attendance on political ideology. This reveals a similar structure but an opposite concentration of positive and negative effects. The effect of attendance on the choice of moderate versus liberal and of conservative versus liberal is negative at the lowest levels of education. But it turns positive for those with higher levels of education (>10.57 and >7.18, respectively). And a very high percentage of the sample is predicted to have a positive effect of attendance on these log odds—89.2% for moderate to liberal and 96.8% for conservative to liberal. The effect of attendance on the conservative to moderate contrast is positive for all levels of education, and as for the other two contrasts, its positive effect increases in magnitude as education increases.

```
****  FOCAL = attend
. est restore mlgtint
. intspec focal(c. attendmonth) main((c.educ, name(Education) range(0(4)20))  ///
>  (c.attendmonth, name(Attend) range(0(1)4))) int2(c.attendmonth#c.educ) ndig(0)
    …
. gfi , ndig(4)

GFI Information from Interaction Specification of
Effect of Attend on g(Pol[Moderate:Liberal]) from Multinomial Logistic Regression
------------------------------------------------------------------------

Effect of Attendmonth =
   -0.4801 + 0.0454*Educ


GFI Information from Interaction Specification of
Effect of Attend on g(Pol[Conservative:Liberal]) from Multinomial Logistic Regression
------------------------------------------------------------------------

Effect of Attendmonth =
   -0.3678 + 0.0512*Educ

 Sign Change Analysis of Effect of Attend
on g(Pol[Moderate:Liberal]), Moderated by Education (MV)
    …
-------------+------------------------
 Sign Changes |   when MV= 10.56797
-------------+------------------------
 % Positive   |        89.2
-------------------------------------
```

```
Sign Change Analysis of Effect of Attend
on g(Pol[Conservative:Liberal]), Moderated by Education (MV)
    ...
------------+-----------------------
Sign Changes |  when MV= 7.176218
------------+-----------------------
% Positive   |        96.8
------------------------------------


. *** 3rd contrast Con:Mod
. est restore mlgt3rd
    ...

. intspec focal(c.attendmonth) main((c.educ, name(Education) range(0(4)20))   ///
>   (c.attendmonth, name(Attend) range(0(1)4))) int2(c.attendmonth#c.educ) ndig(0)
    ...
. gfi , ndig(4)
    ...
GFI Information from Interaction Specification of
Effect of Attend on g(Pol[Conservative:Moderate]) from Multinomial Logistic Regression
-----------------------------------------------------------------------

Effect of Attendmonth =
   0.1123 + 0.0058*Educ
    ...

Sign Change Analysis of Effect of Attend
 on g(Pol[Conservative:Moderate]), Moderated by Education (MV)
    ...
------------+-----------------------
Sign Changes |        Never
------------+-----------------------
% Positive   |       100.0
------------------------------------
```

Factor Change (Odds Ratio) Interpretation of Education Effect

To get the factor change results, I use the *sigreg* command to calculate the odds ratios (factor changes) for a 1 standard deviation change in education as the focal variable. I think that presenting results for a 1 standard deviation change in an interval focal variable usually makes it easier to convey the magnitude of the effect on the outcome. The Stata output includes the boundary value analyses, but I present the formatted significance region table (Table 9.2) saved to Excel by ICALC in lieu of the Stata output version of the table. The ICALC and Stata commands for these results are in boldface in the output; note the repeated use of est restore to switch between using the primary and the supplementary specification of the reference category:

```
****  FOCAL = educ
. est restore mlgtint
    ...
intspec focal(c.educ) main((c.educ, name(Education) range(0(4)20))   ///
>   (c.attendmonth, name(Attend) range(0(1)4))) int2(c.educ#c.attendmonth) ndig(0)
    ...
sigreg , effect(factor(sd)) save(Output\factorchg_educ.xlsx tab) ndig(3)
```

Boundary Values for Significance of Effect of Education on **g(pol[Moderate:Liberal])** Moderated by Attend
Critical value Chi_sq = **3.841** set with p = **0.0500**

	When Attend >=	Sig Changes	When Attend >=	Sig Changes
Effect of educ	2.143	to Not Sig [-8.694]	6.116 (> max)	to Sig [1.278]

Note: Derivatives of Boundary Values in [ ]

Boundary Values for Significance of Effect of Education on **g(pol[Conservative:Liberal])** Moderated by Attend
Critical value Chi_sq = **3.841** set with p = **0.0500**

	When Attend >=	Sig Changes	When Attend >=	Sig Changes
Effect of educ	2.314	to Not Sig [-7.889]	6.617 (> max)	to Sig [1.130]

Note: Derivatives of Boundary Values in [ ]

```
. *** 3rd contrast Con:Mod
. est restore mlgt3rd
    …
intspec focal(c.educ) main((c.educ, name(Education) range(0(4)20))  ///
>   (c.attendmonth, name(Attend) range(0(1)4))) int2(c.educ#c.attendmonth) ndig(0)
    …
sigreg , effect(factor(sd)) save(Output\factorchg_ed_mc.xlsx tab) ndig(3)
```

Boundary Values for Significance of Effect of Education on **g(pol[Conservative:Moderate])** Moderated by Attend
Critical value Chi_sq = **3.841** set with p = **0.0500**

	When Attend >=	Sig Changes	When Attend >=	Sig Changes
Effect of educ	NA	Never	NA	Never

Note: Derivatives of Boundary Values in [ ]

The significance region analyses for the factor change effect of education in Table 9.2 show that education's factor change effects on the conservative to moderate contrast (bottom panel) are all close to 1.0 (no effect) and not significant, unlike education's effect on the other two contrasts. The moderated effect of education on the odds of moderate to liberal (upper panel) and on the odds of conservative to liberal (middle panel) follows the same pattern across the values of attendance. Education has a negative effect—an odds ratio <1—on both contrasts for those who do not attend religious services (*Attend* = 0). This effect lessens in magnitude—moves closer to 1— as attendance increases and then changes to positive—an odds ratio >1.

From the GFI analysis, we know that the sign change occurs at a monthly attendance level of 3.24 for moderates to liberals and 3.50 for conservatives to liberals. The boundary value analysis tells us that the effect of education is significant when monthly attendance is less than 2.1 times for the moderate to liberal contrast and less than 2.3 times for the conservative to liberal contrast. Examining the frequency distribution of attendance for the percentage of cases below, between, and above these cut points indicates that there is a significant negative effect of education for 66% of the sample, a negative but not significant effect of education for 9%, and a positive but not significant effect for 25%.

While the overall pattern and percentages of significant effects are quite similar for both political ideology contrasts to liberal, the magnitudes of the education effect on the conservative to liberal contrast are slightly larger. For the conservative to liberal contrast, a 1 standard deviation increase in education (3.1 years) reduces the

TABLE 9.2 ● FACTOR CHANGE EFFECT OF EDUCATION (1 STANDARD DEVIATION DIFFERENCE) ON POLITICAL IDEOLOGY ODDS MODERATED BY ATTEND

Effect of Educ	Attend				
	0	1	2	3	4
Pol[Moderate:Liberal]					
	0.634*	0.729*	0.839*	0.966	1.112
Pol[Conservative:Liberal]					
	0.574*	0.673*	0.789*	0.924	1.083
Pol[Conservative:Moderate]					
	0.906	0.923	0.940	0.957	0.974
Key					
Plain font, no fill	Pos, Not Sig				
Bold*, filled	Pos, Sig				
Italic, no fill	Neg, Not Sig				
Bold italic*, filled	Neg, Sig				

predicted odds of conservative to liberal identification by a factor of 0.57 among nonattendees, of 0.67 for attendance once a month, and of 0.79 for attendance twice a month. The comparable figures for the moderate to liberal effect of education are 0.63, 0.73, and 0.84, respectively. We could equivalently describe these in terms of the percentage of reduction in the odds, calculated as $(1 - Factor change) \times 100$. Across monthly attendance levels (0, 1, 2), a 1 standard deviation increase in education reduces the predicted odds of conservative to liberal identification by 43%, 32%, and 21%, respectively, and the corresponding moderate to liberal identification reductions are 37%, 27%, and 16%, respectively. Subjectively, I would argue that these are substantial reductions in the odds for a 1 standard deviation change in education.

Discrete Change Interpretation of Attendance Effect

As I did for the factor change effects, I present the discrete change significance region tables saved by ICALC to an Excel file instead of the Stata output. For the discrete change effects, I again use a 1 standard deviation increase in the focal variable. Moreover, the results in Table 9.3 are centered discrete changes (Long & Freese, 2014, p. 259) because this creates an equivalent size of negative and positive discrete changes for moderated coefficients with the same absolute magnitude. I suppress the boundary value analysis with the nobva option in the *sigreg* command because it is not relevant for discrete change effects.

Two quick reminders about using ICALC to generate discrete change results: First, you must have SPOST13 installed. Second, the order of the predictors must be focal

> **TABLE 9.3 ● DISCRETE CHANGE EFFECT OF ATTENDANCE (1 STANDARD DEVIATION CENTERED INCREASE) ON POLITICAL IDEOLOGY, MODERATED BY EDUCATION**

Effect of Attendmonth	Education					
	0	**4**	**8**	**12**	**16**	**20**
pol[Liberal] (18%)						
	0.0422*	0.0335	0.0136	*−0.0226**	*−0.0793**	*−0.1567**
pol[Moderate] (64%)						
	−0.0665	*−0.0652*	*−0.0534**	*−0.0267**	0.0193	**0.0856***
pol[Conservative] (18%)						
	0.0243	0.0316	**0.0398***	**0.0493***	**0.0600***	**0.0711***
Key						
Plain font, no fill	Pos, Not Sig					
Bold*, filled	Pos, Sig					
Italic, no fill	Neg, Not Sig					
Bold italic*, filled	Neg, Sig					

variable, then moderator variable in the *mlogit* command as well as in the *intspec* command for the main( ) and int2( ) options (see the discussion in Chapter 6 of the spost( ) suboption). Failure to do so will produce nonsensical results. Thus, in the syntax above, I reestimated the MNLR model, switching the order of *c.educ* and *c.attendmonth* on the *mlogit* command before running the *intspec* and *sigreg* commands. The ICALC commands used to generate these tables are as follows:

```
. ***   FOCAL = attend
. ***
. mlogit pol c.attendmonth##c.educ age  class female i.race, base(1)
   ...
. intspec focal(c.attendmonth) main((c.attendmonth, name(Attend) range(0(1)4)) ///
>  (c.educ, name(Education) range(0(4)20))) int2(c.attendmonth#c.educ) ndig(0)
   ...
. sigreg , effect(spost(amtopt(am(sd) center) atopt((means) _all))) ///
>  nobva save(Output\Table_9_3)
```

The effect of attending religious services on political ideology exhibits a different pattern of moderation by education across the three outcome categories, as reported in Table 9.3. For the liberal category, attendance has a positive and significant effect at *Education* < 4, and the discrete change is about a 4 percentage point increase in the probability of liberal. But the attendance effect becomes negative and significant at *Education* > 11, steadily increasing in size from a 2 percentage point decline in the probability of liberal to a 16 percentage point drop at *Education* = 20. The pattern of the attendance effect across education for moderates is somewhat the opposite. Attendance has a negative and significant effect on the probability of moderate for education between 5 and 12 years, diminishing from a 5 percentage point decrease in moderate to a 3 percentage point decrease. At *Education* > 17, the effect of attendance becomes positive and significant, predicting a 5 to 9 percentage point increase in the moderate category. Last, attendance has a uniformly positive effect on the probability of conservative, which is significant at *Education* > 7 years. The significant effects of attendance slowly grow from a 4 percentage point increase in conservative to a 7-point increase at *Education* = 20.

How meaningful are the sizes of these discrete change effects, and how do the effect sizes compare across the categories? A simple way to contextualize the effect sizes is to compare them with the sample proportion in the corresponding category shown in the header in Table 9.3 identifying the political ideology category. Not only is this useful for assessing the absolute size of an effect on an outcome category, but it also provides a metric that can be compared across outcome categories. Overall, the effect of attendance appears somewhat more consequential for the probability of liberal identification than for the other two categories, given the magnitude of the negative discrete change effects coupled with the modest representation of liberal in the sample (18%). A midrange effect of −0.08 would reduce the sample proportion of liberal by close to half.

Note that the attendance effect on liberal is almost exclusively negative. Less than 1% of the sample would be predicted to have a significant positive effect of attendance on liberal (*Education* = 0–3), while 84% would have a significant negative effect (*Education* = 12–20). Because the absolute magnitudes of the discrete change effects of attendance on moderate and on conservative are relatively similar, while the conservative category is much smaller in the sample (18% vs. 64%), I would argue that the effect of attendance is more substantial on the conservative outcome. A midrange positive effect of attendance on conservative (0.05) would increase the sample proportion by more than 1/4. In comparison, a midrange negative effect of attendance on moderate (0.05) would *decrease* the sample proportion by about 1/13, and a midrange positive effect (0.07) would *increase* the sample proportion by about 1/9.

The overall picture of how the interaction of attendance at religious services and education affects political ideology seems clearly shown by the discrete change effects. The moderated effect of attendance by education has a varying and slightly complex pattern across the political ideology categories. The simplest is attendance's effect on conservative identification, which is uniformly positive and significant for all but the bottom 3% of the education distribution. This attendance effect steadily grows with education level. The pattern for the liberal category is similar but with a negative effect of attendance that increases in size more sharply across education values in the upper 84% of the distribution. However, attendance also has a more limited positive effect at the bottom 1% of the education distribution. The effect of attendance on the moderate category is more mixed between positive and negative, with a significant negative effect on the probability of moderate for 42% of the education distribution (5–12 years) but a significant positive effect on moderate at the top 11% of education values.

TECHNICAL SIDE NOTE

There is an arguably superior alternative to comparing the discrete change sizes against a category's sample proportion. Instead, compare it with the category's predicted probability used as the base value to calculate the specific discrete change. The "Special Topics" section at the end of the chapter shows how to use the *mchange* command in SPOST13 to calculate this alternative comparison point.

Cautions

There are two cautions about a discrete change effect interpretation of an interaction effect. First, remember that the values of discrete effects will be at least somewhat different depending on the reference values used for the predictors not part of the interaction specification. The results above used sample means, and a supplementary analysis using the "as observed" specification produced nearly identical numbers. But trying several "substantively interesting cases" defined by race–sex group combinations—predictors not part of the interaction—produces some substantial differences in the moderated effects of attendance.

For example, evaluating the discrete change effects for White males who are average on age and social class resulted in attendance having only significant negative effects on the probability of liberal; the positive effects of attendance at low levels of education were no longer significant. For the Other Race–sex combinations used to define substantively interesting cases, the absolute size of the attendance effect on the conservative category was smaller than the size of the attendance effect on the moderate category. And the extent of difference varied across the race–sex groups. That said, contextualizing the effects relative to the outcome category proportions produced the same conclusions, at least in this case.

Second, there is always a concern that the nonlinearity of the link function may understate or exaggerate the inherently nonlinear pattern of moderation for the same reason—the moderated effect is evaluated at different reference values of the moderator. Comparing the discrete change effects for the interacting predictors from the interaction model with the discrete change effects from a main effects model (see Table 9.4) can help disentangle and clarify the moderated effects. In this analysis, the moderated effects of attendance on the liberal and moderate outcomes are slightly enhanced by the nonlinearity of the multinomial logit link function.

Look at the pattern of change of each variable's discrete effect within the row for the liberal and moderate categories in Table 9.4. This reveals that the main effects pattern shows smaller changes with education but parallels the pattern of discrete effect changes from the interaction effects model. But for the conservative outcome, the moderated effect's pattern from the interaction analysis runs counter to the pattern from the main effects analysis. Thus, the effect of attendance on the conservative category is somewhat downplayed in the discrete change effects.

TABLE 9.4 ● DISCRETE CHANGE EFFECTS OF ATTENDANCE, MAIN EFFECTS MODEL

	Effect of Attendance From Main Effects Model					
	Education					
Outcome	0	4	8	12	16	20
Liberal	−0.0171	−0.0238	−0.0324	−0.0429	−0.0546	−0.0663
Moderate	−0.0453	−0.0366	−0.0258	−0.0128	0.0018	0.0171
Conservative	0.0624	0.0603	0.0582	0.0557	0.0528	0.0492

Interpretation Using Displays of Predicted Probabilities

In many ways, the most intuitive approach to understanding effects (including interaction effects) is to produce tabular or visual displays of the predicted probability of each outcome category as it varies with the predictors. For a single-moderator model, a set of 2D tables—one for each outcome category—can effectively present the predicted probabilities. The columns represent the focal variable, the rows represent the moderator, and the cells report the predicted probability of an outcome category for the combination of the focal and moderator values. Reading across a row reveals how the predicted probability of the outcome category changes with the focal variable for a given value of the moderator. Analogously, scanning down a column shows how the predicted probability varies by the moderator for a given value of the focal variable.

It is typically easier to see and discuss the patterns from visual displays of the relationship, but this requires two sets of scatterplots, first treating education as the focal variable and attendance as the moderator and then reversing the roles. I start by examining a plot with curves showing the education–liberal relationship for selected values of attendance, with analogous scatterplots for the moderate and conservative categories. I then interpret a scatterplot whose curves represent the attendance–liberal relationship for selected values of education, with parallel scatterplots for the moderate and conservative categories. Although I focus on the scatterplots for interpretation, it is useful to have (but not necessarily present) the predicted probability tables as a source of specific numbers to use in the interpretation.

ICALC **and Stata Command Sequence**

Using ICALC to produce the predicted probabilities tables usually requires minimal syntax beyond the required definition of the interaction using the *intspec* command. The rows and columns of the table are defined by the ranges specified for the focal and moderator variables, respectively, in the *intspec* command. I define five columns that evenly divide the range of education from its minimum to its maximum. And the six row values (0–4 by increments of 1 plus 4.33) fully span the range of attendance. The *outdisp* command with no options would produce the predicted probabilities using the as-observed option to define reference values for the noninteracting predictors. To save the table to an Excel file, you would add the tab(save(...)) option to the *outdisp* command. Because I prefer to use means for these reference values, I add the out( ) option and its atopt( ) suboption to request this.

```
. intspec focal(c.attendmonth) main((c.educ, name(Education) range(0(5)20))   ///
>  (c., name(Attend) range(0(1)4)4.33) ///
>  int2(c.attendmonth#c.educ) ndig(2)
     ...
. outdisp , out( atopt((means) _all)) ndig(3)
```

Table 9.5 presents the three panels of predicted probabilities—for the liberal, moderate, and conservative outcome categories—produced by these commands.

A quick examination of the rows in each panel of Table 9.5 can help you pick moderator display values for plotting the predicted probabilities. The table shows that education's effect on the liberal and moderate categories is opposite signed for low versus high attendance levels and thus changes direction in the middle of the attendance distribution. The attendance effect on the liberal and moderate categories similarly switches sign in the middle of the education distribution. In this situation, you should select moderator values to represent the initial outcome–focal relationship when the moderator has a low value, a high value of the moderator to capture

TABLE 9.5 ● PREDICTED PROBABILITIES OF POLITICAL IDEOLOGY BY THE INTERACTION OF EDUCATION AND RELIGIOUS SERVICE ATTENDANCE[a]					

	Education				
Attend	**0**	**5**	**10**	**15**	**20**
pol[Liberal] by the Interaction of Attend With Education					
0.00	.031	.064	.128	.240	.403
1.00	.047	.079	.127	.200	.298
2.00	.072	.096	.126	.164	.210
3.00	.109	.116	.124	.133	.141
4.00	.160	.140	.122	.106	.092
4.33	.181	.148	.121	.098	.079
pol[Moderate] by the Interaction of Attend With Education					
0.00	.746	.746	.716	.641	.516
1.00	.714	.712	.694	.653	.585
2.00	.675	.675	.669	.655	.632
3.00	.628	.636	.643	.648	.653
4.00	.572	.594	.614	.633	.651
4.33	.551	.579	.604	.626	.646
pol[Conservative] by the Interaction of Attend With Education					
0.00	.223	.190	.156	.119	.082
1.00	.239	.209	.179	.148	.117
2.00	.253	.229	.205	.181	.158
3.00	.263	.248	.233	.219	.205
4.00	.268	.267	.264	.261	.257
4.33	.268	.272	.275	.276	.275

a. The sample proportion: liberal = 18%, moderate = 64%, and conservative = 18%.

the opposite-signed outcome–focal relationship, and a value preferably equidistant between these two to show the transitional effect. The same selections should work well for an outcome–focal relationship that changes magnitude but not direction, such as the conservative–education effect moderated by attendance and the conservative–attendance effect moderated by education.

Thus, when attendance is the moderator, I specify its reference values as 0, 2, 4; see the first *intspec* command below. And when education is the moderator, I specify its reference values as 4, 12, 20 in the second *intspec* command. I follow my recommended practice of pairing the scatterplots from the interactive model with ones from a main effects model to help identify if there are any concerns with confounded nonlinearities. For this reason, the syntax begins by running and storing the estimates from a main effects model. I next run and store the interaction model estimates. *The predictors must be listed in the same order in these mlogit commands.* This is followed by two sets of *intspec* and *outdisp* commands, the first for education as the focal variable and the second for attendance as the focal variable.

```
. mlogit pol  c.educ c.attendmonth age  class female i.race, base(1)
  ...
. est store mlgtmain

  ...
. mlogit pol c.educ##c.attendmonth age  class female i.race, base(1)
  ...
. est store mlgtint

  ...
. intspec focal(c.educ) main((c.educ, name(Education) range(0(4)20))  ///
>   (c.attendmonth, name(Attend) range(0(2)4))) int2(c.educ#c.attendmonth) ndig(0)
  ...
. outdisp, out(main(mlgtmain) atopt((means) _all)) plot(name(PolEd))  ///
      pltopts(ylab(0(.2).8))
  ...
. intspec focal(c.attendmonth) main((c.attendmonth, name(Attend) range(0(1)4))  ///
>   c.educ, name(Education) range(4(8)20))) int2(c.attendmonth#c.educ) ndig(0)
  ...
. outdisp, out(main(mlgtmain) atopt((means) _all))  plot(name(PolAttend))  ///
      pltopts(ylab(0(.2).8))
```

Interpretation of Predicted Probability Displays

Figure 9.1 presents the scatterplots of the predicted probabilities for each political ideology category against education at low (*solid line*), medium (*small dashed line*), and high (*large dashed line*) levels of monthly attendance. The *black lines* represent predicted probabilities from the interaction model and the *gray lines* the probabilities from a main-effects-only model. Focus first on the interaction model predictions. For both liberal and moderate categories, the outcome category–education relationship changes direction as attendance increases, but in opposite ways. The predicted probability of liberal (top left plot) increases with education at low and medium levels of attendance (*Attend* < 4) but decreases as education rises at high levels of attendance (*Attend* ≥ 4). Relative to the sample proportion of liberal (18%), the predicted change in the probability of liberal across the range of education is fairly substantial—37% for the low-attendance curve, 14% for the medium-attendance curve, and −10% when attendance is high.

Conversely, the predicted probability of moderate (top right plot) decreases as education increases at low and medium levels of attendance but increases with education at high attendance levels. These changes in the probability of moderate with education (−23%, −5%, and 8%, respectively) are smaller than those for the probability of liberal, especially compared with the 64% sample proportion in the moderate category. For the conservative category (bottom plot), the predicted probability curves exhibit a uniform negative relationship with education, and the curves become shallower as attendance increases.[5] The changes in the probability of conservative across education levels (−14%, −8%, and −1%) are smaller than those for moderate and much smaller than for liberal. But comparing them with the sample proportion of conservative (18%) suggests that they are still consequential.

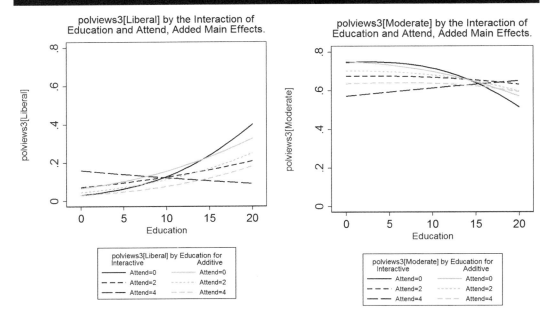

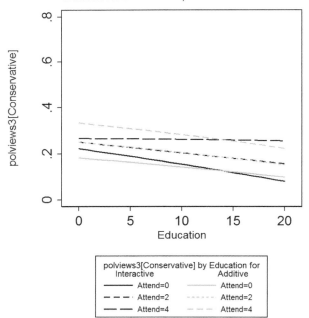

Overall, the scatterplots suggest that attendance moderates the effect of education on political ideology to a fairly substantial degree for the liberal category and to a lesser extent for both the moderate and the conservative category. To what degree is this a result of confounding with the nonlinear link function? We can assess this by comparing the *black prediction curves* from the interaction model with the *gray prediction curves* for the main-effects-only model. The patterns for the liberal and moderate outcomes are slightly enhanced by the nonlinearity because they each heighten a parallel pattern in the main effects model. That is, the slopes of the liberal–education curves become less positive as attendance increases in both the main effects and the interaction effects plots.

Analogously, the slopes of the moderate–education plots are progressively less negative as attendance increases in both plots. But the prediction curves for conservative–education show opposite patterns. In the main effects plots, the slope becomes slightly steeper as attendance increases, while in the interaction effects plot, the slopes become progressively shallower with rising education. This suggests that the effect of education on the conservative outcome is slightly downplayed by the confounding. Not surprisingly, these are the same conclusions about confounding that were evident in the discrete change effects.

Turning to the political ideology–attendance relationship moderated by education (Figure 9.2, *black lines*), the pattern of this moderated relationship is very similar to what I described for the ideology–education relationship for the liberal and the moderate categories but not the conservative category. For the liberal category, the probability rises with attendance at low levels of education, falls slightly with higher attendance at medium education, and sharply declines with increased attendance at high education levels. The changes in probabilities across the range of attendance (10%, −6%, and −32%, respectively) are substantial relative to the 18% sample probability of liberal.

The moderate category shows an opposite pattern: The probability of moderate declines as attendance increases at low and medium education but increases with rising attendance for high education. The changes in the probability of moderate across the span of attendance levels (−20%, −8%, and 13%) are roughly similar to those for the liberal category, but they are in the context of a much larger sample proportion for moderate (64%). The conservative–attendance relationship is positive at all levels of education, and the change in probability across attendance levels grows progressively larger for higher levels of education (7%, 13%, and 19%, respectively). While similar in absolute magnitude to the probability changes for moderate, the probability changes for the conservative category are relative to a much smaller sample proportion (18%).

These results suggest that, like education, attendance is a bit more consequential for the predicted probabilities of the liberal and conservative categories than for the moderate category. Additionally, the effect of attendance is moderated by education to a greater extent in the liberal than in the other two political ideology categories. And again, like the education effect, the pattern of moderation in the liberal and moderate categories is built on a main effects pattern that parallels the interaction effects pattern, which slightly exaggerates the appearance of moderation in the *black lines* for the interaction model predictions.

On the other hand, the moderation of the attendance effect on the conservative category is somewhat greater than what the interactive model scatterplot shows because

FIGURE 9.2 ⬡ PREDICTED PROBABILITY OF IDEOLOGY BY ATTENDANCE MODERATED BY EDUCATION

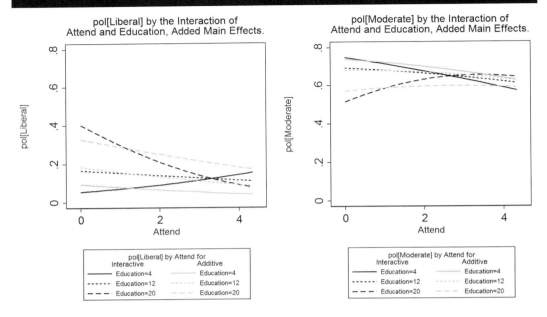

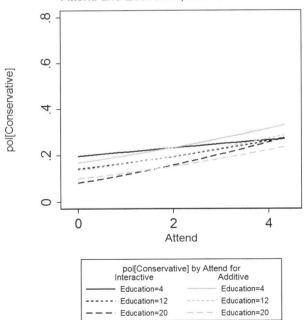

it runs counter to the pattern from the main effects model. That is, the interaction effect increases the slope of the conservative–attendance curve as education rises, while the main effects specification produces prediction curves whose slopes flatten as education increases. Thus, the interaction model has to overcome the opposing tendency to produce the degree of moderation that it shows in the scatterplot.

Cautions

The same two cautions apply to the interpretation of predicted probabilities as the ones I discussed in the wrap-up of the discrete change effects above. I incorporated the impact of the confounded nonlinearities into the interpretation of the pattern of moderation shown by the prediction curves, which created relatively little confounding in either the discrete change interpretation or the prediction curve interpretations.

The other caveat is that the choice of reference values for the noninteracting predictors will have some impact on the shape and location of the prediction curves, the degree of which cannot be known a priori. The same sensitivity analyses (using as-observed or substantively interesting cases defined by race and gender) indicated very little influence of these alternative choices of reference values other than some modest changes in the level of the predicted probabilities across the three outcome categories, but not in the patterns of moderation.

Interpretation Using Displays of Predicted Standardized Latent Outcomes

This last method of interpretation preserves the intuitive appeal of a visual display of the predicted outcome–focal variable relationship without either of the two complications of using predicted probabilities. The trade-off is that the predicted outcomes are the latent outcomes (utilities) underlying the choice among outcome categories, which are only identifiable as the difference in utility between pairs of categories. ICALC creates a plot of the standardized latent outcome (standardized log odds) for each outcome category versus the base category. Standardizing the latent outcomes provides a device for understanding the scale of visual changes and is interpretable in a somewhat abstract metric—the standard deviation of the difference in utilities or, equivalently, the standardized propensity to choose a given category versus the base category. The "Special Topics" section presents the derivation of the formula to calculate the mean and standard deviation of the latent outcome (log odds) for interested readers.

I produced Figures 9.3 and 9.4 with the following ICALC and Stata commands—note the repeated use of the *est restore* and *intspec* commands to switch between using the primary and the supplementary specification of the reference category. To get plots of the standardized latent outcomes, you use the *outdisp* command and within the out( ) option specify the suboptions metric(model) sdy. I also included the atopt((means) _all) suboption to set the reference values for the other predictors and pltopts( ) to customize the formatting.

```
****  FOCAL = educ
. est restore mlgtint
    ...
. intspec focal(c.educ) main((c.educ, name(Education) range(0(5)20))  ///
>  (c.attendmonth, name(Attend) range(0(2)4))) int2(c.educ#c.attendmonth) ndig(0)
    ...
```

```
outdisp, out(metric(model) sdy atopt((means) _all)) plot(name(PolEd_SDY)) ///
>  pltopts(ylab(-1(.5)1) ytit( , size(*.8)) tit( , size(*.9)) leg(cols(1)))
    ...

. est restore mlgt3rd
    ...

. intspec focal(c.educ) main((c.educ, name(Education) range(0(5)20))  ///
>  (c.attendmonth, name(Attend) range(0(2)4))) int2(c.educ#c.attendmonth) ///
>  ndig(0) eqname(Conservative)
    ...

outdisp, out(metric(model) sdy atopt((means) _all)) plot(name(PolEd_SDY3rd)) ///
>  pltopts(ylab(-1(.5)1) ytit( , size(*.8)) tit( , size(*.9)) leg(cols(1)))
    ...

****  FOCAL = attend
. est restore mlgtint
    ...

. intspec focal(c.attendmonth) main((c.attendmonth, name(Attend) range(0(1)4)) ///
>  (c.educ, name(Education) range(0(10)20))) int2(c.attendmonth#c.educ) ndig(0)
    ...

outdisp, out(metric(model) sdy atopt((means) _all)) plot(name(PolAtt_SDY)) ///
>  pltopts(ylab(-1(.5)1) ytit( , size(*.8)) tit( , size(*.9)) leg(cols(1)))
    ...

.est restore mlgt3rd
    ...

.intspec focal(c.attendmonth) main((c.attendmonth, name(Attend) range(0(1)4)) ///
>  (c.educ, name(Education) range(0(10)20))) int2(c.attendmonth#c.educ) ///
>  ndig(0) eqname(Conservative)
    ...

. outdisp, out(metric(model) sdy atopt((means) _all)) plot(name(PolAtt_SDY3rd))
```

The predicted standardized latent outcomes for the choices of moderate to liberal, conservative to liberal, and conservative to moderate in Figure 9.3 highlight how the relationship of ideology to education is moderated by attendance. The patterns are nearly identical for moderate to liberal and conservative to liberal. The education effect is negative until relatively high levels of attendance, about 3.25 times per week for moderate and 3.5 times per week for conservative. For nonattendees (*solid line*), the change in the latent propensity to choose moderate or conservative over liberal across the range of education is fairly substantial. The latent propensities decrease by about 1.5 and 1.8 standard deviations, respectively. At attendance of twice a month, the negative effect of education is much smaller but still notable. The latent propensity declines by more than 1/2 standard deviation for moderate to liberal and by 3/4 standard deviation for conservative to liberal.

Last, for those who attend more than once a week, the latent propensity to choose either moderate or conservative over liberal increases with education, rising by about 1/2 standard deviation for both moderate and conservative. Given the similarities, it is no surprise that the plot of predicted latent outcomes shows only small declines in the propensity to choose conservative over moderate identification across the span of education values. These range from 1/10 to 3/10 standard deviation. The scatterplots clearly show that the moderation of the education effect by attendance is fairly substantial for the moderate to liberal and conservative to liberal contrast and relatively small for the conservative to moderate contrast. Overall, the effect of education is predominantly negative for all three propensities. For the moderate to liberal choice, the effect of education is positive for only 25% of the sample, and the percent positive is smaller yet for the other two propensities (taken from the GFI results).

FIGURE 9.3 ● STANDARDIZED LATENT OUTCOMES BY EDUCATION MODERATED BY ATTENDANCE

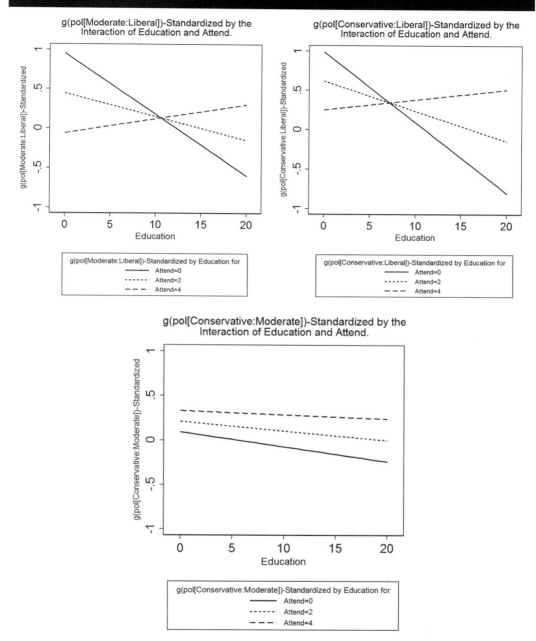

Looking at the political ideology–attendance relationship as moderated by education in Figure 9.4 reveals an initially similar-appearing pattern of moderation. The effect of attendance on the latent propensity to choose moderate or conservative over liberal changes from negative at low levels of education

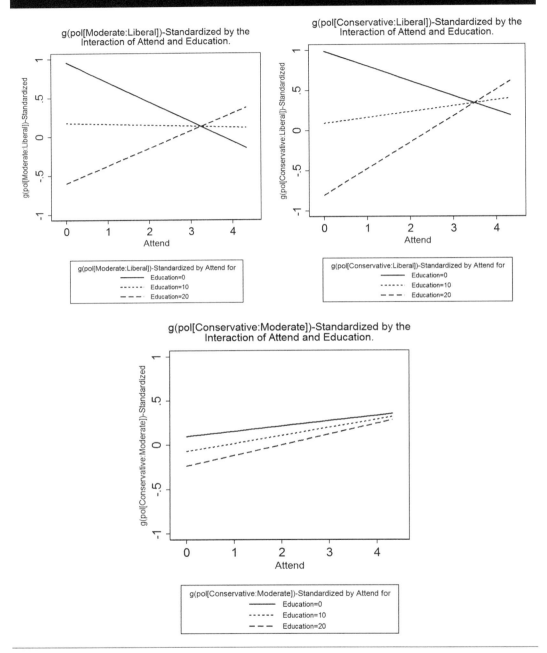

FIGURE 9.4 ● STANDARDIZED LATENT OUTCOMES BY ATTENDANCE MODERATED BY EDUCATION

to positive at higher levels of education. But given the sample distribution of education, the effect of attendance is almost entirely positive in the sample. The effect is positive on the moderate to liberal propensity for 89% of the sample, on the conservative to liberal propensity for 97% of the sample, and on the conservative to moderate propensity for 100% of the sample (see the GFI results).

The change in propensity across the range of attendance is smallest for the conservative to moderate contrast, topping out at 0.5 standard deviation at the highest level of education. The largest change for the moderate to liberal propensity is an increase of 1.1 standard deviations, while for conservative to liberal, it is an increase of 1.4 standard deviations. These maximum changes are a little smaller but similar in magnitude to those for education (-0.6, -1.5, and -1.7, respectively). Also, the extent of moderation of the attendance effect by education is much smaller for the conservative to moderate contrast than for the other two.

Wrap-Up

Like the interpretation of odds ratios, interpreting the latent outcomes requires constantly referring to the contrast categories being discussed. But it avoids the issue of the interpretation being dependent on the (reference) values of the other predictors. And while it is not subject to the confounding of the nonlinearity of the interaction with the nonlinearity of the link function, this comes at the cost of discussing the results in a less intuitive, more abstract metric.

Nevertheless, I think that it provides a clear and understandable picture of the nature of the relationship of the multiple outcomes to the interacting predictors, particularly how the extent of moderation of one of the predictors by the other varies across the outcome contrasts. But it is up to readers to decide for themselves which approach they find the most comprehensible and best to communicate to their audiences. As I have emphasized before, try multiple approaches to interpretation before making up your mind which to use. Remember also that an approach that works best in one analysis may not work as well for another application.

TWO-MODERATOR EXAMPLE (INTERVAL BY TWO NOMINAL)

Data and Testing

This application examines how the interactions of education with sex and education with race/ethnicity predict current occupational class (coded as upper white collar, lower white collar, upper blue collar, and lower blue collar). I analyze these as unordered categories because there is no clear rationale for the relative rank of upper blue collar versus lower white collar. Education is measured at the interval level (years completed, 0–18). Sex is a dichotomous indicator ($1 = $ *women*, $0 = $ *men*), and race/ethnicity (hereafter race for simplicity) is a four-category variable (White, Black, Asian American, Hispanic) coded into dummy variables, using White as the excluded category. The other predictors are age in years (17–83) and a dichotomous measure of marital status ($1 = $ *currently married*, $0 = $ *not married*). The index prediction functions to be estimated for the contrasts of category $m = $ UpperWC, LowerWC, UpperBC with the base category (LowerBC) are as follows:

$$g\left(y_{m,i}\right) = \beta_{m,0} + \beta_{m,1}Education_i + \beta_{m,2}Sex_i + \beta_{m,3}Education_i \times Sex_i$$
$$+ \beta_{m,4}Black_i + \beta_{m,5}Asian_i + \beta_{m,6}Hispanic_i + \beta_{m,7}Education_i \times Black_i$$
$$+ \beta_{m,8}Education_i \times Asian_i + \beta_{m,9}Education_i \times Hispanic_i$$
$$+ \beta_{m,10}Age_i + \beta_{m,11}Married_i$$

The Stata command to run this model uses factor-variable notation to specify the interaction of education by sex (*i.sex##c.educ*) and education by race categories (*i.raceth##c.educ*) to predict current occupational class (*currocc*):

```
mlogit currocc i.sex##c.educ i.raceth##c.educ c.age i.married, base(4)
```

The option base(4) specifies that Category 4 of occupational class be used as the base category in the MNLR. I chose lower blue collar as the reference category because it creates a comparison of the lowest status category with categories that are unambiguously of higher status. In addition, for an MNLR analysis, you often want to examine contrasts other than to the base category for substantive reasons. In this example, the upper-white-collar/lower-white-collar contrast is especially relevant. Thus, for the interpretation with factor changes or predicted latent outcomes, I run a supplementary analysis specifying LowerWC as the base category in the *mlogit* command with the option base(2). The ordering of the interaction terms will sometimes be different in the examples below to correspond to ICALC–SPOST13 interface requirements.

The sample excludes 7,194 respondents who did not have a current occupation and 160 cases in racial/ethnic groups other than White, Black, Asian, or Hispanic. An additional 560 cases were excluded for reasons relevant to the analysis for which these data were originally extracted from the SIPP (see Campbell & Kaufman, 2006, p. 139).[6]

Diagnostic tests of model fit coupled with a sensitivity analysis showed no convincing evidence of problems overall. The tests of the IIA assumption gave mixed results, with the Small-Hsiao test indicating that the assumption is met but the Hausman test suggesting that sometimes it is not. On conceptual grounds, variations of this 2D class categorization (nonmanual vs. manual, each internally ordered by status) have a long history of use in non-Marxist stratification and mobility analyses in the United States, going back at least to Blau and Duncan (1967). This lends credence to considering the classes as conceptually distinct and nonsubstitutable in the eyes of researchers and, hence, adhering to the IIA principle. The Hosmer-Lemeshow statistic and LOWESS plots of the predicted probabilities against the observed probabilities both indicated good fit for the model. Analysis of the residuals found a number of potential outliers, either because their residuals were isolated from the core of the distribution or because they had high leverage values. A sensitivity analysis of the MNLR coefficients and their significance showed that excluding these cases produced inconsequential differences in the results. That is, on the 36 estimated coefficients, the significance of only one changed from barely not significant ($p = .0514$) to significant ($p = .0345$). Thus, I included these cases so as not to bias the results toward larger effects.

Statistical tests of the two sets of interaction terms—education by sex and education by race—support their inclusion in the model but not the addition of their three-way interaction. Global tests of the change in the log likelihood if either or both of these interaction sets are excluded from the model were statistically significant ($p < .001$), and several of the z tests of the individual interaction coefficients for different outcome categories were significant ($p < .001$).

Strategies for Interpreting a Multiple-Moderator Interaction

I recommend organizing the interpretation by discussing each of the predictors as the focal variable in turn, starting with single-moderator interactions. For the current example, this means discussing how the effect of sex is moderated by education, then how the effect of race is moderated by education, and last how the effect of education is moderated by sex and race. To interpret the education effect, you have to decide how to deal with the multiple moderators. As I discussed in Chapter 4, you can either treat each focal–moderator pair individually in calculating moderated effects or predicted values or calculate the moderated effect or predicted value varying simultaneously across both moderators.

The first case is precisely like interpreting a single-moderator interaction except that the second moderator is given a reference value in the same way that you give reference values to other predictors in the model. Applying this to the current example, you would interpret how the effect of education is moderated by sex, setting race to a reference value, and then how the effect of education is moderated by race, setting sex to a reference value. I focus on interpretation using the first strategy but briefly consider the factor change results from the second approach. The difficulty with the simultaneous moderation by race and sex results is that they embody the same patterns as the moderation separately by race and by sex but are harder to recognize, especially with the multiple outcomes (categories) in an MNLR analysis.

I begin the discussion of each moderated effect with a brief GFI analysis of the moderated effect of the focal variable and then demonstrate how to interpret it using selected modes of interpretation. For the effect of sex moderated by education, I use the discrete change effect and predicted probability display approaches for interpretation. For the effect of race moderated by education, I use the factor change effect and the predicted standardized latent outcome methods. Finally, for the moderated effect of education by sex and race, I use the factor change effect and the predicted probability approaches.

The Effect of Sex Moderated by Education

I begin the analysis for this example by running and storing the MNLR estimates for the interaction model using two different outcome categories to define the base category. As needed, I later use the *est restore* command to change the current estimates between these two sets. I specify variations of the *intspec* command shown in the top part of the output to set up the interaction specification to correspond to which predictors are treated as focal and moderating.

INTSPEC **Setup and** GFI **Analysis**

```
. qui mlogit currocc  i.sex##c.educ i.raceth##c.educ  c.age i.married , base(2)
. est store intsex2

. qui mlogit currocc  i.sex##c.educ i.raceth##c.educ  c.age i.married , base(4)
>   est store intsex

. intspec focal(i.sex) main( (i.sex , name(Sex) range(0/1)) ///
>        (c.educ, name(Education) range(0(3)18))) ///
>        int2(c.educ#i.sex ) ndig(0)
  ...
. gfi

GFI Information from Interaction Specification of
Effect of Sex on g(Currocc[upperwc:lowerbc]) from Multinomial Logistic Regression
------------------------------------------------------------------------

Effect of Women =
   -0.0084 + 0.0553*Educ
```

```
GFI Information from Interaction Specification of
Effect of Sex on g(Currocc[lowerwc:lowerbc]) from Multinomial Logistic Regression
----------------------------------------------------------------------

Effect of Women =
    2.0145 - 0.0665*Educ

GFI Information from Interaction Specification of
Effect of Sex on g(Currocc[upperbc:lowerbc]) from Multinomial Logistic Regression
----------------------------------------------------------------------

Effect of Women =
   -3.9549 + 0.1865*Educ
    …
Sign Change Analysis of Effect of Sex
 on g(currocc[upperwc:lowerbc]), Moderated by Education (MV)
    …
-------------+------------------------
 Sign Changes |   when MV= 1.122436
-------------+------------------------
 % Positive   |        99.7
--------------------------------------


 Sign Change Analysis of Effect of Sex
 on g(currocc[lowerwc:lowerbc]), Moderated by Education (MV)
    …
-------------+------------------------
 Sign Changes |     Never
-------------+------------------------
 % Positive   |       100.0
--------------------------------------


 Sign Change Analysis of Effect of Sex
 on g(currocc[upperbc:lowerbc]), Moderated by Education (MV)
    …
-------------+------------------------
 Sign Changes |     Never
-------------+------------------------
 % Positive   |        0.0
--------------------------------------

    …
. est restore intsex2
    …
. intspec …
    …
. gfi
GFI Information from Interaction Specification of
Effect of Sex on g(Currocc[upperwc:lowerwc]) from Multinomial Logistic Regression
----------------------------------------------------------------------

Effect of Women =
   -2.0229 + 0.1218*Educ

Sign Change Analysis of Effect of Sex
 on g(currocc[upperwc:lowerwc]), Moderated by Education (MV)
    …
-------------+------------------------
 Sign Changes |  when MV= 16.61345
-------------+------------------------
 % Positive   |        13.5
--------------------------------------
```

The GFI expression for each of the four selected contrasts has an opposite sign for the baseline effect of sex to that for the sex-by-education interaction coefficient. This raises the possibility that the effect of sex could change sign. Except for the lower-white-collar to lower-blue-collar contrast, the baseline effect of sex is negative. This means that when *Education* = 0,

- women are less likely than men (lower log odds) to be in upper-white-collar or upper-blue-collar jobs than they are to be in lower-blue-collar jobs and

- women are less likely than men to be in upper-white-collar than in lower-white-collar jobs.

Because the coefficient for the interaction of education with sex is positive, the log odds of these contrasts become less negative (possibly positive) as education increases. The sign change analysis output shows that the effect of sex on the upper-white-collar to lower-blue-collar contrast is negative only for *Education* ≤ 1 and is predominantly positive (99.7% of the sample). That is, with rising education, women become increasingly more likely than men to be in upper-white-collar than in lower-blue-collar jobs.

Conversely, for the upper-white-collar with lower-white-collar comparison, the sex effect is primarily negative (86.5% of the sample, *Education* < 17), meaning that women are predicted to be less likely than men to be in upper- versus lower-white-collar jobs. But the sex effect changes to positive—women are more likely than men to be in upper-white-collar jobs—for *Education* > 16 years. For the upper-blue-collar with lower-blue-collar contrast, the effect of sex is always negative while declining in magnitude with increasing education. Finally, for the lower-white-collar with lower-blue-collar comparison, the effect of sex starts and remains positive but diminishes in magnitude with increasing education. This shows that women are predicted to be more likely than men to be employed in lower-white-collar than in lower-blue-collar jobs but the differential is smaller at higher levels of education.

Thinking about how these effects fit together into a larger picture of the occupation–sex–education relationship is relatively straightforward. For the most part, women are predicted to be more likely than men to hold white-collar than blue-collar jobs. But the differential grows with education for upper-white-collar jobs, while it declines in size for lower-white-collar jobs. Within each color-collar type of jobs, men are predicted to be more likely than women to hold upper- than lower-sector jobs, but the disparity decreases with rising education. While this patterning of the effects on the log odds of the occupation outcomes does not directly translate into patterns of effects in the probability metric, it sensitizes us to potential patterns to look for.

Discrete Change Effects

The *sigreg* command is used to calculate the discrete change effects for this example. The syntax begins by restoring the primary set of estimates (stored as "intsex"), which are processed by the same *intspec* command used in the GFI analysis. The *sigreg* command produces the desired discrete change effects and stores the significance region tables in an Excel file saved in the default working directory.

```
. est restore intsex
    ...

. intspec ...

    ...
.

. sigreg, effect(spost(amopt(binary) atopt((means) _all))) ///
>    nobva ndig(3) save(Output\Table_9_6.xlsx tab)
```

Rather than presenting the Stata output tables, I report and discuss the four format-ted significance region tables from the Excel file (one for each outcome category), which I edited together into Table 9.6. These indicate that women have a higher predicted probability of employment in white-collar jobs than men and that this dis-parity first increases with education level and then declines. For upper-white-collar jobs, the sex difference is significant at *Education* > 6 years (98% of the sample), while for lower-white-collar jobs, the sex difference is significant for *Education* < 17 years (87% of the sample).[7] The discrete change in the sex effect is negative and significant for upper-blue-collar jobs at all education levels.

Specifically, the predicted probability that women are employed in upper-blue-collar jobs is less than the probability for men, and this differential initially rises and then declines at higher education levels. In lower-blue-collar jobs, the discrete change effect switches sign as education rises. When *Education* < 9 years, women have a significantly higher probability of employment in lower-blue-collar jobs than men, which declines as education rises (4% of the sample). But for *Education* > 10 years, men have a sig-nificantly higher predicted probability of lower-blue-collar employment than women, which increases and then decreases as education continues to rise (91% of the sample).

As in the prior application example, one way to judge the size of these effects is to com-pare them with the sample proportion in each occupation class. A middle-of-the-range effect of sex on the probability of employment in an upper-white-collar job is about 0.07 (7%), which represents a 20% increase relative to the upper-white-collar sample proportion of 35%.[8] For lower-white-collar jobs, a midrange effect of sex is 0.18 (18%),

TABLE 9.6 ● SIGNIFICANCE REGION FOR DISCRETE CHANGE EFFECTS OF SEX ON CURRENT OCCUPATION, MODERATED BY EDUCATION

	Education						
	0	**3**	**6**	**9**	**12**	**15**	**18**
Upper white collar							
Effect of women	0.000	0.000	0.003	**0.019***	**0.070***	**0.101***	**0.068***
Lower white collar							
Effect of women	**0.036***	**0.079***	**0.160***	**0.262***	**0.271***	**0.105***	*−0.012*
Upper blue collar							
Effect of women	*−0.252**	*−0.280**	*−0.301**	*−0.302**	*−0.250**	*−0.126**	*−0.033**
Lower blue collar							
Effect of women	**0.217***	**0.200***	**0.138***	0.020	*−0.091**	*−0.079**	*−0.023**
Key							
Plain font, no fill	Pos, Not Sig						
Bold*, filled	Pos, Sig						
Italic, no fill	Neg, Not Sig						
Bold italic*, filled	Neg, Sig						

Note: SPOST effect for sex specified as *Amount* = Women vs. Men calculated with at((means) _all).

which is a substantial relative increase of 82% compared with the sample proportion of 22%. The midrange sex effect for upper-blue-collar jobs is similar at 0.20 (20%), but it represents an even larger increase (125%) relative to the sample proportion (16%).

Last, for lower-blue-collar jobs, there was a positive effect of sex in the lower part of the education distribution and a negative effect in the upper part. The midrange positive and significant effect (0.14) is twice as large as the midrange negative and significant effect (0.07) in both absolute size and relative size. Compared with the sample proportion of blue-collar jobs (27%), the midrange positive effect is a notable 52% increase relative to the base proportion, while the midrange negative effect is a 26% decrease relative to the base proportion. In sum, the typical predicted sex gaps in current occupation represent sizable differences in the probability of employment, especially for the lower sector of each color-collar type.

We can again examine the potentially confounding influence of the nonlinear link function from a comparison of the pattern of the moderated sex effect in Table 9.6 with the pattern of the main effects of sex on current occupation as its evaluation point changes with education reference values in Table 9.7. This shows that although the point at which the effect of sex peaks for each occupation class varies between models—as do the magnitudes—the pattern of the difference in probabilities between women and men is virtually identical, whether based on the interaction model or on the main effects model. For example, the predicted upper-blue-collar sex gaps in the two models are as follows:

Model	Education (years)		
	0	**9 (Peak for Both)**	**18**
Interaction model	−0.252	−0.302	−0.033
Additive model	−0.174	−0.257	−0.039

Thus, the underlying shape of the sex effect due to the nonlinear link function contributes substantially to the pattern of apparent moderation. The underlying shape is enhanced by the interaction of sex with education to varying degrees across the occupation classes. In particular, the moderated and main effects are more similar for the two blue-collar classes than for the white-collar classes. The ratio of the moderated to the main sex effect varies from 0.84 to 1.45 for the blue-collar classes, but for the white-collar classes, it ranges from 0.34 to 2.11.[9] Nonetheless, these ratios suggest a substantial degree of moderation of the effect of sex by education, but you would overstate how much if you were to rely solely on the discrete change effects.

Predicted Probability Interpretation

I apply the *outdisp* command to produce visual displays of the predicted probabilities. For an MNLR analysis, this creates a separate display for each category of the outcome variable. For a nominal focal variable like sex, a common choice for a visual display is a bar chart reporting the predicted probability of the outcome for each category of the predictor, the default in *OUTDISP* for a categorical focal variable. The chart is constructed as a set of predicted value bars for the categories of the focal variable, which is repeated for each reference value of the moderator.

I also specify the side-by-side display of the predicted probabilities from the interaction and additive models. As before, this can help distinguish the influence of the

TABLE 9.7 ● DISCRETE CHANGE EFFECT OF SEX WITH CHANGING EDUCATION REFERENCE VALUE, MAIN EFFECTS MODEL

Occupation	Education						
	0	**3**	**6**	**9**	**12**	**15**	**18**
Upper white collar	0.000	0.001	0.006	0.033	0.097	0.086	0.010
Lower white collar	0.017	0.044	0.105	0.202	0.239	0.132	0.049
Upper blue collar	−0.174	−0.202	−0.232	−0.257	−0.244	−0.139	−0.039
Lower blue collar	0.156	0.157	0.121	0.022	−0.092	−0.079	−0.021

moderating predictor from the nonlinearity of the model's link function. I first run and store estimates from a model with only main effects—*with the predictors listed in the identical order as in the interaction effects model*—and then restore the interaction effects to be the active estimates. In the *outdisp* command, I include the suboption main(mainsex) in the out( ) option, where *mainsex* is the name in which I stored the main effects model's estimates.

```
. qui mlogit currocc  i.sex  educ i.raceth age  i.married , base(4)
    …
. qui est store mainsex

    …
. qui est restore intsex

    …
. intspec focal(i.sex) main( (i.sex , name(Sex) range(0/1)) ///
>         (c.educ, name(Education) range(0(3)18))) ///
>         int2(i.sex#c.educ ) ndig(0)
    …
. outdisp , tab(default) plot(name(OccSexEdBar)) out(atopt( ///
> (means)_all) ///
> main(mainsex)) pltopts(ylab(0(.2)1) plotreg(ma(t +2))) ndig(2)
```

The four resulting bar charts in Figure 9.5 provide an effective and intuitive visualization of the pattern of differences in the sex effect on employment as it varies with education for each occupational class. The bar chart for an occupational class consists of horizontal panels for the chosen reference values of education. Each panel has two pairs of bars reporting the predicted probability of employment in the occupational class for women (*gray bar*) and for men (*black bar*). The pair of bars on the left are the interactive model results, and the pair on the right—with a *dashed outline*—are the additive model results. The difference in the height of the paired bars is the sex gap in the probability of employment in the occupational class at a given level of education. Comparing across the panels shows how the sex gap changes as education level changes.

Focus first on the interactive model results. The first bar chart shows that the predicted probability of employment in upper-white-collar jobs increases sharply for both men and women as education levels rise across panels from left to right. Women have a higher probability at all education levels than men, but this predicted gap increases with education levels and then begins to diminish slightly at the top of the education distribution.

FIGURE 9.5 ● PREDICTED PROBABILITY OF OCCUPATION GROUPS BY THE INTERACTION OF SEX AND EDUCATION, INTERACTIVE AND ADDITIVE MODELS

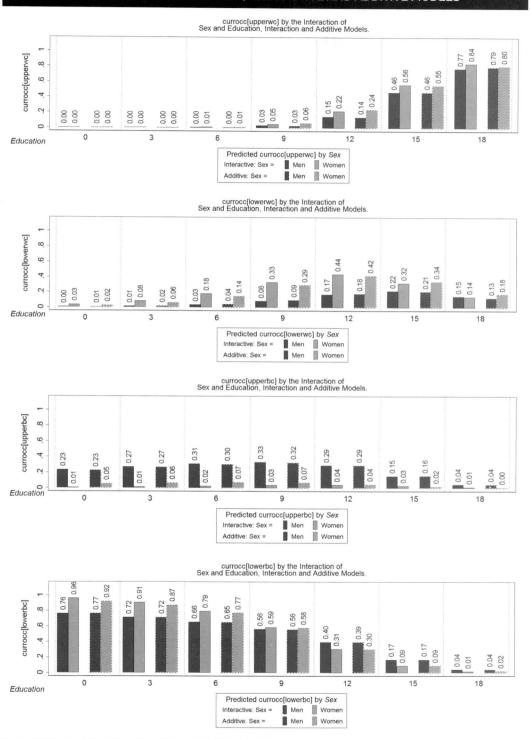

The relationship of the probability of employment in lower-white-collar jobs with sex and education in the second bar chart follows a similar pattern up to *Education* = 12, representing 55% of the sample. The predicted probability of employment for both sexes increases, and the gap in employment probability increasingly favors women. Beyond that point, the sex gap decreases in size as women's employment probability declines from that peak point, while men's employment probability does not peak until *Education* = 15.

The relationship of sex and education to upper-blue-collar employment portrayed by the third bar chart is very different. Although the predicted probabilities of employment for both women and men rise and then fall as education increases, they are relatively stable for *Education* = 3 to 12, as is the sex difference in employment, which favors men.

The last bar chart for lower-blue-collar jobs evidences a steady decline in both women's and men's predicted employment probability as education increases. The sex gap in employment in lower-blue-collar jobs favors women over men for *Education* ≤ 9 (9% of the sample). For the remaining 91% of the sample, with *Education* > 9, men's probability of employment is higher, but that gap diminishes as education continues to increase.

To what extent are these clear-cut patterns the result of interaction between sex and education versus the nonlinearity due to the link function? To address this, contrast the pattern of the interaction model results with the additive model results within the bar chart for each occupational group. That is, compare the pattern of how the sex effect changes with education for the interaction model with that for the additive model. The clear first impression from looking at any of the four bar charts is that the pattern of change in the heights of the bars is nearly identical for the models, with or without the interaction of sex by education. A closer examination of the pattern of the sex gaps (differences in the height of the pair of bars for women and men) in the interaction model with those in the main effects model reveals the influence of the interaction effect on each occupational class.

Note that it can be difficult, as it is in this case, to rely just on a visual perception. I used the value labels on the bars to calculate the gaps in order to accurately summarize the patterns and drew the following conclusions about the interaction effect's import:

- *Upper white collar:* The sex gap is consistently positive (higher probability for women). The sex gap in the interaction effects model changes at a different rate from that in the main effects model across education levels. Between 7 and 14 years of education, the interactive model sex gap grows more slowly, but at 15 or more years, the interactive model sex gap diminishes more slowly. That is, the interaction of sex with education is muting the sex gap in the probability of upper-white-collar employment to a small degree.

- *Lower white collar:* The sex gap is consistently positive. Compared with the main effects model results, the sex gaps in the interaction model favor women and increase more rapidly with education up to 11 years and then decrease more sharply, eventually advantaging men at *Education* = 18. This pattern suggests that the interaction of sex with education is enhancing the sex gap in the probability of lower-white-collar employment.

- *Upper blue collar:* The sex gap is consistently negative (higher probability for men). Although the sex gap is in the opposite direction, the pattern of change with education is similar to that for lower-white-collar employment, with smaller differences. The sex gap in the interactive model increases slightly faster for *Education* < 10 years and then falls marginally more rapidly than in the main effects model. Thus, the sex-by-education interaction heightens the sex gap by a little bit across the range of education.

- Lower blue-collar: The sex gap changes from positive to negative. For lower-blue-collar jobs, the rate of change in the sex gap is nearly identical in the interactive and main effects results. An initial positive sex gap steadily declines until *Education* = 9 years, then turns negative and grows slightly, and finally diminishes in size. The difference is that the positive sex gap at the bottom of the education distribution is larger in the interactive model than in the main effects model. Thus, the sex-by-education interaction somewhat heightens the sex gap but only at lower levels of education.

Not surprisingly, the comparison of predicted values from the interactive model with those from the main effects model indicates the same caution as in the discrete change effect interpretation about the degree to which the sex gap is systematically affected by education. To a fairly large degree, the patterns apparent in the predicted probabilities of each occupational class are due to the nonlinearity of the MNLR model and are somewhat modified by the specification of the sex-by-education interaction. Without a comparison with the main effects model results, the importance of the interaction would almost certainly be overstated.

The Effect of Race/Ethnicity Moderated by Education

This interaction effect illustrates the added complexity of interpreting an interaction that involves multiple comparisons among the categories of the focal variable (race/ethnicity) and among the categories of the outcome (occupational class). To accommodate this, I begin the analysis by estimating and storing MNLR results for two specifications of the outcome base category: lower blue collar (4) and lower white collar (2). And I use the *est restore* command to switch between these estimates as needed. Many of the ICALC commands use the *intspec* command shown below to define and set up the interaction specification. In those cases, it is shown in the syntax only as "intspec ...." In the output from the *gfi* command, I report abbreviated results from the sign change analyses, the two summary rows—"Sign Changes" and "% Positive."

INTSPEC **Setup and** GFI **Analysis**

```
. mlogit currocc  i.raceth##c.educ c.educ##i.sex c.age i.married , base(2)
   …
. est store intrace2
   …
. mlogit currocc  i.raceth##c.educ c.educ##i.sex c.age i.married , base(4)
   …
. est store intrace
   …
. intspec focal(i.raceth) main( (i.raceth, name(RaceEth) range(1/4))  ///
>        (c.educ, name(Education) range(0(3)18)))  ///
>        int2(i.raceth#c.educ ) ndig(0)
   …
. gfi
```

```
GFI Information from Interaction Specification of
Effect of Raceeth on g(Currocc[upperwc:lowerbc]) from Multinomial Logistic Regression
----------------------------------------------------------------------

Effect of Black =
   -1.571 + 0.060*Educ

Effect of AsianAm =
   -2.269 + 0.128*Educ

Effect of Hispanic =
   0.699 - 0.090*Educ
```

```
GFI Information from Interaction Specification of
Effect of Raceeth on g(Currocc[lowerwc:lowerbc]) from Multinomial Logistic Regression
----------------------------------------------------------------------

Effect of Black =
   -2.467 + 0.132*Educ

Effect of AsianAm =
   -0.640 + 0.028*Educ

Effect of Hispanic =
   -0.071 - 0.010*Educ
```

```
GFI Information from Interaction Specification of
Effect of Raceeth on g(Currocc[upperbc:lowerbc]) from Multinomial Logistic Regression
----------------------------------------------------------------------

Effect of Black =
   -2.171 + 0.1190*Educ

Effect of AsianAm =
   -1.097 + 0.053*Educ

Effect of Hispanic =
   -1.085 + 0.067*Educ
   …
```

```
Sign Change Analysis of Effect of RaceEth
  on g(currocc[upperwc:lowerbc]), Moderated by Education (MV)
--------------------------------------------------------------------------------
               |           RaceEth
When           | -------------------------------------------------------------
Education=     |       Black            AsianAm             Hispanic
    …
--------------+-----------------------------------------------------------------
Sign Changes  |       Never        when MV= 17.72514    when MV= 7.770368
--------------+-----------------------------------------------------------------
% Positive    |        0.0              9.9                 2.7
--------------------------------------------------------------------------------


Sign Change Analysis of Effect of RaceEth
  on g(currocc[lowerwc:lowerbc]), Moderated by Education (MV)
--------------------------------------------------------------------------------
               |           RaceEth
When           | -------------------------------------------------------------
Education=     |       Black            AsianAm             Hispanic
    …
--------------+-----------------------------------------------------------------
Sign Changes  |       Never            Never               Never
--------------+-----------------------------------------------------------------
% Positive    |        0.0              0.0                 0.0
--------------------------------------------------------------------------------


Sign Change Analysis of Effect of RaceEth
  on g(currocc[upperbc:lowerbc]), Moderated by Education (MV)
--------------------------------------------------------------------------------
               |           RaceEth
When           | -------------------------------------------------------------
Education=     |       Black            AsianAm             Hispanic

    …
--------------+-----------------------------------------------------------------
Sign Changes  |       Never            Never           when MV= 16.12012
--------------+-----------------------------------------------------------------
% Positive    |        0.0              0.0                 13.5
--------------------------------------------------------------------------------
```

    …

. **est restore intrace2**

    …

. **intspec** …

    …

. **gfi**

```
GFI Information from Interaction Specification of
Effect of Raceeth on g(Currocc[upperwc:lowerwc]) from Multinomial Logistic Regression
------------------------------------------------------------------

Effect of Black =
   0.896 - 0.072*Educ

Effect of AsianAm =
   -1.630 + 0.100*Educ

Effect of Hispanic =
   0.770 - 0.080*Educ
…
Sign Change Analysis of Effect of RaceEth
  on g(currocc[upperwc:lowerwc]), Moderated by Education (MV)
--------------------------------------------------------------------------------
               |           RaceEth
When           | -------------------------------------------------------------
Education=     |       Black            AsianAm             Hispanic
    …
--------------+-----------------------------------------------------------------
Sign Changes  | when MV= 12.42975   when MV= 16.25427   when MV= 9.577981
--------------+-----------------------------------------------------------------
% Positive    |       47.2             13.5                 6.3
--------------------------------------------------------------------------------
```

Usually, the best way to organize the interpretation of a multicategory focal variable in an MNLR analysis is by discussing the difference in the outcome between a given category of the focal variable and the focal variable's reference category. Because there are multiple contrasts among the outcome categories, I discuss the effect of each focal variable contrast in turn on the four outcome contrasts. (*Hint:* To make it easier to see the patterns across the focal variable contrast and for the outcome category contrasts, create a scratch table of the GFI expressions for your own use. For instance, the rows are the race category contrasts, the columns are the outcome category contrasts, and the cells contain the GFI expressions corresponding to the race category and outcome category contrasts.)

Starting with the effect of Black relative to White (reference category), the GFI expression for this moderated effect has the same form for all three contrasts of the other classes with lower blue collar: a negative baseline effect when *Education* = 0, which moves in the positive direction as education level increases. But the effect never changes sign (see the sign change analysis summary rows). Thus, Blacks have a lower chance than Whites of being in any of the occupation classes relative to lower blue collar, but that disparity lessens as education increases. Turning to the within-white-collar contrast (upper to lower), Blacks have a higher, but diminishing, chance compared with Whites of employment in the upper than in the lower sector jobs for education up to 12 years (47% of the sample) and then Blacks have an increasingly lower chance than Whites (53% of the sample).

This same pattern of effects on occupational class, with some minor differences, characterizes the contrast of Asian relative to White. Its effect on the upper-white-collar to lower-blue-collar and on the upper-white-collar to lower-white-collar contrasts turns positive at high levels of education (18 and 17 years, respectively). The other two occupational contrasts—lower white collar to lower blue collar and upper blue collar to lower blue collar—exhibit the same pattern, as do the effects for Black—that is, a negative effect (lower chance of employment than for Whites) that declines in magnitude with rising education.

The Hispanic effects diverge somewhat more from the patterns for the other two race group contrasts. The effect of being Hispanic on the contrasts of upper white collar to lower blue collar and of upper white collar to lower white collar are positive rather than negative at low levels of education (<8 and <10 years, respectively). At higher education levels, the Hispanic effect indicates a lower chance of employment in upper-white-collar relative to either lower-blue-collar jobs or lower-white-collar jobs, and the gap increases with education level. For lower-white-collar relative to lower-blue-collar jobs, the Hispanic effect is negative and increases in size as education levels rise.

Two broader themes emerge from this discussion of the GFI results. First, the effects of the three race categories relative to Whites on the three occupation classes contrasted to lower blue collar are negative for almost all values of education. This indicates that Blacks, Asians, and Hispanics are fairly uniformly more likely than Whites to be employed in lower-blue-collar jobs than in any other category of jobs. Second, looking at the contrasts within occupational class sector—upper to lower blue collar and upper to lower white collar—we can see that non-Whites are generally less likely to be in the upper jobs in each sector. This is almost universal in the blue-collar sector, but there is more deviation from this in the white-collar sector. This is the case especially for Blacks, who are close to a 50–50 split between positive and negative effects given the sample distribution of education values. For Hispanics, only 6% of effects would be positive, and for Asians, only 13%.

Factor Change (Odds Ratio) Interpretation

The discussion of odds ratios builds on the description from the GFI analysis of the effects of race on occupational class and the patterns of moderation of that relationship by education. It is one way to convey a sense of the magnitude of the race effects and how much they vary with education. I use the *sigreg* command to generate tables of odds ratios, running it first on the MNLR estimates with lower blue collar as the base category (*intrace*) and then again on the estimates with lower white collar as the base category (*intrace2*). Note the use of the eqname("upperwc") option added to the end of the second *intspec* command. This limits the calculations and results produced by the second *sigreg* command to a single contrast of an outcome category (*upperwc*) with the reference category (*lowerwc*). The name specified in eqname( ) must be enclosed in double quotes. If your outcome variable has value labels, the name is the value label. Otherwise, it is the numeric value of the outcome category.

```
. est restore intrace

  ...

. intspec ...

  ...

. sigreg, effect(factor) ndig(3)   save(Output\Table_9_8.xlsx tab)
```

Boundary Values for Significance of Effect of RaceEth on **g(currocc[upperwc:lowerbc])** Moderated by Education
Critical value Chi_sq = **3.841** set with p = 0.0500

Effect of RaceEth	When Education >=	Sig Changes	When Education >=	Sig Changes
Black	17.616	to Not Sig [-2.353]	-2.504 (< min)	to Sig [0.210]
AsianAm	14.242	to Not Sig [-1.592]	10.338	to Sig [0.354]
Hispanic	34.961 (> max)	to Not Sig [-0.064]	11.374	to Sig [3.660]

Note: Derivatives of Boundary Values in [ ]

Boundary Values for Significance of Effect of RaceEth on **g(currocc[lowerwc:lowerbc])** Moderated by Education
Critical value Chi_sq = **3.841** set with p = 0.0500

Effect of RaceEth	When Education >=	Sig Changes	When Education >=	Sig Changes
Black	15.841	to Not Sig [-4.464]	38.498 (> max)	to Sig [0.091]
AsianAm	NA	Never	NA	Never
Hispanic	NA	Never	NA	Never

Note: Derivatives of Boundary Values in [ ]

Boundary Values for Significance of Effect of RaceEth on **g(currocc[upperbc:lowerbc])** Moderated by Education
Critical value Chi_sq = **3.841** set with p = 0.0500

Effect of RaceEth	When Education >=	Sig Changes	When Education >=	Sig Changes
Black	15.032	to Not Sig [-4.091]	76.944 (> max)	to Sig [0.012]
AsianAm	NA	Never	NA	Never
Hispanic	12.432	to Not Sig [-2.957]	-830.149 (< min)	to Sig [0.000]

Note: Derivatives of Boundary Values in [ ]

```
. est restore intrace2
. intspec    ...     eqname("upperwc")

  ...

. sigreg, effect(factor) ndig(3)   save(Output\Table_9_8_2.xlsx tab)
```

Table 9.8 presents the significance region tables for the factor change (odds ratio) effects for the four contrasts among the outcome categories. These are the Excel

TABLE 9.8 ● FACTOR CHANGE EFFECTS OF BLACKS, ASIAN AMERICANS, AND HISPANICS COMPARED WITH WHITES ON OCCUPATIONAL CLASS ODDS, MODERATED BY EDUCATION

	Education						
Effect of	**0**	**3**	**6**	**9**	**12**	**15**	**18**
Currocc[Upperwc:Lowerbc]							
Black	*0.208**	*0.249**	*0.298**	*0.357**	*0.427**	*0.511**	*0.612*
AsianAm	*0.103*	*0.152*	*0.223*	*0.327*	*0.480**	*0.705*	1.036
Hispanic	2.011	1.536	1.173	*0.895*	*0.684**	*0.522**	*0.398**
Currocc[Lowerwc:Lowerbc]							
Black	*0.085**	*0.126**	*0.187**	*0.279**	*0.414**	*0.615**	*0.915*
AsianAm	*0.527*	*0.573*	*0.623*	*0.677*	*0.736*	*0.800*	*0.870*
Hispanic	*0.931*	*0.905*	*0.879*	*0.855*	*0.831*	*0.807*	*0.784*
Currocc[Upperbc:Lowerbc]							
Black	*0.114**	*0.163**	*0.233**	*0.333**	*0.475**	*0.679**	*0.970*
AsianAm	*0.334*	*0.391*	*0.458*	*0.536*	*0.628*	*0.735*	*0.860*
Hispanic	*0.338**	*0.413**	*0.506**	*0.619**	*0.758**	*0.927*	1.135
Currocc[Upperwc:Lowerwc]							
Black	2.451	1.974	1.590	1.281	1.031	*0.831*	*0.669*
AsianAm	*0.196*	*0.265*	*0.358*	*0.483*	*0.653*	*0.882*	1.191
Hispanic	2.160	1.697	1.333	1.048	*0.823*	*0.647**	*0.508**
Key							
Plain font, no fill	Pos, Not Sig						
Bold*, filled	Pos, Sig						
Italic, no fill	Neg, Not Sig						
***Bold italic*, filled**	Neg, Sig						

formatted tables saved by ICALC, which I lightly edited to combine. In discussing the odds ratios, I draw on the boundary value analysis results in the output. Notice that it is easy to visually distinguish the negative effects (odds ratios <1) from the positive effects (odds ratios >1) because the odds ratios with values less than 1 are italicized. Looking at the top three panels contrasting the other occupational classes with lower blue collar, it is readily apparent that the three race effects are predominantly negative effects on the odds of each occupational class relative to lower blue collar. That is, the predicted odds of employment in, say, upper to lower blue collar are almost all smaller for each race group than for Whites (odds ratio < 1).

How the effects of Black and Asian American are moderated by education follows the same pattern in these three occupational contrasts. As education increases, the magnitude of the negative effect declines; that is, the odds ratios move along the 0 to 1 continuum in the direction of 1 with rising levels of education. That said, the difference in the odds of employment between Asian Americans and Whites is almost never significant, except for the upper-white-collar to lower-blue-collar contrast and there only for education between 11 and 14 years. Conversely, the difference in the odds of employment between Blacks and Whites is almost always significant, except when education is at or near its highest levels (varying between 16 and 18 years for the three occupational contrasts).

The smallest significant differences in the odds for Blacks versus Whites still represent notable disparities. At 15 years of education, the odds of Blacks being employed in upper-white-collar versus lower-blue-collar jobs are half the size of the odds for Whites (0.511), while for lower white collar or upper blue collar versus lower blue collar, the odds for Blacks are about two thirds of the odds for Whites (0.615 and 0.679, respectively). At low levels of education, the differential is much more substantial. For example, at 6 years of education, the odds for Blacks are about one fifth to one third the size of the odds for Whites. In sum, Blacks are much less likely than Whites to be predicted to be employed in any occupational class other than lower blue collar.

The Hispanic effect on the upper- to lower-blue-collar contrast exhibits the same pattern, a large disparity in the Hispanic–White predicted odds of employment, which lessens as education increases and turns nonsignificant for *Education* > 13 years. More specifically, the Hispanic odds of upper- versus lower-blue-collar employment are about one third the size of the odds for Whites at 0 years of education and diminish in size to three quarters the size of the odds for Whites when *Education* = 12. However, the Hispanic effect shows the opposite pattern for the contrasts of the white-collar classes with lower blue-collar classes. The odds ratios are close to or greater than 1 when *Education* = 0 and become increasingly small fractions (larger negative effects) at higher levels of education. But the Hispanic–White difference is never significant for the lower-white-collar to lower-blue-collar contrast and is only significant for the upper-white-collar to lower-blue-collar contrast when *Education* > 11.

The race effects on the upper- to lower-white-collar contrast exhibit the opposite pattern of moderation by education to that for the upper- to lower-blue-collar contrast. The Black and Hispanic effects change from fairly substantial advantages in the odds of employment (more than twice as large as for Whites) to roughly equal-sized odds ratios in the opposite direction (half to one third the odds for Whites). [*Aside:* Remember that a "positive" odds ratio of 2 is of the same magnitude as a "negative" odds ratio of 0.5.] As education increases from 0 to 18 years, the Asian American effect diminishes from an odds of one fifth the size for Whites, then turns positive (Asian American odds greater than odds for Whites) for *Education* > 13. Despite the fact that many of these odds ratios are substantial in magnitude, the significant effects are limited to the Hispanic–White differences when *Education* > 12.

What is the bigger picture suggested by these effects? First, and consistent with past research on race and socioeconomic inequality (Sakamoto, Goyette, & Kim, 2009), there are very limited differences between Asian Americans and Whites in the predicted odds of employment. Conversely, the Black–White differences in the odds of employment in lower-blue-collar versus the other occupational classes are substantial in size, indicating a predicted concentration of Blacks relative to Whites in lower-blue-collar jobs. These disparities decline in magnitude as education levels increase and are statistically significant, except at the very highest levels of education. The Hispanic effect has a more limited scope of significance, confined to the upper-white-collar to lower-blue-collar contrast for *Education* > 11 years and to the upper- to lower-blue-collar contrast when *Education* < 14 years. Like the effect of Black, the Hispanic effects suggest a greater representation of Hispanics than Whites in lower-blue-collar versus other occupational classes. Finally, and somewhat interestingly, there are limited race effects (at least in terms of statistical significance) on the upper- to lower-white-collar contrast.

Wrap-Up. Odds ratios can be an effective interpretive tool, as evidenced (I hope) by the big-picture summary provided just above. But writing and reading the description of odds ratio results often are somewhat tedious given the need to identify the contrasting categories. It is thus difficult to avoid awkward phrasing and repetitive language using variations of "versus the base category," "relative to the base category," "in contrast to the base category," and so on. Because the focal variable was also categorical and often required referring to the reference category for race, this added to the awkwardness and repetitiveness in the descriptions of the meaning of the odds ratios.

A limitation of using odds ratios for interpretation is that they are not always informative about the importance of the changes in the moderated effects, only about relative differences. Consider the odds ratio of 0.506 for the Hispanic effect on the odds of upper blue collar versus lower blue collar at *Education* = 9 and the odds ratio of 0.758 at *Education* = 12. This means that the effect at *Education* = 9 is two thirds the size of the effect at *Education* = 12. But the meaning of "two thirds the size" could be quite different depending on how large the corresponding probabilities of the upper-blue-collar category are.

Consider the two cases in Table 9.9. Both cases represent odds ratios of 0.5 at *Education* = 9 and 0.75 at *Education* = 12. In Case 1, the change in the percentage (probability) of upper blue collar for Hispanics is 4%, while in Case 2, the change is 0.5%. Given that the base category percentage is 20, the Case 2 difference seems unimportant in this context. But in other substantive contexts, such as severe medical risks, it could be meaningfully large. An additional complication is that there is no one-to-one correspondence between the odds ratio value for a contrast and the

TABLE 9.9 ● ODDS RATIOS AND CHANGES IN PROBABILITY

	Case 1		Case 2	
	Ed = 9	*Ed* = 12	*Ed* = 9	*Ed* = 12
Pr(UpBC|Hispanic)	.08	.12	.01	.015
Pr(LoBC|Hispanic)	.20	.20	.20	.20
Pr(UpBC|White)	.16	.16	.02	.02
Pr(LoBC|White)	.20	.20	.20	.20
Odds ratio	.5 = (.08/.20)/ (.16/.20)	.75 = (.12/.20)/ (.16/.20)	.5 = (.01/.20)/ (.02/.20)	.75 = (.015/.20)/ (.02/.20)
Change in Pr(UpperBC|Hispanic)		.04		.005

probabilities of the two contrasted outcome categories in MNLR results, unlike in binomial logistic regression results (see the boxed *Aside*).

ASIDE: ON THE CORRESPONDENCE OF ODDS AND PROBABILITIES

To show that there is a one-to-one correspondence between the odds of two categories and their probabilities in binomial logistic regression, let p_A = Probability(A) and let p_B = Probability(B), then the odds comparing p_A with p_B is

$$\Omega = \frac{p_A}{p_B}$$

In binomial logistic regression, the probability of the two categories must sum to 1, so

$$p_B = 1 - p_A$$

Substituting this into the expression for the odds of A to B yields a formula for Ω uniquely defined in terms of p_A

$$\Omega = \frac{p_A}{1 - p_A}$$

With some algebra, we can equivalently write p_A and then p_B uniquely in terms of Ω:

$$p_A = \frac{\Omega}{(1+\Omega)} \text{ and } p_B = 1 - p_A = 1 - \frac{\Omega}{(1+\Omega)}$$

But in MNLR, the probabilities of all the categories sum to 1, so the constraint on p_B is

$$p_B = 1 - p_A - \sum_{c \neq A \text{ or } B} p_c$$

Substituting this into the expression for the odds of A to B produces

$$\Omega = \frac{p_A}{1 - p_A - \sum_{c \neq A \text{ or } B} p_c}$$

So there are infinitely many values of the odds of A to B for a given value of p_A corresponding to the possible probabilities of the other categories. Conversely, infinitely many values of p_A can result from a given odds.

Predicted Standardized Latent Outcome Interpretation

Using tables or plots of the predicted values of the standardized latent variables that underlie the occupational class contrasts has two advantages. First, the standardized latent variable for each outcome category contrast is a linear function of the outcome category's MNLR coefficients and the predictors. These visual displays are not confounded by a nonlinear link function to the model coefficients, unlike the predicted probabilities for the outcome categories. Second, the scaling of the latent variable in standard deviation units provides a metric for understanding the magnitude of the changes in predicted values and hence the extent to which the moderator influences the focal variable's effect on the standardized outcome. Because one of the interacting predictors is categorical and the other is interval, there is a choice between two outcome displays:

1. Treat the categorical predictor as the focal variable to create a set of bar charts for each contrast of outcome categories. Each bar chart would report an outcome category's predicted latent value for the categories of the focal variable calculated for the given display value of the moderator. A separate bar chart is constructed for each display value of the moderator, and additional sets are produced for the other outcome category contrasts.

2. To produce scatterplots for each contrast of outcome categories, use the interval predictor as the focal variable, which will define the *x*-axis of the plot. The *y*-axis corresponds to the predicted latent values for the outcome category contrast. Separate prediction lines—latent outcome by focal variable—are plotted for each category of the categorical nominal variable. The vertical height between pairs of lines shows the effect of being in the corresponding pair of categories of the moderator.

When the focal variable has more than two categories, I often find scatterplots easier to interpret and discuss than bar charts, which was true when I looked at both for this example. So I recommend always trying both to see what works best for you.

Consequently, I used the *outdisp* command to generate scatterplots of the four latent outcome variables, with a separate prediction line of the latent variable plotted against education for each race group. The latent outcomes correspond to the four occupational class contrasts discussed in the prior section on factor change effects: contrasts of the three other occupational classes to lower blue collar, plus the upper-white-collar to lower-white-collar contrast.

The vertical height between the lines for pairs of race groups reveals the effects of race, and the changing distance between the lines as education increases shows how education moderates the effect of race. As I noted above, to request a scatterplot, I define the interval predictor (education) as the focal variable and the categorical predictor (race) as the moderator in the *intspec* command. And I specify that the predicted value plotted on the *y*-axis is the standardized latent outcome value by adding the suboptions metric(model) sdy inside the out( ) option. As before, after creating the first three plots, I switch to the MNLR estimates with lower white collar as the outcome reference category and repeat the *intspec* and *outdisp* commands for the fourth plot. I report these plots in Figure 9.6.

```
. est restore intrace
    ...
. intspec focal(c.educ) main((c.educ, name(Education) range(0(3)18)) ///
>       (i.raceth, name(RaceEth) range(1/4)))    int2(c.educ#i.raceth) ndig(0)
    ...
. outdisp , plot(name(RaceEdLatent)) out(metric(model) sdy atopt((means)_all) ///
>   pltopts(ylab(-4.2(1.4)1.4) ytit( , ma(r +3)) tit( , size(*.7)))
    ...
. est restore intrace2
    ...
. intspec focal(c.educ) main((c.educ, name(Education) range(0(3)18)) ///
>       (i.raceth, name(RaceEth) range(1/4))) int2(c.educ#i.raceth) ///
>       ndig(0) eqn("upperwc")
    ...
. outdisp , plot(name(RaceEdLatent2)) out(metric(model) sdy atopt((means)_all) ///
>   pltopts(ylab(-4.2(1.4)1.4) ytit( , ma(r +3)) tit( , size(*.7)))
    ...
```

FIGURE 9.6 ● PREDICTED STANDARDIZED LATENT OUTCOMES BY THE INTERACTION OF EDUCATION AND RACE

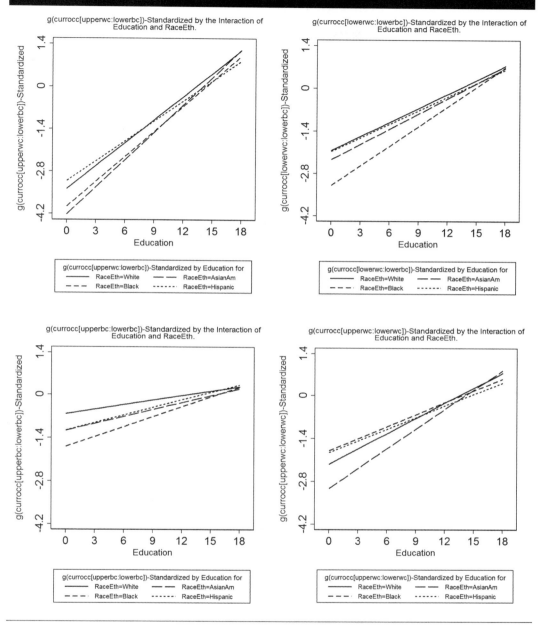

My (and I assume everyone else's) first visual impressions when looking at these scatterplots is that the prediction lines slope upward and the vertical distances between the lines are greatest at the left side of the plots and narrow considerably at the right side of the plots. That is, education has a positive effect on the occupational class contrasts, and the differences between the race groups are greatest when *Education* = 0 and virtually disappear at the highest levels of education. The exceptions are the increasing Hispanic–White differences in the latent propensity of upper white collar versus both lower blue collar and lower white collar at the top of the education distribution. In fact, looking back at the significance region results in Table 9.8, you see precisely that pattern in terms of the significance of the race differences. This comparison is informative because the statistical significance of the factor change effects in the table and the significance of the effects on the latent standardized outcome are identical; both are determined by the statistical significance of the corresponding effects on the log odds of the contrast.[10]

How big are the disparities between Whites and the Other Race groups? For Asian Americans, there are virtually no significant differences with Whites (see Table 9.8). Consequently, I would recommend noting that fact and not discussing the magnitude of their differences from Whites in the plots. Comparing Blacks (*medium dashed lines*) and Whites (*solid lines*), the scatterplots show a clear visual spread between the lines for the contrasts of each occupational class to lower blue collar. The disparity is greatest at *Education* = 0 for all these latent employment contrast propensities. And I would describe it as substantial at about 1 standard deviation lower than Whites' propensities. This diminishes to roughly −1/3 of a standard deviation gap at the median value of education. In the context of the effects of the other (significant) predictors in the model, I would argue that a 1/3 standard deviation difference is meaningfully large. It is 1.5 to 3 times larger than the differences between the marital status categories and the same size as the change in the propensities across the entire range of age values.

Turning to the Hispanic–White differences in employment propensities (*small dashed line* vs. *solid line*), these gaps are visually much smaller in the scatterplots. In particular, for the lower-white-collar versus lower-blue-collar propensity, the prediction lines for Hispanics and Whites nearly overlap, and none of the differences are significant. For the contrasts of upper white collar to both lower white collar and lower blue collar, the Hispanic employment propensities change from larger than those for Whites (positive gap) to smaller than those for Whites (negative gap) as education increases, but only the negative gaps are statistically significant. For the upper- to lower-blue-collar employment contrast, the Hispanic employment propensity is predominantly smaller than that for Whites. But the gap decreases with rising education, eventually turning positive for *Education* > 16. The significant Hispanic–White differences range in size from about 1/7 of a standard deviation at their smallest—around the median level of education—to between 1/3 and 1/2 of a standard deviation at the extrema of education.

In summary, the largest significant Hispanic–White differences in employment propensities are similar in size to the smallest significant Black–White differences in employment propensities. Thus, both visually and numerically, the scatterplots echo the common finding that socioeconomic disparities between Blacks and Whites are greater than those between Hispanics and Whites. Moreover, the predicted employment propensities for Blacks in the three other class categories relative to lower blue collar indicate a concentration of Blacks in lower-blue-collar jobs. The employment propensities for Hispanics also suggest that they are concentrated, in both lower-white-collar and lower-blue-collar jobs, albeit to a lesser degree. Although the scatterplots visually suggest some apparently substantial differences in employment propensity between Asian Americans and Whites (cf. the *large dashed* vs. *solid lines* in the two upper-white-collar employment propensity contrasts), the significance region analyses indicate that almost none of these are significant. As noted before, the findings are consistent with past research, which found the same ordering of groups in the racial hierarchy of socioeconomic outcomes.

Wrap-Up. Interpreting results using the standardized latent outcome has many of the same advantages and disadvantages as a factor change interpretation. It provides a clear understanding of the form of the moderated effect of the focal variable. And there is no confounding of the patterning of the interaction effects with the GLM technique–induced nonlinearity. Additionally, the ability to discuss differences in standard deviation units provides a better sense of the magnitude of the effect and its moderation than a factor change analysis.

On the other hand, there is the same repetitiveness and awkward language in describing the results, created by the fact that each latent outcome represents a contrast between categories. Moreover, there is sometimes a disjuncture between the conceptual model of discrete choice based on utilities, which provides the rationale for a latent outcome and the outcome measure in the analysis. This is partly true for the analysis of occupation classes because workers make employment choices based on the constraint of what jobs they are offered. In this situation, it is unclear what would be meant by a choice based on the relative utility of different occupational classes. For this reason, I avoided using the language of utility and instead used the language of latent employment propensity. A reasonable alternative would be to discuss the results in terms of standardized log odds.

The Effect of Education Moderated by Race/Ethnicity and by Sex

To interpret the moderated effect of education, I begin as usual with a GFI analysis. I then describe how to use factor change effects and predicted probabilities to understand the varying effect of education. I focus the discussion of factor change effects on the effects calculated by treating education moderated by race separately from education moderated by sex. These show the pattern of moderation most clearly. But I also present factor change effects calculated by combining how race and how sex moderate education. This is useful to describe the education effect for each race–sex group. For the predicted probabilities, I use the paired interaction model–main effects model predictions. I do not discuss discrete change effects because, as the predicted probability plots clearly indicate, not only the value but also the sign of the discrete change effects vary across levels of education. Consequently, no single set of reference values can capture this, creating the potential for a very misleading understanding of the interaction effects.

INTSPEC **Setup and** GFI **Analysis**

```
. est restore inted
  ...
. intspec focal(c.educ) main( (c.educ, name(Education) range(0(3)18)) ///
>       (i.raceth, name(RaceEth) range(1/4)) (i.sex , name(Sex) range(0/1))) ///
>       int2(c.educ#i.raceth c.educ#i.sex) ndig(0)
  ...
. gfi, ndig(3)
```

```
GFI Information from Interaction Specification of
Effect of Education on g(Currocc[upperwc:lowerbc]) from Multinomial Logistic Regression
-----------------------------------------------------------------------


Effect of Educ =
    0.659 + 0.060*Black + 0.128*AsianAm - 0.090*Hispanic + 0.055*Women



GFI Information from Interaction Specification of
Effect of Education on g(Currocc[lowerwc:lowerbc]) from Multinomial Logistic Regression
-----------------------------------------------------------------------


Effect of Educ =
    0.350 + 0.132*Black + 0.028*AsianAm - 0.010*Hispanic - 0.066*Women



GFI Information from Interaction Specification of
Effect of Education on g(Currocc[upperbc:lowerbc]) from Multinomial Logistic Regression
-----------------------------------------------------------------------


Effect of Educ =
    0.045 + 0.119*Black + 0.053*AsianAm + 0.067*Hispanic + 0.187*Women
```

. **est restore inted2**

   ...

. **intspec**   ...   **eqn("upperwc")**

   ...

. **gfi, ndig(3)**

```
GFI Information from Interaction Specification of
Effect of Education on g(Currocc[upperwc:lowerwc]) from Multinomial Logistic Regression
-----------------------------------------------------------------------


Effect of Educ =
    0.309 - 0.072*Black + 0.100*AsianAm - 0.080*Hispanic + 0.122*Women
```

I do not report the sign change analysis results because the moderated effect of education on the four employment category contrasts is positive for all the combinations of race and sex. The predominance of positive moderating coefficients (11 of 16) indicates that education generally has a larger positive (or smaller negative) effect on the occupational contrasts for race groups relative to Whites and for women relative to men.

Factor Change Interpretation

I first use the *sigreg* command to calculate the effect of education moderated separately by race and by sex and then simultaneously by race and by sex. Note that the *intspec* command for the separate moderation by race specifies only the education-by-race interaction, which is similarly done for the education-by-sex interaction. Table 9.10 reports the factor change effects (and their significance) for the moderation of education separately by race and sex, and Table 9.11 presents the results for the simultaneous moderation by race and sex. I constructed these tables by combining the formatted significance region tables saved into Excel files. You could instead copy and paste the tables displayed in the Stata Results window (not shown).

**** BY RACE & SEX
. **est restore inted**

   ...

. **intspec** ...

   ...

```
. sigreg, effect(factor) nobva ndig(3) save(Output\Table_9_11.xlsx tab)
  …

. est restore inted2
  …

. intspec   …       eqn("upperwc")
  …

. sigreg, effect(factor) nobva ndig(3) save(Output\Table_9_11_2.xlsx tab)
  …

**** BY RACE
. est restore inted
  …

. intspec focal(c.educ) main( (c.educ, name(Education) range(0(3)18)) ///
>   (i.raceth, name(RaceEth) range(1/4))) ///
>   int2( c.educ#i.raceth ) ndig(0)
  …

. sigreg, effect(factor) nobva ndig(3) save(Output\Table_9_10_1.xlsx tab)      …
. est restore inted2
  …

. intspec   …       eqn("upperwc")
  …

. sigreg, effect(factor) nobva ndig(3) save(Output\Table_9_10_2.xlsx tab)
  …

**** BY SEX
. est restore inted
  …

. intspec focal(c.educ) main( (c.educ, name(Education) range(0(3)18)) ///
>   (i.sex , name(Sex) range(0/1))) ///
>   int2( c.educ#i.sex) ndig(0)
  …

.sigreg, effect(factor) nobva ndig(3) save(Output\Table_9_10_3.xlsx tab)
  …

. est restore inted2
  …

. intspec       …           eqn("upperwc")
  …

. sigreg, effect(factor) nobva ndig(3) save(Output\Table_9_10_4.xlsx tab)
```

All of the education effects on the four reported occupational class contrasts are positive, and all but one (for Asian Americans comparing upper with lower blue collar) are statistically significant. This means that the predicted odds of employment in the three occupation classes compared with lower blue collar increase with education for every race–sex group. And similarly, the employment odds for upper white collar compared with lower white collar rise with education for each group. But the extent to which these predicted employment odds increase with education varies by race and by sex.

The right-hand vertical panel in Table 9.10 reports how sex moderates the effect of education on the employment odds of the occupational class contrasts. In the comparisons of the upper with the lower sector of each color-collar type, the effect of education on the employment odds is significantly larger for women than for men.[11] Education increases the odds of upper- to lower-white-collar employment by a factor of about 1½ for women and 1⅓ for men for a 1-year change in education. In the blue-collar sector comparison, the increases are by a factor of 1.26 for women and 1.05 for men.

Education has essentially the same effect for men and women on the upper-white-collar to lower-blue-collar contrast, doubling the odds of relative employment for both sexes. Last, education increases the employment odds for lower white collar

TABLE 9.10 ● FACTOR CHANGE EFFECTS OF EDUCATION MODERATED SEPARATELY BY RACE AND SEX

	Moderated by Race					Moderated by Sex		
	Raceeth						Sex	
	White	Black	AsianAm	Hispanic			Men	Women
upperwc:lowerbc						upperwc:lowerbc		
Effect of Educ	**1.932***	**2.051***	**2.196***	**1.766***		Effect of Educ	**1.932***	**2.042***
lowerwc:lowerbc						lowerwc:lowerbc		
Effect of Educ	**1.419***	**1.619***	**1.459***	**1.405***		Effect of Educ	**1.419***	**1.328***
upperbc:lowerbc						upperbc:lowerbc		
Effect of Educ	**1.046***	**1.179***	1.103	**1.119***		Effect of Educ	**1.046***	**1.261***
upperwc:lowerwc						upperwc:lowerwc		
Effect of Educ	**1.362***	**1.267***	**1.505***	**1.257***		Effect of Educ	**1.362***	**1.538***
Key						Key		
Plain font, no fill	Pos, Not Sig					Plain font, no fill		Pos, Not Sig
Bold*, filled	Pos, Sig					**Bold*, filled**		Pos, Sig
Italic, no fill	Neg, not Sig					*Italic, no fill*		Neg, Not Sig
Bold italic*, filled	Neg, Sig					***Bold italic*, filled***		Neg, Sig

versus lower blue collar somewhat more for men than for women; the employment odds are 1.4 times higher for men and 1.3 times higher for women. From a bigger-picture standpoint, these factor changes suggest that as education decreases, both men and women have higher odds of employment in the lower-blue-collar class than in the other three classes. But this difference is notably greater for women than for men for the employment odds of lower blue collar versus upper blue collar.

The factor change effects of education moderated by race are presented in the left-hand vertical panel of Table 9.10, and they portray a varied pattern of moderation of the education effect by race group. On all four occupational class contrasts, the education effect is larger for Asian Americans than for Whites, but none of the differences in the education effect between these two groups are significant. For Hispanics compared with Whites, there is a significantly larger effect of education on the predicted employment odds of upper- versus lower-blue-collar jobs. Both are modest in size, with the education effect increasing the odds by a factor of 1.05 for Whites and 1.12 for Hispanics. But none of the other effects for Hispanics—all negative—are significantly different than for Whites.

The effect of education is higher for Blacks than for Whites in the three contrasts to lower blue collar and is smaller for the upper- to lower-white-collar contrast. Only two of these differences are statistically significant: (1) lower white collar to lower blue collar, in which education increases the employment odds by a factor of

1.62 for Blacks and 1.42 for Whites, and (2) upper blue collar to lower blue collar, where education increases the odds by 1.12 for Blacks and by 1.05 for Whites.

One implication of this set of findings is that higher levels of education provide pathways to jobs other than lower blue collar for all race groups. Blacks, Hispanics, and Asian Americans mainly have pathways facilitated by education equivalent to those for Whites, with a limited number of enhanced pathways for Blacks and for Hispanics relative to Whites. On the flip side, this indicates that lower education is associated with greater odds of lower-blue-collar employment relative to all the other classes for all the ethnoracial groups. And in some cases, the concentration in lower blue collar is greater for Blacks and Hispanics than for Whites.

The same patterns of moderation by sex and by race can be gleaned from Table 9.11, which presents the factor changes calculated for all combinations of race–sex groups. But it requires more work to see those patterns when the two-way interaction of

TABLE 9.11 ● FACTOR CHANGE EFFECTS OF EDUCATION MODERATED SIMULTANEOUSLY BY RACE AND SEX

| | Effect of Educ | | | |
| | Raceeth | | | |
Sex	**White**	**Black**	**AsianAm**	**Hispanic**
upperwc:lowerbc				
Men	1.932*	2.051*	2.196*	1.766*
Women	2.042*	2.168*	2.321*	1.866*
lowerwc:lowerbc				
Men	1.419*	1.619*	1.459*	1.405*
Women	1.328*	1.515*	1.365*	1.315*
upperbc:lowerbc				
Men	1.046*	1.179*	1.103	1.119*
Women	1.261*	1.420*	1.329*	1.349*
upperwc:lowerwc				
Men	1.362*	1.267*	1.505*	1.257*
Women	1.538*	1.431*	1.700*	1.419*
Key				
Plain font, no fill	Pos, Not Sig			
Bold*, filled	Pos, Sig			
Italic, no fill	Neg, Not Sig			
***Bold italic*, filled**	Neg, Sig			

education by sex is overlaid with the two-way interaction of education by race. Indeed, you can use the factor changes separately by race and sex in Table 9.10 to calculate the factor changes in Table 9.11. For each employment contrast, the rows of factor change effects by race for men in Table 9.11 are equal to the factor change effects by race in Table 9.10. To get the factor changes for the row for women in Table 9.11, multiply the factor changes by race in Table 9.10 for an occupation contrast by the ratio of women's factor change over men's factor change for the same occupation contrast. For example, the factor change for Hispanic women for upper white collar versus lower blue collar is

$$1.766 \quad \times \quad 2.042 \quad / \quad 1.932 \quad = 1.8665$$

| Hispanic odds ratio | Women's odds ratio | Men's odds ratio |

Consequently, these results are primarily useful for reporting the combined effect of the two two-way interactions with education—factor changes in education for race–sex groups.

Wrap-Up. For the interaction of the focal variable with multiple moderators, this example demonstrates that it is simpler to see and interpret the patterns of moderation by examining each moderator's effect separately. And there is no advantage in the simultaneous examination of the moderating effects because the patterns are identical to those from considering the moderators' influence one at a time. If there had been a three-way interaction, however, we can only do a simultaneous assessment because the patterns are by definition different for the two-way from the three-way calculations.

Interpretation Using Predicted Probabilities

For an interval focal variable, the *outdisp* command produces sets of scatterplots for each outcome category showing the predicted probability of the category plotted against the focal variable (education), with separate prediction lines for each category of the first moderator (race). An individual scatterplot is drawn for every category of the second moderator (sex). Typically, I would run this twice, first without a comparison with the main effects model predictions. The second time, I would specify a comparison of the interactive predictions with the main effects predictions (main( ) suboption), with a separate scatterplot showing each interactive prediction line paired with its corresponding main effects prediction line (single(1) suboption). Such a disaggregated comparison is needed to assess the differences between the interactive and main effects predictions because they are subtle and most of them would be difficult to discern if the plots were not disaggregated.

An alternative to the scatterplot is a stacked area chart, which is sometimes an easier-to-interpret visual display. I illustrate its use for this example. A stacked area chart shows the predicted probabilities for each outcome category plotted against education on the same plot in a specified order, with the probability for each outcome stacked on top of (added to) the probability of the prior outcomes. The area between each plotted line is shaded, and the vertical gap between the lines represents the probability of the top outcome. This plot type is not included in the Stata graph commands and is not currently available directly from ICALC. But you can create a stacked area with the following steps:

1. Using the SPOST13 *mgen* command to generate plotting variables and then concatenate a set of Stata two-way area plots

2. Using ICALC to save the plotting data to an Excel file to create a stacked area chart in Excel or to export to another platform

In the "Special Topics" section at the end of the chapter, I describe how to implement both of these steps to create the kind of stacked area charts shown in Figure 9.7. I use these to interpret the relationship between the occupational class categories and the education-by-race and education-by-sex interactions. I then discuss the paired interactive–main effects prediction plots to assess the extent to which the portrayal of the relationship in the stacked area charts reflects the interactive relationship versus the nonlinearity of the MNLR.

The stacked area charts in Figure 9.7 reveal the overall patterns of the effect of education on occupational class fairly readily, but I find it useful to also look at—but not present—the scatterplots for each outcome category to clarify some of the less visually distinct patterns. There are dramatic and opposite changes in the probability of employment in upper-white-collar (*black fill*) and lower-blue-collar jobs (*medium gray shading*) as education increases. Not surprisingly, the probability of lower-blue-collar employment declines with education, while the probability of upper-white-collar employment increases. There is also a trade-off in the probabilities of the upper-blue-collar (*white shading*) versus lower-white-collar (*light gray*

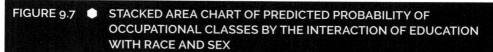

FIGURE 9.7 ● STACKED AREA CHART OF PREDICTED PROBABILITY OF OCCUPATIONAL CLASSES BY THE INTERACTION OF EDUCATION WITH RACE AND SEX

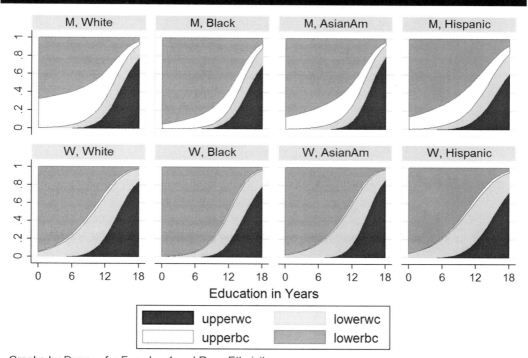

Graphs by Dummy for Female =1 and Race Ethnicity

shading) classes. The probability of upper blue collar declines with rising education as the probability of lower white collar increases. This is most visible for men in the upper panel of the charts.

This trade-off also demonstrates why I decided not to use discrete change effects to interpret the moderated education effect on occupational class. For a given race and sex, the discrete change effect of education is evaluated such that it relies on the central tendency of the sample distribution of education, unless you choose to use a substantively interesting cases approach for setting reference or display values for education (see the boxed "Aside" below). For upper-blue-collar and lower-white-collar predicted probabilities, the slope of the predicted values by education changes from positive to negative across the range of education values. This would be obscured by interpreting the discrete change effects unless you use multiple reference values for education, making the interpretation of an already complicated-to-interpret technique even more complex.

ASIDE

It is not well recognized how the calculation of a discrete change effect relies directly or indirectly on the central tendency of the distribution of a predictor depending on how you set the at( ) option used by the Stata *margins* command or the SPOST13 *mchange* command. If you specify at((means) _all), the discrete change effect for *X* is calculated as the difference in the predicted probabilities as *X* changes from its mean to its mean + 1. You can see this in the standard output produced by the *mchange* command, which lists the at values used in the calculation in the section titled "Base Values of Regressors." If you specify at((asobs) _all), then the discrete change for *X* is the average across the sample of the difference in the predicted probabilities when *X* changes from its observed value to its observed value + 1. By averaging the predicted difference across the sample values of *X*, it is indirectly dependent on the central tendency of *X*. Either case will result in obscuring the fact that the discrete change effect changes sign across the range of *X*'s values.

The charts also portray a number of ways in which the occupational class–education relationship varies by race and by sex. Comparing the charts for men in the top panel with those for women in the bottom panel, we can see the following:

1. The probability of upper-blue-collar employment (*white shading*) is small and hardly changes with education for women. Men have a much higher employment probability, which initially increases and then declines with education.

2. For lower-blue-collar jobs (*medium gray fill*), the probability falls more rapidly with increasing education for women than for men. Consequently, women's initially higher employment probability at lower levels of education changes to a higher probability for men between 9 and 11 years of education. Given the sample distribution of education, men's probability is higher for close to 90% of the sample.

3. The trade-off between upper- and lower-blue-collar employment probabilities (*black* and *medium gray shading*, respectively) is larger for women than for men. That is, upper-white-collar employment rises more

quickly and lower-blue-collar employment falls more rapidly as education increases for women.

A comparison of the charts by column provides insight into the ethnoracial group variation in the class–education relationships. This indicates that Whites and Blacks are the most different, with the other two groups in between but somewhat more similar to Whites:

1. For Whites, the rate of decline in lower-blue-collar employment probability as education increases is notably less than for the other three groups, while for Blacks, the probability changes are the largest.

2. For Blacks, the lower-white-collar probability rises and falls more slowly than for the other three groups. Asian Americans and Whites have similarly faster increases and decreases in employment probability as education levels rise.

3. The upper-blue-collar probability increases more slowly and then declines more quickly as education increases for Whites than it does for the other ethnoracial groups. For Blacks, the probabilities rise more quickly than for the other groups and fall more slowly than for all other groups except Hispanics.

4. The increase in the probability of upper-white-collar employment with rising education is greatest for Whites, similar to that for Asian Americans, and least for Blacks and Hispanics.

How Much Is Education Moderated by Race and Sex? To explore the possible confounding of the moderation of education's effect by the link function's nonlinearity, I again use plots comparing the interaction model's and the additive model's predictions. To best see the differences, I create separate scatterplots for an occupation's paired education prediction lines (interaction and additive model lines) for each combination of race and sex. The syntax for this is shown in the outdisp command below. The name of the previously stored estimates from the main effects model, mainedrace, is listed in the main( ) suboption of the out( ) option, and I include the single(1) suboption in the plot( ) option to disaggregate the plots by the values of the first moderator.

```
. est restore inted
    ...
. intspec focal(c.educ) main( (c.educ, name(Education) range(0(3)18)) ///
>      (i.sex , name(Sex) range(0/1)) (i.raceth, name(RaceEth) range(1/4))) ///
>          int2(c.educ#i.sex c.educ#i.raceth) ndig(0)
    ...
. outdisp, plot(name(EdRaceSex) save(Output\OccEdRaceSex_Phat.xlsx) single(1)) /// >
   out(main(mainedrace) atopt((means) _all)) pltopts(ylab(0(.2)1))
```

Figure 9.8 provides a visual representation of the divergence between the predictions from the interactive effects model and the main effects model. Each of the 32 scatterplots shows a pair of predictions—one from the interactive and one from the noninteractive model results—for each combination of outcome category, race group, and sex group. Each page presents the scatterplots for an occupational class category, with the scatterplots for the four ethnoracial groups in the first row for men

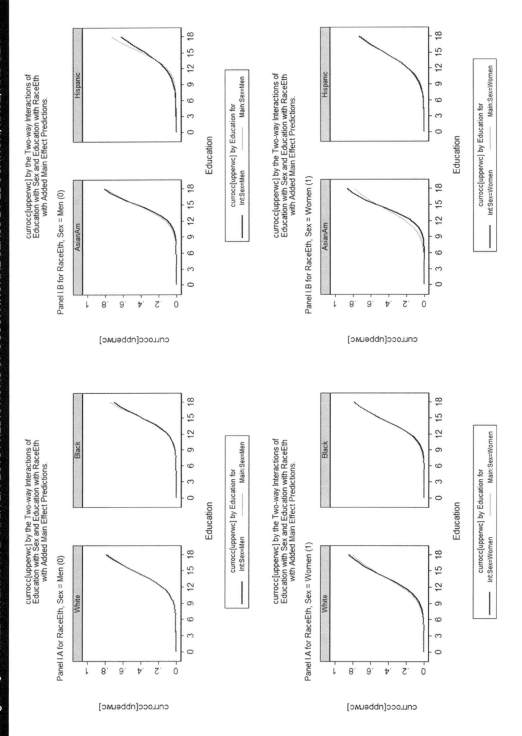

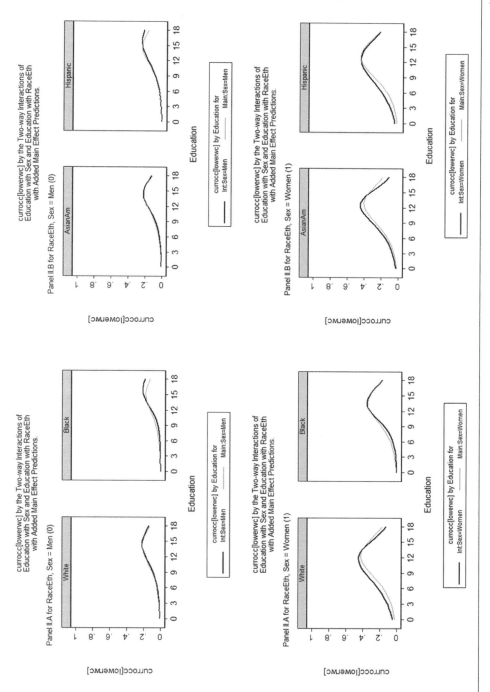

(Continued)

FIGURE 9.8 (CONTINUED)

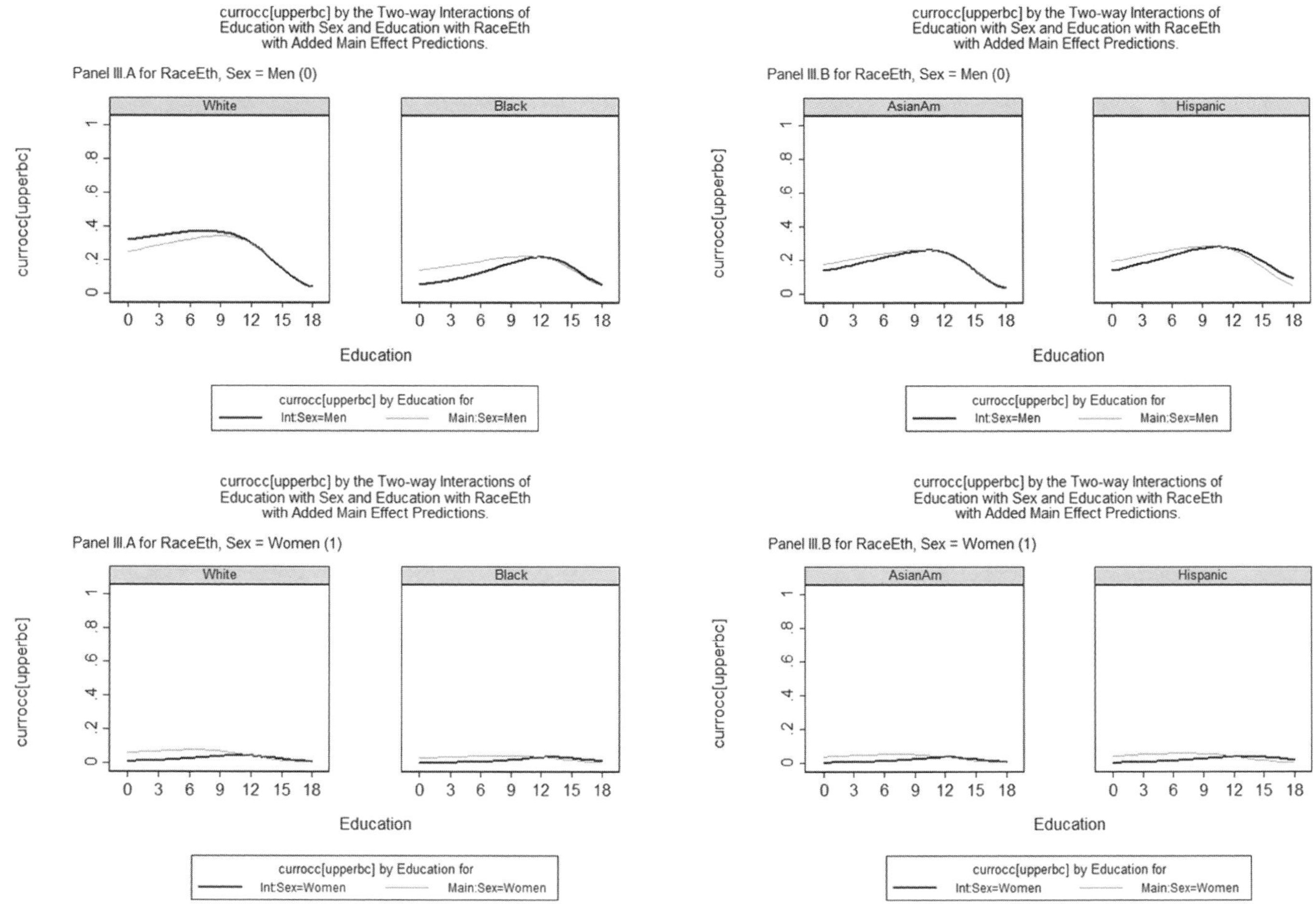
currocc[upperbc] by the Two-way Interactions of Education with Sex and Education with RaceEth with Added Main Effect Predictions.
Panel III.A for RaceEth, Sex = Men (0)
White
Black
currocc[upperbc]
Education
currocc[upperbc] by Education for
Int:Sex=Men
Main:Sex=Men
currocc[upperbc] by the Two-way Interactions of Education with Sex and Education with RaceEth with Added Main Effect Predictions.
Panel III.B for RaceEth, Sex = Men (0)
AsianAm
Hispanic
currocc[upperbc]
Education
currocc[upperbc] by Education for
Int:Sex=Men
Main:Sex=Men
currocc[upperbc] by the Two-way Interactions of Education with Sex and Education with RaceEth with Added Main Effect Predictions.
Panel III.A for RaceEth, Sex = Women (1)
White
Black
currocc[upperbc]
Education
currocc[upperbc] by Education for
Int:Sex=Women
Main:Sex=Women
currocc[upperbc] by the Two-way Interactions of Education with Sex and Education with RaceEth with Added Main Effect Predictions.
Panel III.B for RaceEth, Sex = Women (1)
AsianAm
Hispanic
currocc[upperbc]
Education
currocc[upperbc] by Education for
Int:Sex=Women
Main:Sex=Women

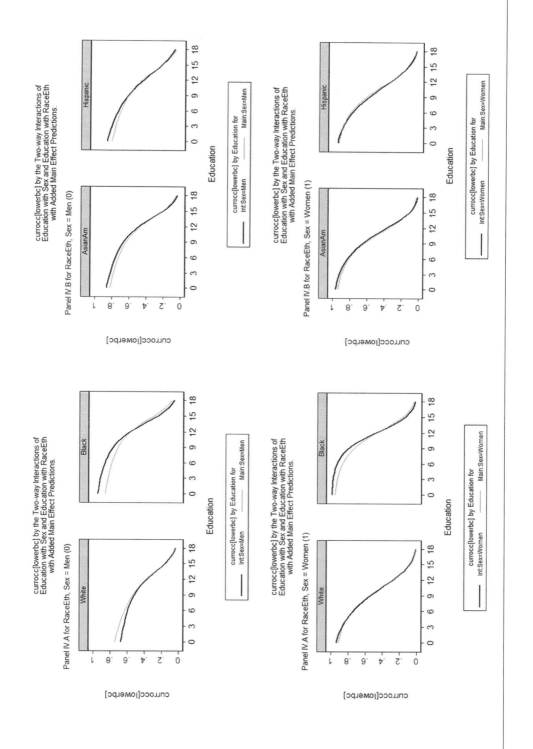

and in the second row for women. Greater visual (and numeric) divergence between the paired predictions indicates a larger degree of moderation. If the pair of lines were to show the same direction of effect, this would suggest that the apparent moderation is somewhat overstated due to confounding with the nonlinearity of the GLM link function. But if the lines were to run in opposite directions, the apparent moderation would be understated.

The clear impression from examining these scatterplots is that visual comparisons of the education effect on occupational class from the stacked area charts are dominated by the MNLR models' nonlinear link function connecting the predicted probabilities and the predictors. That is, the *black lines* for the interactive model predictions are not that different from the *gray lines* for the main effects model predictions. They are barely distinguishable for upper-white-collar jobs, with a little more variation in the other occupational classes across sex or race:

1. For the lower-white-collar category, the interactive education effect for women increases and then decreases a little more quickly than in the main effects model, with the reverse characterizing the difference for men. Similarly, for Whites and Asian Americans, the education effect rises and falls slightly faster than is shown by the main effects predictions, balanced by the opposite for Blacks and Hispanics.

2. The interactive education effect on the upper-blue-collar job probability changes more quickly for men than in the main effects model and correspondingly more slowly for women. The variation by race is between Whites and the other three groups, with the White interactive education effect changing more slowly than in the main effects model.

3. For the lower-blue-collar class, the moderated effect of education declines more rapidly for men than it does in the main effects model and correspondingly changes more slowly for women. The interactive education effect for Blacks shows the greatest divergence from the education main effects, first declining more slowly and then more quickly than it does in the main effects predictions, balanced by small differences for the other ethnoracial groups.

Wrap-Up. In sum, as we just saw for the interpretation of the effect of sex moderated by education and for the effect of race moderated by education, reliance on predicted probabilities to assess the degree of moderation would be misleading in this case. As this discussion of the paired scatterplots reiterated, the changes in the effect of education with race or sex in the interactive model prediction are similar to those in the main effects model prediction. Consequently, a varying and sometimes large portion of the apparent moderation is due to the visual confounding of the interactive effect and the nonlinearity of the MNLR in generating predicted probabilities. On the other hand, the predicted probability portrayal of the relationship between the occupational class outcome and education, race, and sex is much clearer and more easily explained to most audiences than are the factor changes or plots of the standardized latent outcome.

A good approach is to draw on both. Focus the discussion on the predicted probabilities, and refer to the factor change results (or the standardized latent outcome plots) to comment on the degree of moderation. Although the paired interaction–main effects prediction scatterplots also provide information about the degree of moderation, in this example, they are both numerous and more complicated to work with than factor changes or standardized latent outcome plots.

As always, I recommend trying all of the options to determine which one allows you to best tell the story of the relationship in a particular analysis.

SPECIAL TOPICS

Getting the Base Probability for a Discrete Change Effect From SPOST13

The *mchange* command in SPOST13 calculates a discrete change for the effect of X in an MNLR by taking the difference between the predicted probability of an outcome category when $X = x^{start}$ and the predicted probability when $X = x^{end}$ (Long & Freese, 2014, p. 416). The first predicted probability is referred to as the base probability as it defines the base from which the change is calculated. ICALC uses the *mchange* command to calculate discrete changes but does not access or save the base probability information. Thus, you will need to specify and run an appropriate *mchange* command to get this information. The basic syntax is

```
mchange focalvar if e(sample) [weights], contents-of-amtopt ///
    at( contents-of atopt moderator1 = value)
```

where amtopt and atopt are what you specified on the *sigreg* command and *value* is the value of the moderator in the *sigreg* table corresponding to the discrete change effect for which you are trying to find the base probability.

Thus, to find the base probabilities for the three outcome categories for the discrete change effect of education when *Attendance* = 2 (twice a month), you would use

```
mchange educ, am(sd) center at((means) _all attendmonth=2)
```

If you wanted to find the base probabilities for a discrete change effect value not shown in the significance region table, you would need to first determine what value of the moderator produces the discrete change value of interest. For example, to find the base probability of the liberal category for a midrange education effect of 0.05, you need to determine the value of attendance (the moderator) at which the education effect for the liberal category is 0.05. By trial and error (running the *mchange* command for attendance values between 0.9 and 1.0 by 0.01 increments), you can determine that at *Attendance* = 0.93 the education effect is 0.05. You then get the base probability for liberal by running

```
mchange educ , am(sd) center at((means) _all attendmonth=.93)
```

Finding the Standard Deviation and Mean of the Latent Outcomes (Utilities)

For an overview of the discrete choice model and its application to develop the MNLR model, see the discussion in Long (1997, pp. 155–156), which includes references to more detailed presentations. The underlying premise is that each person chooses the category of the outcome that has the highest utility for him or her, where the utility has both a systematic component based on the person's characteristics (predictors) and a random component. Correspondingly, the choice between a specific pair of categories depends on the difference in utilities between those two categories. We can conceptualize the latent outcome $y^*_{m:0}$ for the choice between the given category m and the reference category 0 as that difference in utilities.

To develop the formula ICALC uses to estimate the mean and the standard deviation of the latent outcome, I work from several formulas and assumptions from Long (1997, chap. 6):

1. Let u_{im} be the utility of choosing category m for person i, defined as the mean utility of that choice for that person plus a random error:

$$\mu_{im} = \mu_{im} + \epsilon_{im}, \text{ where } \mu_{im} = \text{mean utility and } \epsilon_{im} = \text{random error}$$

2. The mean utility is a linear combination of a set of $J + 1$ predictors (x_j):

$$\mu_{im} = \sum_{j=0}^{J} b_{mj} x_{ij}, \text{ where for convenience } x_{i0} = 1 \text{ and } b_{m0} \text{ is the intercept}$$

3. Each random error is uncorrelated with the predictors and is independently and identically distributed as a Type I extreme-value distribution with the density function (NIST/SEMATECH, 2017)

$$f(\epsilon) = e^{-\epsilon - e^{-\epsilon}}, \text{ with mean} = 0.5772 \text{ and variance} = \frac{\pi^2}{6}$$

4. The set of parameters b_{mj} are identified by the constraint that the parameters for the reference category $m = 0$ are all set to zero: $b_{0j} = 0$ for all j.

Thus, we can write the utility of category m for person i as

$$u_{im} = \sum_{j=0}^{J} b_{mj} x_{ij} + \epsilon_{im}$$

Then write the latent outcome for the choice between category m and the reference category 0 as the difference in utilities between those categories:

$$y^*_{m:0} = u_{im} - u_{i0} = \sum_{j=0}^{J} (b_{mj} - b_{0j}) x_{ij} + (\epsilon_{im} - \epsilon_{i0})$$

By assumption, $b_{0j} = 0$, so

$$y^*_{m:0} = \sum_{j=0}^{J} b_{mj} x_{ij} + (\epsilon_{im} - \epsilon_{i0})$$

Now the predicted log odds of category m to category 0 is given by

$$\log \widehat{Odds}_{i,m:0} = \sum_{j=0}^{J} b_{mj} x_{ij}$$

So we can rewrite the latent outcome as

$$y^*_{m:0} = \log \widehat{Odds}_{i,m:0} + (\epsilon_{im} - \epsilon_{i0})$$

Then the mean of the latent outcome is

$$\mathcal{E}\left(y^*_{m:0}\right) = \mathcal{E}\left(\log\widehat{Odds}_{i,m:0}\right) + \mathcal{E}\left(\epsilon_{im} - \epsilon_{i0}\right)$$

The expected values of the errors, ϵ_{im} and ϵ_{i0}, are both 0.5772, so the latent outcome's mean is

$$\mathcal{E}\left(y^*_{m:0}\right) = \mathcal{E}\left(\log\widehat{Odds}_{i,m:0}\right)$$

which can be estimated by the sample mean of $\log\widehat{Odds}_{i,m:0}$. Because the covariance of the errors and the predicted log odds is 0,[12] the variance of the latent outcome is then given by

$$Var\left(y^*_{m:0}\right) = Var\left(\log\widehat{Odds}_{i,m:0}\right) + Var\left(\epsilon_{im} - \epsilon_{i0}\right)$$

Furthermore, the errors are uncorrelated, so

$$Var\left(\epsilon_{im} - \epsilon_{i0}\right) = Var\left(\epsilon_{im}\right) + Var\left(\epsilon_{i0}\right) = \frac{\pi^2}{6} + \frac{\pi^2}{6} = \frac{\pi^2}{3}$$

Then,

$$Var\left(y^*_{m:0}\right) = Var\left(\log\widehat{Odds}_{i,m:0}\right) + \frac{\pi^2}{3}$$

Thus, to estimate the standard deviation of the latent outcome $y^*_{m:0}$, ICALC gets the observed variance of the predicted log odds for category m to category 0, adds $\frac{\pi^2}{3}$, and then takes the square root of that result.

Creating a Stacked Area Chart

Option 1: Using *mgen* in Stata

You specify what values of your predictors are used to calculate the predicted value by how you specify the at( ) option in Line 1. I used (means) _all to set the default calculation to a predictor's mean and then listed the range of numbers for each predictor in the interaction that defines a calculation point. stub(occ) defines a prefix that will be used to name the plotting variables. Lines 2 and 3 apply value labels to the nominal plotting predictors. Lines 4 and 5 specify the first probability to be plotted (occpr1) and the focal variable (occeduc) against which to plot the probability, the variables for which to create separate plots (sorted as first variable within second variable), the color with which to fill the area, and the labels for the x- and y-axes. Lines 6 to 8 define the plots to be stacked on and their fill colors. *Note the difference in the name of the y-axis variables compared with Line 4.* The uppercase C in the name indicates that it is the cumulative probability; for example, occCpr2 is the cumulative probability of Category number 2 or less. Line 9 formats the legend for the plot to name each shaded area by its corresponding category. Stata would otherwise use the default variable names created by *mgen* and would name the shaded area as cumulative probability outcomes—for example, lowerwc/upperwc.

1. mgen , at((means) _all educ=(0/18) sex=(0/1) raceth=(1/4)) stub(occ)
2. lab val occsex mf
3. lab val occraceth RE
4. twoway area occpr1 occeduc , by(occsex occraceth, rows(2)) color(black) ///
5. xlab(0(6)18) ylab(0(.2)1) ///
6. || rarea occCpr2 occpr1 occeduc , by(occsex occraceth) color(gs13) ///
7. || rarea occCpr3 occCpr2 occeduc , by(occsex occraceth) fc(white) lc(gs4) lw(vvthin) ///
8. || rarea occCpr4 occCpr3 occeduc , by(occsex occraceth) color(gs9) ///
9. leg(label(1 "upperwc") lab(2 "lowerwc") label(3 "upperbc") lab(4 "lowerbc"))
** use legend label not labelling by varnames. varnames creates label for lowerwc as lowerwc/upperwc etc.

In Excel

When you request that plotting data from *outdisp* be saved into an Excel file, ICALC saves the plotting data into separate worksheets for each outcome category. The worksheets are named *plotdata_categorylabel*. It is easier to create the stacked area chart from the data if you reorganize the data in the following order in a single worksheet:

moderator1   moderator2   focal variable   outcome probabilities

The order in which you list the outcome probabilities determines the order of stacking in the area chart. You may need to do some experimenting to see what order creates the most readable charts. You should list the variable names in the first row of the worksheet, *with the exception that you must leave the cell blank for the focal variable's name*. If you put in the focal variable's name, then Excel will treat the focal variable as a variable to be stacked rather than as the labels for the x-axis. For example, the top rows for Black women (*raceeth* = 2 and *sex* = 1) are as follows:

	AJ	AK	AL	AM	AN	AO	AP
1	raceeth	sex		upperwc	lowerwc	upperbc	lowerbc
2	2	1	0	0.0000317	0.004502	0.001023	0.994443
3	2	1	0.72	0.0000552	0.006061	0.001314	0.99257
4	2	1	1.44	0.0000961	0.008153	0.001688	0.990063

Note the blank cell in row 1 of column AL, where I edited out the variable name for the focal variable, education. There are always 26 rows of data beneath the variable names. The steps to create a stacked area chart for Black women are as follows:

1. Select (highlight) the cells for the focal variable and the outcome probabilities from the top row with the variable names down to the 27th row—AL1 to AP27 in this example.

2. Click *Insert* on the menu bar, click *Area* in the *Charts* box, and then click the icon for *Stacked Area* in the *2-D Area* box.

3. Customize the format to suit your preferences. I set the maximum value on the y-axis from 0 to 1, showed every fourth x-axis label with one decimal

place, changed the area shading to a black–white–gray scheme, and added a chart title identifying the race–sex group.

4. Use Steps 1 and 2 to create the charts for the other combinations of the moderators. You can easily copy the formatting from the first chart you created to the other charts by the following steps:

   a. Click on *Home* on the menu bar.

   b. Click on the first chart, and then click on the *copy* icon in the *Clipboard* section of the *Home* tab.

   c. Click on another chart, then in the *Clipboard* section click on the *Paste* drop-down menu and click on *Paste Special*. In the dialog box that opens, click on *Formats*. This should apply the formatting from the first chart to the second one.

You can also download a do-file named Create_stacked_area_data.do from www.icalcrlk.com to reorganize the data. It creates another tab named *plotdata_combined* in your saved Excel file, with the plotting data reorganized to easily create the stacked area charts. For each combination of categories of the two moderators, there is a separate set of columns with variable names in the first row followed by the data in the next 26 rows for that combination of moderators' values. Edit Lines 5, 6, 9, and 12 to adapt them to your analysis.

CHAPTER 9 NOTES

1. When all of the predictors are categorical, log-linear analysis is sometimes used, especially in older applications. Log linear is closely related to MNLR, and it is possible to specify a log-linear model that is identical to MNLR, both theoretically and empirically (Fox, 2008, p. 400). But the reverse does not hold true; not all log-linear models have an MNLR equivalent.

2. The original GSS ordinal variable was first recoded into an estimated annual frequency based on the values and then divided by 12 to provide a monthly metric.

3. Technically, subjective social class is an ordinal variable, but testing its inclusion as a categorical versus an interval variable suggests that treating it as interval is the preferred option. Thus, I treated it as interval for simplicity.

4. Only one of the six test results was significant for the liberal category on the Small-Hsaio test.

5. At the maximum value of attendance (4.33 times per month), the education effect turns positive.

6. Excluded were certain types of interracially married couple households ($N = 27$), 326 cases missing immigration data, 207 cases who were married but had missing spouse information on key variables, and one case that was identified as likely to have major data errors.

7. I determined these significance region boundaries by respecifying the reference values for education to single-year increments on the *intspec* command and then rerunning the *sigreg* command.

8. I calculated the midrange effects in two different ways, and then averaged those two results. The first was the average of the discrete change effects that were significant at 1-year increments of education from 0 to 18. The second was a weighted average, using the sample proportion in each year of education as the weight.

9. This comparison conservatively sets aside an extreme low ratio and an extreme high ratio for the white-collar classes.

10. For factor changes (odds ratios), I follow the Stata procedure for testing the significance of the odds ratio by using the significance test on the model coefficient "in the natural

space of the model (H_0: $b = 0$)." This is asymptotically equivalent to testing the odds ratio in the transformed parameter space—that is, "in the OR space (H_0: $e^b = 1$)" (see http://www.stata.com/support/faqs/statistics/delta-rule/). The effect on the standardized latent variable is a multiplicative transformation of the model coefficient, so its standard error is the same multiplicative transformation of the standard error for the model coefficient. Hence, it has the same test statistic—a function of the estimated parameter divided by its standard error—used to test the model coefficient.

11. I determined the statistical significance of the differences between the factor change values I discuss in this section directly from the MNLR results. The significance of the difference between the education factor change effect at moderator category m and the factor change at its reference category is the same as the significance of the coefficient for the interaction of education and moderator category m.

12. The predicted log odds is a linear combination of the predictors, and the errors are uncorrelated with the predictors by assumption. Hence, the errors are uncorrelated with the predicted log odds.

ORDINAL REGRESSION MODELS

OVERVIEW

Properties and Use of Ordinal Regression Models

Unlike outcomes with unordered categories analyzed with MNLR, these models require ordinal-level outcomes. Given their predominance in applications, I cover what Long and Freese (2014) call ordinal regression models, which estimate the cumulative probabilities of the outcome categories. I focus on ordinal logistic regression, with a brief application to ordinal probit analysis. Both ordinal logistic and ordinal probit predict the probability that a case is in category k of the C category of an ordinal outcome using an index function of the predictors multiplied by their coefficients $(x_i'\beta)$ and a set of $C+1$ cut points (τ_k). The value of the index function relative to the values of the cut points determines the probability that Y_i is in category k (see Equation 10.3) below. Note that β specifies that the coefficients do not vary with the cut points; this is known as the parallel regression assumption. As is true for binary outcomes, the ordinal logistic and ordinal probit models usually provide similar results once you take into account the different scaling of the coefficients (Long & Freese, 2014, p. 318). There are also a variety of other ordered regression models with different assumptions (Fullerton & Xu, 2016; Liu, 2016; Long, 1997, pp. 145–147; Long & Freese, 2014, pp. 370–382). Fullerton and Xu (2016, pp. 9–10) present a useful two-factor typology of ordered regression models: (1) the degree of constraints on the coefficients to a parallel regression assumption and (2) the type of comparison across categories.

The modes of interpretation of interaction effects for ordinal regression models are nearly identical to those for the logit and probit models of binary outcomes in Chapter 8. The one notable difference is that the meaning of the odds ratios (factor changes) for the ordinal logistic model is somewhat different. It describes changes in the cumulative odds—not the base odds—for a given outcome category. For readers not already familiar with the basics of interpretation of the ordinal logistic regression model, I would recommend reading Chapter 4 of Liu (2016) for an introduction and Chapter 7 of Long and Freese (2014) for a more extensive discussion of interpretation.

Data and Circumstances When Commonly Used

Ordinal regression models are increasingly the preferred technique to use for an ordinal-level outcome measure. That is, $Y_i = Category_k$, where the index $k = 1, 2, 3, \ldots, C$. And there is some property of the categories that increases monotonically with the index k such that the categories are in the same order as the property:

$$Property_1 < Property_2 < Property_3 < \cdots < Property_C \qquad (10.1)$$

The classic example is a Likert-type scale for which the response categories are *strongly disagree*, *disagree*, *agree*, and *strongly agree*, ordered by the property of strength of agreement. Technically, any numeric values that maintain the category order can be used. For convenience, I assume that Y_i is coded numerically by the category index $(1, 2, 3, \ldots, C)$, which is common practice but does not mean that the outcome can or should be treated as interval. As Long and Freese (2014) note,

> Perhaps because of this coding, it is tempting to analyze ordinal outcomes with the linear regression model (LRM). However, an ordinal dependent variable violates the assumptions of the LRM, which can lead to incorrect conclusions, as demonstrated strikingly by McKelvey and Zavoina (1975: 117) and Winship and Mare (1984: 521–523). (p. 309)

In part, the difficulty with using OLS for an ordinal outcome arises from the assumption that the categories are equidistant in terms of the real underlying metric. However, an outcome whose categories exhibit an ordering property is not necessarily better analyzed assuming ordered rather than unordered categories, as the political ideology example in Chapter 9 demonstrated.

Published Examples.

- Cech et al. (2011) used ordinal logistic regression of intention to persist in engineering majors (*very unlikely* to *very likely*) to test the interactive effect of students' gender and their family plans. They interpreted the significant interaction effect from plots of predicted persistence by family plans separately for women and for men.

- Yen, Coats, and Dalton (1992) compared the results from ordinal probit and ordinal logistic regression predicting an election's degree of competitiveness by the interaction between district homogeneity and past performance of incumbents. For interpretation, they briefly discussed the meaning of the coefficients.

- In an ordinal logistic regression analysis, Visser, Mills, Heyse, Wittek, and Bollettino (2016) assessed whether trust in management can buffer the negative effects of job autonomy on satisfaction with work–life balance (7-point Likert-type scale). They briefly discussed the coefficients and interpreted how the moderated effect of autonomy differs by trust level.

- Pisljar, van der Lippe, and den Dulk (2011) studied the health of hospital workers in a set of European countries using an ordinal logistic regression of self-reported health (*excellent* to *poor* in four categories). They found two-way interactions between working in eastern Europe and workers' overtime work and their job autonomy. They interpreted the interactions by discussing the coefficients and plotting predicted health against overtime and against autonomy separately for eastern and for western European hospital location.

GLM Properties and Coefficient Interpretation for Ordinal Regression Models

Ordinal logistic and ordinal probit regression are typically developed starting from the assumptions that an interval latent variable y^* underlies the C observed categories and that y^* is a linear function of the predictors and an error term:

$$y_i^* = \underline{x}_i' \underline{\beta} + \varepsilon_i \tag{10.2}$$

Moreover, the latent variable is divided into the observed categories according to a set of $C + 1$ ordered cut points τ_k between $+\infty$ and $-\infty$. Given this, the probability p_{ik} of being in category k conditional on the predictors is

$$p_{ik} = \Pr\left(Y_i = k \mid \underline{x}_i\right) = F\left(\tau_k - \underline{x}_i' \underline{\beta}\right) - F\left(\tau_{k-1} - \underline{x}_i' \underline{\beta}\right) \tag{10.3}$$

where $F(\ )$ is a cumulative distribution function, $\tau_0 = -\infty$ and $\tau_C = +\infty$. For ordinal probit, $F(\ )$ is the cumulative normal distribution with $Mean = 0$ and $Variance = 1$. For ordinal logistic, $F(\ )$ is the cumulative logistic distribution with $Mean = 0$ and $Variance = \dfrac{\pi^2}{3}$.

Equivalently (see the boxed *Aside*), you can write the relationship between the observed outcome and the predictors in terms of the cumulative probabilities cp_{ik} as

$$cp_{ik} = \Pr\left(Y_i \leq k \mid x_i\right) = F\left(\tau_k - \underline{x}_i' \underline{\beta}\right) \tag{10.4}$$

This form is convenient for considering ordinal logistic and ordinal probit as GLMs. Both specify that the probability distribution of the ordered categories, conditional on the predictors, follows a multinomial distribution. The link function for the relationship between the observed and modeled outcomes applies on a category-by-category basis, with the observed outcome for a category defined as its cumulative probability and the link function as the inverse of the cumulative distribution function, $F^{-1}(\ )$. For ordinal logistic regression, the link function for category k simplifies to $F^{-1}(\ ) = \ln\left(\dfrac{cp_k}{1 - cp_k}\right)$. There is no equivalent simplifying expression for ordinal probit models.

ASIDE: ORDERED REGRESSION MODEL WRITTEN AS CUMULATIVE PROBABILITIES

Define the kth cumulative probability for case i as

$$cp_{ik} = \sum_{c=1}^{k} p_{ic}$$

(Continued)

(Continued)

Substituting for p_{ik} from Equation 10.3,

$$cp_{ik} = \sum_{c=1}^{k}\left[F\left(\tau_c - \underline{x}_i'\underline{\beta}\right) - F\left(\tau_{c-1} - \underline{x}_i'\underline{\beta}\right)\right]$$

$$= \left[F\left(\tau_1 - \underline{x}_i'\underline{\beta}\right) - F\left(\tau_0 - \underline{x}_i'\underline{\beta}\right)\right] + \left[F\left(\tau_2 - \underline{x}_i'\underline{\beta}\right) - F\left(\tau_1 - \underline{x}_i'\underline{\beta}\right)\right] + \cdots +$$

$$\left[F\left(\tau_c - \underline{x}_i'\underline{\beta}\right) - F\left(\tau_{c-1} - \underline{x}_i'\underline{\beta}\right)\right]$$

Add the first term in each pair of terms in square brackets to the second term in the next pair, and recognize that they sum to zero. Repeat across the set of terms. This leaves the second term in the first bracketed pair and the first term in the last bracketed pair:

$$cp_k = -F\left(\tau_0 - \underline{x}_i'\underline{\beta}\right) + F\left(\tau_k - \underline{x}_i'\underline{\beta}\right)$$

But $\tau_0 = -\infty$, thus $F\left(\tau_0 - \underline{x}_i'\underline{\beta}\right) = 0$ because the cumulative distribution between $-\infty$ and $-\infty - \underline{x}_i'\underline{\beta}$ is equal to 0. This leaves the expression in Equation 10.4:

$$cp_k = F\left(\tau_k - \underline{x}_i'\underline{\beta}\right)$$

Two restrictions on the parameters are usually made to identify these models. Like binomial logistic regression and probit analysis, the variance of $F(\ )$ is fixed (at 1.0 for probit and at $\frac{\pi^2}{3}$ for logistic). The second constraint can be set in different ways. Some software, including Stata, fix the value of the intercept at 0. There are other mathematically equivalent parameterizations (e.g., setting one of the τ_ks to 0 instead of the intercept); see Long and Freese (2014, pp. 235–238) for a discussion and an example of how to translate results between them.

Interpretation of Interaction Effects

The approach to the interpretation of interaction effects in ordinal regression models closely resembles the options available for binomial logistic regression and probit analysis. A key difference is that you must interpret multiple outcomes (categories) for most of the approaches. This chapter demonstrates how to interpret results:

1. As effects on the latent interval variable underlying the categories (optionally standardized)

2. As factor changes (odds ratios) in the cumulative odds—only applicable for ordinal logistic regression

3. As a marginal effect on the probability of a category—typically, as a discrete change, but a marginal change (instantaneous) can be used for interval predictors

4. Using tabular or visual displays of the predicted probabilities of the categories or the predicted latent outcome

Diagnostic Tests and Procedures

Diagnostic testing focuses on the parallel regressions assumption, which is explicit in the formula in Equation 10.4 describing the relationship between the cumulative probability of each category and the predictors. The only parameter that changes across the categories is the cut point, τ_k, which serves as the intercept for the index function for the kth category of the ordered outcome. Other than this intercept term, the regression parameters are assumed to be equal across the categories. Consequently, an ordinal regression model is not appropriate for the data if they cannot support the assumption of equal regression parameters across categories. You can formally test this assumption using Wald, likelihood ratio, or score tests; for details of the tests, see Fullerton and Xu (2016, pp. 109–119). Long and Freese (2014, pp. 328–331) describe how to calculate these tests in Stata, which requires installing one or more user-written commands. It is good practice, especially when analyzing large samples, to also consider informal diagnostics comparing results from parallel and nonparallel models—Bayesian information criterion (BIC) and/or Akaike information criterion (AIC) statistics, coefficients, odds ratios (if applicable), or predicted probabilities—to avoid rejecting the assumption when differences are substantively small or due to other types of model misspecification (Fullerton & Xu, 2016, pp. 119–130).

For example, Long and Freese (2014, p. 331) caution that the data for ordinal regression frequently fail the Brant test, yet in some instances the results do not differ notably from those estimated by alternative models (e.g., multinomial logit or generalized ordered logit). They also caution that model misspecification is always an alternative explanation for a significant diagnostic test. In this regard, an advantage of the formal test using the Brant criterion (an approximate Wald test) is that you can also examine the set of binary analyses from which the Brant criterion is calculated. This lets you make a judgment of the degree of difference in the coefficients and to potentially identify misspecifications of the functional form of predictors.

Data Source for Examples

The first example is a single-moderator interaction (education by sex) predicting a four-category ordinal outcome (frequency of purchasing chemical-free food). The second example analyzes subjective class identification with two interval moderators of the same nominal predictor—race by education and race by income. Note that this includes neither a three-way interaction among the predictors nor the two-way interaction of education by income. Both examples analyze data from the 2010 GSS (GSS_2010.dta). This data set and the example do-files can be downloaded at www.icalcrlk.com.

ONE-MODERATOR EXAMPLE (INTERVAL BY NOMINAL)

Data and Testing

In this example, I interpret how education and sex interact to predict how often respondents report that they purchase pesticide-free produce. I focus on interpretations for an ordinal logistic regression model, but I end this section with a brief consideration of the ordinal probit results. The response categories for the frequency of purchase of chemical-free produce are

$$1 = Never \quad 2 = Sometimes \quad 3 = Often \quad 4 = Always$$

Years of education completed (education) is an interval scale ranging from 0 to 20, and sex is a dummy variable coded 1 = *female* and 0 = *male*. Two of the remaining predictors are interval measures: age (18–89) and size of place of residence in millions (0–8.01). Race is a three-category measure (1 = *White*, 2 = *Black*, and 3 = *Other*), with Other used as the reference category. Political views also consists of three categories (1 = *Conservative*, 2 = *Moderate*, and 3 = *Liberal*), and Conservative is the reference category. The index function for estimating the latent outcome takes the following form:

$$y_i^* = \beta_1 Education_i + \beta_2 Female_i + \beta_3 Education_i \times Female_i + \beta_4 Age_i \\ + \beta_5 Size_i + \beta_6 White_i + \beta_7 Black_i + \beta_8 Moderate_i + \beta_9 Liberal_i \quad (10.5)$$

Note that there is no β_0 parameter for an intercept because its value is fixed at 0 to identify the other parameters. Because the analysis uses ordinal logistic regression, we can write an equivalent expression for the log of the odds of being in a response category $>k$ to being in a response category $\leq k$ (see the boxed *Aside*):

$$\ln\left(\frac{1-cp_k}{cp_k}\right) = \beta_1 Education_i + \beta_2 Female_i + \beta_3 Education_i \times Female_i \\ + \beta_4 Age_i + \beta_5 Size_i + \beta_6 White_i + \beta_7 Black_i + \\ \beta_8 Moderate_i + \beta_9 Liberal_i - \tau_k \quad (10.6)$$

The sample of $N = 1,340$ excludes 640 respondents with missing data on the outcome, another 40 with missing data on political views, and 5 more with missing information on one or more of the other predictors.

ASIDE: MODEL WRITTEN AS A LOG ODDS OF CUMULATIVE CATEGORIES

Let $\Omega_{\leq k|>k}$ represent the odds of being in a response category $\leq k$ to being in a response category $>k$. According to Long and Freese (2014, p. 335),

$$\Omega_{\leq k|>k} = e^{\tau_k - \underline{x}'\underline{\beta}}$$

We want the inverse of this odds, contrasting being in a response category $>k$ to being in a response category $\leq k$:

$$\Omega_{>k|\leq k} = e^{-\left(\tau_k - \underline{x}'\underline{\beta}\right)}$$

Taking the log gives

$$\ln\left(\Omega_{>k|\leq k}\right) = \ln e^{-\left(\tau_k - \underline{x}'\underline{\beta}\right)} = -\left(\tau_k - \underline{x}'\underline{\beta}\right) = \underline{x}'\underline{\beta} - \tau_k$$

The Brant test of the ordinal logistic results is not significant either globally ($p = .065$) or for any of the individual predictors (smallest $p = .093$). This suggests that the data are consistent with the parallel regression assumption, as in the other four global tests produced by the *oparallel* command. I also compared the BIC and AIC statistics with those from alternative analytic techniques: ordinal probit, multinomial logit, and two generalized ordinal logistic regression models (Williams, 2016)—one with no predictors constrained to meet the parallel regression lines assumption and one with selected predictors constrained as chosen by the autofit option. The BIC statistics suggested the use of ordinal logistic regression, providing positive support vis-à-vis ordinal probit and strong support relative to the other models. The AIC results were similar, except that the AIC had positive support for the auto-fit generalized ordinal logistic regression. Alternative statistical tests of the inclusion of the interaction term—$\beta_3 Education \times Female_i$—gave consistent results. The change in the log likelihood if the term is excluded from the model was significant ($p = .040$), as was the z test of the coefficient ($p = .045$), which provided justification to proceed with the inclusion and interpretation of the interaction.

In the following sections, I apply the GFI, SIGREG, and EFFDISP tools to interpret first how sex moderates the effect of education on the frequency of purchasing chemical-free produce and then how education moderates the effect of sex on purchase. For each moderated effect, I demonstrate how to create and interpret three types of effects: (1) on the latent variable, (2) on the odds of more frequent to less frequent purchasing, and (3) on the probability of each purchase category. I then illustrate several of the OUTDISP tool options that permit interpreting the effects of education and sex simultaneously. I conclude this example with a brief discussion of the interpretation of the interaction effect from the ordinal probit analysis.

Education Moderated by Sex

The output for the example begins with the estimation command for *ologit*, followed by the *intspec* command specifying education as the focal variable and sex as the moderator and setting their display names and ranges. The factor option in the *gfi* command requests the algebraic expressions for the moderated effect of education on the latent outcome (the log odds of more frequent to less frequent purchase) as well as the factor change effect on the odds of more to less frequent purchase. Subsequent applications of the *sigreg* command produce significance region tables for the three types of effects, and the *effdisp* command illustrates the corresponding effect display plot for the effect on the latent variable.

```
. ologit chemfree2 c.educ##i.female ib3.race age i.libcon size
    ...

. intspec focal(c.educ) main((c.educ, name(Education) range(0(4)20)) ///
    (i.female, name(Sex) range(0/1))) ndig(0) int2(c.educ#i.female)

Interaction Effects on Chemfree2 Specified as

    Main effect terms: educ  1.female
    Two-way interaction terms: i.female#c.educ

  These will be treated as: Focal variable = educ ("Education")
    moderated by interaction(s) with
        1.female ("Sex")

. gfi, factor

GFI Information from Interaction Specification of
Effect of Education on g(Chemfree2) from Ordinal logistic Regression
----------------------------------------------------------------------
```

```
Effect of Educ =
   -0.0125 + 0.0673*Female

Factor Change Effect (1 unit change in) Educ =
   e^-0.0125 * e^( 0.0673*Female) =
   0.9876 * 1.0696^Female
```

We can use the fact that "female" is a dummy variable to rewrite the GFI expressions to convey the separate effects of education for men and for women:

$$Effect\ for\ men\ (female = 0) = -0.0125 + 0.0673 \times female$$
$$= -0.0125 + 0.0673 \times 0 = -0.0125$$
$$Effect\ for\ women\ (female = 1) = -0.0125 + 0.0673 \times female$$
$$= -0.0125 + 0.0673 \times 1 = 0.0548$$

We see that men and women have opposite-signed effects of education on the latent propensity to purchase pesticide-free produce. Men's propensity to purchase declines by 0.0125 for each 1-year difference in education, while women's propensity increases by 0.0548. Similarly, the factor change expression for the effects of education for men and for women is given by

$$Factor\ effect\ for\ men\ (female = 0) = 0.9876 \times 1.0696^{female} = 0.9876 \times 1.0696^{0}$$
$$= 0.9876 \times 1 = 0.9876$$
$$Factor\ effect\ for\ women\ (female = 1) = 0.9876 \times 1.0696^{female} = 0.9876 \times 1.0696^{1}$$
$$= 1.0563$$

These numbers tell us that the odds of more frequent purchases relative to less frequent purchases decreases for men by a factor of 0.988 for a 1-year difference in education but increases for women by a factor of 1.056. While both of these sets of calculations indicate the nature and direction of the moderated effect of education, neither tells us much about the size (meaningfulness) of the effects.

We can use the SIGREG tool not only to determine if the moderated effects are significant but also to rescale the calculated effects to provide a better sense of the magnitude of the effect of education. I demonstrate three different ways to accomplish this, in part by using a 1 standard deviation difference in education rather than a 1-year difference:

1. Calculate the standardized change in the latent outcome for a standard deviation change in education. Specify the *sigreg* option effect(b(sdyx)).

2. Calculate the factor change in the odds of higher versus lower frequency purchase corresponding to a standard deviation change in education. Use the option effect(factor(sd)).

3. Calculate the discrete change in the probability of each purchase category predicted by a standard deviation change in education. Apply the option spost(amtopt(am(sd) center) atopt((means) _all)).

Standardized Change in the Latent Outcome

```
.sigreg , effect(b(sdyx))
   Significance Region for Effect of Education (1 s.d. difference)
   on g(Chemfree2)-standardized at Selected Values of Sex
```

Effect of		At Sex= Male	Female
Educ		-0.0206	**0.0903***

Key: Plain font = Pos, Not Sig   **Bold font*** = Pos, Sig

Italic font = Neg, Not Sig   *Italic font** = Neg, Sig

The significance region table indicates that the effect of education is negative and not significant for men but is positive and significant for women. More specifically, a 1 standard deviation increase in education predicts almost a 1/10 of a standard deviation increase in women's propensity to purchase pesticide-free produce. To me, this indicates a small but not inconsequential effect of education.[1]

Predicted Change in the Odds of More Frequent Purchase

```
. sigreg , effect(factor(sd))
    …
```

Significance Region for Factor Change Effect of Education (1 s.d. difference) on Chemfree2 at Selected Values of Sex

Effect of		At Sex= Male	Female
Educ		0.9624	**1.1825***

Key: Plain font = Pos, Not Sig   **Bold font*** = Pos, Sig

Italic font = Neg, Not Sig   *Italic font** = Neg, Sig

Not surprisingly, the effect of education on the odds of more frequent purchase is negative and nonsignificant for men but positive and significant for women. For women, a 1 standard deviation change in education predicts that the odds of more frequent purchases will increase by a factor of 1.18—that is, by 18%. Like the effect on the latent outcome, this suggests a relatively small change.

Discrete Change in the Probabilities of Each Purchase Category

```
.sigreg, effect(spost(amtopt(am(sd) center) atopt((means) _all)))
    …
```

Significance Region for SPOST Change Effect of Education on chemfree2[Never] at Selected Values of Sex

Effect of		At Sex= Male	Female
Educ		0.0087	*-0.0322**

    …

Significance Region for SPOST Change Effect of Education on chemfree2[Sometimes] at Selected Values of Sex

Effect of		At Sex= Male	Female
Educ		*-0.0008*	*-0.0078**

Key: Plain font = Pos, Not Sig   Bold font* = Pos, Sig

    …

```
Significance Region for SPOST Change Effect of Education
on chemfree2[Often] at Selected Values of Sex
                              At Sex=
      Effect of      |    Male      Female
--------------+----------------------------
         Educ  |   -0.0051     0.0224*


Significance Region for SPOST Change Effect of Education
on chemfree2[Always] at Selected Values of Sex
                              At Sex=
      Effect of      |    Male      Female
--------------+----------------------------
         Educ  |   -0.0028     0.0176*
  Key: Plain font = Pos, Not Sig    Bold font* = Pos, Sig
       Italic font = Neg, Not Sig   Italic font* = Neg, Sig
```

```
Spost Effect for  educ specified as amount =    +SD_centered  calculated with
   at(  (means) _all )
```

Again, we see that there is no significant effect of education on purchase frequency for men but there is for women. While I could summarize the education effect for women as positive, it is more appropriate to consider the details provided by the separate effects on each outcome category. A 1 standard deviation increase in women's education has a negative effect on the probabilities of never and of sometimes purchasing pesticide-free produce, decreasing the probability of the categories by 3% and <1%, respectively. On the other hand, education has a positive effect on women often and always buying pesticide-free produce, increasing these probabilities of buying by about 2%. You can judge the size of these effects relative to the observed probability of each category. The effect on the "sometimes" category is the smallest, at slightly more than 2%, relative to its category's probability. The effects on the "never" and "often" categories are four times larger, at 11% and 9% of their category's observed probability, respectively. The effect of education for women on the "always" category represents a predicted change of about 16% of the observed category probability. Like the other two types of effects, these results suggest modest but meaningful effects of education on purchase and frequency for women but not for men.

Effect Displays

Given the simplicity of the significance region tables in this case, I do not think that the effect displays are necessarily preferable to the significance region tables, and to present both would be redundant. I provide one example for the effect on the latent propensity to purchase pesticide-free produce by way of illustration. Because I want the default error bar plot for a nominal moderator, the only option necessary is the same one used in the *sigreg* command to specify the effect type, effect(b(sdyx)). I also use the pltopts( ) option to adjust the size of the title and ndig( ) to format the *y*-axis labels on the plot in Figure 10.1:

```
. effdisp , effect(b(sdyx)) pltopts(tit(, size(*.8))) ndig(2)
```

Figure 10.1 shows the same information as the significance region table. The effect of education for men is not significant since the error bars bracket (include) an effect size of zero. The effect of education for women is significant because zero does not lie within its confidence interval. Clearly, this would be a good alternative to presenting the significance region table, depending on your preference for visual versus tabular displays.

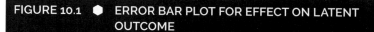

FIGURE 10.1 ● ERROR BAR PLOT FOR EFFECT ON LATENT OUTCOME

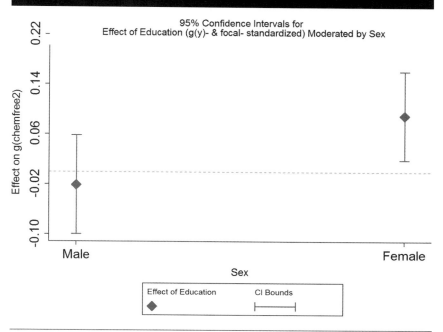

Sex Moderated by Education

Because I use SPOST13 to calculate discrete change effects, I begin by rerunning the *ologit* command to reorder the predictors in the factor-variable specification of the two-way interaction. The variable for sex is listed first, followed by the variable for education (i.female##c.education) (see the discussion in Chapter 6 for details on the predictor-ordering requirements when invoking SPOST13). I also revise the *intspec* command to set sex as the focal variable and education as the moderator, as well as specifying their display names and ranges. As before, I list the factor option in the *gfi* command to display the factor change effects. Although I create significance region tables for the same three types of effects, I change the specification of the amount of change in the focal variable for which the effect is calculated. With an interval measure such as education as the focal variable, I like to specify scaling the effects for 1 standard deviation changes in the focal variable. But this is not sensible when treating a nominal measure such as sex as the focal variable. Instead, I use the default for effects scaled for a one-unit change in sex (from 0 to 1, the difference between men and women) in the *sigreg* and *effdisp* commands below.

```
. ologit chemfree2 i.female##c.educ ib3.race age i.libcon size

    ...

. intspec focal(i.female) main((c.educ, name(Educ) range(0(2)20)) (i.female,
>   name(Sex) range(0/1))) ndig(0) int2(i.female#c.educ)

Interaction Effects on Chemfree2 Specified as
    Main effect terms: 1.female   educ
      Two-way interaction terms: i.female#c.educ
```

```
      These will be treated as: Focal variable = 1.female ("Sex")
         moderated by interaction(s) with
            educ ("Educ")
. gfi, factor
GFI Information from Interaction Specification of
Effect of Sex on g(Chemfree2) from Ordinal logistic Regression

Effect of Female =
   -0.4789 + 0.0673*Educ


Factor Change Effect (1 unit change in) Female =
   e^-0.4789 * e^( 0.0673*Educ) =
   0.6195 * 1.0696^Educ
```

The GFI results reveal a negative effect of female when *Education* = 0 (−0.4789, because $0.0673 \times 0 = 0$), which moves in the positive direction as education increases. The sign change analysis table (not shown) reports that the sign changes to positive when *Education* > 7.11. That is, women have a lower propensity than men to purchase pesticide-free produce if *Education* ≤ 7 years but a higher propensity if *Education* > 7 years. Analogously, the factor change GFI expression indicates that when *Education* = 0, women's odds of more frequent to less frequent purchasing are less than two thirds that of men (0.6195, because 1.0696^0 = 1). Women's odds remain lower than men's until *Education* > 7 years. Next, I elaborate the meaning of this moderated relationship using three different ways to define the statistical significance and magnitude of the effect of sex on the purchase of pesticide-free produce.

Standardized Change in the Latent Outcome

```
. sigreg , effect(b(sdy)) ndig(2) concise
    ...
Boundary Values for Significance of Effect of Sex on g(Chemfree2) Moderated by Educ
    Critical value Chi_sq = 3.841 set with p = 0.0500


 +--------------------------------------------------------------------------------+
 |                   | When Educ >= | Sig Changes | When Educ >= | Sig Changes |
 |-------------------+--------------+-------------+--------------+-------------|
 | Effect of Female |   NA (< min) |     NA      |    10.99     |   to Sig    |
 +--------------------------------------------------------------------------------+


    Significance Region for Effect of Sex
     on g(Chemfree2)-standardized at Selected Values of Educ
                                       At Educ=
Effect of  |   0     2     4     6     8    10    12    14    16    18    20
-----------+-----------------------------------------------------------------------
  Female   | -.26  -.19  -.11  -.04   .03   .10   .18*  .25*  .32*  .39*  .47*
-----------+-----------------------------------------------------------------------
     Key: Plain font   = Pos, Not Sig    Bold font*    = Pos, Sig
          Italic font  = Neg, Not Sig    Italic font*  = Neg, Sig
```

The sdy specification produces effects scaled to the standard deviation of the latent outcome *y* and a default one-unit change in the focal variable; the option ndig(2) formats the significance region table results with two digits, and concise suppresses details in the boundary values table. The boundary value analysis reports that the female–male difference in the propensity to purchase pesticide-free produce is only significant when *Education* ≥ 11 years. The significant differences are positive, meaning that women have a higher propensity to purchase such produce than men, and range in size from about 1/5 of a standard deviation difference to nearly 1/2 of a standard deviation difference in the propensity. I would consider these differences somewhat similar to but smaller in magnitude than the significant education effect (for women), which predicted a difference in the propensity of 6/10 of a standard deviation from the minimum to the maximum education value.

Predicted Change in the Odds of More Frequent Purchase

```
. sigreg , effect(factor(1)) ndig(2)
   ...
   Significance Region for Factor Change Effect of Sex (1 unit difference)
      on g(Chemfree2) at Selected Values of Educ
```

							At Educ=					
Effect of		0	2	4	6	8	10	12	14	16	18	20
Female		.62	.71	.81	.93	1.06	1.21	**1.39***	**1.59***	**1.82***	**2.08***	**2.38***

```
        Key: Plain font  = Pos, Not Sig      Bold font*   = Pos, Sig
             Italic font = Neg, Not Sig     Italic font* = Neg, Sig
```

As they should, the factor change effect results parallel what we just saw for effects on the latent outcome. In particular, as discussed in Chapter 3, the boundary values analysis for effects on $g(y)$ also apply to factor changes. Thus, the factor change effect of sex is only significant for *Education* ≥ 11 years, where it has a positive effect. When *Education* = 12, women have a 39% higher odds of more frequent relative to less frequent purchases than do men. This steadily increases with education and reaches a maximum at 20 years of education where women's odds are more than twice as high as men's. The significant factor change effects are again in the ballpark but somewhat smaller than the significant education factor change effect (for women). The factor change in the odds from the maximum value of education to its minimum is 2.81, showing that women's odds of more frequent to less frequent purchasing almost triples between *Education* = 0 and 20.

Discrete Change in the Probabilities of Each Purchase Category

```
. sigreg, effect(spost(amtopt(am(bin)) atopt((means) _all))) ndig(3)
   Significance Region for SPOST Change Effect of Sex
      on chemfree2[Never] at Selected Values of Educ
```

							At Educ=					
Effect of		0	2	4	6	8	10	12	14	16	18	20
Female		0.111	0.079	0.047	0.017	-0.013	-0.042	-0.070*	-0.097*	-0.123*	-0.148*	-0.172*

```
   Significance Region for SPOST Change Effect of Sex
      on chemfree2[Sometimes] at Selected Values of Educ
```

							At Educ=					
Effect of		0	2	4	6	8	10	12	14	16	18	20
Female		-0.014	-0.008	-0.004	-0.001	0.000	-0.000	-0.003	-0.007	-0.013*	-0.020*	-0.028

```
   Significance Region for SPOST Change Effect of Sex
      on chemfree2[Often] at Selected Values of Educ
```

```
                                                          At Educ=
Effect of  |     0        2        4        6        8        10       12       14       16       18       20
-----------+-----------------------------------------------------------------------------------------------------
   Female  |  -0.063   -0.046   -0.028   -0.010    0.008    0.026    0.045*   0.063*   0.081*   0.098*   0.114*
-----------+-----------------------------------------------------------------------------------------------------


 Significance Region for SPOST Change Effect of Sex
 on chemfree2[Always] at Selected Values of Educ
_____
                                                          At Educ=
Effect of  |     0        2        4        6        8        10       12       14       16       18       20
-----------+-----------------------------------------------------------------------------------------------------
   Female  |  -0.034   -0.025   -0.016   -0.006    0.005    0.016    0.028*   0.041*   0.055*   0.070*   0.086*
-----------+-----------------------------------------------------------------------------------------------------

    Key: Plain font  = Pos, Not Sig    Bold font*    = Pos, Sig
         Italic font = Neg, Not Sig    Italic font*  = Neg, Sig

Spost Effect for  female specified as amount =    Female vs Male  calculated with

  at(  (means) _all )
```

The discrete change results take a little more work to interpret, simply because you need to understand and discuss the effect of the focal variable (sex) on the probabilities of each outcome category. These results exhibit the same pattern of significant effects of sex (differences between men and women) only when education is in the upper part of its distribution, with a turning point varying from greater than 11 to greater than 14 across the purchase categories. The significant effects of sex are negative in the "never" and "sometimes" categories, meaning that women have a lower predicted probability than men for these categories. But the significant effects are positive in the "often" and "always" categories, indicating that women have a higher predicted probability than men in these categories. That is, there is an overall positive effect of sex on the frequency of purchasing pesticide-free produce because women are predicted to be less likely in the less frequent purchasing categories and more likely in the more frequent purchasing categories.

How big are these effects? Comparing the smallest significant effect in a category with the observed probability of the category suggests that the sex effect is not trivial except for the "sometimes" category. In the other categories, the smallest significant sex effects range from 15% of the size of the observed probability in the category up to 23%, while the largest sex effect is between 44% and 80% of the size. Keeping in mind that these are only the significant effects, the comparison still suggests a fairly substantial predicted difference in the category probabilities between men and women. A similar calculation for the significant effect of education (significant only for women) shows that it is 17% to 31% larger than the largest significant sex effect.[2]

Effect Displays

Given the greater complexity of the discrete change effects—changing sign and significance across the four categories of the outcome—I demonstrate the use of an effect display to interpret them. I use the same specification of the effect() option as for the *sigreg* command and force the same labeling of the *y*-axis on the plot for each outcome category with the pltopts( ) option (see the discussion of how to use this option in the "Special Topics" section of Chapter 7).

```
. effdisp, effect(spost(amtopt(am(bin)) atopt((means) _all))) ///
> pltopts(ylab( -.39(.13).39,form(%8.2f) labsize(*.75)))
```

I think the first impression of the confidence bounds plots in Figure 10.2 conveys the main message that the effect of sex differs both with education and with the particular outcome category. The significance boundary lines also draw attention to the fact that the sex effect is only significant in the upper part of the education distribution. But do not overlook the bigger picture that among the more educated part of the population, the overall effect of sex across the ordered categories is a predicted

FIGURE 10.2 ◆ CONFIDENCE BOUNDS PLOT FOR DISCRETE CHANGE EFFECTS

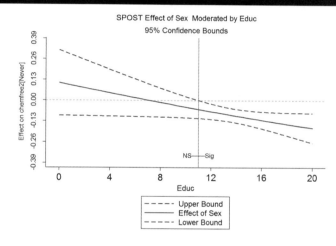

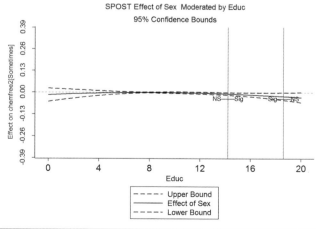

(Continued)

FIGURE 10.2 ● (CONTINUED)

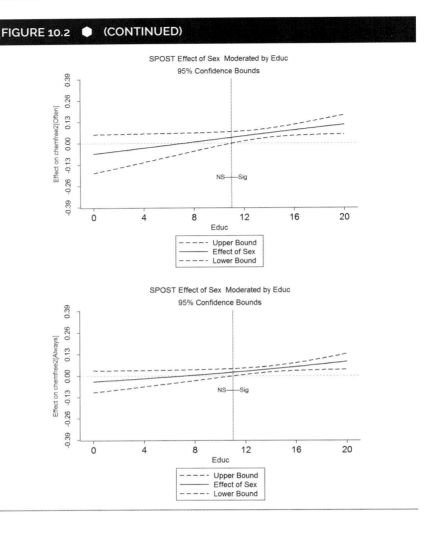

increase in the frequency of purchasing pesticide-free produce for women compared with men. That is, the predicted probability of never or sometimes purchasing is smaller for women than for men and the difference increases with education.

At the same time, the predicted probability of often or always purchasing is larger for women than for men, with the difference growing with education. Together, these paint a picture of women's distribution across the categories as shifted relative to men's, with women shifted out of the lower frequency categories (never and sometimes) and into the higher frequency categories (often and always). If you choose to present plots instead of significance region tables, I would add a discussion of the size of the sex effect, as I did in discussing the significance region table above.

OUTDISP **for the Effects of Education and Sex Simultaneously**

When one of the predictors in the interaction is interval and the other is nominal, you need to choose between using a bar chart or a scatterplot. In this example, the bar chart would report a set of bars repeated for the display values of education, where each set consists of a bar for men and a bar for women showing the value of an outcome predicted by sex and by the display value of education. The scatterplot would show separate prediction lines for men and women, each line plotting a predicted outcome against education. I think the scatterplot is easier to interpret and explain than the bar chart, mainly because it facilitates interpreting the effects of both predictors from the same visual display. I start with a plot and a table of the predicted latent outcome, which you can also conceptualize as the log odds of more frequent to less frequent purchases. Then, I illustrate the use of tables and plots of predicted probabilities and show how a comparison with the results from a main effects model can help untangle the confounded nonlinearities. Because I want to do this latter comparison, I begin by running and storing the estimates from the main effects model and the interaction effects model. I use the quiet prefix to suppress showing the estimation results. The *intspec* command is then run while the interaction effects are the active (last run) estimates, and it must specify education as the focal variable to get scatterplots of an outcome against education:

```
quiet ologit chemfree2 c.educ i.female ib3.race age i.libcon size
...
est store mainmoded

...
quiet ologit chemfree2 c.educ##i.female ib3.race age i.libcon size
...
est store intmoded

...
intspec focal(c.educ) main( (c.educ, name(Education) range(0(4)20)) ///
    (i.female, name( Sex) range(0/1))) ndig(0) int2(c.educ#i.female)
```

Displays of the Predicted Latent Outcome

In the out( ) option for the *outdisp* command, I indicate that the outcome metric is the model (latent outcome) and use the sdy suboption to scale the latent outcome in standard deviation units. For consistency, I also use the atopt( ) suboption to set the noninteracting predictors to their means for the calculations.[3] I specify the save( ) suboption in the table( ) option to save the Excel formatted table of the predicted latent propensity and the abs suboption to make font sizes proportional to the absolute value of the predicted value—thus, large negative and large positive predicted values both are in larger font sizes. Last, the name( ) suboption in the plot( ) option stores the plot as a memory graph:

```
. outdisp, plot(name(Latent)) tab(save(Output\table_10_1.xlsx abs) out(sdy
>   metric(model) atopt((means) _all))
...
```

The patterns in Table 10.1 are easy to see, but remember that the font size is proportional to the magnitude of the effect in either the positive or the negative direction. For men, the predicted propensity to purchase chemical-free food becomes increasingly small as education rises (larger and larger negative magnitudes). Conversely, the predicted propensity for women to purchase increases at a sharper rate with higher levels of education (changing from a large negative magnitude to a large positive magnitude). Comparing the sex difference within levels of education reveals that women have a lower predicted propensity to purchase than men at the lowest levels of education, but this reverses for those with 8 or more years of education. And the difference—a higher predicted propensity for women than for men—becomes significant at 11 or more years of education. (Note how I incorporated details from the earlier sign change and boundary value analyses into this discussion.)

TABLE 10.1   ●   *g*(Chemfree2) STANDARDIZED BY THE INTERACTION OF EDUCATION WITH SEX

Education	Sex	
	Male	Female
0	−0.0429	−0.3008
4	−0.0699	−0.1827
8	−0.0969	−0.0647
12	−0.1238	0.0533
16	−0.1508	0.1713
20	−0.1778	0.2893

Although the table is effective, I think the scatterplot in Figure 10.3 is even better. It is immediately clear that the effect of education on the propensity is positive for women and negative for men and that the sex gap changes from a higher propensity for men at low education levels to a higher propensity for women as education rises. How big are these differences? Consider how much the propensity changes across the

FIGURE 10.3   ●   SCATTERPLOT OF STANDARDIZED LATENT PROPENSITY TO PURCHASE PESTICIDE-FREE PRODUCE

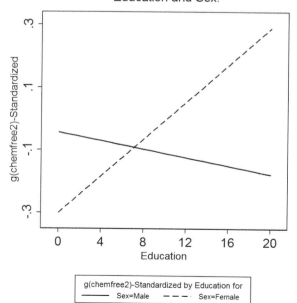

g(chemfree2)-Standardized by the Interaction of Education and Sex.

g(chemfree2)-Standardized by Education for
——— Sex=Male   – – – · Sex=Female

range of education. For men, the difference is relatively small—0.13 of a standard deviation—but for women it is more than 1/2 a standard deviation (0.59). If you look at the sex differences, they vary from 1/4 of a standard deviation higher for men than for women (at *Education* = 0) to almost 1/2 a standard deviation (0.47) higher for women than for men (at *Education* = 20). Overall, I would consider such effects meaningfully large but not overwhelming.

Displays of the Predicted Outcome Category Probabilities

```
. outdisp, plot(name(PredProb) tab(save(Output\table_10_2.xlsx)) ///
>    out(atopt((means) _all) metric(obs)) pltopts(ylab(.05(.1).45))
```

Many analysts and audiences prefer a more concrete outcome measure than the abstraction of the standardized latent propensity, such as plotting or creating tables of the predicted probability of each category as it varies with the interacting predictors' values. To create these displays, I revised the *outdisp* command options. In particular, the out( ) option now lists the metric as observed. I also rename the memory graph and the file name for saving the table of predicted values. I start by interpreting the scatterplots in Figure 10.4 first, because I think they are more useful than the set of tables in Table 10.2 in this instance.

The first thing I notice about the prediction lines in Figure 10.4 is that they show different relationships for men (*solid lines*) and for women (*dashed lines*) between education and the predicted probability of each category, adding up to opposite relationships overall. For men, increasing education predicts a higher probability of being in the "never" category, no differences in the "sometimes" category, and lower probabilities for the "often" and "always" categories—that is, an overall negative relationship between education and purchasing for men. In contrast, for women, higher levels of education predicts a lower probability of being in the "never" category, a very shallow inverted-U-shaped pattern in the "sometimes" category, and higher probabilities for the "often" and "always" categories. These add up to an overall positive relationship of higher frequency of purchasing pesticide-free produce with education for women.

We can also glean the relationship between sex and the frequency of purchasing from these plots. The "sometimes" category shows small male–female differences that change very little with education. In the other three categories, the sex difference in the predicted probability switches direction between the lower end of the education distribution and the middle and upper parts. That is, for the "never" category women are predicted to have a higher probability than men when *Education* < 8 years, but women have an increasingly lower predicted probability than men beyond that threshold. In the "often" and "always" categories, there are parallel patterns in which women's probability is less than men's for education less than 8 years but switches to an increasingly larger probability than men's as education continues to rise.

I present the predicted probability table for each category without the different-sized fonts to distinguish effect sizes. The differential font sizing is applied to each table independently—this makes it easy to see within-table differences, but it masks the differences between categories. That said, you can see the same patterns I just described for the plots.

Based on the visual patterns, you could be inclined to rank the categories from highest to lowest in terms of the degree of change in the sex effect as never, often, always, and sometimes. And the same would hold true for the steepness of the education effect, especially for women. Because the levels of probability at which we see the plotted lines differ, some part of these differences is likely due to the nonlinear link

FIGURE 10.4 ⬡ SCATTERPLOT OF PREDICTED CATEGORY PROBABILITIES

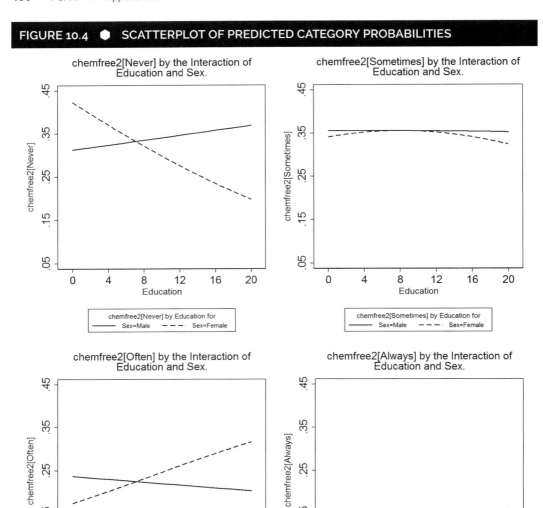

function. As before, I want to distinguish between describing the degree of moderation of the interacting predictors and the expressed relationship between those predictors and the outcome. In the next section, I demonstrate how a plot with superimposed main effects predictions helps parse out the nonlinearity.

TABLE 10.2 ● PREDICTED CATEGORY PROBABILITIES

Education	Sex		Education	Sex	
	Male	Female		Male	Female
Never			Often		
0	.3131	.4239	0	.2373	.1744
4	.3240	.3715	4	.2305	.2024
8	.3350	.3219	8	.2237	.2318
12	.3463	.2760	12	.2169	.2617
16	.3577	.2344	16	.2102	.2907
20	.3693	.1974	20	.2036	.3175
Sometimes			Always		
0	.3558	.3414	0	.0938	.0603
4	.3559	.3522	4	.0896	.0739
8	.3557	.3559	8	.0856	.0904
12	.3550	.3522	12	.0818	.1102
16	.3540	.3413	16	.0781	.1335
20	.3525	.3241	20	.0746	.1610

Displays of the Predicted Outcome Category Probabilities, With Superimposed Main Effects

```
. outdisp , plot(name(PrHat_Super)) out(atopt((means) _all) main(mainmoded))
> pltopts(ylab(.05(.1).45) tit( , size(*.85)))
```

To superimpose the main effects predictions on the predicted probabilities plot, presented in Figure 10.5, I added the main( ) suboption to indicate that the main effects estimates are stored under the name "mainmoded." I also changed the name of the memory graph inside the plot( ) option and removed the table option, as the prior run had already saved the predicted probabilities tables. My reading of the superimposed main effects predictions suggests several conclusions. First, the apparent effect magnitude differences across the outcome categories are a result of where along the cumulative probability function the predictions for each category are made—reflecting the reality of the observed distribution of cases across the categories. That is, the main effects differences in magnitude (e.g., the vertical gap between the education prediction lines for men and for women) exhibit similar-magnitude differences across the categories, including the lack of differences in the "sometimes" category.

At the same time, the comparison also shows that the opposite sign of the education effect for men and women and the change in direction of the sex effect with education in the "never," "Often," and "always" categories are not a result of the link function nonlinearity. None of the main effects predictions suggest such patterns, and

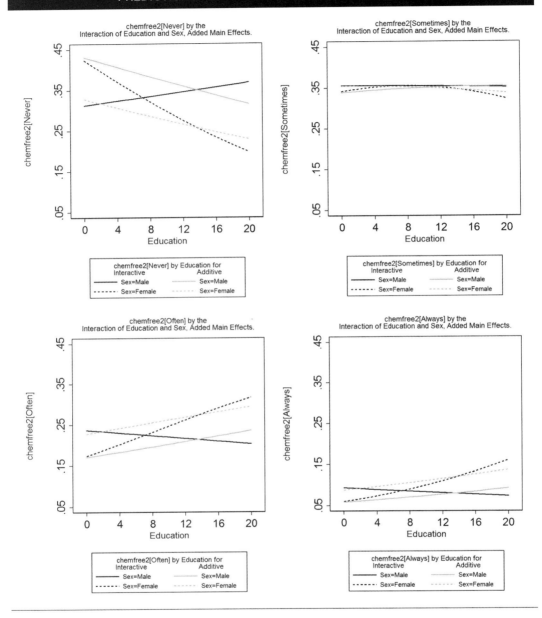

the moderated interaction effects appear proportional to the size of the main effects prediction differences in each of these categories. Indeed, the fact that the education effects are of opposite sign for women and men suggests that the visual display understates the degree of moderation of education by sex.

In sum, this analysis suggests minimal problems with interpreting how much education moderates the sex effect, and vice versa, within the outcome categories. But it does suggest caution in suggesting differences in the magnitude of moderation across the outcome categories, except for the absence of moderation in the "sometimes" category.

Ordinal Probit Results

The primary difference between the interpretation of ordinal probit and ordinal logistic results is that you have fewer options for interpretation. Most important, you cannot calculate or interpret factor change effects. While you can calculate the effects on or predict a value for a latent outcome measure, you cannot interpret these in terms of a log odds of more frequent to less frequent purchase of pesticide-free produce. But the coefficient estimates, and hence calculation of the moderated effects or predicted values, typically are quite similar if you adjust for the different scaling of the coefficients (Long & Freese, 2014, pp. 317–318). In terms of the ICALC commands, you could run an ordinal probit model and then apply exactly the same commands as used above. Any options that do not apply to an ordinal regression model would generate a warning message and be ignored.

As an example of the similarity between ordinal logistic and ordinal probit results, compare the significance-region table for the ordinal probit-estimated effect of sex moderated by education shown below with the table for ordinal logistic in the output above.

```
     Significance Region for Effect of Sex (sdy unit difference)
        on q(Chemfree2)-standardized at Selected Values of Educ
                                        At Educ=
  Effect of |   0     2     4     6     8    10    12    14     16     18     20
  ----------+-----------------------------------------------------------------
    Female  | -0.29 -0.21 -0.13 -0.05  0.03  0.11  0.19* 0.27*  0.35*  0.43*  0.51*
  ----------+-----------------------------------------------------------------
       Key: Plain font  = Pos, Not Sig    Bold font*    = Pos, Sig
            Italic font = Neg, Not Sig    Italic font*  = Neg, Sig
```

Both sets of effects are scaled in standard deviation units of the latent outcome, which makes them appropriate for contrasting the probit and logit results. If we consider the size of the significant moderated effects between the results, the probit-estimated effects are 8% to 10% larger than the logit-estimated effects. But the patterns of increase in the size of the moderated sex effect between adjacent education categories presented in Table 10.3 are nearly identical for the probit and logit results. Furthermore, the boundary value analyses (not shown) calculate close to the same value of education beyond which the sex effect becomes significant (10.97 vs. 10.99).

Consequently, the interpretation of how education moderates sex as an effect on the latent propensity to purchase would be the same.

TABLE 10.3	●	RATIO OF MODERATED SEX EFFECTS FOR ADJACENT EDUCATION VALUES			
	$\dfrac{Ed = 12}{Ed = 10}$	$\dfrac{Ed = 14}{Ed = 12}$	$\dfrac{Ed = 16}{Ed = 14}$	$\dfrac{Ed = 18}{Ed = 16}$	$\dfrac{Ed = 20}{Ed = 18}$
Probit	1.69	1.41	1.29	1.23	1.18
Logit	1.73	1.42	1.30	1.23	1.19

Similarly, look at the predicted probability plots for the "never" category drawn from the ordinal probit and ordinal logistic results and reported in Figure 10.6 (the other categories show comparable differences). The difference between the logit and probit results is barely detectable visually. The slope of the education prediction line for men is a little steeper in the probit results (0.08 vs. 0.06), while the line for women is a little shallower in the probit results (−0.21 vs. −0.23). Keep in mind that the discrete change effect analysis above showed no significant effect of education for men in this or any category. The differences in the predicted probabilities by sex (the vertical gap between the lines) are even smaller. Within each plot, the average of the absolute value of the sex gap is 0.084. Comparing the sex gap at each level of education across the logit and probit results, the average absolute difference in the sex gap between the

FIGURE 10.6	●	PREDICTED PROBABILITY PLOTS FOR THE "NEVER" PURCHASE CATEGORY FROM ORDINAL PROBIT AND ORDINAL LOGIT

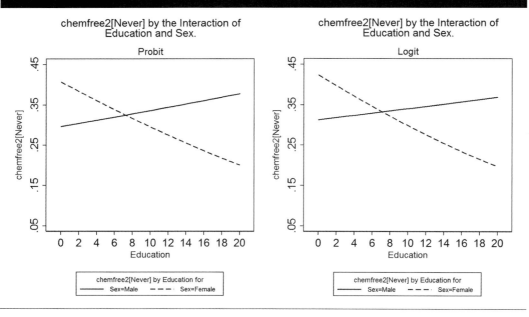

plots is only 0.002. Thus, the interpretation of the predicted probability plots from the ordinal logistic results discussed above applies equally well to the ordinal probit results.

TWO-MODERATOR INTERACTION EXAMPLE (NOMINAL BY TWO INTERVAL)

Data and Testing

This example analyzes the relationship between subjective class identification predicted by the interaction of race by log income and race by education; also included as predictors are sex, occupational status, and age. Class is a four-category ordinal measure (1 = *lower*, 2 = *working*, 3 = *middle*, 4 = *upper*). Years of education completed is an interval indicator (0–20, $\bar{X}$ = 13.7, s = 3.1), and log income is the natural log of annual household income (−2.996 to 3.147, $\bar{X}$ = 1.34, s = 1.14).[4] Sex is a dummy-coded variable (1 = *female*, 0 = *male*). Occupational status (17.1–97.2, $\bar{X}$ = 49.4, s = 19.2) and age (18–89, $\bar{X}$ = 47.5, s = 16.6) are interval measures. The corresponding index function for the latent outcome is

$$
\begin{aligned}
g(y_i) = y_i^* = {} & \beta_1 Black_i + \beta_2 Other_i + \beta_3 Education_i + \beta_4 \log(Income)_i \\
& + \beta_5 Black_i \times Education_i + \beta_6 Other_i \times Education_i \\
& + \beta_7 Black_i \times \log(Income)_i + \beta_8 Other_i \times \log(Income)_i \\
& + \beta_9 OccStatus_i + \beta_{10} Female_i + \beta_{11} Age_i
\end{aligned}
\tag{10.7}
$$

Equivalently, for an ordinal logistic regression, we can express the log odds of a response category $>k$ to a response category $\le k$ using the cumulative probabilities of the ordered categories (cp_k) as

$$
\begin{aligned}
\ln\left(\frac{1-cp_k}{cp_k}\right) = {} & \beta_1 Black_i + \beta_2 Other_i + \beta_3 Education_i + \beta_4 \log(Income)_i \\
& + \beta_5 Black_i \times Education_i + \beta_6 Other_i \times Education_i \\
& + \beta_7 Black_i \times \log(Income)_i + \beta_8 Other_i \times \log(Income)_i \\
& + \beta_9 OccStatus_i + \beta_{10} Female_i + \beta_{11} Age_i - \tau_k
\end{aligned}
\tag{10.8}
$$

The sample size after excluding cases with missing information is $N = 1{,}659$, which excludes 239 respondents with missing information for log income, another 134 missing on occupational status, and an additional 12 missing on one or more of the other variables analyzed.

The initial diagnostic testing of the parallel regressions assumption using the Brant test indicated that the data did not meet the assumption and in particular suggested that the income effect is problematic. Long and Freese (2014, chap. 7) analyzed the same outcome (subjective social class) as an example, albeit with different predictors and for a different sample (multiple years) of GSS data. They also noted the role of income in violating the assumption and speculated that income might be more consequential for lower class identification than for higher thresholds. This speculation led me to try using the natural log of income, which in essence models a diminishing effect of untransformed income at higher levels of income.

Using this transformation of income, I reran the diagnostic testing and concluded that it provided mixed but acceptable support for proceeding with an ordinal logistic regression analysis. Although the global tests of the parallel regression assumption were all significant ($p < .001$), none of the Brant tests for individual predictors were significant at $p = .05$. Two were marginally significant at $p = .10$ (the dummy for Black and the product term for the Other Races dummy by log income). I then compared the BIC and AIC statistics from the ordinal logistic regression with alternatives that did not require making the parallel regression assumption for all predictors:

- The multinomial logit model
- A generalized ordinal logistic regression model (Williams, 2016) with no predictors constrained to meet the parallel regression assumption

The BIC statistics favored the ordinal logistic regression over both alternatives, whereas the AIC supported the ordinal logistic only over the multinomial logit.

I also tried a specification with selected predictors constrained as chosen by the autofit option. This confirmed that the parallel regression assumption was correct for all but two terms, the main effects term for log income and the squared term for age. But it is not sensible to test these predictors separately since they are part of a multiple-variable functional form. The proper test is of all the terms that are part of the same multivariable functional form. I reran the generalized ordinal logistic with and without all the components of the race-by-income interaction and with and without both the linear and the quadratic terms for age. The BIC statistics for all of the combinations of constrained/unconstrained sets of predictors favor the ordinal logistic regression analysis that constrains all the predictors to meet the parallel regression assumption. The AIC favored the ordinal logistic when testing the unconstrained age specification but supported the alternative when testing the interaction of race by log income. With mixed overall support leaning toward the ordinal logistic regression and the fact that the detailed Brant test was not significant for any of the predictors individually, I decided that this provided adequate grounds for using the unconstrained ordinal logistic regression results.

Last, statistical tests of the coefficients for the two-way interactions of race by education and race by log income—$\beta_5 Black_i \times Education_i + \beta_6 Other_i \times Education_i + \beta_7 Black_i \times \log(Income)_i + \beta_8 Other_i \times \log(Income)_i$—support their inclusion in the model. A global test of the change in the log likelihood if these four terms are excluded from the model is statistically significant ($p < .001$), and the exclusion of either the pair of race-by-education or the pair of race-by-log income terms is similarly significant ($p < .002$). Additionally, z tests for one of the individual two-way coefficients in each pair are significant ($p < .05$).

Approaches to Interpreting the Two-Moderator Interaction

I begin by interpreting each component predictor in turn and how its effect is moderated by the other predictor(s). For each, I briefly discuss the GFI results and then use significance-region tables and/or effects displays to describe the nature of each moderated effect and to lay a foundation for understanding the results of tables and plots of the predicted outcome(s). I discuss the single-moderator effects first (how race moderates the education effect, how race moderates the income effect) and then consider how education and income moderate the effect of race. For the single-moderator effects, I present only significance region table results for the moderated effect of education and discuss only effect display results for the moderated effect of income to avoid repetitive interpretations and still show how to work with each type of display. To demonstrate the range of effect and interpretive options, I discuss the following:

1. The effect of education moderated by race on standardized subjective class identification

2. The factor change effect of income moderated by race on the odds of a higher class versus a lower class identification

3. The effect of race moderated by education and income, both as a factor change effect of race on the odds of higher to lower class identification and as a discrete change effect on the probability of each subjective class identification category

I then illustrate how to use different types of outcome displays to provide a more integrated portrayal of the interaction effects as a whole.

Throughout this example, I use the same Stata *ologit* command—without having to reorder the interacting predictors within the list of independent variables—because I only request SPOST13 marginal effects for the race effect moderated by education and by income, not for the effect of education moderated by race or income moderated by race (see Chapter 6 for an explanation of the ordering requirements when getting SPOST13 calculated effects):

```
. ologit class i.race##c.educ i.race##c.loginc i.female c.age
```

On the other hand, the ICALC *intspec* command to define the interaction is revised when the focal variable changes and when a moderator's display values are changed. Thus, each application of an ICALC tool reports the corresponding *intspec* command.

Moderated Effect of Education by Race on Standardized Class Identification

```
. intspec focal(c.educ) main((c.educ, name(Education) range(0(4)20)) ///
>            (i.race, range(1/3) name(Race)) ) int2(c.educ#i.race) ndig(0)
   ...
. gfi

GFI Information from Interaction Specification of
Effect of Education on q(Class) from Ordinal logistic Regression

Effect of Educ =
   0.1703 + 0.0343*BLACK - 0.1251*OTHER
```

Because the moderator (race) is nominal, it is easy and informative to write out the effect of education for each category by plugging the appropriate values of 0 or 1 into the dummy-coded indicators for race to specify each of the race groups in turn (or you could just get them from the sign change table):

- *Whites:* Specify Black = 0 and Other = 0 to get the education effect of Whites:

$$0.1703 = 0.1703 + 0.0343 \times 0 - 0.1251 \times 0$$

- *Blacks:* Specify Black = 1 and Other = 0 to get the education effect of Blacks:

$$0.2047 = 0.1703 + 0.0343 \times 1 - 0.1251 \times 0 = 0.1703 + 0.0343$$

- *Other Races:* Specify Black = 0 and Other = 1 to get the education effect of Other Races:

$$0.0452 = 0.1703 + 0.0343 \times 0 - 0.1251 \times 1 = 0.1703 - 0.1251$$

The effect of education for all three race groups is positive, predicting that their latent class identification increases with education levels. The predicted increase is similar for Blacks and Whites—the predicted increase in identification is 0.20 and 0.17 per year of education, respectively—with a much smaller moderated effect for Other Races (0.05). To better understand how big these effects are, let us look at the significance region table in which the moderated effects are fully standardized. That is, they represent the effect of a 1 standard deviation increase in education on latent class identification scaled in standard deviation units. Use the sdyx suboption in the effect( ) option to specify this:

```
. sigreg , effect(b(sdyx))
    ...

Significance Region for Effect of Education (1 s.d. difference)
on g(Class)-standardized at Selected Values of Race
```

		At Race=	
Effect of	WHITE	BLACK	OTHER
Educ	**0.2414***	**0.2901***	0.0641

```
Key: Plain font  = Pos, Not Sig    Bold font*   = Pos, Sig
     Italic font = Neg, Not Sig    Italic font* = Neg, Sig
```

These results tell us that for Whites and Blacks, education has a significant effect on the latent class identity, and a 1 standard deviation increase in education predicts about 1/4 of a standard deviation increase toward higher class identification. The moderated effect of education for the Other Races group is small—less than 1/10 of a standard deviation—and is not significantly different from an effect of 0.

The Moderated Effect of Log Income by Race as Factor Changes in the Cumulative Odds

```
. intspec focal(c.loginc) main((i.race, range(1/3) name(Race)) ///
>  (c.loginc, name(Income) range(-3(1)3))) int2(c.loginc#i.race) ndig(0)
    ...
. gfi, factor
    ...
GFI Information from Interaction Specification of
Effect of Income on g(Class) from Ordinal logistic Regression

Effect of Loginc =
  0.5744 - 0.5153 *BLACK - 0.2955 *OTHER
  Factor Change Effect (1 unit change in) Loginc =
    e^ 0.5744 * e^(- 0.5153*BLACK) * e^(- 0.2955*OTHER) =
    1.7761 * 0.5973^BLACK * 0.7442^OTHER
```

The GFI expression for the factor change effect of education for each race group is calculated by substituting 0 or 1 as appropriate into the race dummy indicators in the formulas defining the powers of *e*. Remember that any number raised to the power of 0 is equal to 1.

- *Whites:* Specify Black = 0 and Other = 0 to get the income factor change effect for Whites:

$$1.7761 = 1.7761 \times 0.5973\wedge0 \times 0.7442\wedge0 = 1.7761 \times 1 \times 1$$

- *Blacks:* Specify Black = 1 and Other = 0 to get the income factor change effect for Blacks:

$$1.0549 = 1.7761 \times 0.5973\wedge1 \times 0.7442\wedge0 = 1.7761 \times 0.5973 \times 1$$

- *Other Races:* Specify Black = 0 and Other = 1 to get the income factor change effect for Other Races:

$$1.3143 = 1.7761 \times 0.5973\wedge0 \times 0.7442\wedge1 = 1.7761 \times 1 \times 0.7442$$

These results tell us that log income has a positive effect on the odds of higher to lower class identification, which is greatest for Whites and lowest for Blacks. Rather than interpreting these factor change effects, let's focus on the effects display for the factor change effects. I also get significance region results to provide exact values for the moderated effects to use in the discussion. Because the metric for log income (log dollars) does not convey much meaning, I request factor changes for a 1 standard deviation change in log income with the factor(sd) suboptions for effect( ), and I tidy up the appearance of the plot with the specifications in pltopts( ):

```
. sigreg, effect(factor(sd))
    ...

    Significance Region for Factor Change Effect of Income (1 s.d. difference)
    on Class at Selected Values of Race
                                At Race=
    Effect of       |     WHITE       BLACK       OTHER
    ----------------+----------------------------------
        Loginc      |    1.9207*     1.0694      1.3728
    Key: Plain font  = Pos, Not Sig   Bold font*   = Pos, Sig
         Italic font = Neg, Not Sig   Italic font* = Neg, Sig
    ...

. effdisp, effect(factor(sd)) ndig(1) pltopts(ylab(.5(.5)3) ti(,size(*.9)))
```

Figure 10.7 presents an error bar plot showing the factor change effect of log income—the *diamond* marker—with confidence intervals drawn around the effect. The *horizontal dotted line* at the value of 1 on the *y*-axis represents the no-effect reference line for a factor change because the value of 1 means that the ratios of the odds are equal. You can see that the factor change effect of log income is positive for all three race groups (values >1.0) but is significant only for Whites. That is, the White confidence interval does not contain the no-effect value of 1.0, but the confidence interval for Blacks and Other Races does. For Whites, the odds of higher to lower class identification are predicted to almost double (1.92) when log income increases by 1 standard deviation. For Other Races, the odds are predicted to increase by much

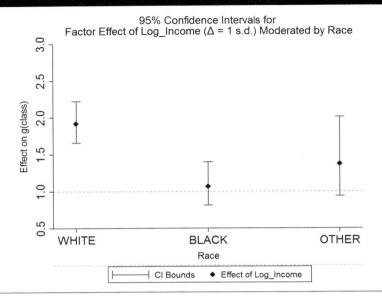

FIGURE 10.7 ● ERROR BAR PLOT OF FACTOR CHANGE EFFECT OF INCOME MODERATED BY RACE

less, a factor of 1.37 (a 37% increase), and even less for Blacks, a 25% increase; both factor changes do not significantly differ from a null effect.

While the significant factor change for Whites appears fairly substantial, it is important to keep in mind that it may not be absolutely large in terms of the probability changes. One way of assessing this is to calculate for each class category the cumulative odds and probabilities that would result from increasing the observed odds by the amount of the factor change. Then compare this with the observed cumulative odds and probabilities of higher to lower class identification for each class category. For example, the observed odds of upper class is 0.0279 (probability = 2.71%). How much would that observed odds change if log income increased by 1 standard deviation for Whites, given that log income's factor change effect is 1.92 for Whites? The formula for calculating the resulting probability of being in category k or higher, $\Pr(\geq k\ after)$, when the cumulative odds Ω is now Ω multiplied by a factor change value of FC is as follows (this uses the formula for calculating the probability from the odds shown in Chapter 9, "Aside: On the Correspondence of Odds and Probabilities"):

$$\Pr(\geq k, after) = \frac{\Omega \times FC}{1 + \Omega \times FC}$$

(10.9)

and

$$Odds(\geq k\ \text{to} < k, after) = \frac{\Pr(\geq k, after)}{\Pr(< k, after)}$$

Table 10.4 reports these cumulative odds and probabilities using the factor change in the log income for Whites ($FC = 1.92$). At the three thresholds, the factor change

TABLE 10.4 ● CUMULATIVE ODDS AND PROBABILITIES, OBSERVED AND WITH FACTOR CHANGE OF 1.92

	Cumulative Odds, At Class or Higher to Lower	Pr(At/ Higher) (%)	Pr(Lower) (%)
Observed odds			
Upper	0.0279	2.71	97.29
Middle	0.7819	43.88	56.12
Working	12.4771	92.58	7.42
Factor change × Observed odds			
Upper	0.0535	5.08	94.92
Middle	1.5012	60.02	39.98
Working	23.9560	95.99	4.01

corresponding to a 1 standard deviation change in log income predicts what are arguably substantial absolute changes in the cumulative probabilities. The probability of upper class increases by 2.4 percentage points (5.08 − 2.71), starting from a base of 2.7 percentage points; the probability of middle class or higher increases by 16 percentage points, starting from 44 percentage points; and the probability of working class or higher increases by 3.4 percentage points.

The discussion of the factor changes by race shown in Figure 10.7 described whether the log income factor change for each race is significantly different from no effect. But I sidestepped the issue of how to determine, for example, that the factor change effect of income for Blacks is different from the factor change effect for Other Races. There are several different approaches to this issue, which I address in the "Special Topics" section at the end of this chapter.

The Moderated Effect of Race by Education and Race by Log Income

I start with a brief description of what the GFI expressions tell you about this two-moderator interaction to provide an overview of the patterns. I then illustrate two approaches to probing the details of the interaction effects, one using factor change effects on the cumulative odds and the other using discrete change effects on the category probabilities.

```
. intspec focal(i.race) main((i.race, range(1/3) name(Race)) ///
>   (c.educ, name(Education) range(0(2)20)) (c.loginc, name(Log_Income)
>   range(-3(1)3))) int2(i.race#c.educ i.race#c.loginc) ndig(0)
    ...
. gfi, factor
```

```
GFI Information from Interaction Specification of
Effect of Race on q(Class) from Ordinal logistic Regression

Effect of BLACK =
    -0.3445 + 0.0343*Educ - 0.5153*Loginc

    Factor Change Effect (1 unit change in) BLACK =
        e^-0.3445 * e^( 0.0343*Educ) * e^(- 0.5153*Loginc) =
        0.7086 * 1.0349^Educ * 0.5973^Loginc

Effect of OTHER =
    2.1156 - 0.1251*Educ - 0.2955*Loginc
    Factor Change Effect (1 unit change in) OTHER =
        e^ 2.1156 * e^(- 0.1251*Educ) * e^(- 0.2955*Loginc) =
        8.2946 * 0.8824^Educ * 0.7442^Loginc
```

The effect of Black race is the expected Black–White difference in latent class identification; a negative effect means that the predicted White identification is higher
(more toward the upper class) than Black identification, while a positive effect indicates that the predicted Black identification is higher. The GFI expression shows that
the Black–White gap in latent class identification increases with education (0.0343)
but decreases with log income (−0.5153). The GFI for the factor change effect of
Black race analogously indicates that the odds of higher to lower class categories
increase relative to the odds for Whites as education increases (by a factor of 1.0349)
but diminish as log income rises (by a factor of 0.5973).

The effect of the Other Races group is also a comparison with Whites and exhibits a dissimilar pattern of moderation by education—that is, a negative effect.
Specifically, the GFI indicates that the Other Races–Whites difference in latent
identification changes in the negative direction (higher class identification for
Whites than for Other Races) with increasing education (−0.1251) and with
increasing log income (−0.2955). In terms of the factor change effect expression,
the cumulative log odds of higher to lower class categories decreases for Other
Races relative to Whites by a factor of 0.8824 for education and by a factor of
0.7442 for log income.

These brief descriptions highlight how the Black effect and the Other Races effect
change with the moderators but not whether the effects are always positive or always
negative, or if and where they change sign. While we could look at the sign change
analysis produced by the *gfi* command, it is usually more efficient (less redundant)
to use the significance region tables or effects displays to provide this information, as
well as for determining when the effect significantly differs from the no-effect value.
However, if the patterns of change are complicated, then I would suggest spending
time understanding the sign change analyses. In the next section, I demonstrate the
use of factor change effects on the cumulative odds for probing the interaction effect.
And in the following section, I consider the interpretation of discrete change effects
on the predicted probabilities of the categories.

Factor Change Effect of Race on the
Cumulative Odds (Higher Versus Lower Class)

```
. intspec focal(i.race) main((i.race, range(1/3) name(Race)) ///
>  (c.educ, name(Education) range(0(5)20)) (c.loginc, name(Log_Income) ///
>  range(-3(1)3))) int2( i.race#c.educ i.race#c.loginc) ndig(0)

. sigreg , effect(factor) save(Output\table_10_5.xlsx tab mat)
```

My initial specification of the display values for the moderators in the *intspec* command (not shown) requested a more detailed division of the range of values for the moderators—every 2 years for education and every 0.5 unit for log income. This enabled me to determine how to condense the set of display values and still capture thresholds where the effect changed significance. Given the size of the table, I present the formatted significance region table saved by ICALC to an Excel file, which better highlights sign and significance changes. Table 10.5 reports the resulting significance region table for the effect of the two race dummy indicators, Blacks and Other Races, with White as the reference group. Note that the factor change effects with a value less than 1 conceptually represent "negative" effects and hence are formatted as such in the table. Another important point to keep in mind is that there is no three-way interaction between race, education, and income. Consequently, each row in a panel exhibits the same multiplicative pattern of how the effect of race changes with education, and every column has the same pattern of changes with respect to log income.

Take a minute before you read on to study the results in Panel A for the effect of Black race in order to identify the three patterns of how this effect changes with education regardless of income, with income regardless of education, and with the combinations of values of income and education.

What you see should be the following:

- The factor change effect of Black race numerically increases (moves in the positive direction) as education increases at every income level—
  read across the rows. Specifically, between adjacent education categories (5 years apart), the effect of being Black increases by a factor of 1.0349^5 = 1.187 (look at the GFI to see why this is so). That is, the odds of higher versus lower class identification for Blacks compared with Whites increase by almost 19% ($100 \times [1.187 - 1]$). Note that at higher levels of log income, the factor change effect becomes a smaller negative effect as education rises.

- Conversely, at any education level, the factor change effect of Black race diminishes (moves in the negative direction) by a factor of 0.597 as log income increases by one unit. This means that the odds of higher versus lower class for Blacks compared with Whites drop by 40% ($100 \times [0.597 - 1]$).

- The combined moderation by education and log income creates two distinct regions of the effect of Black, divided by the level of log income. When log income is less than a value varying between −0.5 and 0.5, the factor change effect of Black is positive. Thus, Blacks relative to Whites have a larger predicted odds of higher versus lower class identification. But for income above that boundary level, the effect of Black is negative, and Blacks have a smaller expected odds than Whites of higher versus lower class identification. Around 80% of the sample falls in the upper log income region, so cases are predominantly in the region where the odds of higher versus lower class identification are predicted to be (much) smaller for Blacks than for Whites.

More briefly, the effect of Other Races follows a different pattern of moderation by education and income. The factor change effect declines (moves in the negative direction) as either education or log income increases. This creates distinct regions in the

TABLE 10.5 ⬡ SIGNIFICANCE REGION TABLE FOR THE FACTOR CHANGE EFFECT OF RACE ON CLASS

Log Income	Education				
	0	5	10	15	20
Panel A: Factor change effect of Black race on class, moderated by education and log income					
−3	3.3247	**3.9476***	**4.6872***	**5.5654***	**6.6081***
−2	1.9859	2.3580	**2.7998***	**3.3244***	**3.9472***
−1	1.1863	1.4085	1.6724	1.9857	2.3578
0	*0.7086*	*0.8413*	*0.9990*	1.1861	1.4084
1	*0.4233*	*0.5026*	***0.5967****	*0.7085*	*0.8413*
2	*0.2528*	***0.3002****	***0.3564****	***0.4232****	*0.5025*
3	*0.1510*	***0.1793****	***0.2129****	***0.2528****	***0.3002****
Panel B: Factor change effect of Other Races on class moderated by education and log income					
−3	**20.1296***	**10.7699***	**5.7622***	3.0830	1.6495
−2	**14.9790***	**8.0142***	**4.2879***	2.2941	1.2274
−1	**11.1463***	**5.9636***	**3.1907***	1.7071	*0.9134*
0	**8.2943***	**4.4377***	**2.3743***	1.2703	*0.6797*
1	**6.1720***	**3.3022***	**1.7668***	*0.9453*	*0.5058*
2	**4.5928***	2.4573	1.3147	*0.7034*	***0.3763****
3	3.4176	1.8285	*0.9783*	*0.5234*	***0.2801****
Key					
Plain font, no fill	Pos, Not Sig				
Bold*, filled	Pos, Sig				
Italic, no fill	Neg, Not Sig				
Bold italic*, filled	Neg, Sig				

diagonal corners of Panel B. The upper-left quadrant, roughly bounded by *Education* ≤ 12 and log *Income* ≤ 1, has positive and significant effects of Other Races, which decline with both education and log income. For the 20% of the sample in this region, the odds of higher versus lower class identification are predicted to be larger—often many times larger—for the Other Races than for the Whites category. The opposite holds true in the lower-right quadrant (28% of the sample) bounded by *Education* ≥ 15 and log *Income* ≥ 1.5. Here, the Other Races effect is negative and significant, meaning that the expected odds of higher to lower class identification are smaller for Other Races than for Whites.

TABLE 10.6 ●	CUMULATIVE ODDS AND PROBABILITIES, OBSERVED AND WITH FACTOR CHANGE OF 0.300		
	Cumulative Odds, At Class or Higher to Lower	**Pr(At/Higher) (%)**	**Pr(Lower) (%)**
Observed odds			
Upper	0.0279	2.71	97.29
Middle	0.7820	43.88	56.12
Working	12.4939	92.58	7.42
Factor change × Observed odds			
Upper	0.0084	0.83	99.17
Middle	0.2346	19.00	81.00
Working	3.7431	78.92	21.08

To provide a sense of the magnitude of the factor change effects of Black race and of Other Races, we can use Equation 10.9 to calculate how much the probability of higher class and lower class identification would alter as a result of the factor change. We saw in Table 10.4 that a factor change of 1.92 produced notable changes in the probabilities, and almost all of the significant positive factor change effects of Black and Other Races are larger than 2. What about factor changes in the other direction, reducing the cumulative odds at each threshold? Table 10.6 reports the results for a factor change of 0.300, the effect of Black on the cumulative odds when *Education* = 20 and log *Income* = 3; this is the median significant factor change in the negative direction in Table 10.5. This produces even more substantial absolute changes in the predicted probabilities, although in the opposite direction, for two of the three thresholds. At the threshold for upper class versus middle/working/lower class, the probability of upper class decreases by only about 2 percentage points. But this is a two-thirds reduction away from an observed probability of 2.71%, where there is limited room for this probability to decrease, so I would suggest this is still a notable change.

Theoretically, effects displays could also be used to report the factor change effects of race moderated by education and income, but plots of factor changes do not visually represent the values well if there are both positive factor changes, which can range from 1 to ∞, and negative factor changes, which can range from 0 to 1. A plot presents negative factor changes with a fraction of the vertical distinction given to positive factor changes. You could, of course, instead plot the logarithm of the factor change, which is identical to plotting the moderated effect on the latent outcome.

Discrete Change Effects of Race on the Probability of Each Class Category

Interpreting any given moderated discrete change effect of race on the probability of an outcome category is usually easier and more accessible to a wide range of audiences than a factor change interpretation. But the amount of information—numerous discrete change values presented in a number of tables and/or plots for each category of the outcome—can feel somewhat overwhelming to manage, especially when the focal variable is nominal level with more than two categories. For this example, I present and discuss a set of significance region tables separately for each of the two effects of race. This requires interpreting four significance region tables for each race effect, one for each outcome category. Although it requires some concentration, this is not as difficult as it may sound because the significance region tables are formatted to highlight sign and significance changes. After some experimentation starting with 7 to 10 display values, I decided to use 5 display values for the two moderators defining the rows and columns of the table because this captured the patterns well and using fewer display values did not.

However, for ordinal regression analyses like this example—multiple effects of the focal predictor with two or more moderators—I would not try to use effects displays. Even if you limit the display values for the second moderator to four, you would have to interpret 16 confidence bounds plots for each effect of race as it varies with education—that is, one for each combination of an outcome category (four) and display value of log income (four). Furthermore, it would be difficult to see the pattern of moderation by log income without producing a second set of plots of the effect of race as it varies with income, repeated for the display values of education.

To create the significance region tables for discrete change effects, it is important that the terms comprising the interaction are in the same order in the Stata *ologit* command and the ICALC *intspec* command (see Chapter 6). To emphasize this, the syntax listing starts with the *ologit* and *intspec* commands with the appropriately ordered predictors. In the *sigreg* command, inside the effect( ) option, the suboptions specify the following:

- spost( ): An SPOST13 calculated marginal effect

- amtopt( ): SPOST13 option for the amount of predictor change—"bin" means from 0 to 1

- atopt( ): How to set reference values for noninteracting predictors

- save( ): The path and file name in which to save the Excel formatted table

- ndig( ): The number of digits in significance table results

```
. ologit class i.race##c.educ i.race##c.loginc c.sei i.female c.age
  ...
. intspec focal(i.race) main((i.race, range(1/3) name(Race)) ///
>   c.educ, name(Education) range(0(5)20)) (c.loginc, name(Log_Income) ///
>   range(-3(1.5)3))) int2( i.race#c.educ i.race#c.loginc) ndig(1)
  ...
. sigreg, effect(spost( amtopt(am(bin)) atopt((means) _all))) ///
>   save(Output\table_10_6.xlsx tab) ndig(4)
```

Tables 10.7 and 10.8 present the significance tables first for the effect of Black on the four class categories and then for the effect of Other Races. In the subtables for the effect of Black, there is a clear but varying vertical separation (log *Income* ≤ −1.5 vs. log *Income* ≥ 1.5) of the predicted difference between Blacks and Whites in the probability of identifying in the class category. Given this, the best way to tell the story is to discuss the effect of Black when log income is low across the class categories and then the effect of Black when log income is high across the outcome categories. Note that there are no significant effects of Black when log *Income* = 0.

TABLE 10.7 ⬡ DISCRETE CHANGE EFFECTS OF BLACK ON CLASS CATEGORY PROBABILITIES, MODERATED BY EDUCATION AND LOG INCOME

Log Income	Education				
	0	5	10	15	20
Lower class					
−3.0	−.2115	−.3249*	−.3350*	−.2436*	−.1390*
−1.5	−.0936	−.1475	−.1414*	−.0948*	−.0517*
0.0	.0845	.0386	.0001	−.0115	−.0099
1.5	.2694	.1719	.0766*	.0269*	.0079
3.0	.4004	.2402	.1087*	.0416*	.0147
Middle class					
−3.0	.0132	.0376	.0987	.2206*	.3704*
−1.5	.0071	.0242	.0697	.1585	.2492
0.0	−.0090	−.0105	−.0001	.0354	.0768
1.5	−.0469	−.0867*	−.1310*	−.1374*	−.0811
3.0	−.1267*	−.2244*	−.3110*	−.2939*	−.1515

Log Income	Education				
	0	5	10	15	20
Working class					
−3.0	.1981	.2865*	.2339*	.0160	−.2513
−1.5	.0863	.1227	.0699*	−.0694	−.2146
0.0	−.0753	−.0278	−.0000	−.0256	−.0749
1.5	−.2215	−.0829	.0591*	.1196*	.0888
3.0	−.2705	−.0086	.2184*	.2866*	.2063*
Upper class					
−3.0	.0003	.0008	.0024	.0070	.0199
−1.5	.0002	.0006	.0018	.0057	.0171
0.0	−.0002	−.0003	−.0000	.0016	.0079
1.5	−.0011	−.0023*	−.0047*	−.0090*	−.0156
3.0	−.0032*	−.0072*	−.0160*	−.0344*	−.0695*

Key

Plain font, no fill	Pos, Not Sig
Bold*, filled	Pos, Sig
Italic, no fill	Neg, Not Sig
Bold italic*, filled	Neg, Sig

TABLE 10.8 ● DISCRETE CHANGE EFFECTS OF OTHER RACES ON CLASS CATEGORY PROBABILITIES, MODERATED BY EDUCATION AND LOG INCOME

Log Income	Education						Log Income	Education				
	0	5	10	15	20			0	5	10	15	20
	Lower class							Working class				
-3.0	-.6264*	-.5280*	-.3650*	-.1887	-.0592		-3.0	.5254*	.4112*	.2383*	.0689	-.0171
-1.5	-.5544*	-.3870*	-.2072*	-.0752	-.0041		-1.5	.4143*	.2332*	.0566	-.0354	-.0080
0.0	-.4057*	-.2229*	-.0886*	-.0156	.0157		0.0	.2214*	.0380	-.0596	-.0369	.0797
1.5	-.2361*	-.0999*	-.0253	.0075	.0188		1.5	.0134	-.0890	-.0701	.0434	.1759*
3.0	-.1092	-.0333	.0007	.0132	.0162		3.0	-.1230	-.1068	.0047	.1359	.2211*
	Middle class							Upper class				
-3.0	.0987	.1140	.1236	.1166	.0739		-3.0	.0023	.0028	.0031	.0032	.0023
-1.5	.1366	.1498*	.1464	.1070	.0116		-1.5	.0034	.0039	.0042	.0036	.0005
0.0	.1794	.1795*	.1431*	.0502	-.0891		0.0	.0049	.0054	.0050*	.0023	-.0063
1.5	.2158	.1821	.0909	-.0472	-.1693		1.5	.0069	.0068	.0045	-.0037	-.0255*
3.0	.2232	.1329	-.0050	-.1274	-.1657		3.0	.0090	.0072	-.0004	-.0216*	-.0716*

Key

Plain font, no fill	Pos, Not Sig
Bold*, filled	Pos, Sig
Italic, no fill	Neg, Not Sig
Bold italic*, filled	Neg, Sig

Effect of Black When Log Income ≤ −1.5. The discrete change effects of lower class identification are negative, indicating a lower predicted probability of identifying as lower class for Blacks than for Whites. But notice that the gap between Blacks and Whites rises and then diminishes as education increases. We see the same pattern of the rising and falling effect of Black on the probability of working class, but the effect is largely positive (only the positive effects are significant). The predicted Black–White gap in middle class is consistently positive—higher probability for Blacks and Whites—but only significant for the highest levels of education. And, unlike for lower and working classes, the effect magnitude steadily increases with education. Last, there are no significant effects of Black on upper-class identification, and the difference in probabilities between Blacks and Whites is small. This pattern suggests that Blacks and Whites differentially weight the importance of income and education in making a subjective assessment of their social class. That is, when income is low and education is high, Blacks appear more likely than Whites to identify as middle class than as lower or working. And when education is not the lowest but not high, Blacks are more likely to identify as working class than as lower or middle class.

Effect of Black When Log Income ≥ −1.5. Focusing only on significant effects of Black when income is high, there is a simple pattern. The predicted Black–White difference in the probability of class identification is positive for lower- and working-class identification—Blacks have a higher predicted probability than Whites. Conversely, it is negative for middle and upper class, meaning that Blacks have a lower predicted probability than Whites. These gaps primarily increase in magnitude (whether positive or negative) as income rises—that is, larger positive effects on working-class identification and larger negative effects on middle- and upper-class identification. But the effects for lower class move in the negative direction with rising education, resulting in smaller positive magnitudes for the effect of Black. This pattern is the flip side of the speculation above that Blacks give less credence to income in making class identification than Whites. That is, when income is high, Blacks are less likely to choose higher class identifications, just as when income is low they are less likely to choose the lowest class identification.

There is also a pair of coefficient regions for the effect of Other Races in Table 10.8, quite distinct from those for the effect of Black. The first region is the upper-left quadrant, where log income and education both range from their lowest to middle values, and the second is the lower-right quadrant, where both education and log income are high. As I just did, I discuss each region in turn across the four classes and then integrate them into a narrative or theme.

Effect of Other Races When Education and Log Income Are Both Low to Middle. This region is populated by significant negative effects of Other Races on lower class identification, which smoothly decline in size along both dimensions. This shows a (much) smaller predicted lower class identification for Other Races than for Whites, but the disparity diminishes as either education or log income increases. For working class, this area consists of positive significant effects—a higher predicted probability for Other Races than for Whites—declining in size with both education and log income. But it excludes several of the middle-valued education and log income cells. For both middle and upper classes, the region is primarily middle-valued cells that contain positive significant effects, showing a larger expected probability for the Other Races than for the Whites category.

This progression of effects again suggests a differential evaluation process in which Other Races translate their education and income into higher subjective class identification than Whites if education and income are below the highest levels. If their education and income are in the lower part of the range, they appear to identify as working class over lower class. But if education and income are more in the middle range, they identify as middle class.

Effect of Other Races When Education and Log Income Are Both High. This region has significant effects in only two of the class categories. The significant effects are positive for working class but negative for upper class. This pattern suggests that the highest levels of education and log income are undervalued by Other Races relative to Whites in choosing class identification.

Overall, the effect of Black and the effect of Other Races suggest a more limited range of class identification for these groups than for Whites but in somewhat different ways. When education and log income are in the lower part of their ranges, the predicted class identification for Other Races relative to Whites is shifted away from lower class and toward working or middle class. But for high education and log income, their class identification is shifted away from upper class and more toward working class. For Blacks, the compressed range of identification appears more a result of giving less weight to income than to education when making a subjective class identification.

Keep in mind that discrete change effects are subject to confounded nonlinearity due to the link function. So you need to be careful not to attribute differences in the patterns of race effects to moderation by education or log income. For this example, if you were to construct a table akin to Table 10.7 calculated from the main-effects-only model, you would see the same pattern for the lower class category of the discrete change effects rising and falling with education at low levels of log income. This suggests some degree of confounding and hence the need for caution in how you discuss these results. I return to this issue in the next section, illustrating the use of predicted probabilities to interpret the interaction effects.

OUTDISP **for the Effects of Race, Education, and Income Simultaneously**

As I have noted repeatedly, visual displays of the predicted outcome(s) are often the preferred interpretive technique. If you do work with the observed metric, I strongly recommend that you examine how the pattern of the relationship differs between predictions from the interaction model and predictions from a main effects model to guard against misinterpreting the nature of the moderated effects. Ordinal regression models are particularly challenging to interpret in the observed metric because you must interpret multiple outcomes, the probability of each category.

Alternatively, you can interpret your results in the model metric, which has a single outcome—an underlying latent interval variable—and is not subject to the confounded nonlinearities problem. If you do so, I advise standardizing the latent outcome's predicted values to provide a metric for understanding the size of the effects. I illustrate the use of predicted value tables and plots in the next sections—first the standardized latent outcome (model metric) and then the probabilities of each category (observed metric). For each table or plot, I present but do not discuss the *intspec* command used to define the interaction because they differ only in the minor details of display values, the format of labels, and whether education or income is treated as the first moderator. I do briefly note the features and specifications in the *outdisp* command used to define the desired table or plot of predicted values.

Predicted Values in the Model Metric (Standardized Latent Outcome)

Table of Predicted Values

The ICALC syntax that follows for the tables and plots of the standardized latent outcome has the same specifications in the out( ) option of the *outdisp* command: The atopt( ) suboption sets the other predictors' reference values to their means, the metric( ) suboption specifies the predicted values in the model metric, and sdy indicates that the predicted values are standardized. The rest of the syntax below is used to define the table of predicted values (Table 10.9). The tab( ) option with the save( ) suboption generates a predicted value table in the results window and saves it in an Excel file, and ndig(3) formats the table values with three decimal places. Because the later plots have different display values, the plot option is not specified.

```
. intspec focal(i.race) main((i.race, range(1/3) name(Race))(c.educ, ///
>    name(Education) range(0(5)20))(c.loginc, name(Log_Income) range(-3(1.5)3))) ///
>    int2( i.race#c.educ i.race#c.loginc) ndig(1)
    ...
. outdisp , out(atopt((means) _all) metric(model) sdy ) ///
>    tab(save(Output\Race_2Mods_table_10_9.xlsx) ) ndig(3)
```

It is easy to see the patterns of how latent class identification is predicted to vary with the education-by-race group or the income-by-race group in Table 10.9. Examine the change across a row to determine the effect of education for a race group, and the change down a column for a race group to see its income effect. Keep in mind that the education effect for each race category does not vary with log income, and neither do the income effects vary by education because there is no three-way interaction of race by education by income.

Every row shows a predicted increase in latent identification as education increases, which is manifested from the font sizes changing from smallest to largest as education increases. The predicted increase toward higher latent identification across education's range is much larger for Blacks and Whites (1.90 and 1.58 standard deviations, respectively) than for Other Races (0.41). How big are these differences? If we translate these predicted changes across the range of education into a change per standard deviation of education (3.05 years)—a fully standardized effect—class identification is predicted to change by 0.25 standard deviation or more for Whites and Blacks but by less than 0.1 standard deviation for Other Races. By most researchers' rules of thumb, the education effect for Blacks and Whites would be considered moderate.

Analogously, read down a column to see that the income effect is positive—again note the increasing font size—but varies by race. Each column has sets of three lines. Compare the top line in each triplet down a column to see the positive income effect for Whites, the middle line for the positive effect for Blacks, and the third line for the positive effect for Other Races. Across the range of income's values, the predicted change in latent identification for Whites (1.60 standard deviations) is comparable in magnitude with the education effect for Whites. It is more than double the predicted change for Other Races (0.77 standard deviation) and nearly 10 times the predicted change for Blacks (0.17 standard deviation). In fully standardized terms, the income effect for Whites (0.30) is sizable, but those for Other Races (0.15) and especially Blacks (0.03) are not. This echoes what we saw in the earlier analysis of the moderated effects of income and education by race on class identification. Income and education appear equally salient to Whites' class identification, while education

TABLE 10.9 ● *g*(Class) STANDARDIZED BY THE TWO-WAY INTERACTIONS OF RACE WITH EDUCATION AND WITH LOG INCOME

Log Income	Race	Education				
		0	**5**	**10**	**15**	**20**
−3.0	White	−2.229	−1.834	−1.438	−1.042	−0.647
	Black	−1.671	−1.196	−0.720	−0.245	0.231
	Other	−0.834	−0.729	−0.624	−0.519	−0.414
−1.5	White	−1.829	−1.433	−1.038	−0.642	−0.246
	Black	−1.630	−1.155	−0.679	−0.204	0.272
	Other	−0.640	−0.535	−0.430	−0.325	−0.220
0.0	White	−1.429	−1.033	−0.637	−0.242	0.154
	Black	−1.589	−1.113	−0.638	−0.162	0.313
	Other	−0.446	−0.341	−0.236	−0.131	−0.026
1.5	White	−1.028	−0.633	−0.237	0.159	0.554
	Black	−1.548	−1.072	−0.597	−0.121	0.354
	Other	−0.251	−0.146	−0.041	0.064	0.169
3.0	White	−0.628	−0.232	0.163	0.559	0.955
	Black	−1.506	−1.031	−0.556	−0.080	0.395
	Other	−0.057	0.048	0.153	0.258	0.363

is much more consequential than income for Blacks' identification; and conversely, income matters more than education for Other Races' identification.

It is not as easy to pull out the race differences in identification and how they vary with education and income just from examining the table. But it is quite possible to do so using head or hand calculations of the race differences in predicted values and how they change. Let's begin with the Black minus White differences in the table. Reading across the top row (lowest income level), the difference is positive and slightly increasing with education, and essentially the same pattern but with smaller Black–White differences characterizes the next two levels of log income. As income continues to rise, the difference switches sign, and Whites have a higher predicted latent identification than Blacks, but the gap decreases slightly with education. The overall pattern is that Black–White differences in class identification change with rising income from a higher predicted latent identification for Blacks than for Whites to a higher identification for Whites than for Blacks. Increasing education levels narrow the gap somewhat at lower income levels but widen it slightly at higher income levels.

Apply the same comparative process to the Other Races minus White differences in latent identification to uncover the pattern: The Other Races category generally has a higher latent class identification than the Whites category except for combinations of high education and high income levels. The positive gap is predicted to narrow with education at lower income levels but remains positive. At higher incomes, the gap switches from positive to negative as education rises. How big are the race differences in class identification? One way to assess this is the average of the absolute value of the race gap for a pair of groups. For each pair of groups, the average absolute gap in class identification is around 0.25 standard deviation for Other Races to Whites or to Blacks (0.21 and 0.25, respectively), to almost 0.5 standard deviation for Blacks to Whites (0.45), which I think most readers would consider a reasonably large difference.

In sum, working with the results in the predicted value table is an effective and relatively accessible way to present and interpret these interaction effects, which can be done with a single table of predicted values. In the next section, I illustrate the use of plots of the predicted values against the interacting predictors. These usually provide an even more accessible portrayal of the relationship, which requires less work on the part of both the researcher and the audience to understand. The minor drawback is that in a case like this with 2 two-way interactions, it often requires two sets of plots to easily see and interpret the patterns, as I do in the next section.

Plots of Predicted Values

For a nominal focal variable, the default plot is a bar chart. But if one of the moderators is interval, then a scatterplot of the predicted values against the interval moderator, with separate prediction lines for each category of the nominal focal variable, is much easier to interpret. To specify this, you use the plot( ) option with the type(scat) suboption. The first pair of *intspec* and *outdisp* commands use the standardized latent outcome as the *y*-axis and education as the *x*-axis to plot the predicted latent outcome for each race group on the same graph, with a separate graph created/calculated using each of log income's display values (Figure 10.8). The second pair switches the roles of education and log income to produce graphs for each education display value, each graph plotting the predicted latent outcome for the three race groups against the values of log income (Figure 10.9).

```
. intspec focal(i.race) main((i.race, range(1/3) name(Race)) (c.educ, ///
>   name(Education) range(0(4)20)) (c.loginc, name(Log_Income) range(-3(2)3))) ///
>     int2( i.race#c.educ i.race#c.loginc) ndig(0)
  ...
. outdisp, out(atopt((means) _all) metric(model) sdy) ///
>    plot(type(scat) name(EdClass)) pltopts(by( , tit( , size(*.8))) leg( rows(1)))
```

```
. intspec focal(i.race) main((i.race, range(1/3) name(Race))   ///
>   (c.loginc, name(Log_Income) range(-3(2)3)) (c.educ,  name(Education) ///
>   range(2(6)20))) int2( i.race#c.loginc i.race#c.educ) ndig(0)
    ...
. outdisp , out(atopt((means) _all) metric(model) sdy)  plot(type(scat ///
>   name(IncClass)) pltopts(by( , tit( , size(*.8))) leg( rows(1)))
```

Examining the scatterplots of predicted latent class identification against education in Figure 10.8 readily reveals with less mental effort the same patterns that I discussed above. But you cannot as accurately talk about the numeric differences—such as the change in class identification across the range of education—directly from the figure, which is why I suggest producing both the plots and the predicted values table so that you can supplement your interpretation of the plots with those numeric differences, as I do in the following discussion.

The slope of latent identification predicted by education is similar for Blacks (*short dashed lines*) and Whites (*solid lines*). Latent identification changes by about 1/4 of a standard deviation per standard deviation increase in education, which I would classify as a moderate effect. Also, recall that the interaction term for Black by education represents the difference in the education slope and is not significant. The Black–White gap in predicted identification is positive and grows slightly as education increases when log income is low (top two graphs), but the difference is negative and converges somewhat with higher education when log income is high (bottom two graphs).

The *long dashed line* for the predicted latent identification by education for the Other Races group is much shallower (less than 1/10 of a standard deviation per standard deviation of education) and is significantly different from the Whites slope. The Other Races–Whites gap in latent class ID is predominantly positive—a higher level of identification for Other Races than for Whites—except when both education and log income are not at low levels. You can see this in the bottom two graphs for higher income levels from the crossing of the *solid* and *long dashed lines* toward the right-hand side for higher education values. Overall, the differences among the race groups in latent identification shown in Figure 10.8 are notable; for each pair of groups, the average of the absolute value of the difference in identification is between almost 1/4 of a standard deviation (0.22) and 1/3 of a standard deviation (0.32).

To see how log income affects latent class identification and how it moderates the effect of race on latent identification, look at Figure 10.9. This presents scatterplots of each racial group's latent identification against log income for different display values of education. It is immediately clear that the slope of the class identification prediction line against log income is much steeper for Whites (*solid line*) than for the other two groups. Recall from the discussion of how race moderates log income that the effect of log income is not significantly different from 0 for either Blacks or Other Races. The magnitude of the income effect on class identification for Whites is comparable with this group's education effect, 3/10 of a standard deviation increase in latent identification for 1 standard deviation rise in log income.

You can determine the race gaps in latent class identification by comparing the changing vertical gap between the lines for pairs of race groups. The Black–White differences change with income in an identical pattern regardless of level of education. At low to moderate levels of income, Blacks' predicted class identification is higher than that for Whites. But at moderate to high levels of income, Blacks' predicted identification is lower than that for Whites. The same is not true for how the Other Races–White gap changes with income levels. When education

FIGURE 10.8 ● SCATTERPLOT OF STANDARDIZED LATENT CLASS
BY EDUCATION

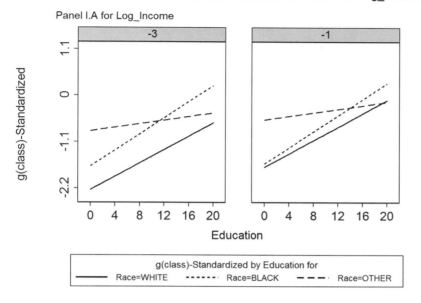

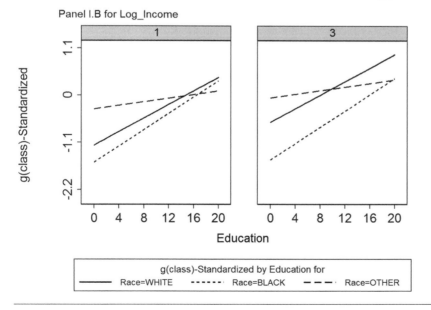

is low, Other Races' predicted latent class identification is higher than that for Whites, but the differences diminish as log income rises. On the other hand, when education is high, the difference changes from positive to negative as log income increases. That is, the predicted latent identification for Other Races changes from

FIGURE 10.9 ● **SCATTERPLOT OF STANDARDIZED LATENT CLASS BY INCOME**

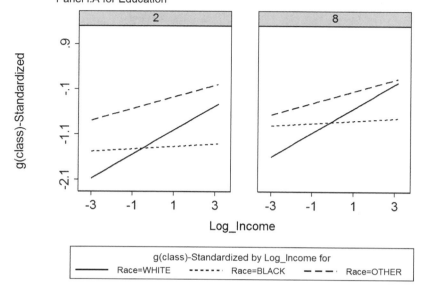

g(class)-Standardized by the
Two-way Interactions of Race with Log_Income and Race with Education.

Panel I.A for Education

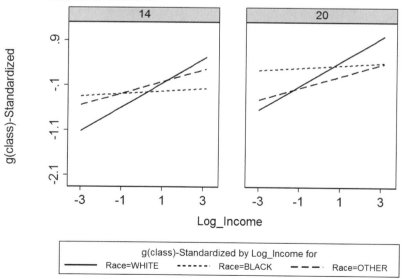

g(class)-Standardized by the
Two-way Interactions of Race with Log_Income and Race with Education.

Panel I.B for Education

higher than Whites' to lower than Whites' as the level of log income rises. As described above, the differences in latent class identification between pairs of race groups are fairly sizable, ranging from 1/4 of a standard deviation to 1/3 of a standard deviation on average.

In conclusion, regardless of whether you decide to use a predicted values table or a visual display of predicted values to interpret and explain the nature of an interaction effect, I would highly recommend that you always get both. If you decide to interpret the predicted values table, the plots will be useful for helping you see the patterns of how the interacting predictors affect the (latent) outcome. Indeed, this is what I did in writing up this example. And if you use visual displays instead, the predicted values table will be useful to help you add specific numeric illustrations to your interpretation. The same is true if you decide to use and interpret predicted probabilities of the categories, which I discuss next.

Predicted Values in the Observed Metric (Probability of Outcome Categories)

Interpreting interaction effects on the probability of the outcome categories is complicated by the need to both discuss the pattern of interaction for a given category and construct an overall integrative story about the interactive relationship. I illustrate this interpretation with a primary discussion of the predicted value plots, followed by a brief consideration of the predicted value tables. I then revisit the plots to show how the interpretation of the predicted probabilities from the interactive model are influenced by the confounded nonlinearities.

Predicted Value Plots

I use two sets of plots to interpret the effects of the interaction of race by education and race by income on the predicted probabilities of each class. The first set plots the predicted probabilities against education, with separate prediction lines for each race and a separate graph for the display values of income (Figure 10.10). Because it is difficult to determine the effect of income from this set of plots, I also produce scatter-plots of the predicted probabilities against income repeated for the display values of education (Figure 10.11). These plots also superimpose the main effects predictions onto the interactive effects plots. Consequently, I run and store the results of the interactive model and the corresponding main effects model for each set of plots. I show these to remind readers that the predictors in the main effects model must be listed in the same order as in the corresponding interactive effects model.

```
    quietly {
        ologit class  i.race##c.educ i.race##c.loginc c.sei  i.female c.age
        est store intmoded
        ologit class  i.race c.educ c.loginc c.sei  i.female c.age
        est store mained
    }
    . est restore intmoded
        …
    . intspec focal(i.race) main((i.race, range(1/3) name(Race)) ///
    >   (c.educ, name(Education) range(0(4)20)) (c.loginc, name(Income) ///
    >    range(-3(2)3))) int2(i.race#c.educ i.race#c.loginc) ndig(0)
        …
    . outdisp, out(atopt((means) _all) main(mained)) plot(type(scat) name(RaceEd)) ///
    >   tab(save(Output\Table_10_10.xlsx)) ndig(3) pltopts(ylab( , format(%5.2f)))
        …

    quietly {
        ologit class i.race##c.loginc  i.race##c.educ c.sei  i.female c.age
        est store intmodinc
        ologit class i.race c.loginc  c.educ c.sei  i.female c.age
        est store maininc
    }
        …
```

```
. est restore intmodinc
intspec focal(i.race) main((i.race, range(1/3) name(Race)) ///
>   (c.loginc, name(Income) range(-3(2)3)) (c.educ, name(Education) ///
>   range(0 6.33 12.67 20))) int2(i.race#c.educ i.race#c.loginc) ndig(0)
    ...
outdisp, out(atopt((means) _all) main(maininc)) plot(type(scat) name(RaceInc)) ///
>   ndig(2) pltopts(ylab( , format(%5.2f)))
```

Rather than describing the results in Figure 10.10 for each class in turn, I find that discussing each interacting predictor's effects across the set of classes is more effective in telling a holistic story. Focus for now on the *black lines* for the interaction model predictions. Let's start with the relationship between education and the predicted probability of each class, which, not surprisingly, describes a mechanism that shifts people's identification to higher classes at higher levels of education but to a varying degree across race. Look at the probability of lower class identification in Panel I. The *monotonically declining curves*—the larger probabilities of lower class identification for low education and smaller ones for high education—show a negative relationship between education and lower class identification. You can also see that the predicted probability changes relatively little across education levels for the Other Races group (*large dashed line*). In contrast, the predicted probability starts higher for the other two groups and declines more rapidly for Blacks (*small dashed line*) than for Whites (*solid line*). This indicates that their identification is predicted to shift to higher class categories.

Turning to Panel II for working-class identification, there is again much less of a relationship between education and working-class identification for the Other Races group. Note the relatively high probabilities, which diminish as education rises to a small but varying degree, suggesting a small possible shift into higher class identification. For the Black and White groups, the effect of education on working-class identification shows a parabolic pattern, rising to a peak and then falling as education continues to increase, which is the epitome of a transfer mechanism. That is, the rising working-class probabilities from low education to middle levels are the counterpart of the initially higher but decreasing probabilities of lower class identification. As education increases from its lowest values, identification is predicted to shift to the working class. And the predicted decline in working-class probabilities at high education levels reflects identity being transferred to higher classes.

But the parabolic pattern is incomplete for Whites, in the sense that it shows a lesser drop-off in the probability of working-class identification with higher education when income is low but more of a drop-off with education when income is high. This is indicative of a greater shift in identification to higher classes more broadly. Panel III shows a continuation of this transfer process, with increased probabilities of middle-class identification, which accelerate at higher levels of education. The last panel (Panel IV for upper-class identification) shows the final piece of the transfer mechanism. The predicted probability of identification for all groups remains relatively low until the highest levels of education (especially when coupled with higher income levels), at which point it rises more sharply for Blacks than for the Other Races category but not for Whites.

FIGURE 10.10 ● **PREDICTED PROBABILITIES OF CLASS BY EDUCATION, INTERACTION MODEL WITH ADDED MAIN EFFECTS MODEL PREDICTIONS**

class[LOWER_CLASS] by the
Two-way Interactions of Race with Education and Race with Income
with Added Main Effect Predictions

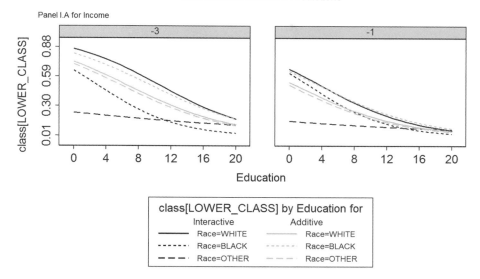

class[LOWER_CLASS] by the
Two-way Interactions of Race with Education and Race with Income
with Added Main Effect Predictions

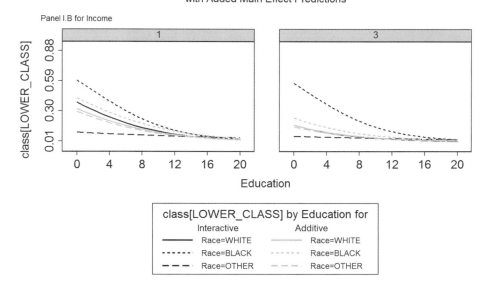

(Continued)

FIGURE 10.10   ●   (CONTINUED)

class[WORKING_CLASS] by the
Two-way Interactions of Race with Education and Race with Income
with Added Main Effect Predictions

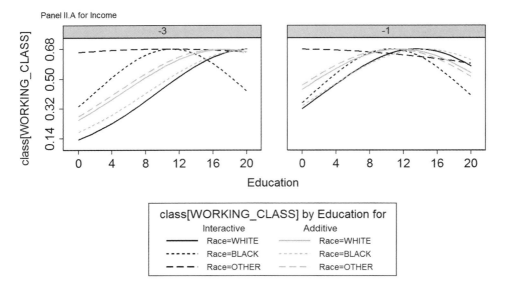

class[WORKING_CLASS] by the
Two-way Interactions of Race with Education and Race with Income
with Added Main Effect Predictions

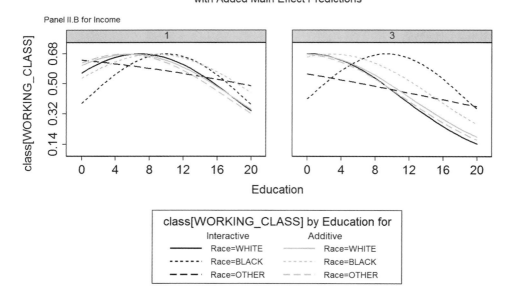

class[MIDDLE_CLASS] by the
Two-way Interactions of Race with Education and Race with Income
with Added Main Effect Predictions

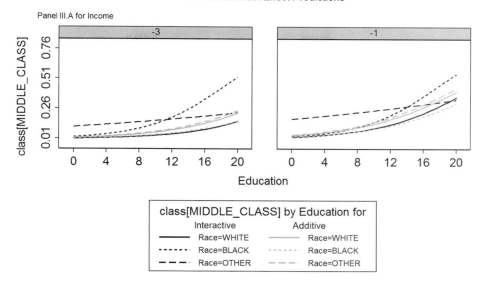

class[MIDDLE_CLASS] by the
Two-way Interactions of Race with Education and Race with Income
with Added Main Effect Predictions

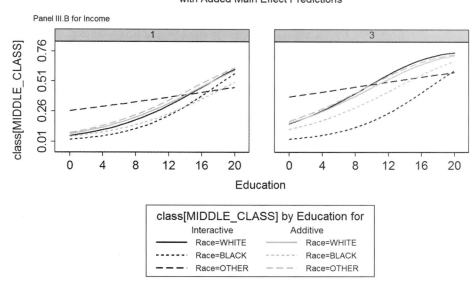

(Continued)

FIGURE 10.10 ● (CONTINUED)

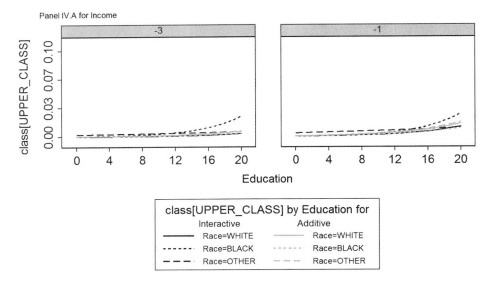

class[UPPER_CLASS] by the
Two-way Interactions of Race with Education and Race with Income
with Added Main Effect Predictions

Panel IV.A for Income

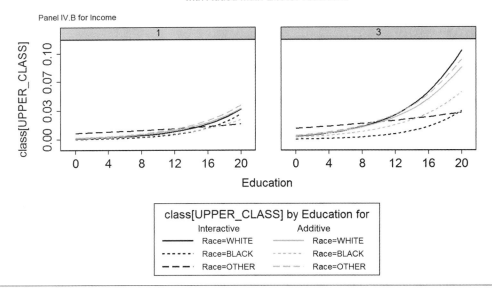

class[UPPER_CLASS] by the
Two-way Interactions of Race with Education and Race with Income
with Added Main Effect Predictions

Panel IV.B for Income

I purposely minimized in this interpretation a description of how the education effects change with the level of income because these differences are not the result of the interaction specification. That is, there is no interaction between education and log income. Consequently, the change in the effect of education for Whites, for example, with income shows the influence of the link function nonlinearity. The following discussion of the effect of log income similarly shows how the income effect varies with education.

I use the plots in Figure 10.11 to interpret the patterns of predicted probability by log income because it is very difficult to discern them in Figure 10.10. We see the same type of transfer mechanism as for education, with the key difference that income is less consequential for Blacks' predicted class identification than education. Panel I of Figure 10.11 portrays both points quite clearly. The uniformly negative effect of log income on the probability of lower class identification shows the first transfer process step—the declining probability of lower class identification requires an increasing probability of higher class identification. And Blacks' predicted probability of lower class identification has a shallower slope (*small dashed line*) than the Other Races prediction line (*large dashes*) and an even shallower slope than for Whites (*solid line*), showing a lesser predicted influence of income for Blacks.

The working-class predicted probability plots in Panel II portray a parabolic relationship between income and class identification for Whites and Other Races. This supports the transfer process imagery in which increasing income levels initially predict a falling probability of lower class identification offset by a higher probability of working-class identification. The following diminishing probability of working-class identification as income rises is offset by a predicted increase in higher class identification. In contrast, for Blacks, higher levels of income always slightly increase the predicted probability of working-class identification when education is low but the opposite when education is high. The predicted probability plots for both middle-class and upper-class identification (Panels III and IV, respectively) complete the picture of the shifting identification process. Low probabilities of identification in these two classes correspond to higher predicted probabilities of identification as working class, while high predicted probabilities here correspond to lower predicted probabilities for working-class identification. And you continue to see little predicted influence of income on Blacks' class identification.

Last, we need to interpret how race is related to class identification, which we can do from either Figure 10.10 or Figure 10.11 by examining the vertical gap between the plotted lines for pairs of race groups. Using Figure 10.10, how the gap changes with education levels within a scatterplot portrays the interactive effect of education on race, while comparing the gap between scatterplots for different levels of income reveals the interactive effect of income on race. The Black–White gap in the probability of lower class identification (Panel I) changes with education in a different way depending on low versus high income, as we saw in the prior discussion of discrete change effects.

At low income levels, Whites have a higher predicted probability of lower class identification than Blacks. Both groups' predicted probability declines with education,

FIGURE 10.11 ● PREDICTED PROBABILITIES OF CLASS BY LOG INCOME, INTERACTION MODEL WITH ADDED MAIN EFFECTS MODEL PREDICTIONS

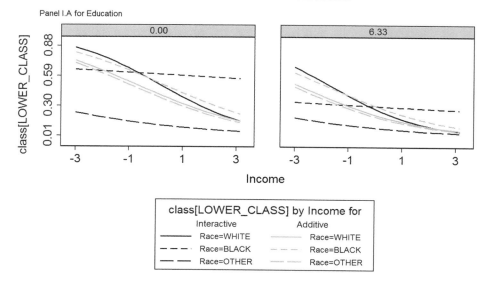

class[LOWER_CLASS] by the Two-way Interactions of
Race with Income and Race with Education
with Added Main Effect Predictions

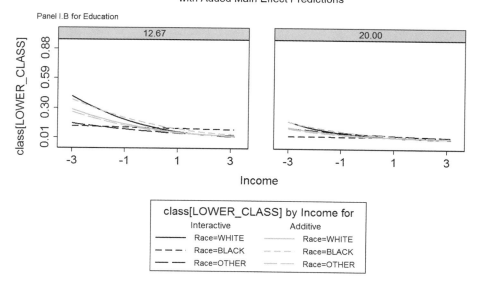

class[LOWER_CLASS] by the Two-way Interactions of
Race with Income and Race with Education
with Added Main Effect Predictions

class[WORKING_CLASS] by the Two-way Interactions of
Race with Income and Race with Education
with Added Main Effect Predictions

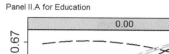

Panel II.A for Education

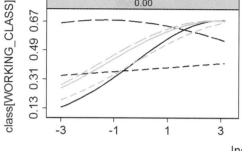

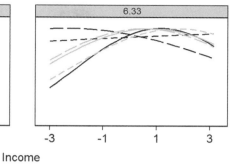

Income

class[WORKING_CLASS] by Income for

Interactive	Additive
Race=WHITE	Race=WHITE
Race=BLACK	Race=BLACK
Race=OTHER	Race=OTHER

class[WORKING_CLASS] by the Two-way Interactions of
Race with Income and Race with Education
with Added Main Effect Predictions

Panel II.B for Education

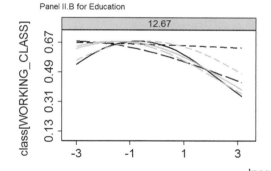

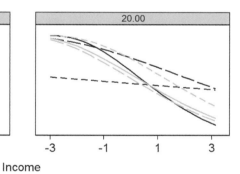

Income

class[WORKING_CLASS] by Income for

Interactive	Additive
Race=WHITE	Race=WHITE
Race=BLACK	Race=BLACK
Race=OTHER	Race=OTHER

(Continued)

FIGURE 10.11 ⬡ (CONTINUED)

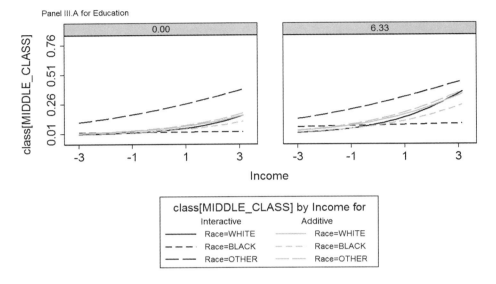

class[MIDDLE_CLASS] by the Two-way Interactions of
Race with Income and Race with Education
with Added Main Effect Predictions

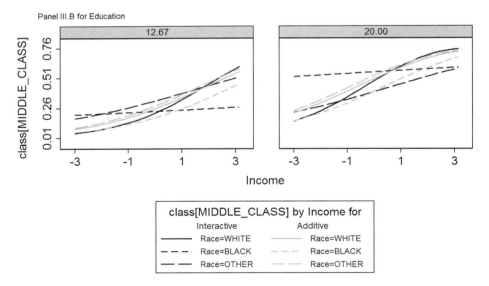

class[MIDDLE_CLASS] by the Two-way Interactions of
Race with Income and Race with Education
with Added Main Effect Predictions

class[UPPER_CLASS] by the Two-way Interactions of
Race with Income and Race with Education
with Added Main Effect Predictions

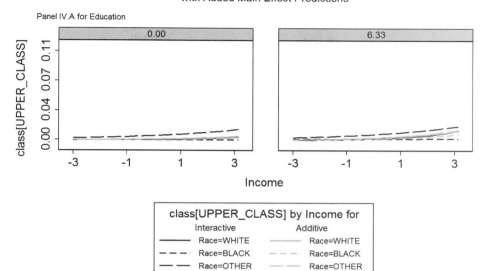

class[UPPER_CLASS] by the Two-way Interactions of
Race with Income and Race with Education
with Added Main Effect Predictions

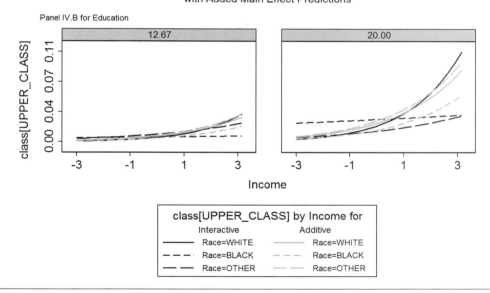

but the difference between the groups initially increases and then begins to drop. At high income levels, Blacks have a higher predicted probability of lower class identification, and the gap between the groups declines as education rises. This suggests that Whites' class identification shifts less into higher classes with rising education when income is low (higher probability of lower class) but Blacks identification shifts less into identifying with higher classes when income is high (higher probability of lower class). But in either case, the probabilities converge to very small values when education rises to its highest levels.

Turning to Panel II for working-class identification, Blacks' predicted identification is higher than Whites' except for the combination of high education and low income or low education and high income. This indicates that at low income, Blacks have a lower predicted chance to transfer to higher class identification than Whites (higher predicted working class) for low- to middle-range education. But Blacks have a greater predicted shift to higher class identification than Whites when education is high. Conversely, when income is high, Whites have a smaller expected chance of transfer to higher class identification when education is low. The pattern of identification shift is simpler for the middle-class and upper-class categories. When income is low, the predicted probability of Blacks' identification as middle class is greater than Whites' and rises more with education, corresponding to a greater transfer out of working class. In contrast, when income is high, Whites' predicted probability of identification starts and remains higher with increasing education than Blacks'. This again corresponds to an identity transfer out of working class. Part of what underlies these patterns of transfers is that Blacks' predicted probabilities for class identification are less affected by income than Whites'.

The Other Races–White gap exhibits a somewhat simpler pattern because, as the earlier effects analysis demonstrated, education and income exert relatively little influence on the class identification probabilities for the Other Races group. Thus, the patterns of the gap are primarily determined by the effects of education and income for Whites. Looking at Panel I, the low and relatively unchanging probability of lower class identification for the Other Races group means that this group has a higher predicted probability of identifying as one of the higher classes. But the Other Races–White differences converge at the highest values of education.

For working-class identification, the Other Races category has a greater and invariant predicted probability compared with the Whites category except at the maximum value of education. As income rises, the point at which Whites' predicted probability becomes greater than the Other Races probability shifts to lower values of education. Eventually, a second transition value emerges at which the higher probability reverses back to Other Races. The overall picture is that Other Races have a relatively high and stable predicted probability of working-class identification, which is greater than that for Whites when income and education are either equally low or equally high. How the Other Races–Whites gap varies with income and education is the same for middle-class and upper-class identification. When income and education are both low, the expected probability of identification is greater for the Other Races category. As income rises, the probability for Whites becomes greater than that for Other Races when education exceeds a threshold point, with the threshold point at a lower value of education as income increases.

Behind these detailed patterns of change is a relatively straightforward summary of the relationship between class identification and race, varying with education and income. For all three groups, increasing education and income predict changes in the probability that shift the probability of identification from lower to higher classes. For Whites, the probability of class identification depends more on education and income than it does for Blacks and Other Races. Thus, the shift of the probability of identification from lower to higher classes is smaller than for the other two groups when income and education are both low but larger when income and education are both high. For Blacks, the predicted probability of identification changes primarily as education increases and relatively little with income, while the opposite is true for Other Races.

Predicted Value Tables

Note: The same *intspec* and *outdisp* commands produced Table 10.10 and Figure 10.10. They are repeated here to clearly document how to produce this table:

```
. intspec focal(i.race) main((i.race, range(1/3) name(Race)) ///
>   (c.educ, name(Education) range(0(4)20)) (c.loginc, name(Income) ///
>    range(-3(2)3))) int2(i.race#c.educ i.race#c.loginc) ndig(0)
    …
. outdisp, out(atopt((means) _all) main(mained)) plot(type(scat) name(RaceEd)) ///
>   tab(save(Output\Table_10_10.xlsx)) ndig(3) pltopts(ylab( , format(%5.2f)))
```

Table 10.10 reports the probabilities predicted by the interaction of race, education, and log income for the four class categories. The predicted probabilities in the subtable for each class are formatted with font sizes proportionate to their relative values within that subtable (class category)—not relative to the table as a whole—to highlight the patterns within each class. Given the extensive discussion above of the predicted value plots, I comment only briefly on how to read the tables to see some of the effects already discussed. The organization of the table makes it particularly easy to see the relationship between education and the predicted probabilities of class as they vary by race. Reading across a row shows the relationship between education and the predicted probability for a given race; comparing the rows shows how the education effect varies by race.

In Panel I for lower class, each row shows the predicted probability increasing from left to right across the row with higher values of education—that is, a negative effect. And the comparison among race groups shows that the change in probability across the range of education is much less for Other Races than it is for Whites or Blacks. Moreover, the change in predicted probability is higher for Whites than for Blacks when log income is low but higher for Blacks than for Whites when log income is high. It is also straightforward to determine the pattern of the education effect for working class—relatively high probabilities with little change by education levels for Other Races, predicted probabilities that rise and fall with education for Blacks, and predicted probabilities for Whites whose pattern changes with increasing income levels from strictly increasing to increasing, then decreasing, to strictly decreasing. I leave it to the readers to convince themselves of these patterns and then to examine the patterns in the subtables for middle class and upper class to verify that they come to the same conclusion about the patterns as described above.

Note that you can read the income effect for race groups from these tables the same way we did in Table 10.9. Reading down a column, compare the top cell in each triplet of rows for the White effect, the middle cell for the Black effect, and the bottom cell for the Other Races effect. The key difference is that you must do that with these tables for each class category. As we just saw comparing the prediction lines, it is less straightforward but possible to derive the effect of race for each class category

TABLE 10.10 ● PREDICTED PROBABILITIES OF CLASS, INTERACTION MODEL WITH ADDED MAIN EFFECTS MODEL PREDICTIONS

		Education					
Loginc	Race	0	4	8	12	16	20
class[LOWER_CLASS] by the Two-Way Interactions of Race With Education and With Loginc							
−3	White	.860	.756	.610	.442	.286	.169
	Black	.648	.448	.264	.136	.065	.030
	Other	.233	.202	.175	.150	.129	.110
−1	White	.660	.495	.332	.201	.113	.060
	Black	.621	.419	.241	.123	.058	.027
	Other	.148	.127	.108	.092	.078	.066
1	White	.381	.237	.136	.074	.039	.020
	Black	.592	.391	.220	.111	.052	.024
	Other	.091	.077	.065	.055	.046	.039
3	White	.163	.090	.048	.025	.013	.006
	Black	.564	.363	.201	.100	.047	.021
	Other	.054	.045	.038	.032	.027	.023
class[WORKING_CLASS] by the Two-Way Interactions of Race With Education and With Loginc							
−3	White	.135	.232	.367	.514	.630	.679
	Black	.333	.509	.644	.676	.592	.428
	Other	.660	.672	.679	.679	.673	.662
−1	White	.322	.469	.600	.673	.665	.578
	Black	.358	.533	.656	.671	.571	.402
	Other	.679	.673	.661	.643	.621	.593
1	White	.563	.658	.676	.613	.487	.339
	Black	.383	.556	.665	.663	.550	.376
	Other	.642	.619	.591	.559	.524	.487
3	White	.679	.641	.531	.385	.247	.144
	Black	.409	.577	.673	.653	.526	.351
	Other	.556	.521	.484	.445	.405	.366

Loginc	Race	Education					
		0	**4**	**8**	**12**	**16**	**20**
class[MIDDLE_CLASS] by the Two-Way Interactions of Race With Education and With Loginc							
−3	White	.006	.011	.022	.043	.081	.148
	Black	.019	.042	.090	.183	.333	.519
	Other	.105	.123	.143	.167	.193	.222
−1	White	.018	.035	.067	.124	.217	.350
	Black	.021	.047	.100	.201	.359	.545
	Other	.169	.195	.225	.257	.293	.330
1	White	.055	.102	.183	.304	.456	.606
	Black	.024	.053	.111	.220	.385	.571
	Other	.260	.296	.334	.373	.415	.456
3	White	.154	.262	.407	.562	.685	.746
	Black	.027	.059	.124	.241	.412	.595
	Other	.377	.418	.460	.501	.542	.581
class[UPPER_CLASS] by the Two-Way Interactions of Race With Education and With Loginc							
−3	White	.000	.000	.000	.001	.002	.004
	Black	.000	.001	.002	.005	.011	.024
	Other	.002	.003	.003	.004	.005	.006
−1	White	.000	.001	.001	.003	.006	.011
	Black	.000	.001	.002	.005	.012	.026
	Other	.004	.005	.006	.007	.009	.010
1	White	.001	.002	.005	.009	.018	.035
	Black	.001	.001	.003	.006	.013	.030
	Other	.007	.009	.011	.013	.015	.018
3	White	.004	.007	.015	.028	.055	.103
	Black	.001	.001	.003	.007	.015	.033
	Other	.013	.015	.018	.022	.026	.031

by comparing the Black with the White entries and the Other Races with the White entries in a triplet. It takes time and effort to work out the patterns of how those differences change with education and log income.

Evaluating Confounded Nonlinearities in the Interactive Effect Model's Predicted Probabilities

The prior description of the relationship between race and the predicted probabilities of the class categories as it varies by education and by income did not take into account to what degree the patterns reflect the interactive effects of these predictors versus the nonlinearity due to the link function. You can directly assess this with two types of comparisons of the main effects model prediction lines. First, the extent to which the main effects prediction lines are not parallel reveals the potential for influence from the link function nonlinearity because the link function is the only source of nonlinearity in the main effects predicted probabilities. Second, the degree of divergence between a main effects prediction line and its corresponding interaction effects prediction line indicates the degree of moderation. The plots in Figures 10.10 and 10.11 show both the interactive model predictions (in *black*) and the main effects model predictions (in *light gray*) using the same line style for each race group in both sets (*solid line* for Whites, *short dashed line* for Blacks, *long dashed line* for Other Races).

Overall, the main effects prediction lines are fairly parallel to one another in terms of their slopes by education or by income in each of the class categories. They do differ in terms of the vertical gaps between pairs of prediction lines for race groups for a given class category. For the lower class predicted probabilities, the main effects model predicted gaps are smaller for high income than for low income and for high education than for low education. This indicates that the patterns identified in the discussion above would be somewhat of an exaggeration of the degree of moderation of the race gaps by education and by income because they are enhancing what would be predicted solely from the nonlinearity of the model.

For the working-class main effects model predictions, the patterns in the race gaps are more complicated. The gaps are small at low education levels regardless of income but change from small to increasingly large at high education when income changes from low to high. The gaps by income are larger at low income and smaller at high income when education is low but reverse and become smaller at low income and larger at high income when education is high. This suggests that the racial gaps in the interactive model for working class may be understated to some degree when education is low for any level of income, when education is high and income is low, or when income is high and education is low. For the middle- and upper-class predicted probabilities, the magnitude of the racial gaps is small at low education levels for all incomes, and the gap is larger at high education and increasingly so as income rises. When income is low, the gaps are small but increase with education levels, while the gaps are larger when income is high, and they grow and then diminish as education increases. This suggests that the racial gaps at low education for all income levels may be understated but the gaps at low income levels and higher education levels may be overstated.

If after reading the last section on interpretations using predicted probabilities, you come to the conclusion that this is too complicated a process for understanding the interactive effects of race, education, and income on class identification—let alone for trying to parse out the interactive effects from the nonlinearities of an ordinal regression model—then you are in agreement with my perspective. I think that working with the predicted latent outcome in ordinal regression models, especially for interaction effects, is the best course of action, and I highly recommend it. You have interaction effects on a single outcome to interpret rather than interpreting the effects on the multiple outcome categories. You can standardize the latent outcome to provide a metric for interpretation, which while still abstract conveys some meaning about the magnitude of the effects and differences. And by definition there is only a single source of nonlinearity in the predicted values, that which you modeled by how you specified your interaction.

SPECIAL TOPICS

Testing the Equality of Factor Change Effects for Different Moderator Values

The notation for variable names for the focal and moderating variables and for the values of the moderating variables used in the calculating options below are as follows:

Variable Type	Variable Name	Variable Value	Name in Factor-variable Notation
Focal	Fvar	—	c.Fvar
Moderator 1	Mvar1	m1	m1.Mvar1
Moderator 2	Mvar2	m2	c.Mvar2
⋮	⋮	⋮	⋮
Product of Focal and Moderator 1	FvarByMvar1		c.Fvar #m1.Mvar1
Product of Focal and Moderator 2	FvarByMvar2		c.Fvar #m2.Mvar2
⋮	⋮		⋮
Three-way product	FvarByMvar1ByMvar2		c.Fvar#m1.Mvar1#m2.Mvar2

The equation name used to define the names of coefficients is notated by *eqname*. The fourth column provides an example of how you would use factor-variable names if Moderator 1 is categorical and the other predictors are interval. Each option is applied to two examples:

1. Is the factor change effect of log income equal for the Black and Other Races categories?

2. Is the factor change effect of Black when *Education* = 12 and log *Income* = 1 equal to the effect when *Education* = 16 and log *Income* = 3? This is the type of question you might ask if you were using ideal types to define reference values for your moderators.

Option 1: Stata *test* Command

You can use the *test* command to perform a Wald test that the moderated effect of the focal variable *Fvar* at specific values of its moderators—*Mvar1* = *m1*, *Mvar2* = *m2*, . . .—is equal to its moderated effect at different specific values of its moderators—*Mvar1* = *m1*$_a$, *Mvar2* = *m2*$_a$, ... Technically, this tests the equality of the effects on the log odds and not on the odds, but the tests are asymptotically equivalent (Sribeny & Wiggins, n.d.). It is unclear which has better finite sample properties, although the Stata FAQ makes an argument in favor of the first test in the "natural perimeter space." Conceptually, the generic Stata *test* command assesses the expression as follows:

> test   Moderated effect value at (Mvar1 = m1, Mvar2 = m2, . . .) =
>
> Moderated effect value at (Mvar1 = m1$_a$, Mvar2 = m2$_a$, . . .)

To run this command, you need to write the moderated effects in terms of the estimated coefficients and the specified values of the moderators. For a single-moderator specification, the Stata command for this test is

```
test _b[eqname:Fvar] + _b[eqname:FvarByMvar1] * m1 = ///
    _b[eqname:Fvar] + _b[eqname:FvarByMvar1] * m1ₐ
```

For a two-moderator specification, the command is

```
test _b[eqname:Fvar]+ _b[eqname:FvarByMvar1]*m1 + ///
    _b[eqname:FvarByMvar1]*m2 = ///
    _b[eqname:Fvar]+ _b[eqname:FvarByMvar1]* m1ₐ ///
    _ +b[eqname:FvarByMvar2]* m2ₐ
```

For a three-way interaction specification, the command is

```
test _b[eqname:Fvar]+ _b[eqname:FvarByMvar1]*m1 + ///
    b[eqname:FvarByMvar1]*m2 ///
    _b[eqname:FvarByMvar1ByMvar2]*m1*m2  =       ///
    _b[eqname:Fvar]+ _b[eqname:FvarByMvar1]*m1ₐ + ///
        _b[eqname:FvarByMvar2]*m2ₐ ///
    _b[eqname:FvarByMvar1ByMvar2]* m1ₐ * m2ₐ
```

You must modify the above expressions if you use factor-variable notation to define your interactions. Replace the variable name with the factor-variable notation name shown above, and do not multiply by the value of the factor variable if it is categorical. For instance, the two-moderator command for *Mvar1* (categorical) and the other two variables (interval) would be

```
test _b[eqname:c.Fvar]+_b[eqname:c.Fvar#m1.Mvar1]+ ///
    _b[eqname:c.Fvar#c.Mvar2]*m2 ///
= _b[eqname:c.Fvar]+_b[eqname:c.Fvar#m1ₐ.Mvar1]+_ ///
    b[eqname:c.Fvar#c.Mvar2]*m2ₐ
```

Test the equality of the income factor change effect for Blacks and Other Races:

```
. test _b[class:c.loginc]+_b[class:c.loginc#2.race] =
_b[class:c.loginc]+_b[class:c.loginc#3.race]

 ( 1)  [class]2.race#c.loginc - [class]3.race#c.loginc = 0

        chi2(  1) =    1.11
      Prob > chi2 =    0.2920
```

Conclude that the factor change effect of log income is the same for Blacks and Other Races.

Test the equality of the factor change effect of Black at different education and log income values:

```
. test _b[class:2.race]+_b[class:2.race#c.educ]*12+_b[class:2.race#c.loginc]*1= ///
>       _b[class:2.race] +_b[class:2.race#c.educ]*16 +_b[class:2.race#c.loginc]*3

 ( 1)   - 4*[class]2.race#c.educ - 2*[class]2.race#c.loginc = 0

              chi2(  1) =      9.81
           Prob > chi2 =    0.0017
```

Conclude that the factor change effect of Black differs for the pair of education–income values.

Option 2: Stata *testnl* Command

This option tests the equality between two factor change effects by using a Wald test of a nonlinear expression of the estimated coefficients and applying the Delta method to calculate the variance of the nonlinear expression (*Stata Base Reference Manual Release 14*). The factor change effect of the focal variable is mathematically equal to the exponentiated value of the moderated effect of the focal variable on the log odds:

$$Factor\ change = e^{Moderated\ effect\ on\ ln\ Odds}$$

Thus, the nonlinear expression we specify for the *testnl* command is the exponentiated value of the moderated effect on the log odds at the first set of moderator values equals the exponentiated value of the moderated effect at the second set of moderator values, where exp( ) calculates the exponentiated value:

testnl   exp(Moderated effect value at [Mvar1 = m1, Mvar2 = m2, . . .]) = ///
exp(Moderated effect value at [Mvar1 = m1$_a$, Mvar2 = m2$_a$, . . .])

To apply this, you need to write the moderated effects inside the term exp( ), using the estimated coefficients and the specified values of the moderators just as we did above. For example, the two-moderator interaction expression would be

testnl exp(_b[eqname:Fvar] + _b[eqname:FvarByMvar1] * m1 + ///
_b[eqname:FvarByMvar1] * m2) = ///
exp(_b[eqname:Fvar] + _b[eqname:FvarByMvar1] * m1$_a$ + ///
_b[eqname:FvarByMvar2] * m2$_a$)

Applying this to the same two examples as above,

```
. testnl exp(_b[class:c.loginc]+_b[class:c.loginc#2.race]) =  ///
>    exp(_b[class:c.loginc]+_b[class:c.loginc#3.race])

  (1)    exp(_b[class:c.loginc]+_b[class:c.loginc#2.race]) =
     >       exp(_b[class:c.loginc]+_b[class:c.loginc#3.race])

              chi2(1) =        1.01
           Prob > chi2 =     0.3137
```

```
. testnl exp( _b[class:2.race] +_b[class:2.race#c.educ]*12 ///
>    +_b[class:2.race#c.loginc]*1) =  ///
>    exp(_b[class:2.race] +_b[class:2.race#c.educ]*16 ///
>    +_b[class:2.race#c.loginc]*3)

 (1)     exp( _b[class:2.race] +_b[class:2.race#c.educ]*12
     >      +_b[class:2.race#c.loginc]*1) = exp(_b[class:2.race]
     >      +_b[class:2.race#c.educ]*16 +_b[class:2.race#c.loginc]*3)

             chi2(1) =         11.27
         Prob > chi2 =        0.0008
```

These results are quite similar for the two options, which is typically but not necessarily the case. The single-moderator test statistic value from Option 1 is 1.11 ($p = .2920$), while the value from Option 2 is 1.01 ($p = .3137$). For the two-moderator example, the Option 1 test statistic is 9.81 ($p = .0017$), and the Option 2 test statistic is 11.27 ($p = .0008$).

How to Calculate the Average Standardized Latent Outcome by Race Group

Both sets of Stata commands to do this calculation use the last estimation results, so be careful that no other estimation commands have been run in between (ICALC always restores your last estimation results when it finishes, so it can be run in between). The first set of code uses the *predict* command to create a variable named ylat, which has valid values only for cases in the estimation sample (if e(sample)) and whose content is the predicted latent outcome—the sum of the coefficients times the predictors (xb). The *replace* command standardizes the predicted latent outcome by subtracting its mean (4.6378) and dividing the result by its standard deviation (2.1522), both of which are reported in the ICALC output when you create an outcome display for the standardized latent outcome. The *tabstat* command calculates the mean of the latent outcome for each race group.

```
predict ylat if e(sample), xb
replace ylat = (ylat-4.6378)/2.1522
tabstat ylat , by(race)
```

The second set of code automates the process but requires that you have SPOST13 installed to use the *fitstat* command, which stores a number of fit statistics, including the estimated variance of the latent outcome, saved as r(Vystar). The *margins* command calculates the expression inside the exp( ) option, which is defined as the predicted latent outcome predict(xb) minus its mean `latmn` divided by the square root of its variance `latsd`—with the predictions calculated separately by race group as specified with the option over(race). Note that the single quote symbol preceding the term for the mean and square root of the latent variance is not the usual single quote symbol found on the keyboard; it is a backtick (`), often found on the keyboard as the leftmost key in the row of numbers/symbols, paired with the tilde (~) symbol.

```
quiet {
    predict ylat if e(sample), xb
    sum ylat, meanonly
    loc latmn = r(mean)
    fitstat
    loc latsd = r(Vystar)^.5
    drop ylat
}
margins if e(sample), exp((predict(xb)-`latmn')/`latsd') over(race) nose
```

CHAPTER 10 NOTES

1. You could also calculate the change in the latent propensity across the range of education's values. Specify the option effect(b(sdy)) to get the standardized change in propensity for a 1-year difference in education (0.0295), and multiply that value by 20 (0.59). The difference between the least possible and the most possible educated person would be 6/10 of a standard deviation change in the propensity.

2. That is, the discrete change effect of education is calculated across the range of education values (0–20).

3. You get the same results if you use the as-observed strategy for setting reference values when you work in the model metric, so specifying the suboption is not really necessary. But since I prefer using the mean to set reference/display values when working in the observed metric, I include the suboption for consistency and clarity.

4. The original income measure does not have zeros or negative numbers, so it is possible to calculate the log. The prelogged measure of income was in categories, which I converted to numeric values using the midpoints for the closed-ended categories and a Pareto estimate for the top open-ended category. There were no zeros because the bottom category had a midpoint value of $500.

COUNT MODELS

OVERVIEW

Properties and Use of Count Models

A count outcome Y_i is an interval-level measure of the number of events experienced by a unit of analysis: $Y_i = 0, 1, 2, \ldots$ Although the outcome is interval level, OLS regression is likely to be, at a minimum, an inefficient estimator and may be a biased and inconsistent estimator when applied to count data (Long, 1997; Long & Freese, 2014). The reason is the common presumption that the higher the count, the less likely its occurrence. This suggests that the value of the count increases at a decreasing rate as a predictor increases or decreases. That is, there is a nonlinear relationship between the count and the predictors, leading to the conclusion that OLS is an inefficient and biased estimator.

I cover a set of four related models commonly used to analyze count outcomes: (1) Poisson regression, (2) negative binomial regression, (3) zero-inflated Poisson regression (ZIP), and (4) zero-inflated negative binomial regression (ZINB). As in other chapters, I presume that readers already have a foundation of knowledge concerning count models and their interpretation in the absence of interaction effects. If not, I would suggest reading Long (1997, chap. 8) for more technical details and Long and Freese (2014, chap. 9) for an excellent introduction to their interpretation before you proceed.

Data and Circumstances When Commonly Used

A count model predicts the probability that a case has a count of events equal to Y_i derived from a model of the set of probabilities that $Y_i = 0, 1, 2, \ldots$ Typically, count data have a disproportionate number of zeros and a nonlinear relationship between the predictors and the expected count—a smaller rate of change in the expected count as a predictor increases for a positive effect or decreases for a negative effect. Count models assume that units of analysis have fixed and equal exposure to the process generating the count of events. In many situations, this means that the time frame during which events are counted is the same for all cases—for example, the number of poor mental health days a worker experienced during the past 30 days or the number of labor actions initiated against a company during the past year.

In other situations, the number of members constituting a unit of analysis may determine the level of exposure of the unit of analysis to the process creating events; for example, the number of daily student absences reported by homeroom teachers in schools depends on homeroom class size. If a variable measuring exposure is available, unequal exposure can be incorporated into a count model. For the first situation, you would need a measure of the varying exposure time for each case, and for the second situation, you would need a measure of the number of members of each unit of analysis.

Published Examples.

- For an analysis of the spatial density of Tea Party organizations, McVeigh et al. (2014) used a negative binomial regression of the number of organizations in a county on a set of county socio-demographic characteristics. They tested whether the effect of educational residential segregation is moderated by returns to education and income inequality. Their interpretation of these significant interactions describes the effect on the number of organizations of education segregation at the mean of each moderator (returns, inequality) and how that effect changes at higher levels of each moderator.

- He (2012) used a negative binomial regression to study the effects of the implementation of policy on behavior; specifically how the implementation of a market-based regulation that shops in Guiyang, China, charge for plastic bags affected consumers' use of plastic bags. The author interprets the coefficients for several significant interactions between the dummy indicator of before/after the policy implementation and a set of individual and contextual predictors.

- To study how neighborhood composition moderates the effect of individual self-control on juvenile delinquency, Jones (2017) estimated a hierarchical ZINB of the number of nonserious delinquent acts on individual and neighborhood predictors. A central focus is the interpretation of the coefficients for a significant interaction between individual low self-control and neighborhood (aggregate) low self-control.

- Ailshire, Karraker, and Clarke (2017) analyzed how cognitive functioning among older adults is related to air pollution and neighborhood social stress. Specifically, they estimated a negative binomial regression of the number of cognitive errors on the interaction between perceived neighborhood stress and a measure of fine particulate matter air pollution to assess the role of stress as a vulnerability factor. They interpreted the interaction by discussing the coefficients from a plot of predicted cognitive errors against fine particulate matter levels, separately for low and high neighborhood stress.

GLM Properties and Coefficient Interpretation for Count Models

Poisson Regression. This provides the starting point for the other three models, which relax one or more of the assumptions of the Poisson model. Poisson regression specifies that the distribution of counts conditional on the predictors follows a Poisson distribution, with a log link function defining the relationship between the observed and modeled outcomes. Specifically, the conditional probability that the count equals c is given by

$$\Pr\left(y_i \mid \underline{x}_i = c\right) = \frac{\left(\underline{x}_i'\underline{\beta}\right)^c e^{-\underline{x}_i'\underline{\beta}}}{c!}$$

and the conditional mean of the observed outcome is

$$\mu_i = e^{x_i' \underline{\beta}} \tag{11.1}$$

The log link function defines the modeled outcome as the log of the count, and thus the coefficients represent effects on the log count:

$$\eta_i = \ln(\mu_i) = \underline{x}_i' \underline{\beta} \tag{11.2}$$

A key property of a Poisson distribution is that its variance is equal to its mean, which specifies that the variance of the counts should be equal to their mean. This is frequently at odds with the empirical reality of overdispersion in count data, with the variance larger than the mean.

Negative Binomial Regression. It relaxes the assumption that the variance equals the mean by adding a parameter δ_i for the variance (Long, 1997; Long & Freese, 2014) to the log link expressions in Equations 11.1 and 11.2:

$$\eta_i = \underline{x}_i' \underline{\beta} + \ln(\delta_i) \quad \text{because } \mu_i = \delta_i e^{x_i' \underline{\beta}} \tag{11.3}$$

To identify this model, the usual assumption is that δ_i is distributed as $\Gamma(v)$ with $\mathcal{E}(\delta_i) = 1$. Note that most estimation programs (including Stata) report $\alpha = \dfrac{1}{v}$ rather than v. With this assumption, the negative binomial regression model has the same conditional mean structure as the Poisson regression model. But the predicted probability of a given count is now

$$\Pr(y_i \mid \underline{x}_i = c) = \frac{\Gamma(c+v)}{c!\,\Gamma(v)} \left(\frac{v}{v + e^{x_i'\underline{\beta}}} \right)^v \left(\frac{e^{x_i'\underline{\beta}}}{v + e^{x_i'\underline{\beta}}} \right)^c \tag{11.4}$$

ZIP and ZINB. They modify Poisson regression and negative binomial regression, respectively, to take into account that in practice they often underpredict the number of zeros. The ZIP and ZINB models add a second process to the model that can also generate a count of 0. The underlying logic is that some cases will have a count of 0 as a result of the general process that generates differential counts. And some cases will always have a count of 0, with a case's probability of being always zero conditional on a set of predictors that may or may not overlap with the predictors used to model the general process.

Consider predicting the number of poor mental health days experienced by a worker. Some workers might be taking medications that prevent them from feeling poorly, and hence they would always be zero on the count outcome. Some workers who do not take such medications might also have a value of 0 on the count outcome as a result of the Poisson or negative binomial process generating the distribution of counts. The ZIP model uses Poisson regression to estimate the general process and logistic regression (optionally probit analysis) to estimate the probability that a case is always zero. Analogously, the ZINB model uses negative binomial regression for the general process and logistic (or probit) for the always-zero process.

Interpretation. Despite these variations among the models, the principles for interpreting the effects of predictors on the distribution of the counts are identical (Long & Freese, 2014, chap. 9), although some formulas and details differ. Interaction and other effects can be interpreted

1. as effects on the log number of events (optionally standardized);

2. as factor changes in the number of events, sometimes called an incidence–rate ratio;

3. as a marginal effect on the number of events—either a marginal change (instantaneous) or a discrete change in the number of events; and

4. using tabular or visual displays of the predicted number of events, the predicted log number of events, or the predicted probability of specific numbers of counts (predicted probabilities are considered in the "Special Topics" section at the end of the chapter).

Diagnostic Tests and Procedures

Although it is not a formal test, comparison of the observed distribution of the counts with the predicted probability distributions of the counts from each model can be informative about model fit and model selection. Four diagnostic tests or statistics for count models are used to guide the choice between the four models for a particular analysis (Long & Freese, 2014, pp. 544–549). The BIC and the AIC can be applied to any pair of models and are general (nonspecific) tests of model fit. Two additional tests assess specific pairs of models. If the likelihood ratio chi-square test of the overdispersion parameter (H_0: $\alpha = 0$, with the constraint that $\alpha \geq 0$) is significant, this provides evidence for selecting the negative binomial regression model over the Poisson or the ZINB model over the ZIP model. If not significant, it suggests making the other choice in each pair. Until quite recently, the Vuong test was used to compare a zero-inflated model with its non–zero-inflated counterpart. But Wilson (2015) demonstrated that that it is not an appropriate application of the Vuong test. Finally, checking for collinearity among the predictors—apart from the necessary functional collinearity among interaction terms—is standard.

Data Source for Examples

Both application examples analyze GSS data. The single-moderator example data are from the 2010 survey (GSS_2010.dta), and the three-way interaction example uses 1987 data (GSS_1987.dta). Both data sets and example do-files can be downloaded at www.icalcrlk.com.

ONE-MODERATOR EXAMPLE (INTERVAL BY NOMINAL)

Data and Testing

This example illustrates the interpretation of an interval-by-nominal interaction effect, how work–family conflict and occupational status interact in predicting the number of poor mental health days a worker experienced during the past month. Work–family conflict is coded as a three-category variable (1 = *never*, 2 = *sometimes*, and 3 = *often*), with "never" used as the reference category. Occupational status (SEI) is an interval scale ranging from 17.6 to 97.2. Three of the other predictors are interval measures: age (18–88 years), education (0–20 years), and number of children (0–8). Female is dummy coded (1 = *female*, 0 = *male*). The index function to be estimated takes the form

$$g(y_i) = \beta_0 + \beta_1 SEI_i + \beta_2 Sometimes_i + \beta_3 Often_i + \beta_4 SEI_i \times Sometimes_i + \beta_5 SEI_i$$
$$\times \, Often_i + \beta_6 Age_i + \beta_7 Educ_i + \beta_8 Female_i$$

Keep in mind that $g(y_i)$ in this example is the logged number of days of poor mental health. The sample excludes 890 respondents not in the labor force, another 178 with missing data on the outcome, and 26 more with missing information on one or more of the predictors in the analysis.

The BIC and AIC statistics both suggest the choice of the ZINB model for this analysis, as do the targeted tests for overdispersion and adding a zero-inflated component to the model. The likelihood ratio test for overdispersion is significant ($p < .001$) for both the negative binomial regression versus Poisson regression and the ZINB versus ZIP, which suggests using either negative binomial regression or ZINB. An examination of the distribution of the observed and predicted counts provides support for choosing the ZINB. The ZINB predicted probabilities are closer to the observed than are the other three models' predictions, both overall and for most specific counts. The sum of the absolute difference between the observed and predicted probabilities is only 0.108 for ZINB, compared with 0.154 for the negative binomial, 0.288 for the ZIP, and 1.109 for the Poisson predictions.

Statistical testing of the interaction terms in the ZINB model of work–family conflict by occupational status—$\beta_5 SEI_i \times Sometimes_i + \beta_6 SEI_i \times Often_i$—provides consistent results. A global test of the change in the log likelihood if the two terms are excluded is significant ($p = .029$), as are the z tests of each coefficient individually ($p = .011$ and $.007$, respectively), which provides ample grounds for including and interpreting the interaction.

An important point to keep in mind for this example is that the meaning of the coefficients and predicted outcomes is slightly different in the zero-inflated models from their meaning in the non–zero-inflated models. In the ZIP and ZINB models, the effects and the predicted outcomes apply only to the group that is not always zero. In the following sections, I illustrate how to use the GFI, SIGREG, and EFFDISP tools first to interpret the effects of work–family conflict moderated by occupational status and then to apply them to occupational status moderated by conflict. I then demonstrate the interpretation of the results from several OUTDISP tool options, which show the effects of both work–family conflict and occupational status simultaneously. The "Special Topics" section at the end of the chapter includes a brief consideration of how to use the ICALC commands for an interaction effect in the zero-inflated part of the model—namely, the logistic regression (or probit) predicting whether a case is in the always-zero group or not.

Work–Family Conflict Moderated by Occupational Status

Following the estimation command for ZINB in the following output, the *intspec* command defines conflict as the focal variable and occupational status as the moderator and sets the display names and ranges. The factor option in the *gfi* command requests the algebraic expressions for the moderated effect of conflict on logged number of days as well as the factor change effect on the number of days.

```
.zinb pmhdays i.wfconflict3##c.sei age educ childs i.sex ///
> if wrkstat <=4 , inf(age i.sex)
    ...


.intspec focal(i.wfconflict3) main((i.wfconflict3, name(WorkVsFam) range(1/3)) ///
c.sei, name(OccStatus) range(17(10)97))) int2(c.sei#i.wfconflict3) ndig(0)
    ...


. gfi, factor ndig(3)
GFI Information from Interaction Specification of
Effect of Workvsfam on g(Pmhdays) from Zero-Inflated Negative Binomial Regression

Effect of Sometimes =
   -0.937 + 0.027*Sei


   Factor Change Effect (1 unit change in) Sometimes =
      e^-0.937 * e^( 0.027*Sei) =

   0.392 * 1.027^Sei

Effect of Often =
   -0.815 + 0.036*Sei


   Factor Change Effect (1 unit change in) Often =
      e^-0.815 * e^( 0.036*Sei) =

   0.443 * 1.037^Sei


Sign Change Analysis of Effect of Workvsfam
on g(Pmhdays), Moderated by Occstatus (MV)


    ------------------------------------------------------------
            |      Workvsfam
When        | -------------------------------------------------
Occstatus=  |     Sometimes              Often
------------+------------------------------------------------
         17 | Neg b =    -0.473    Neg b =    -0.197
         27 | Neg b =    -0.200    Pos b =     0.167
         37 | Pos b =     0.073    Pos b =     0.530
         47 | Pos b =     0.346    Pos b =     0.894
         57 | Pos b =     0.619    Pos b =     1.257
         67 | Pos b =     0.892    Pos b =     1.621
         77 | Pos b =     1.165    Pos b =     1.985
         87 | Pos b =     1.438    Pos b =     2.348
         97 | Pos b =     1.711    Pos b =     2.712
------------+------------------------------------------------
Sign Changes |  when MV= 34.32263    when MV= 22.41374
------------+------------------------------------------------
% Positive  |       70.8                 99.3
    ------------------------------------------------------------
```

The first thing to notice about the GFI expression is that the two conflict categories have the same structure of moderation by SEI. The main effects term is negative—each conflict category has fewer logged poor mental health days than the reference category "never" if occupational status were to equal zero (which is not a possible value). But the moderating effect of SEI is positive. The sign change analysis shows that the effect of the conflict categories turns positive at a relatively low value of SEI (34 and 22 for the two categories in order). This indicates that above roughly the first quartile of status, the more frequent conflict categories are expected to have a higher (logged) number of poor health days reported relative to workers who never experience such conflict.

The factor change expressions show an analogous pattern and make explicit that the moderating variable (SEI) acts as a multiplier to turn the initial effects of conflict on poor health days (*not* logged days) from "positive" to "negative." Keep in mind that a factor effect <1 means that a predictor reduces the outcome by that factor, which is conceptually a "negative" effect, while a factor effect >1 predicts an increase in the outcome by that factor, a "positive" effect. Consider, for example, the expression for the effect of the "sometimes" category: 0.392×1.027^{SEI}. The 0.392 tells us that workers in the "sometimes" category would be expected to have about 40% of the number of poor mental health days of those in the "never" category if *SEI* = 0. For a given value of SEI, the factor change effect of "sometimes" increases from 0.392 by a factor of 1.027 raised to a power equal to the value of SEI. For example, at *Occupational status* = 57, the factor change effect of "sometimes" is 1.790 (0.392×1.027^{57}). This means that the number of poor health days is expected to be 1.79 times as high for workers in the "sometimes" as in the "never" category when *Status* = 57.

SIGREG Results

For the SIGREG tool, I specify the effect (factor) option to demonstrate the use of factor change effects for interpreting the moderated effect of a predictor, the save(filepath tab) option to create the Excel formatted significance region results shown in Table 11.1, and the concise option to produce a briefer boundary value analysis table. Although the boundary value analysis is calculated for the moderated effects of the estimated coefficients, it also applies asymptotically to the factor effect coefficients (Feiveson, 2017). In practice, I would not present the boundary values table when writing up or presenting results but would use the exact boundary values to supplement the discussion of a significance region table or an effects display.

```
. sigreg, effect(factor) save(Output\Table_11_1.xlsx) concise
Boundary Values for Significance of Effect of Workvsfam on g(Pmhdays) Moderated by Occstatus
    Critical value Chi_sq = 3.841 set with p = 0.0500
+--------------------------------------------------------------------------------------+
| Effect of Workvsfam | When Occstatus >= | Sig Changes | When Occstatus >= | Sig Changes |
|---------------------+-------------------+-------------+-------------------+-------------|
|           Sometimes |      NA (< min)   |     NA      |      47.9057      |   to Sig    |
|               Often |      NA (< min)   |     NA      |      36.4809      |   to Sig    |
+--------------------------------------------------------------------------------------+
```

Table 11.1 makes the effect of work–family conflict and how it is moderated by occupational status quite clear. The effect of more frequent conflict relative to no conflict turns positive and then significant as status increases; that is, more frequent conflict is expected to increase the number of poor mental health days, and the effect of each conflict category increases with status. Note that there is a lower threshold for significance for the category with the more frequent conflict; "Sometimes" becomes significant at *Status* = 47.9, but "Often" turns significant at 36.5. Moreover, the expected difference in the number of days among the conflict categories rises with occupational status. At *Status* = 47 (a little below the mean), experiencing conflict

TABLE 11.1 ● FACTOR CHANGE EFFECT OF WORKVSFAM (ONE-UNIT DIFFERENCE) ON PMHDAYS MODERATED BY OCCSTATUS, FORMATTED TO HIGHLIGHT SIGN AND SIGNIFICANCE

Effect of	Occstatus								
	17	**27**	**37**	**47**	**57**	**67**	**77**	**87**	**97**
Sometimes	*0.6232*	*0.8188*	1.0758	1.4134	**1.8570***	**2.4398***	**3.2056***	**4.2116***	**5.5333***
Often	*0.8213*	1.1814	**1.6995***	**2.4446***	**3.5164***	**5.0582***	**7.2760***	**10.4663***	**15.0552***

Key	
Plain font, no fill	Pos, Not Sig
Bold*, filled	Pos, Sig
Italic, no fill	Neg, Not Sig
***Bold italic*, filled**	Neg, Sig

Note: Workvsfam = work versus family; Pmhdays = number of poor mental health days; Occstatus = occupational status.

sometimes predicts a 41% increase in the number of poor mental health days, while experiencing conflict often predicts nearly 2.5 times as many poor mental health days as for workers with no conflict. At *Status* = 67 (just below 1 standard deviation above the mean), the effect of "Sometimes" is to more than double the expected number of days, and the effect of "Often" predicts five times as many poor mental health days as for those who experience no conflict.

The preceding discussion points out a disadvantage of presenting results with display values for the moderating variable equally spaced from its minimum to its maximum. The display values may not be especially meaningful without some distributional information. In such a case, I would recommend rerunning the ICALC commands after redefining the display values with the option range(meanpm2) in the *intspec* command. This defines the display values as the mean, the mean plus or minus 1 standard deviation, and the mean plus or minus 2 standard deviations; if an upper or lower limit lies outside the sample range, it is truncated to the maximum or minimum, respectively. The new syntax is as follows:

```
.intspec focal(i.wfconflict3) main((i.wfconflict3, name(WorkVsFam) range(1/3)) ///
> (c.sei, name(OccStatus) range(meanpm2)) )) ndig(0) int2(c.sei#i.wfconflict3)
    …
```

```
. sigreg, effect(factor) save(Output\Table_11_2.xlsx tab)
```

The results in Table 11.2 now make it straightforward to describe the changing effect of work–family conflict at the mean of occupational status, and at 1 and 2 standard deviations above the mean—status values of 51, 70, and 89, respectively.

EFFDISP **Results**

An effects display is a graphical alternative for presenting the information in the significance region table. I create a confidence bounds plot for the effect of each

TABLE 11.2 ● FACTOR CHANGE EFFECT OF WORKVSFAM (ONE-UNIT DIFFERENCE) ON PMHDAYS MODERATED BY OCCSTATUS, REVISED DISPLAY VALUES

Effect of	Occstatus				
	18	32	51	70	89
Sometimes	*0.6335*	*0.9327*	**1.5738***	**2.6556***	**4.4810***
Often	*0.8394*	1.4052	**2.8208***	**5.6627***	**11.3675***
Key					
Plain font, no fill	Pos, Not Sig				
Bold*, filled	Pos, Sig				
Italic, no fill	Neg, Not Sig				
Bold italic*, filled	Neg, Sig				

Note: Workvsfam = work versus family; Pmhdays = number of poor mental health days; Occstatus = occupational status.

category of work–family conflict, which is the default type for an interval-level moderator. Because I want to use the same display values as in Table 11.2 to label the *x*-axis, I do not need to run a new *intspec* command. For the *effdisp* command, I use the effect( ) option to specify the effect type as factor change and the ndig( ) option to control the axis-label numeric formatting:

```
. effdisp , effect(factor) ndig(0)
```

The plots in Figure 11.1 show the factor change effect of each work–family conflict category as a *solid line* that slopes upward as occupational status increases. The factor change effect increases more quickly for the "often" conflict category. A *vertical gray line* marks the change between nonsignificance and significance on each plot to show, as we saw before, that the effect of the "often" experience conflict category becomes significant at lower levels of status. I find the confidence bounds plot in this instance to be not very informative. The problem is that the maximum value for the upper bound on the plot for the "often" category overshadows the display of the moderated effect, making it difficult to see the values of the factor change effect. A line plot works better as an effects display because it displays the moderated effect with significance change markers but does not show the confidence bounds. The plot(type( )) and sigmark options in the following *effdisp* command create the desired line plot shown in Figure 11.2.

```
. effdisp , plot(type(line)) effect(factor) ndig(0) sigmark
```

This better shows the magnitudes of the factor change effects for each category of conflict and how they increase with occupational status at different rates. If I presented these plots instead of the significance region table, I would supplement the discussion with the numeric factor change values from the significance region table.

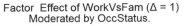

FIGURE 11.1   ⬡   FACTOR CHANGE EFFECT CONFIDENCE BOUNDS PLOT

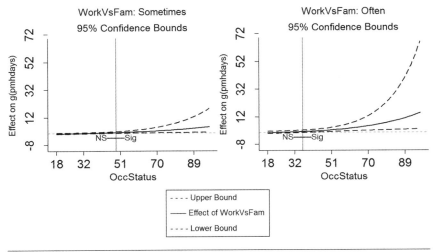

FIGURE 11.2   ⬡   LINE PLOTS FOR FACTOR CHANGE EFFECTS

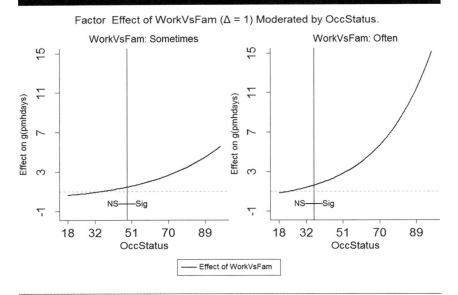

For example, at mean occupational status (51 days), workers sometimes or often experiencing conflict would be expected to have significantly more days of poor mental health than workers who never experienced conflict, 1.5 times and 2.8 times as many, respectively. The increase in the number of days is predicted to be much larger for workers whose status is 1 standard deviation above the mean (70). Those who sometimes have work–family conflict expect more than 2.65 times as many poor health days as those in the "never" category, and those who often experience conflict can expect 5.7 times as many days of poor mental health. Last, those with very high status—2 standard deviations above the mean (89)—would expect considerably more days with poor mental health, 4.5 and 11.4 times as many, respectively, for those who sometimes or often have conflicts.

Before considering how predicted outcome displays assist in interpreting the effects of work–family conflict and job status, let's see how status's effect is moderated by work–family conflict.

Occupational Status Moderated by Work–Family Conflict

I again start with a GFI analysis following a revised *intspec* command, changing the focal( ) option to "status" and reordering the predictors specifying the term in the int2( ) option. Because I plan to get SPOST13 results, I first reestimate the ZINB results so that the two-way interaction term in the Stata ZINB model command has *c.sei* and *i.wfconflict3* in the same order as in the int2( ) option of the ICALC *intspec* command (see Chapter 6, for a discussion of the ordering requirements when getting SPOST13 results):

```
. zinb pmhdays c.sei##i.wfconflict3 age educ childs female if wrkstat <=4
>    inf(age female)
   …
. intspec  focal(c.sei) main((c.sei, name(OccStatus) range(17(10)97))
> (i.wfconflict3, name(WorkVsFam) range(1/4))) ndig(0) int2(c.sei#i.wfconflict)
   …
. gfi, factor ndig(4)

GFI Information from Interaction Specification of
Effect of Occstatus on g(Pmhdays) from Zero-Inflated Negative Binomial Regression

Effect of Sei =
   -0.033 + 0.027*Sometimes + 0.036*Often

   Factor Change Effect (1 unit change in) Sei =
      e^-0.033 * e^( 0.027*Sometimes) * e^( 0.036*Often) =

      0.968 * 1.027^Sometimes * 1.037^Often

Sign Change Analysis of Effect of Occstatus
on g(Pmhdays), Moderated by Workvsfam (MV)
-------------------------------------
             |        Occstatus
When         |    -------------------
Workvsfam=   |
-------------+-----------------------
      Never  | Neg b =     -0.0326
   Sometimes | Neg b =     -0.0053
      Often  | Pos b =      0.0037
-------------+-----------------------
Sign Changes |       Sometimes
-------------+-----------------------
% Positive   |          7.4
-------------------------------------
```

The GFI expression for the effect of occupational status on the log number of days of poor mental health tells us that status has a negative effect (−0.033) for workers with no conflict, the reference category. The negative moderated effect of SEI is smaller for the "sometimes" category because we add a positive interaction coefficient to the main effects coefficient, which is not quite big enough to change it from positive to negative. For the "often" category, the interaction coefficient is substantial enough to make the moderated effect of status positive. I want to call readers' attention to the GFI expression for the factor change effects on the number of days. Because at most one of the dummy variables for conflict can equal 1 for a given case, the multiplier term for the dummy variable equal to 1 will multiply the main effect by a value other than 1. The other multiplier terms raise their coefficients to a power of 0, which means that they multiply the main effect by a value of 1. For the "never" category, all of the multiplier terms are equal to 1, so the factor change effect of status for this category is 0.968. That is, for workers with no work–family conflict, the expected number of poor mental health days is smaller by a factor of 0.968 for a one-unit increase in occupational status. Equivalently, we could say that the expected number of days declines by 3.16% (100 × (1 − 0.9684)).

SIGREG **Results**

I get significance region results for the discrete change effect of occupational status using SPOST13 to illustrate this option for interpretation. Because the significance region table is small and the results are not complex, I show the significance region table from the Stata output rather than the one saved to and formatted in Excel. To get the SPOST13 discrete change effect, I use the amtopt( ) suboption to define the amount of discrete change as a one-unit change in status and to request that it be a centered change. The atopt( ) suboption specifies the at( ) option that is passed to the Stata *margins* command to do calculations by setting all predictors other than the interacting variables to their means:

```
. sigreg, effect(spost(amtopt(am(one) center)) atopt((means) _all)) nobva
  …

  Significance Region for SPOST Change Effect of Occstatus
  on Pmhdays at Selected Values of Workvsfam
  -----------------------------------------------------
                                At Workvsfam=
   Effect of       |     Never    Sometimes     Often
  ---------------+-------------------------------------
          Sei    |   -0.0728*    -0.0188       0.0234
  -----------------------------------------------------
    Key: Plain font  = Pos, Not Sig    Bold font*   = Pos, Sig
         Italic font = Neg, Not Sig    Italic font* = Neg, Sig


  Spost Effect for  sei specified as amount =    +1_centered   calculated with
     at(  (means) _all )
```

These results show that the effect of occupational status on the number of poor mental health days is significant only for workers who never experience work–family conflict. And in that case, status's effect appears to buffer workers. For each unit increase in status, the predicted number of days decreases by 0.0731. Because a one-unit increase in status is both small and in a nonintuitive metric, I think that examining the effect of a standard deviation difference in status provides a better sense of the magnitude of the status effect. You can request this by changing the am( ) suboption to specify "sd" instead of "one," which produces the following results:

```
. sigreg, effect(spost(amtopt(am(sd) center) atopt((means) _all))) nobva ndig(2)
  ...

  Significance Region for SPOST Change Effect of Occstatus
  on Pmhdays at Selected Values of Workvsfam
  -------------------------------------------------------
                             At Workvsfam=
    Effect of    |     Never    Sometimes    Often
  ---------------+---------------------------------------
            Sei  |     -1.42*      -0.36       0.45
  -------------------------------------------------------

  Spost Effect for  sei specified as amount =   +SD_centered   calculated with
      at(  (means)  _all )
```

This tells us that a 1 standard deviation increase in status for those who never experience conflict predicts almost a 1.5-day decrease per month (−1.42) in the number of poor mental health days per month. Given that the number of poor mental health days has a mean of 3.8 and a standard deviation of 7.3, a reduction of 1.4 days appears substantial enough to be meaningful. Notice that the effect of status at other levels of conflict is much smaller, between 1/3 and 1/2 of a day, and is not significant.

How, if at all, are the discrete change results subject to the problem of confounded nonlinearities? To assess this, I used SPOST13 to evaluate the discrete change effect of occupational status at each category of work–family conflict from a main effects model. These would all be equal if there were no confounding.[1] The discrete change effects are relatively similar in absolute magnitude for the "never" and "sometimes" categories (−0.33 and −0.38, respectively), suggesting that these evaluation points are not creating any confounding because the main effects model shows them to be equal. But the discrete change effect of status for "often" is almost twice as large (−0.67) as for the other categories, which would suggest that its discrete change effect from the interaction model is increased by confounding. But in this instance, this is not especially consequential since the status effect for "often" is not significant.

EFFDISP **Results**

If you prefer a graphical representation of the moderated effect of occupational status, an error bar chart (Figure 11.3) provides a quite accessible portrayal of these results. It is the default effect display for a categorical moderator, so the plot( ) option is not needed. The effect( ) option requests the same discrete change effect as in the significance region table, and the ndig( ) option formats the *y*-axis. On the initial plot, the lower bound of the error bar for the "never" category overlapped with the *x*-axis, so I used the pltopts( ) option to extend the *y*-axis slightly.[2] The interpretation of this figure is identical to that for the significance region table. The confidence interval for the status effect in the "never" category of work–family conflict is the only one that does not include the value 0. Thus, this status effect is the only one that is significant. And the *diamond marker* for the effect shows that a 1 standard deviation difference in status predicts a reduction in poor mental health days of about 1.5 days per month (−1.42).

```
. effdisp, effect(spost( amtopt(am(sd) center)) atopt((means) _all)) ndig(0)
>   pltopts(ylab(-3.75 " ", add custom notick))
```

FIGURE 11.3 ● ERROR BAR CHART FOR THE EFFECT OF STATUS MODERATED BY CONFLICT

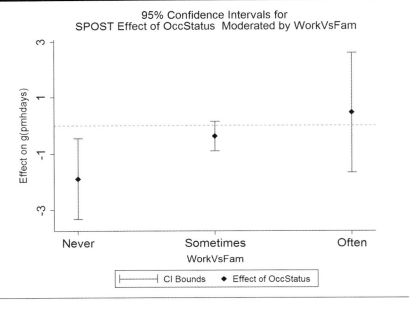

OUTDISP **for Work–Family Conflict and Occupational Status Simultaneously**

In addition to accessibility, tables or visual displays of the predicted outcome often have the advantage that you can use a single display to present and explain how each variable in the interaction moderates the effect of the other. When one of the interacting predictors is interval and the other is nominal, the default choice is a scatterplot of a predicted outcome plotted against the interval predictor, with repeated prediction lines corresponding to the categories of the nominal predictor. I find this easier to interpret than a bar chart of the predicted outcome for the categories of a predictor, with repeated sets of charts for the display values of the interval predictor. I start with a plot and a table of the predicted observed outcome, the number of days of poor mental health, and then consider options to parse out the confounded nonlinearities. I specify the save( ) suboption in the table( ) option to produce and save the Excel formatted table of the predicted number of days and the name( ) suboption in the plot( ) option to name and store the plot as a memory graph:

```
intspec focal(c.sei) main((c.sei, name(OccStatus) range(17(16)97)) ///
>     (i.wfconflict3, name(WorkVsFam) range(1/3))) ndig(0) int2(c.sei#i.wfconflict3)
    ...
```

```
outdisp, tab(save(Output\table11_3.xlsx)) plot(name(obs)) ndig(1)
```

When looking at these results, remember that they describe predicted outcomes for workers in the not-always-zero group. It is especially easy to see the effect of occupational status and how it is moderated by conflict in Table 11.3 by comparing the columns. For workers who often experience conflict, the number of days is predicted to rise with occupational status. The opposite characterizes the status effect for those who never or sometimes experience conflict—status predicts a decline in the number of days that is particularly rapid for the "never" category.

TABLE 11.3 ● PMHDAYS BY THE INTERACTION OF OCCSTATUS WITH WORKVSFAM

Occstatus	Workvsfam			Average Absolute Difference
	Never	Sometimes	Often	
17	6.8	4.2	5.5	1.70
33	4.0	3.9	5.9	1.35
49	2.4	3.5	6.2	2.58
65	1.4	3.3	6.6	3.48
81	0.8	3.0	7.0	4.13
97	0.5	2.7	7.5	4.64

Note: Pmhdays = number of poor mental health days; Occstatus = occupational status; Workvsfam = work versus family.

Comparing the rows shows two patterns in the moderated effect of work–family conflict:

1. The degree of difference among the categories initially declines with status and then rises; see the rightmost column, which reports the average absolute differences among the row entries.

2. There is a reordering of which types of conflict experience the most and the fewest days of poor mental health with increasing occupational status. At the bottom of the status distribution, the ordering is volatile, with the "never" category changing from the most days to the fewest days, the "sometimes" category switching from the least to the middle, and the "often" category going from the middle to the most. But across most of the status distribution (>33), the number of days is highest for the "often" category, followed by the "sometimes" category, and lowest for the "never" category.

The predicted values plot in Figure 11.4 consists of three prediction lines of the relationship between occupational status and the number days of poor mental health, one line for each of the categories of work–family conflict. The patterns just described for the predicted outcome table characterize the moderated effects of status and of conflict, so I will not repeat them.

An alternative to using predicted counts to interpret the interaction effects would be to use the predicted probabilities of different values of the count outcome—that is, $Pr(y = 0)$, $Pr(y = 1)$, $Pr(y = 2)$, ... Because ICALC does not have the functionality to work with predicted probabilities, I do not discuss that option here. Instead, I provide an example in the "Special Topics" section at the end of the chapter, demonstrating how to use SPOST13 and Stata commands to create tables and plots of predicted probabilities as they vary with the interacting predictors. Note that like the predicted counts, the predicted probabilities reflect both sources of nonlinearity.

An observed outcome plot with superimposed main effects is one approach to assessing whether the link function nonlinearity is confounding the interpretation of the

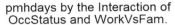

FIGURE 11.4 ● SCATTERPLOT OF OBSERVED OUTCOME (DAYS OF POOR MENTAL HEALTH)

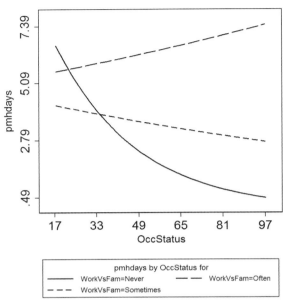

interaction effect. This compares the interaction model's predictions of the observed outcome with the predictions from the corresponding main effects model. The main effects model predictions only embody the nonlinearity of the link function, so the comparison can help illuminate the extent to which each interacting predictor moderates the other. ICALC superimposes the main effects model predictions on the same plot as the interaction model predictions if you include the suboption main(estimates_name) in the out( ) option. To make this work, you must first store the estimation results from the main effects model, rerun or restore the estimates from the interaction effects model, and then run the *outdisp* command to produce the plot presented in Figure 11.5:

```
. zinb pmhdays c.sei i.wfconflict3 age educ childs female if wrkstat <=4,
>       inf(age female)
    ...
.est store mainsei
    ...
.zinb pmhdays c.sei##i.wfconflict3 age educ childs female if wrkstat <=4 ,
>       inf(age female)
    ...

.outdisp, out(main(mainsei) model(obs) atopt((means) _all)) plot(name(obsMain))
>       ndig(2) tab(def) pltopts(tit( , size(*.8)))
```

The *black lines* are the predicted number of poor mental health days from the interaction model; the *solid line* is for the workers who never experience conflict, the

FIGURE 11.5 ● MAIN EFFECTS PREDICTIONS SUPERIMPOSED ON INTERACTION MODEL PREDICTIONS

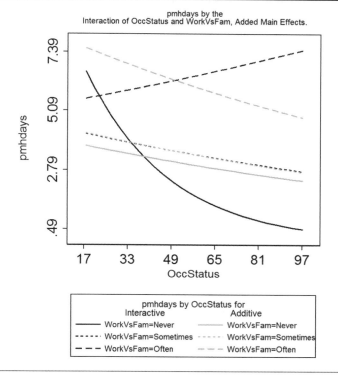

pmhdays by the
Interaction of OccStatus and WorkVsFam, Added Main Effects.

small dashed line is for "sometimes"—almost perfectly overlaid with its main effects prediction—and the *larger dashed line* is for "often." The main effects model predictions are drawn with *gray lines*, with the same line pattern for the categories of work–family conflict as the one used for the interaction model. The main effects prediction lines reveal that the effect of status for the "often" conflict category has a noticeably steeper slope than the main effects prediction lines for the other two conflict categories. This suggests that the "often" conflict effect is evaluated in a region where the link function "enhances" the magnitude of the prediction curve, while the "never" conflict effect is in an area that "mutes" its magnitude.

Thus, the link function nonlinearity appears to somewhat heighten the appearance of the moderated differences in the status effect between the "often" and "sometimes" conflict categories but to understate the moderated difference between the "never" and "sometimes" categories.

What about the moderation of work–family conflict by status? The main effects predictions suggest that the link function creates a smaller difference in the predicted outcome among the work–family conflict categories—compare the spread between the *gray prediction lines* at the left side and the right side of the plot. Because these are main effects predictions, the changing spread must be due to the link function nonlinearity and not the interaction. This means that the link function slightly diminishes the apparent degree of moderation of the work–family conflict effect by status.

Overall, the comparison of the interaction model and main effects model predictions shows evidence of a limited confounding of the interaction effects' nonlinearity and the nonlinearity of the link function. Consequently, the plot of the predicted number of days (observed outcome metric) by the interaction of occupational status and work–family conflict provides a reasonably good guide to the nature of the relationship. That is, occupational status has a significant negative effect on the number of poor mental health days only for workers who never experience work–family conflict. Across the range of status, the predicted number of days declines from almost 7 days per month for the lowest status workers to 1/2 a day for the highest status workers. The difference in the expected number of days among the work–family conflict categories changes from nonsignificant when occupational status is relatively low (<33) to an increasingly large gap with a consistent ordering of categories, the highest number of days for those who often experience conflict and the lowest number for those who never experience conflict. The disparity between "often" and "never" is 1.75 days when status is 1 standard deviation below its mean. The difference is 2.29 at mean occupational status, and the gap is 5.74 at 1 standard deviation above the mean status.

A model metric outcome plot (Figure 11.6) avoids the problem of confounding the nonlinearity of the model with the nonlinearity of the interaction effect altogether. For count models, this means interpreting effects on the log count (log number of days of poor mental health), with the qualifier that for zero-inflated models this applies only to those who are not in the always-zero group. You can improve the interpretability of results in the model metric—log *Count*—with the dual suboption. This adds a second *y*-axis on the right-hand side labeled with the values of the expected count corresponding to selected model metric values. The sdy suboption to calculate a standardized model metric outcome is not available for zero-inflated count models but is illustrated in the next example for a negative binomial regression model.

```
outdisp,  out(metric(model) dual) plot(metricDual) ndig(2)  tab(def)
>         pltopts(ti( , size(*.9)))
```

FIGURE 11.6 ●   PREDICTED LOG NUMBER OF DAYS WITH DUAL AXES

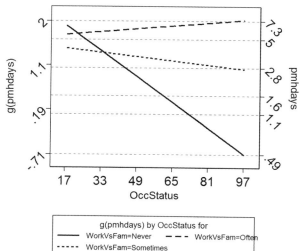

g(pmhdays) by the
Interaction of OccStatus and WorkVsFam, Dual Outcome Axes.

This plot unambiguously portrays the nature of the interaction effects. We know that the status effect among those with no work–family conflict (*solid black line*) is significant, and it exhibits the steepest change in the log number of days. Across the range of occupational status, the predicted value drops by 2.61 logged days. To provide some context for the size of this effect, the right-hand side y-axis tells us that it corresponds to a decline of around 6 days (6.25 days to be exact if we look at the predicted values table). And we know with certainty that the difference between the conflict categories in the magnitude of the status effects is solely a function of the interaction. Similarly, the changing effect of work–family conflict with occupational status is made manifest. Looking at the top-left corner of the plot, the conflict categories differ in the expected log number of days by at most 1/2 a logged day at low status (about 2.5 days). But at the highest status at the right side of the plot, the conflict category predictions are spread much farther apart (by 2.7 logged days), equating to almost 7 days. And for the vast majority of the status distribution, the "never" category has the fewest predicted days of poor mental health, and the "often" category has the most.

In the end, I think that either of the alternative plots makes it clear that the magnitude of change in the moderated effects of work–family conflict and of occupational status is not primarily due to the confounded nonlinearities. And they both provide a basis for arguing that the "plain-vanilla" plot of the predicted number of days of poor mental health (the observed metric) provides an acceptable picture of the relationship among work–family conflict, occupational status, and the frequency of reporting poor mental health days. In practice, I personally might prefer the last chart, the dual axis. The reason for trying all the options is to find what best suits your particular analysis and your interpretation style.

THREE-WAY INTERACTION EXAMPLE (INTERVAL BY INTERVAL BY NOMINAL)

Data and Testing

This example revisits in more detail Example 4 from Chapter 5, analyzing the relationship between the number of voluntary association memberships and the interaction of age by education by sex as well as its relationship to perceived social class and race. The number of memberships ranges from 0 to 16 (the maximum possible). Sex is a binary indicator, with 57.4% of the cases coded as 1 (female). Race is a three-category indicator represented by dummy variables for Black (29.9%) and for Other Races (2.9%), with White as the reference category (67.2%). The remaining predictors are interval level. Age and education are measured in years (with means = 44.1 and 12.5 and standard deviations = 17.1 and 3.1, respectively). Perceived class is a 10-point scale running from lowest (1) to highest (10)—mean = 5.8 and standard deviation = 1.9. The index function we want to estimate is

$$g(y_i) = \beta_0 + \beta_1 Education_i + \beta_2 Age_i + \beta_3 Sex_i + \beta_4 Education_i \times Age_i$$
$$+ \beta_5 Education_i \times Sex_i + \beta_6 Age_i \times Sex_i + \beta_7 Education_i \times Age_i$$
$$\times Sex_i + \beta_8 Class_i + \beta_9 Black_i + \beta_{10} Other_i$$

The data are from the 1987 GSS, and the sample excludes 322 respondents with missing information on perceived social class and an additional 10 cases with missing information on one or more of the other variables in the analyses.

The diagnostic testing provides consistent evidence for choosing either the negative binomial regression model or the ZINB model for this analysis but a mixed assessment of the choice between these two. The likelihood ratio test for overdispersion is significant for the two paired contrasts, indicating the choice either of negative binomial regression over Poisson regression or of ZINB over ZIP, with the BIC, AIC, and fit to the observed proportions also supporting this conclusion. Comparing the negative binomial regression and the ZINB models, the AIC statistic suggests selecting the ZINB. But the BIC statistic favors the negative binomial regression model, as does a consideration of the absolute differences between the observed and predicted probabilities for the two models. The negative binomial regression predictions are closer to the observed probabilities than the ZINB for counts = 0 to 4, and virtually indistinguishable beyond that. Overall, the average absolute difference for the negative binomial regression predictions (0.0036) is half the size of the average for the ZINB (0.0065). Given this mixed assessment, and the lack of a strong substantive argument for the existence of an always-zero group, I choose the negative binomial regression model because of its parsimony relative to the ZINB model.

Statistical tests of the coefficients constituting the terms for the three-way interaction of education by age by female,

$$\beta_4 Education_i \times Age_i + \beta_5 Education_i \times Sex_i +$$
$$\beta_6 Age_i \times Sex_i + \beta_7 Education_i \times Age_i \times Sex_i$$

indicate their inclusion in the negative binomial regression model. A global test of the change in the log likelihood if these four terms are excluded from the model is statistically significant ($p = .045$). And the z test of the three-way coefficient, β_7, is similarly significant ($p < .05$).

Approaches to Interpreting the Three-Way Interaction

I organize the interpretation of the interaction effects by focusing in turn on each component predictor and how its effect is moderated by the other two. In each case, I first briefly discuss the GFI results and then focus on the significance region table results, in some cases supplemented by an effects display, to provide a foundation of understanding of the nature of each moderated effect. To demonstrate the range of options, I discuss the effect of the following:

1. Age on the logged number of memberships as it changes with education and sex

2. Education as a factor change in the number of memberships and how it varies with age and sex

3. Sex as a discrete change in the number of memberships varying with age and education

I then illustrate how to use outcome displays to provide a more integrated portrayal of the interaction effects as a whole. The initial Stata *nbreg* command to estimate the model and the ICALC *intspec* command to define the interaction are

```
. nbreg memnum c.age##c.ed##i.sex i.race class
        …
. intspec focal(c.age)  main((c.age , name(Age) range(18(14)88))  ///
>        (c.ed, name(Education) range(0(2)20)) i.sex, name(Sex) range(0/1)))  ///
>        int2(c.age#c.ed c.age#i.sex  c.ed#i.sex) int3(c.age#c.ed#i.sex)  ndig(0)
```

If you sometimes use SPOST13 to calculate marginal effects for ICALC calculations—for example, for a significance region table or an effects display—you need to ensure that the variables are referenced in both commands in the same relative order (see Chapter 6). In this application, the information in the *intspec* command will change across the example as the focal variable changes or to modify the display values of a predictor (i.e., the range( ) suboption).

Moderated Effect of Age on Log Number of Memberships

```
. gfi, ndig(6)

GFI Information from Interaction Specification of
Effect of Age on g(Memnum) from Negative Binomial Regression
---------------------------------------------------------------------

Effect of Age =
-0.000235 + 0.000518*ed + 0.023569*Women - 0.002102*ed*Women
```

Rather than focus on the numeric details from the GFI expression of how the sign and magnitude of the effect of age change with education and sex, I instead describe the overall pattern and then use the significance region results to provide the details:

- For men, substitute *Women* = 0 into the GFI to get the effect of age:

  $-0.000235 + 0.000518 \times Ed + 0.023569 \times 0 - 0.002102 \times Ed \times 0$
  $= -0.000235 + 0.000518 \times Ed$

- This means that the baseline effect of age for men (with zero education) is negative, reducing the number of voluntary association memberships as age increases, but the age effect becomes positive as education increases.

- For women, substitute *Women* = 1 into the GFI to get the effect of age:

  $-0.000235 + 0.000518 \times Ed + 0.023569 \times 1 - 0.002102 \times Ed \times 1 =$
  $0.023334 - 0.001584 \times Ed$

- Thus, the baseline effect of age for women (with zero education) is positive but moves in the negative direction with rising education.

```
. sigreg, save(Output\table_11_4 tab)
    …
. sigreg, save(Output\table_11_4_sdyx tab)  effect(b(sdyx))
    …
. effdisp , plot(type(errbar)) effect(b(sdyx)) ///
> pltopts(xlab(0(4)20) msym(d)) ndig(1)
    …
```

Table 11.4 presents the significance region table for the moderated effect of age. The left-hand panel shows the effect of a 1-year difference in age. This shows that the effect of age for men increases with education and is positive and significant for education between 11 and 17 years. In contrast, for women, the effect decreases with education and is positive and significant for education between 0 and 11 years. Note that the significant effects of age for women are generally larger than the significant effects for men.

It is harder to get a read on the magnitude of these effects given that the outcome is the logged number of memberships. The right-hand panel of the table shows the

| TABLE 11.4 ● EFFECT OF AGE MODERATED BY THE INTERACTION OF SEX AND EDUCATION ON *g(Memnum)*, EFFECT OF 1-YEAR DIFFERENCE AND 1 STANDARD DEVIATION DIFFERENCE |

1-Year Difference			1 *SD* Difference		
	Sex			Sex	
Education	Men	Women	Education	Men	Women
0	−0.0002	**0.0233***	0	−0.0036	**0.3558***
2	0.0008	**0.0202***	2	0.0122	**0.3075***
4	0.0018	**0.0170***	4	0.0280	**0.2592***
6	0.0029	**0.0138***	6	0.0438	**0.2109***
8	0.0039	**0.0107***	8	0.0596	**0.1626***
10	0.0049	**0.0075***	10	0.0754	**0.1143***
12	**0.0060***	0.0043	12	**0.0913***	0.0660
14	**0.0070***	0.0012	14	**0.1071***	0.0177
16	**0.0081***	−0.0020	16	**0.1229***	−0.0306
18	0.0091	−0.0052	18	0.1387	−0.0789
20	0.0101	−0.0083	20	0.1545	−0.1272
Key					
Plain font, no fill	Pos, Not Sig				
Bold*, filled	Pos, Sig				
Italic, no fill	Neg, Not Sig				
***Bold italic*, filled**	Neg, Sig				

Note: Pmhdays = number of poor mental health days; Occstatus = occupational status; Workvsfam = work versus family.

effects of a 1 standard deviation change in age scaled to represent the change in the log number of memberships in standard deviation units. These values convey the sense that the magnitude of the age effect is meaningful. For a standard deviation difference in age (17.1 years), the significant age effects for men are about 1/10 of a standard deviation increase in logged memberships, while the significant effects for women range between 1/10 and 1/3 of a standard deviation increase.

An error bar chart is also effective in this particular example, more so than the default confidence bounds plot for an interval moderator. The display values for education define the spacing of the error bars in Figure 11.7. The pattern of the age effect, rising with education for men but falling with education for women, is obvious, as is the fact that only the positive effects of age are significant (error bars do not bracket the zero-effect line) but at different levels of education for men and for women. This would be a good alternative to presenting the significance region table. Note the use of the pltopts( ) option in the *effdisp* command shown above to change the *x*-axis labels for education from the display values every 2 years to every 4 years.

FIGURE 11.7 ERROR BAR CHART FOR MODERATED EFFECT OF AGE

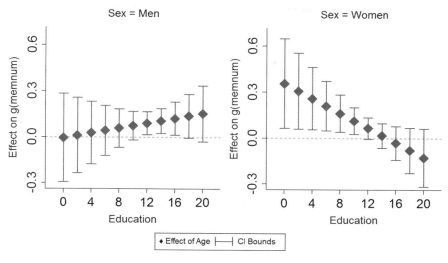

95% Confidence Intervals for Effect of Age (g(y)- & focal- standardized) Moderated by the Interaction of Education and Sex.

Moderated Effect of Education as a Factor Change in Number of Memberships

```
. intspec focal(c.ed)  main((c.age , name(Age) range(18(14)88))  ///
>   (c.ed, name(Education) range(0(2)20)) (i.sex, name(Sex) range(0/1)))  ///
>   int2(c.age#c.ed c.ed#i.sex c.age#i.sex) int3(c.ed#c.age#i.sex)ndig(0)
   ...
. gfi , factor

GFI Information from Interaction Specification of
Effect of Education on g(Memnum) from Negative Binomial Regression
--------------------------------------------------------------------

Effect of Ed =
  0.0753 + 0.0005*Age + 0.1351*Women - 0.0021*Age*Women


  Factor Change Effect (1 unit change in) Ed =
     e^ 0.0753 * e^( 0.0005*Age) * e^( 0.1351*Women) * e^(- 0.0021*Age*Women) =

  1.0782 * 1.0005^Age * 1.1447^Women * 0.9979^(Age*Women)
```

Let's start by finding and interpreting the GFI expression for the factor change effect of education for men and for women. Keep in mind that a number raised to the power of 0 is equal to 1. For men, we set *Women* = 0 in the factor change GFI, which yields a factor change effect of education of

$$1.0782 \times 1.0005^{\wedge}Age \times 1.1447^{\wedge}0 \times 0.9979^{\wedge}(Age \times 0)$$
$$= 1.0782 \times 1.0005^{\wedge}Age \times 1 \times 1$$
$$= 1.0782 \times 1.0005^{\wedge}Age$$

This expression will always have a value greater than 1, meaning that education always has a positive effect on the number of memberships and that effect increases with age. Substituting *Women* = 1 to get the factor change effect of education for women gives us

$$1.0782 \times 1.0005^{\wedge}Age \times 1.1447^{\wedge}1 \times 0.9979^{\wedge}(Age \times 1)$$
$$= (1.0782 \times 1.1447) \times 1.0005^{\wedge}Age \times 0.9979^{\wedge}Age$$
$$= 1.2342 \times (1.0005 \times 0.9979)^{\wedge}Age$$
$$= 1.2342 \times 0.9984^{\wedge}Age$$

This tells us that the factor change effect of education for women moves toward a less positive (more negative) effect as age increases. Rather than doing further calculations on this expression to provide specifics, let's move on to examining the significance region results for the factor change effect of education. Because the results have a straightforward pattern, I show the significance region table from the results window rather than the one saved to Excel.

```
. sigreg, effect(factor) nobva ndig(3)

  Significance Region for Factor Change Effect of Ed (1 unit difference)
  on Memnum at Selected Values of the Interaction of Age and Sex
-----------------------------------------------------------------------
                                  At Age≡
  At Sex=   |     18       32       46       60       74       88
------------+----------------------------------------------------------
     Men    |   1.088*   1.096*   1.104*   1.112*   1.120*   1.128*
   Women    |   1.199*   1.173*   1.147*   1.122*   1.098*   1.074*
-----------------------------------------------------------------------
  Key: Plain font  = Pos, Not Sig    Bold font*    = Pos, Sig
       Italic font = Neg, Not Sig    Italic font*  = Neg, Sig
```

This shows that the effect of education on the number of memberships is positive and significant at all ages for both men and women. For men at the youngest age, a 1-year increase in education predicts that the number of memberships will be larger by a factor of 1.088—that is, almost a 9% increase. This effect slowly increases with age such that at 88 years, the number of memberships would increase by about 13% for a 1-year difference in education. For women, the expected factor change in the number of memberships declines from 1.199 at *Age* = 18 to 1.074 at *Age* = 88. That is, for the youngest women, memberships are expected to rise by 20% for 1 year's difference in education, but by only 7% for the oldest women. For a visual display, the confidence bounds plot in Figure 11.8 shows this pattern quite well. It could be presented instead of the significance region table, but I would still get the significance region table to report specific effect values in the discussion of the figure.

```
. effdisp, plot(type(cbound)) effect(factor) ndig(0) pltopts(ylab( , form(%6.2f)))
```

Moderated Effect of Sex as a Discrete Change in Number of Memberships

```
. intspec focal(i.sex)  main((c.age , name(Age) range(20(10)80))  ///
>  (c.ed, name(Education) range(0(2)20)) (i.sex, name(Sex) range(0/1)))  ///
>  int2(c.age#i.sex c.ed#i.sex c.age#c.ed) int3(i.sex#c.age#c.ed) ndig(0)
   ...
. gfi

GFI Information from Interaction Specification of
Effect of Sex on g(Memnum) from Negative Binomial Regression
-----------------------------------------------------------------------

Effect of Women =
   -1.7303 + 0.0236*Age + 0.1351*Ed - 0.0021*Age*Ed
```

FIGURE 11.8 ● CONFIDENCE BOUNDS PLOTS FOR MODERATED EFFECT OF EDUCATION

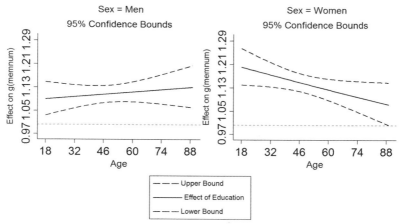

The GFI expression for the moderated effect of sex is not as simple to understand as were the expressions for the moderated effects of age and education, because both of the moderators are now interval. We can still describe the general pattern of how the effect of sex changes with age and education. Remember that a negative effect of sex means that the model predicts that women would have fewer memberships than men. An initial negative effect of sex (−1.7303) changes toward a positive (less negative) effect as either age or education increases, but at a slower rate for higher valued combinations of age with education. Determining more precisely where the effect changes and by how much is best done from the significance region table that follows.

```
. sigreg, effect(spost(amtopt(am(binary))) atopt( (means) _all))) ///
>   save(Output\table_11_5 tab) ndig(2)
```

Table 11.5 makes the pattern of the moderated effect of sex relatively easy to see and understand. The effect of sex is negative for most combinations of age and educa-tion. It is positive only for high education values paired with relatively low age val-ues (bottom-left corner) and, conversely, low education values paired with relatively high age values (top-right corner). Moreover, the significant negative effects are fairly similar in magnitude, with the number of memberships for women ranging from about 0.25 to 0.5 lower than for men; and they are roughly bounded in a region defined by education between 0 and 14 and age between 18 and 68 years. The effect of sex is positive—women are expected to have more memberships than men—and significant only for education between 17 and 20 years and ages between 18 and 36 years. Furthermore, these expected differences favoring women range from almost 1 membership to 2.4 memberships, which are much larger than the significant advan-tages for men (at most 0.5 membership) that characterize most combinations of age and education.

TABLE 11.5 ● SPOST CHANGE EFFECT OF WOMEN MODERATED BY THE INTERACTION OF AGE AND EDUCATION ON MEMNUM, FORMATTED TO HIGHLIGHT SIGN AND SIGNIFICANCE

	Age							
Education	**18**	**28**	**38**	**48**	**58**	**68**	**78**	**88**
0	*−0.37*	***−0.33****	***−0.28****	***−0.22****	*−0.15*	*−0.06*	0.06	0.20
2	***−0.40****	***−0.36****	***−0.31****	***−0.25****	*−0.18*	*−0.08*	0.03	0.17
5	***−0.43****	***−0.39****	***−0.34****	***−0.29****	*−0.22*	*−0.13*	*−0.03*	0.09
8	***−0.41****	***−0.38****	***−0.35****	***−0.31****	***−0.26****	*−0.21*	*−0.14*	*−0.07*
11	*−0.27*	***−0.28****	***−0.29****	***−0.30****	***−0.31****	***−0.32****	*−0.33*	*−0.35*
14	0.09	*−0.01*	*−0.11*	*−0.23*	***−0.36****	***−0.49****	*−0.64*	*−0.80*
17	**0.88***	0.58	0.28	*−0.05*	*−0.39*	*−0.75*	*−1.12*	*−1.53*
20	**2.44***	**1.74***	1.03	0.33	*−0.39*	*−1.12*	*−1.87*	*−2.66*
Key								
Plain font, no fill	Pos, Not Sig							
Bold*, filled	Pos, Sig							
Italic, no fill	Neg, Not Sig							
***Bold italic*, filled**	Neg, Sig							

Note: Memnum = number of memberships.

You could also try an effects display for the sex effect, but I find it less informative than the significance region table for this example. The difficulty with the confidence bounds plot in Figure 11.9 is that one of the plots (*Education* = 20) has a much bigger range of values for both the moderated effect and its confidence bounds. Consequently, the information on the other plots is compressed and hard to read because the same scaling is used for every plot for comparability. Readers can judge this for themselves by examining these five plots.

```
. effdisp , effect(spost(amtopt(am(binary)) atopt((means) all))) ///
> plot(type(cbound)) ndig(0)
```

In the next section, I draw on what we learned from studying the three separate moderated effects to demonstrate how to provide a more integrated explanation of the relationship of the three-way interaction to the number of memberships.

OUTDISP **for the Effects of Age, Education, and Sex Simultaneously**

Presenting a table or plot of the predicted outcome in the observed metric and discussing how it varies with the interacting predictors is often the most user-friendly strategy for a wide range of audiences. In addition, you can sometimes use a single table or a set of plots to describe the effect of all three interacting predictors. But you need to address the confounded nonlinearities in some manner to provide an accurate interpretation of the degree to which the predictors moderate one another's effect on the outcome. As discussed in the previous chapters, this can be done in

FIGURE 11.9 ● CONFIDENCE BOUNDS PLOT FOR MODERATED DISCRETE CHANGE EFFECT OF WOMEN

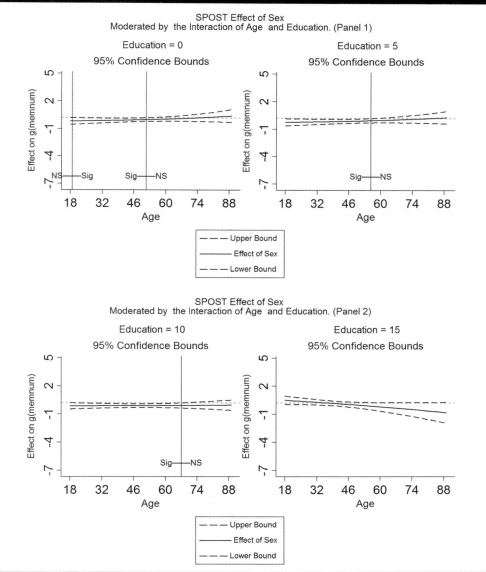

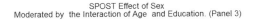

FIGURE 11.9 ● (CONTINUED)

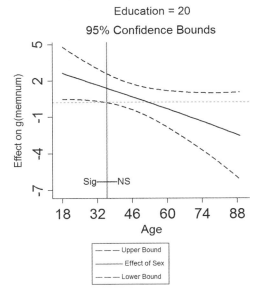

two basic ways. You can base your interpretations on the predicted outcome in the model metric rather than the observed metric, or you can compare the patterns in the observed predictions from the interactive model with those in the corresponding main effects model. I apply these strategies to the count model results first with tables and then with plots.

Predicted Values Tables

I start with Table 11.6 reporting the predicted values in the observed metric (number of memberships), followed by Table 11.7 with the predicted model metric outcome (logged number of memberships). I use the same *intspec* command to set up the definition of the interaction model and then run the *outdisp* command twice, once to get the predictions of the observed outcome and then for the modeled outcome predictions (adding the out(metric(model)) option). Note the specification of the save( ) suboption to produce the Excel formatted predicted values tables with the font size proportional to the magnitude of the effects presented in Tables 11.6 and 11.7. You control the organization of the table by your specification of the focal and moderating variables. The focal variable (*Ed*) defines the rows in the table, the first moderator (*Age*) defines the columns, and the second moderator (*Sex*) splits the table into horizontal subtables. I find that using the predictor with the fewest display values as the second moderator to split the table into subtables usually works best. I tried several alternative organizations, and this is the easiest to interpret.

```
. intspec focal(c.ed)  main((c.ed , name(Education) range(0(5)20))  (c.age ,    ///
> name(Age) range(20(20)80))   (i.sex, name(Sex) range(0/1)))  ///
> int2(c.age#c.ed c.ed#i.sex c.age#i.sex) int3(c.ed#c.age#i.sex) ndig(0)
   ...
. outdisp , tab(row(focal) save(Output\Pred_Memnum_sex.xlsx)) ndig(2)
>. out(atopt(means) _all)
   ...
. outdisp , tab(row(focal) save(Output\Pred_log_Memnum_sex.xlsx) ) ndig(2)
> out(metric(model) sdy)
```

Comparing the columns within the subtable for men and then within the subtable for women shows the relationship between education and memberships and how it changes with age for each sex. The differential font size makes it apparent that the number of memberships is always expected to rise with education levels (read down each column within a subtable). Similarly, it is clear that the education–membership

TABLE 11.6 ● MEMNUM BY THE THREE-WAY INTERACTION OF EDUCATION, AGE, AND SEX

Sex	Education	Age			
		20	**40**	**60**	**80**
Men	0	0.50	0.50	0.50	0.49
	5	0.77	0.81	0.85	0.89
	10	1.18	1.30	1.44	1.59
	15	1.81	2.11	2.45	2.85
	20	2.78	3.41	4.17	5.11
Women	0	0.14	0.23	0.36	0.58
	5	0.35	0.47	0.65	0.88
	10	0.85	0.99	1.15	1.33
	15	2.08	2.06	2.04	2.03
	20	5.08	4.30	3.64	3.08
Men–women	0	0.36	0.27	0.13	−0.08
	5	0.42	0.33	0.20	0.01
	10	0.33	0.32	0.29	0.26
	15	−0.27	0.05	0.41	0.82
	20	−2.30	−0.89	0.53	2.03

Note: Memnum = number of memberships.

TABLE 11.7 ● g(Memnum)-STANDARDIZED BY THE THREE-WAY INTERACTION OF EDUCATION, AGE, AND SEX					
Sex	**Education**	**Age**			
		20	**40**	**60**	**80**
Men	0	−1.10	−1.10	−1.10	−1.11
	5	−0.71	−0.67	−0.63	−0.59
	10	−0.33	−0.24	−0.15	−0.07
	15	0.05	0.19	0.32	0.45
	20	0.43	0.61	0.79	0.98
Women	0	−2.22	−1.80	−1.39	−0.97
	5	−1.42	−1.15	−0.87	−0.60
	10	−0.62	−0.49	−0.36	−0.22
	15	0.17	0.17	0.16	0.15
	20	0.97	0.82	0.67	0.52
Men−women	0	1.12	0.70	0.28	−0.14
	5	0.71	0.48	0.24	0.01
	10	0.29	0.25	0.20	0.16
	15	−0.12	0.02	0.16	0.30
	20	−0.54	−0.21	0.12	0.45

relationship varies with age in opposite patterns for men and for women. For men, the predicted change in membership with education increases in magnitude as age increases—the change in membership across the range of education values is 2.28 at $Age = 20$, rising to 4.62 at $Age = 80$. Women show the opposite pattern as the membership difference is expected to drop from 4.94 at $Age = 20$ to 2.50 at $Age = 80$. Given that the mean membership is 1.7 with a standard deviation of 1.9, these changes are all substantively large.

Looking across the rows within the male and female subtables indicates how the age–membership relationship varies with education differently for women and for men. This is easy to see by looking at the predicted change in membership across the age range shown, from 20 to 80 years. For women, the age–membership relationship is positive at lower levels of education; initially, it becomes larger in size, then declines as education rises from 0 to 10 years (predicted differences of 0.44, 0.53, and 0.48, respectively), and then changes to an increasingly negative effect when education ≥15 (−0.05 and −2.00, respectively). For men, the age–membership relationship starts as effectively no age differences at *Education* = 0 (−0.01), but then becomes increasingly positive as education increases (expected differences of 0.12, 0.41, 1.04, and 2.33, respectively). Note that the age differences are generally smaller than the education differences, suggesting that the education–membership relationship is more substantial.

It would take more head calculation and concentration to see the changing pattern of the sex–membership relationship because you need to compare rows (or columns) between the male and female subtables. Instead, I calculated and added a subtable for the men–women values in the saved Excel file (or you could easily respecify the organization of the table). This shows a generally higher predicted number of memberships for men than for women—positive gaps—except when education is very low and age is very high, or vice versa. Look at the upper-right and the lower-left corners of the men–women subtable. Outside those areas, the predicted membership difference between men and women rises and then falls with education at younger ages but only increases at older ages. However, the gap decreases with age when education levels are low and increases with age at higher education levels. Note that most of the gaps are modest in size, less than half a membership difference between men and women.

As usual, in the preceding discussion I avoided talking about the effects or the moderated effects but, rather, interpreted the table in terms of the relationships between the predictors and the outcome. I want to emphasize the point that predictions in the observed metric for nonlinear link functions are not always accurate guides to the pattern or the degree of moderation. Looking at the predicted values table in the model metric in Table 11.7 demonstrates this, although none of the discrepancies are particularly large. Consider how the age–membership relationship for women varies with education according to Table 11.7—there is an initial positive effect of age on membership (*Education* = 0), which declines in magnitude as education rises, then changes sign to negative when education is slightly smaller than 15 and continues to increase in magnitude.

In contrast, in Table 11.6, the initial positive relationship between age and memberships first increases in magnitude before beginning to drop, then changes to an increasingly negative effect. The pattern of moderation of the sex effect in the interior of the table—excluding the two corners in which women have higher predicted memberships than men—is simpler and more clear-cut. At all but the oldest age, the degree of male advantage in the number of memberships consistently declines with education level, whereas this was inconsistent in Table 11.6. Moreover, the age at which the men–women differences in membership change from always decreasing with education to always increasing with education is different—between 60 and 70 instead of between 50 and 60.

As I have noted before, whether you should use a table of predicted values in the observed or the modeled metric depends on the importance of discussing the pattern and degree of the moderated effects. You have just seen that the predicted observed values do not accurately portray the pattern of moderation for this example. My inclination in principle would be to focus instead on interpreting the table in the modeled outcome metric. Note that Table 11.7 reports predicted value differences in standard deviation units for the log number of memberships to provide a sense of the magnitude of the predicted differences. For example, it is clear in this table, as in the prior one, that the differences in membership across the range of education values are fairly substantial, between 1.5 and 3 standard deviations in the log number of memberships.

Predicted Values Plots

Many analysts and readers find a graphic display of the information in the predicted values tables easier to explain and to understand. With a three-way interaction, there are many different configurations of plotting variables and their display values that you could choose. I recommend trying several options to see what works best for you in a particular analysis. The basic structure is a display of the outcome against a (focal) predictor, with the same graph repeating that display for different values of the first moderator and the entire set repeated for different values of the second moderator. When one of the interacting predictors is categorical, I find that the plots are easiest to interpret if you use the nominal-level measure as your second moderator.

In this example, I use a scatterplot of predicted membership against age, with the prediction line repeated on the same graph for different values of education and with separate graphs for men and women. (I did examine alternatives, such as switching the roles of age and education and trying a bar chart with sex as the focal variable and age and education as moderators.) I present and discuss three different scatterplots of this relationship: (1) a scatterplot of the predicted number of memberships from the interaction model, (2) a scatterplot of the predicted memberships from the interaction model with superimposed main effects model prediction lines, and (3) a scatterplot of the logged membership in standard deviation units. To construct these plots, you need to first run and store the main effects and the interaction effects models' estimation results. The *intspec* command specifies age as the focal variable—what membership is plotted against—as well as display names and display values for age, education, and sex. The *outdisp* command requests the default plot type—in this case a scatterplot—and specifies that the reference values for the predictors not part of the interaction are set to their means.

```
. nbreg memnum c.age c.ed i.sex i.race class
      …
. est store mainmod
      …
. nbreg memnum c.age##c.ed##i.sex i.race class
.     …
. est store intmod
      …
. intspec focal(c.age)  main((c.age , name(Age) range(18(14)88))  ///
> (c.ed , name(Education) range(0(5)20)) (i.sex, name(Sex) range(0/1))) ///
> int2(c.age#c.ed c.age#i.sex  c.ed#i.sex) int3(c.age#c.ed#i.sex) ndig(0)
      …
outdisp ,  plot(def) out(atopt((means) _all))  ndig(3)
```

The plot of predicted number of memberships in Figure 11.10 exhibits the same patterns as Table 11.6 for the relationship of membership with the three interacting predictors. But the plotted prediction lines for membership by age visually focus the reader's attention more on the age–membership relationship. You can still determine the pattern for the education–membership relationship by examining the vertical gap between the plotted lines—which shows the expected difference in membership between education levels—and how that gap varies with age and how it differs between the plots for men and women. For example, comparing the lines for any two education levels shows that the predicted number of memberships is greater for higher levels of education and that the membership difference by education changes with age. For men, the predicted membership differences by education increase with age, while for women the membership differences decrease with age. In a similar fashion, you can compare the prediction line for a given education level between the plot for men and the plot for women to assess the sex–membership relationship. This generally indicates more predicted memberships for men than for women, except for those who have either high education levels paired with relatively young ages or low education levels with relatively old ages.

The unsurprising fact that the predicted values plot shows the same patterns as the predicted values table means that the plot has the same problem of confounded

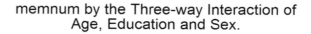

FIGURE 11.10 ● PREDICTED OUTCOME PLOT IN OBSERVED METRIC

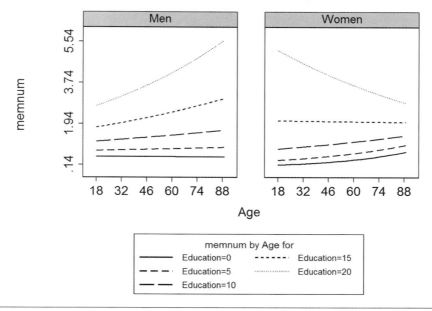

nonlinearities when interpreting the interaction effect. The first option for dealing with this is to supplement the predicted values plot by adding prediction lines from a corresponding main effects model. You produce this by adding the main( ) suboption with the name of the stored main effects estimation results inside the out( ) option:

```
. outdisp , out(main(mainmod) atopt((means) _all)) ndig(3) plot(def)
```

The main effects prediction lines in Figure 11.11 provide a reference point for parsing out the nonlinearity due to the model from the nonlinearity of the interaction effect. The *black lines* are the prediction lines from the interactive model, and the *gray lines* are the predictions from the main effects model. This portrays a very different sense of the extent to which a predictor in the interaction is moderated by the other two. The divergence of the main effects prediction lines from parallel shows the impact of the model-induced nonlinearity—steeper slopes for higher levels of education. At the same time, we can see that education moderates the age effect differently for men than for women. For men, higher levels of education intensify the positive effect of age on membership, while for women education substantially diminishes the positive effect of age, turning it increasingly negative at high levels of education.

It is difficult to use Figure 11.11 to separate out how education (or sex) is moderated by the other two predictors. It is much easier to create a similar plot reversing the roles of age and education to separate out the moderation of education with age and sex from the nonlinearity of the link function. This requires a revised *intspec* command to declare education as the focal variable and to change the display values for

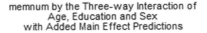

FIGURE 11.11 ● PREDICTED OUTCOME PLOT IN OBSERVED METRIC WITH SUPERIMPOSED MAIN EFFECTS MODEL PREDICTIONS

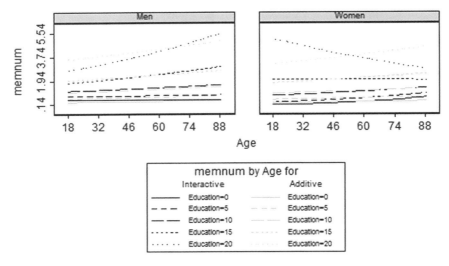

age. In the *outdisp* command, I add the single(1) suboption in the plot( ) option to deconstruct the plot by age. This creates a separate plot for each of the age display values to facilitate comparing the education interaction model predictions with the main effects predictions:

```
.  intspec focal(c.ed) main((c.ed,name(Education)range(0(5)20)) ///
>  (c.age , name(Age) range(18(23)87))(i.sex , name(Sex) range(0/1))) ///
>  int2(c.ed#c.age c.ed#i.sex c.age#i.sex)int3(c.ed#c.age#i.sex)ndig(0)
   ...
.  outdisp , out(main(mainmod)  atopt((means) _all)) plot(sing(1)) ndig(2)
```

Figure 11.12 presents such a plot with prediction lines for the membership–education relationship at several ages, separately for men and for women. The prediction lines in the nondeconstructed plot are so overlaid with one another as to make them difficult to distinguish well; readers should try running this to see for themselves. Divergences between the *black* and the *gray lines* represent how and to what degree the effect of education is moderated by age and sex. Let's start by looking at how the effect of education changes with age for each sex. Compare plots in the same column from top to bottom. For men, the positive effect of education is smaller at younger ages and increases in magnitude at older ages for both the interactive model results (black lines) and the main effect model results (gray lines). Also note that the black lines (interactive model effects) have a shallower slope than the gray lines (main effects model effects) at younger ages which reverses as age increases. Thus the moderation of the education effect by age for men is overstated because it builds on (enhances) the main effect model's pattern of change with age. For women, the reverse pattern holds—the positive effect of education on memberships is largest for women at younger ages and declines with age. And thus, the interaction predictions understate how much age moderates the education effect for women since it runs counter to the main effects model's pattern.

FIGURE 11.12 ⬡ DECONSTRUCTED PREDICTED OUTCOME IN OBSERVED METRIC

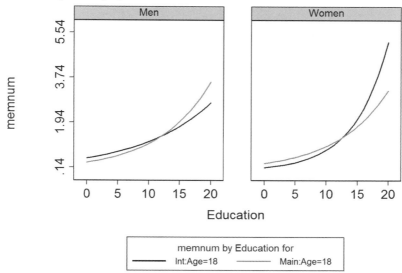

memnum by the Three-way Interaction of
Education, Age and Sex
with Added Main Effect Predictions.

Age = 18 (18)

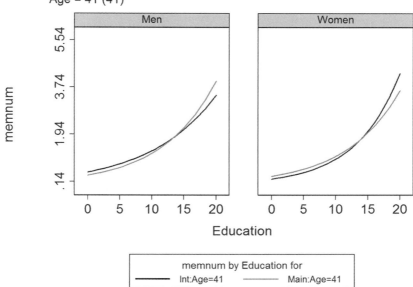

memnum by the Three-way Interaction of
Education, Age and Sex
with Added Main Effect Predictions.

Age = 41 (41)

(Continued)

FIGURE 11.12 ● (CONTINUED)

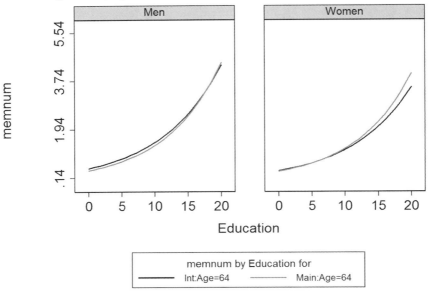

memnum by the Three-way Interaction of
Education, Age and Sex
with Added Main Effect Predictions.

Age = 64 (64)

memnum by Education for
Int:Age=64   Main:Age=64

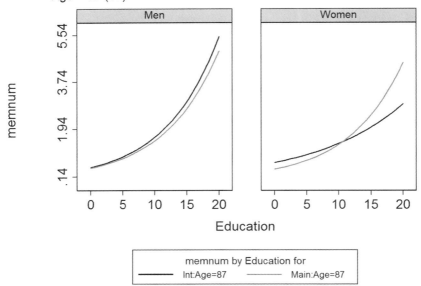

memnum by the Three-way Interaction of
Education, Age and Sex
with Added Main Effect Predictions.

Age = 87 (87)

memnum by Education for
Int:Age=87   Main:Age=87

The second, and I think more readily interpretable, option to parse out the modeled nonlinearity from the interaction effect is to interpret the predicted outcome in the model metric. The predicted log counts are standardized to provide a more meaningful scale; ICALC reports the mean and standard deviation values used for the standardization (see section "Special Topics" at the end of the chapter for a discussion of the calculation of the standardized model metric outcome for count models). I also like to create dual *y*-axis labeling: the left-side *y*-axis labeled in the model metric (standardized log memberships) and the right-hand *y*-axis labeled in observed metric values (number of memberships) corresponding to the selected model metric values. Note that I use the same *intspec* command for these plots in Figure 11.13 as for Figure 11.10, but in the *outdisp* command in the out() option, I specify the suboption metric(model) and the keywords sdy and dual:

```
. intspec focal(c.age)  main((c.age , name(Age) range(18(14)88)) ///
> (c.ed , name(Education) range(0(5)20)) (i.sex , name(Sex) range(0/1))) ///
> int2(c.age#c.ed c.age#i.sex  c.ed#i.sex) int3(c.age#c.ed#i.sex) ndig(0)
   ...
.outdisp , out(metric(model) dual sdy atopt((means) all) plot(def) ndig(3)
   ...

Standard deviation of g(Memnum) =   1.1202

            Mean =   0.5378
```

The plots in Figure 11.13 reveal the moderation of the age effect by education and sex quite straightforwardly. For men, the age effect on membership changes from zero to increasingly positive as education levels rise—the largest effect is almost 2/3 of a standard deviation change in logged memberships at *Education* = 20 across age's

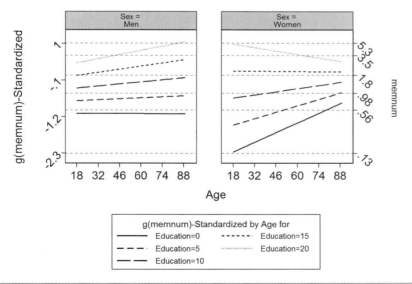

FIGURE 11.13 ⬡ PREDICTED OUTCOME PLOT IN MODEL METRIC

g(memnum)-Standardized by the Three-way Interaction of Age, Education and Sex, with Dual Outcome Axes.

range (or, equivalently, 2.8 more memberships). In contrast, for women, the age effect changes from positive at low levels of education to negative at the top of the education distribution. The biggest age effects for women are at the bottom of the education distribution—a 1.46 standard deviation increase in logged memberships at *Education* = 0 and a 0.96 increase at *Education* = 5 (0.56 and 0.66 more memberships, respectively). Moreover, women's age effect at the highest education level is as large as any of the age effects for men—a 0.52 standard deviation decline in logged memberships (2.3 fewer memberships).

With some concentration, you can also see the moderation of the effect of education by age and sex from this figure. Compare the vertical gaps between the plotted lines as age increases to interpret the effect of education. The prediction lines for men show a fan-shaped opening to the right. This means a larger expected difference in logged membership between education levels as age increases, and the vertical order of the lines—increasing from low to high education—indicates a positive education effect. That is, there is a positive effect of education that is largest at the oldest ages. Conversely, women's prediction lines exhibit a fan-shaped opening to the left, indicating a smaller difference in the expected logged membership between education levels as age increases—that is, a positive education effect diminishing with age. Comparing the prediction lines between the plots for women and men shows the effect of sex. The top two lines (*Education* = 15, 20) for women lie above the lines for men at the youngest ages but below them as age increases. The reverse characterizes the lower two lines (*Education* = 0, 5). Overall, the prediction lines for men are above those for women—men are expected to belong to more organizations—except when education and age are at opposite ends of their respective ordered distributions.

The patterns of the moderated effects of education and sex would be easier to identify if you revise and rerun the ICALC commands, once to make education your focal variable and then to make sex your focal variable. I would not present these figures but, rather, use them to help better understand and then explain the patterns in Figure 11.13. Similarly, for any plot of predicted values, I would always get the corresponding table of predicted values to get exact figures to provide details in the interpretation of the plot.

SPECIAL TOPICS

Using Predicted Probabilities for Interpretation

How the predictors affect the distribution of the predicted probabilities of the outcome count is sometimes used to interpret the results from count models. ICALC does not currently have functionality for this mode of interpretation, so I use the first example results (reported days of poor mental health) to briefly illustrate how to produce tables and plots showing how the distribution of the counts' probabilities vary with the interacting predictors. I assume that readers are familiar with the SPOST13 commands *mtable* and *mgen*, which I apply to produce the desired results; so I will provide little explanation of the meaning of the options in these commands.

The basic strategy is to calculate the probability that the count outcome = 0, 1, 2, . . . for each combination of values of your interacting predictors and describe how the patterns of the probabilities change with the predictors. To avoid drowning in the amount of data produced, you need to decide the maximum count value for which you want the predicted probability. At some point, there will be very little difference in the predicted probability across the combinations of your predictors, and that is the ideal point to pick as the maximum value you will use. This obviously depends on the data you're analyzing, but I would recommend initially picking a value between 5 and 9 as your maximum and seeing if at that maximum value there are still perceptible differences in the predicted probabilities.

One quick note on which $Pr(y = 0)$ you want to analyze: For Poisson or negative binomial, there is only one $Pr(y = 0)$, calculated for all cases. For ZIP or ZINB, there are three $Pr(y = 0)$ that can be calculated:

1. For all cases

2. For those in the always-zero group

3. For those in the not-always-zero group

If your interaction is in the zero-inflated component, then you will want $Pr(y = 0)$ for those in the always-zero group, while if your interaction is in the count model portion, then you will want $Pr(y = 0)$ for those in the not-always-zero group. Given the options available through the Stata *margins* command and the SPOST13 commands, you can get the $Pr(y = 0)$ for the always-zero group and for all cases. Subtract the latter quantity from the former to get $Pr(y = 0)$ for the not-always-zero group. The following example is for a ZINB regression, which illustrates how to do this for an interaction in the count model component.

Table of the Predicted Probability Distribution of Counts

You need to run the *mtable* command twice and combine the results. The first command will produce results for $Pr(y = 0)$ for the always-zero group by specifying the predict(pr) option. The option pr(0/9) in the second command calculates $Pr(y = 0)$ for all cases and $Pr(y = c)$ for $c = 1, 2, . . . ,$ max. The second *mtable* command is specified to place the second set of results beside the first by using the right option. The at( ) option in both *mtable* commands is used to specify the display (calculating) values for each of the predictors in the interaction, and the atmeans option sets the reference values for the other predictors to their means. The atvars( ) option in the first *mtable* command specifies that the variables to which it refers—the interacting predictors—will be listed to identify the table rows. In the second *mtable* command, the atvars(_none) option suppresses the repetition of the values of the interacting predictors in the combined subsequent table. In both *mtable* commands, the norownumbers option suppresses row numbering, and the *wide* command ensures the compatibility of the tables' setup/formatting for combining. The model estimation command and the *mtable* commands used are as follows:

```
.zinb pmhdays i.wfconflict3##c.sei c.age educ childs i.sex if wrkstat <=4 , ///
>       inf(c.age##i.sex)
...

. mtable , at(wfconflict3=(1/3) sei=(17(20)97)) atmeans atvars(wfconflict3 ///
  sei) predict(pr) clear norownumbers wide
...

. mtable , at(wfconflict3=(1/3) sei=(17(20)97)) atmeans    atvars(_none)   ///
>       pr(0/9) right norownumbers wide
```

2.

3.

wfconf~t	sei	PrAll0	PrAny0_0	PrAny0_1	PrAny0_2	PrAny0_3	PrAny0_4	PrAny0_5	PrAny0_6	PrAny0_7	PrAny0_8	PrAny0_9
0	17	.467	.523	.042	.036	.031	.028	.025	.023	.021	.019	.018
0	37	.467	.557	.064	.052	.043	.036	.031	.027	.023	.020	.018
0	57	.467	.606	.091	.066	.050	.039	.030	.024	.019	.015	.012
0	77	.467	.672	.114	.072	.047	.031	.021	.014	.009	.006	.004
0	97	.467	.749	.124	.061	.031	.016	.009	.005	.002	.001	.001
1	17	.467	.546	.057	.047	.040	.034	.030	.026	.023	.021	.018
1	37	.467	.552	.065	.050	.042	.036	.031	.027	.023	.020	.018
1	57	.467	.559	.070	.052	.043	.037	.031	.027	.023	.020	.017
1	77	.467	.566	.070	.055	.045	.038	.032	.027	.023	.020	.017
1	97	.467	.574	.074	.058	.047	.038	.032	.027	.022	.019	.016
0	17	.467	.531	.048	.040	.035	.031	.027	.024	.022	.020	.018
0	37	.467	.528	.045	.038	.033	.030	.026	.024	.022	.020	.018
0	57	.467	.524	.043	.036	.032	.028	.026	.023	.021	.019	.018
0	77	.467	.521	.041	.035	.031	.027	.025	.023	.021	.019	.017
0	97	.467	.518	.039	.033	.029	.026	.024	.022	.020	.019	.017

I lightly edited the table of predicted probabilities produced by *mtable* in the results window to improve readability. The three panels correspond to the categories of work–family conflict—never, sometimes, often—whose values are shown in the first two columns, and the rows are the display values for SES shown in the third column. The remaining columns report the probabilities of different numbers of counts. To get $\Pr(y = 0)$ in the not-always-zero group, you have to subtract $\Pr(y = 0)$ for the always-zero group—PrAll0 in the table—from $\Pr(y = 0)$ for all cases—PrAny0_0. Looking across the columns of probabilities, the differences by SES and work–family conflict have largely disappeared for counts ≥6, except for the combination of "never" for work conflict and higher levels of SES. Even if you concentrate on the columns for counts <6, it requires some work to figure out the patterns.

> For a Poisson or negative binomial regression, you would need only one *mtable* command to get the predicted probabilities:
>
> mtable , at(wfconflict=(1/3) sei=(17(20)97)) atmeans atvars(wfconflict /// sei) pr(0/9) clear norownumbers wide

Plotting the Predicted Probability Distribution of Counts

A more accessible approach is to plot the predicted probabilities against SEI for each category of work–family conflict. You generate the plotting variables using the *mgen* command. As I did with *mtable*, I use two separate *mgen* commands—once to create $\Pr(y = 0)$ for the always-zero group and again in the full sample to create $\Pr(y = 0)$ for the not-always-zero group; the *gen ZZpr0* . . . command creates this variable for plotting. (*Note:* You need only the first *mgen* command if you are not analyzing a zero-inflated model.) The next several commands apply value and variable labels to improve the readability of the plots. If I plotted $\Pr(y = 0)$ on the same scale and axis as the other probabilities, it would compress the other plots. Thus, I format the predicted probabilities by SES plot to show the prediction line for $\Pr(y = 0)$ separately in the top area of the graph and on a scale different from that of the other predicted probabilities plotted in the bottom three quarters of the graph area. This is primarily controlled by the specifications in the options for yaxis( ), ylab( ), and ysc( ) for each scatterplot. I use the by( ) option to create the separate scatterplots shown in Figure 11.14 for each of the categories of work–family conflict (wfconflict3).

```
. zinb pmhdays i.wfconflict3##c.sei c.age educ childs i.sex if wrkstat <=4 ,
>   inf(c.age##i.sex)
   …
. mgen , at(wfconflict3=(1/3) sei=(17(20)97)) atmeans stub(ZZ) pr(0/9) noci replace
   …
. mgen , at(wfconflict3=(1/3) sei=(17(20)97)) atmeans stub(ZZaz) predict(pr) noci
>   replace
   …
. gen ZZpr0 = ZZprany0 -ZZazprall0
   …
. lab val ZZwfconflict3 often
. forvalues i=0/9 {
  2.        lab var ZZpr`i' "Pr(y=`i')"
  3.  }
. lab var ZZsei "SES"
. scatter  ZZpr1 ZZpr2 ZZpr3 ZZpr4 ZZpr5  ZZsei , by(ZZwfconflict, cols(3)) ///
>   scheme(s1mono) conn(l l l l l) name(PrSES, replace) ///
>   yaxis(1) ylab(0(.04).12, axis(1)) ysc(r(.17)  axis(1)) aspect(1.5) ysize(5.75) ///
>   xsize(6.5) leg( symy(*.7) symx(*.7) size(*.7)) ///
>   || scatter ZZpr0 ZZsei, conn(l) by(ZZwfconflict3) yaxis(2) ylab(.05(.1).25, ///
>   axis(2)) ysc(r(-.72) axis(2))
```

FIGURE 11.14 ● PREDICTED PROBABILITY DISTRIBUTION SCATTERPLOT FOR TWO-WAY INTERACTION

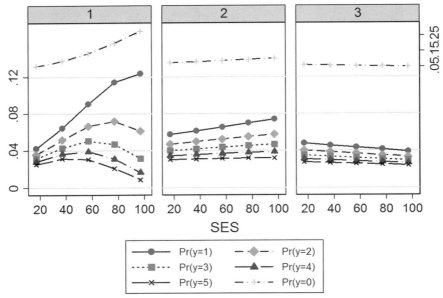

Graphs by Work Family Conflict 3 Category

These plots make it clear that the relationship between SES and poor mental health days is quite different across the categories of work–family conflict. The "never" category exhibits the largest predicted changes in probabilities with SES. Specifically, the probability of 0 or 1 day of poor mental health increases with SES, while the probabilities of higher counts (>1) initially rise with SES and then decline. This indicates that the negative relationship of SES and the predicted number of poor health days for those workers who never experience conflict is generated not only by an increase in the probability of 0 or 1 day of poor mental health but also by declines in the probability of higher counts at higher levels of SES. In the "sometimes" and "often" conflict categories, the predicted probabilities change much less with SES—recall that the SES moderated effect is not significant in these categories—but in opposite directions from each other. The predicted probabilities increase with SES in the "sometimes" category but decrease with SES in the "often" category.

You can also discern the relationship between work–family conflict and poor mental health days. The difference in the predicted probabilities for 0 or 1 day across categories is smallest at low SES but increases as SES rises due to the magnitude of change in the "never" category.

You could repeat this process for a main effects model for comparison with these plots to assess the degree of confounding of the two sources of nonlinearity. Just change the *zinb* command to replace the interaction specification—i.wfconflict##c.sei—with a main effects specification—i.wfconflict 3c.sei. The plots in Figure 11.15 show a decreasing magnitude of the predicted probabilities from the "never" to the "sometimes," to the "often" conflict category. This suggests that some of the differences we see in the interaction model plots are enhanced by the nonlinearity of the link function.

FIGURE 11.15 ⬡ PREDICTED PROBABILITIES FROM A MAIN EFFECTS MODEL

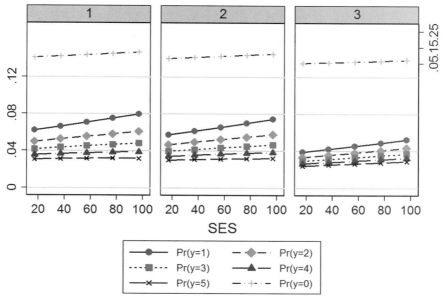

Graphs by Work Family Conflict 3 Category

Working With Interaction Effects in the Zero-Inflated Model Component

If a count model with a zero-inflated component contains an interaction effect, you would interpret it in the same way that you would any binary outcome estimated by logistic regression (or optionally probit analysis). The outcome analyzed in this component of the model is whether or not a case is in the always-zero group. Because the independent variables predict the probability of being always zero, it is common for predictors to have opposite-signed effects in the zero-inflated versus the count component of the model. The observed outcome metric is the probability of being in the always-zero group, and the model outcome metric can be considered the latent propensity to be in the always-zero group. All the approaches and tools covered in Chapter 8 can be applied to interpret an interaction in the zero-inflated model results, so I do not include an example of the interpretation.

But there are two important points about how to use the ICALC commands to do the analyses that I illustrate with a hypothetical example. Suppose the ZINB example had included an interaction between age and female in the zero-inflated component:

```
zinb pmhdays i.wfconflict3##c.sei c.age educ childs i.sex if wrkstat <=4, ///
inf(c.age##i.sex)
```

To get the ICALC results for this age-by-female interaction, you must tell ICALC that you want to use the zero-inflated model component's results, because the default is to use the results predicting the counts for those who are not in the always-zero group. To request this, include in the *intspec* command eqname(inflate) as an option, as well as the usual options to describe the interaction:

```
intspec focal(c.age) main( (c.age, name(Age) range(18(14)88)) ///
(i.sex, name(Sex) range(0/1))) ndig(0) int2(c.age#i.sex) eqname(inflate)
```

Second, any predictors listed using factor-variable notation that are in both the count model and the zero-inflated model must be specified in the same way. For example, I sometimes do not use "i." notation for a two-category nominal variable such as sex unless it is part of an interaction. Suppose I had listed the variable *sex* (already dummy coded as 1 = *sex*) and not *i.sex* in the count component of the ZINB model (see Line 1 of the command below) but used *i.sex* with factor-variable notation to specify the interaction in the zero-inflated component (see Line 2 of the command):

```
zinb pmhdays i.wfconflict3##c.sei c.age educ childs female if wrkstat <=4, ///
inf(c.age##i.sex)
```

ZINB would produce the same results as if I had used *i.sex* in both places. But this would cause confusion (and an error message) when subsequently running the *margins* command (which ICALC does for various calculations). Any reference to "sex" would be ambiguous because Stata will not know if you are referring to *sex*, *1.sex*, or *0.sex*. So you must specify *i.sex* in both places to avoid this confusion.

Standardized Log Count for Poisson and Negative Binomial Models

Let Y be the outcome measure of the observed count and Z represent the natural log of the count; that is, $Z = \ln(Y)$. The standardized log *Count* is defined as

$$\frac{Z - \hat{\mathcal{E}}(Z)}{\widehat{Var}(Z)^{\frac{1}{2}}}$$

To standardize the predicted values of log *Count*, we need an estimate of the mean and the variance (standard deviation) of Z. Because Z is not defined when $Y = 0$, we cannot just calculate Z and then find its mean and variance. But we can use the Delta method, which approximates the mean and the variance of a transformation of a random variable, to estimate the mean and variance of Z using the transformation function and the mean and variance of Y (Feiveson, 2017, Stata FAQ page). If $g(\ )$ is the differentiable function defining the transformation $Z = g(Y)$, and μ and σ^2 are the mean and variance of Y, then the estimates are

$$\mathcal{E}(Z) \approx g(\mu) \quad Var(Z) \approx g'(\mu)^2 \times \sigma^2$$

In this case, $g(\mu) = \ln(\mu)$, so

$$g'(\mu) = \frac{dg}{d\mu} = \frac{d\ln(\mu)}{d\mu} = \frac{1}{\mu}$$

Substituting these into the generic expressions above gives

$$\mathcal{E}(Z) \approx \ln(\mu) \quad Var(Z) \approx \frac{1}{\mu^2} \times \sigma^2$$

Applying these to our example, plug in the mean and variance of the number of memberships to estimate the mean and variance of the logged number of memberships as

$$\mathcal{E}(Z) \approx \ln(\mu) = \ln(1.712172) = 0.53776$$

$$Var(Z) \approx \frac{1}{\mu^2} \times \sigma^2 = \frac{1}{1.712172^2} \times 3.678876 = 1.2549325$$

Standard deviation $= 1.12024$

These estimated values are reported in the results window.

Getting the Count Value Equivalent to a Standardized Log Count Value

You can calculate this by hand with the mean and standard deviation reported by ICALC (0.5378 and 1.1202, respectively, from the three-way interaction example):

$$Count = e^{Stanardized \; \ln(Count) \times SD \, + \, mean}$$

For example, the count value corresponding to a standardized log *Count* value of 1.0478 (predicted values for *Sex* = men, *Age* = 87, and *Education* = 20) is

$$e^{1.0478 \times 1.1202 \, + \, 0.5378} = e^{1.7115} = 5.538$$

The easier and more accurate way is to let ICALC do the work for you. First, if you do not already have a table of predicted values in the model metric, add the table( ) option to your *outdisp* command and rerun it. Then, revise the out( ) option in the *outdisp* command to delete the sdy and the metric(model) suboptions; keep the table( ) option. Rerun the *intspec* command (if needed), followed by this revised *outdisp* command. The count value equivalent of a standardized log *Count* in the first table is the value in the same cell location (same values of the interacting variables) in the second table.

CHAPTER 11 NOTES

1. After running the main effects model, I used the SPOST13 *mchange* command to calculate the discrete effects:

    ```
    zinb pmhdays c.sei i.wfconflict3 age educ childs female if wrkstat <=4 , ///
        inf(age female)
      mchange sei , am(sd) center at(wfconflict3=1 (means) _all) brief
      mchange sei , am(sd) center at(wfconflict3=2 (means) _all) brief
      mchange sei , am(sd) center at(wfconflict3=3 (means) _all) brief
    ```

2. The contents of pltopt( ) specify adding a label consisting of a blank space to the *y*-axis at a value of −3.75 and that no tick mark is drawn only for that added label.

12

EXTENSIONS AND FINAL THOUGHTS

This is the way the world ends

Not with a bang but a whimper

—T. S. Eliot (1925)

In this concluding chapter, I first briefly discuss some extensions to several modeling situations that I did not cover in the applications chapters: how to interpret the effect of a polynomial function of a predictor interacted with another predictor; interaction effects when analyzing censored dependent variables such as Tobit or Heckman selection models; and interaction effects when analyzing duration data such as Cox proportional hazards models. In each of these cases, the calculations and results needed for the complete interpretation of interaction effects can only be obtained in part using the current version of the ICALC commands. In the following sections, I illustrate for these situations what calculations and output can be produced using ICALC and overview how to generate those that ICALC cannot produce. I end by revisiting two cautionary points, on the use and interpretation of standardized effects and on the consequences of model misspecification.

EXTENSIONS

Interaction of a Polynomial Function of a Predictor With Another Predictor

For this illustration, I use an OLS regression of respondents' education on the interaction between a quadratic function of age and a three-category nominal indicator of race, with additional predictors for parental education and a dummy variable for female. In essence, this interaction specifies that the quadratic effect of age on education is different for each race group. It also means that race differences in education vary as a quadratic function of age. To interpret this interaction, I first use the *margins* command to produce the equivalent of a significance region table for the moderated effects of race and age. I then apply the *margins* and *marginsplot* commands to demonstrate how to create a predicted values table and plots to show how race and age interact in predicting education.

In line with the principle of marginality, your prediction function should include product terms for each component of the polynomial function of age with the dummy indicators for race along with the main effect terms for each predictor. Using factor-variable notation in Stata to specify polynomial effects such as the quadratic effect of age and the dummy variables for race and their interaction greatly simplifies the calculation of the interpretive aids. And it ensures that the values specified for calculations involving age are properly substituted into its squared term; for example, if age is set to 40, age-squared will be set to $40^2 = 1,600$.

Thus, I use *i.race* to specify dummy indicators for the race categories and *c.age##c .age* to specify a quadratic function of age as a predictor, and then I include *i.race##c .age##c.age* in the regression model:

regress educ i.race##c.age##c.age i.female pared if wrkstat != 6

This properly represents the interaction between race and the age quadratic function because it includes all the relevant higher and lower order terms:

- Main effects of the race dummies, age, and the square of age (*c.age#c.age*)

- Interactions between the race dummies and age

- Interactions between the race dummies and the squared term for age

The relevant part of the OLS regression results is as follows:

$$
\begin{aligned}
Educ = &-1.76544\ Black + 2.54562\ Other + 0.10186\ Age \\
&- 0.00078\ c.Age \# c.Age + 0.08286\ Black \times Age \\
&- 0.00100\ Black \times c.Age \# c.\ Age - 0.10662\ Other \times Age \\
&+ 0.00085\ Other \times c.Age \# c.Age + ...
\end{aligned}
\tag{12.1}
$$

ASIDE

You could use ICALC to get some of the calculations for the effect of race moderated by age. But to get properly calculated results from the current version of ICALC, you cannot use factor-variable notation to specify the quadratic term for age but must use a constructed predictor—*agesq*—containing the squared values of age. In using the *intspec* command to define the interaction specification, you would then specify the display values for *agesq* to correspond to the square of the display values for age. You would square the first display value for age (18) and specify that as the first display value for *agesq* (324); the second pair of display values would be 36 and its square (1,296), and so on. You could then get some correct results from the *gfi* and the *sigreg* commands, but they would be mixed in with meaningless results; for example, the diagonal cells in the significance region table correctly report the moderated effect value and significance of Black or Other Races, but the remainder of the table would not be sensible.

Moderated Effect of Race

Let's start by getting the algebraic expression for the moderated effect of the race dummy variables as they change with age. Applying the GFI principle to Equation 12.1, we get

$$\text{Effect of } Black = -1.76544 + 0.08286 \, Age - 0.00100 \, Age^2$$
$$\text{Effect of } Other = 2.54562 - 0.10662 \, Age + 0.00085 \, Age^2$$

(12.2)

We can use the *margins* command with the dydx( ) option to calculate the value of these effects and their significance at selected values for age:

```
margins , dydx(race) at(age=(18(18)90) (means) _all)

Conditional marginal effects                    Number of obs    =      1,815
Model VCE    : OLS

Expression   : Linear prediction, predict()
dy/dx w.r.t. : 2.race 3.race

1._at        : 1.race         =      .7779614 (mean)
               2.race         =      .1443526 (mean)
               3.race         =       .077686 (mean)
               age            =            18
               0.female       =      .4341598 (mean)
               1.female       =      .5658402 (mean)
               pared          =      12.10028 (mean)

2._at        : 1.race         =      .7779614 (mean)
               2.race         =      .1443526 (mean)
               3.race         =       .077686 (mean)
               age            =            36
               0.female       =      .4341598 (mean)
               1.female       =      .5658402 (mean)
               pared          =      12.10028 (mean)

3._at        : 1.race         =      .7779614 (mean)
               2.race         =      .1443526 (mean)
               3.race         =       .077686 (mean)
               age            =            54
               0.female       =      .4341598 (mean)
               1.female       =      .5658402 (mean)
               pared          =      12.10028 (mean)

4._at        : 1.race         =      .7779614 (mean)
               2.race         =      .1443526 (mean)
               3.race         =       .077686 (mean)
               age            =            72
               0.female       =      .4341598 (mean)
               1.female       =      .5658402 (mean)
               pared          =      12.10028 (mean)

5._at        : 1.race         =      .7779614 (mean)
               2.race         =      .1443526 (mean)
               3.race         =       .077686 (mean)
               age            =            90
               0.female       =      .4341598 (mean)
               1.female       =      .5658402 (mean)
               pared          =      12.10028 (mean)
```

```
--------------------------------------------------------------------------------
            |                Delta-method
            |      dy/dx   Std. Err.        t    P>|t|     [95% Conf. Interval]
------------+-------------------------------------------------------------------
2.race      |
       _at  |
         1  |   -.5968727   .5235969    -1.14    0.254    -1.623793    .4300473
         2  |   -.0741665   .2191955    -0.34    0.735    -.5040703    .3557372
         3  |   -.1973244   .2507767    -0.79    0.431    -.6891677     .294519
         4  |   -.9663463    .366903    -2.63    0.009    -1.685946   -.2467469
         5  |   -2.381232   .9479652    -2.51    0.012    -4.240457   -.5220071
------------+-------------------------------------------------------------------
3.race      |
       _at  |
         1  |    .900715   .6866055     1.31    0.190    -.4459105     2.24734
         2  |   -.1954809   .290825     -0.67    0.502    -.7658702    .3749083
         3  |   -.7429687   .3605119    -2.06    0.039    -1.450033    -.035904
         4  |   -.7417483   .6492477    -1.14    0.253    -2.015105    .5316082
         5  |   -.1918197    1.74035    -0.11    0.912    -3.605132    3.221493
--------------------------------------------------------------------------------
Note: dy/dx for factor levels is the discrete change from the base level.
```

The table at the end of the Stata output has the moderated effects shown in the first column with the significance level shown in the fourth column. But to decipher the meaning of the _at labels, we have to look at the list before the table results and determine the pattern and correspondence to the values of the moderator, age. This is straightforward with a single moderator, but it can become awkward and hard to use for multiple moderators. Nevertheless, we can see from the bottom panel of this table that the Other Race–White difference in education is significant only for ages mostly in the 50s to early 60s; the effect is negative in this range, indicating a lower level of predicted education for Other Races compared with Whites.

In contrast, the Black–White difference in education turns significant for ages 65 and above, where it is negative and increasing in magnitude, suggesting a larger and larger deficit in predicted education relative to Whites. Note that this would be much easier to interpret by adding the display values of age to the _at column to create a table like the significance region tables that ICALC produces for models without polynomial interactions. And as we will see shortly, plots of predicted values work especially well to show the effects.

Moderated Effect of Age

For a polynomial function, the algebraic rearrangement principles of GFI do not apply. Instead, you must take the partial derivative of the regression in Equation 12.1, which gives[1]

$$\text{Effect of } Age = 0.10186 - 2 \times 0.00078 \, Age + 0.08286 \, Black - 2 \times 0.00100$$
$$Black \times Age - 0.10662 \, Other + 2 \times 0.00085 \, Other \times Age$$

Rearranging the terms and simplifying yields

$$\text{Effect of } Age = \left(0.10186 + 0.08286 \, Black - 0.10662 \, Other\right) +$$
$$\left(-0.00156 - 0.00200 \, Black + 0.00170 \, Other\right) \times Age \tag{12.3}$$

We can make this easier to work with and understand by writing out the effect of age for each race category and plugging the appropriate values into the Black and Other dummy variables; then simplifying,

$$\text{Effect of } Age_{\text{for White }(Black=0,\,Other=0)} = (0.10186 + 0.08286 \times 0 - 0.10662 \times 0)$$
$$+ (-0.00156 - 0.00200 \times 0 + 0.00170 \times 0) \times Age$$
$$= 0.10186 - 0.00156 \times Age$$

$$\text{Effect of } Age_{\text{for Black }(Black=1,\,Other=0)} = (0.10186 + 0.08286 \times 1 - 0.10662 \times 0)$$
$$+ (-0.00156 - 0.00200 \times 1 + 0.00170 \times 0) \times Age$$
$$= 0.18472 - 0.00356 \times Age$$

$$\text{Effect of } Age_{\text{for Other }(Black=0,\,Other=1)} = (0.10186 + 0.08286 \times 0 - 0.10662 \times 1)$$
$$+ (-0.00156 - 0.00200 \times 0$$
$$+ 0.00170 \times 1) \times Age \quad (12.4)$$
$$= -0.00476 + 0.00014 \times Age$$

Looking at these expressions, we can see that the effect of age on education may be initially positive for Whites and Blacks at younger ages and then turn negative, and vice versa for Other Races. We can calculate the value of age at which the sign changes as -1 times the ratio of the constant divided by the coefficient of age in Equation 12.4:

$$Age_{\text{for White}} = -\frac{0.10186}{-0.00156} = 65.3$$
$$Age_{\text{for Black}} = -\frac{0.18472}{-0.00356} = 51.9 \quad (12.5)$$
$$Age_{\text{for Other}} = -\frac{-0.00476}{0.00014} = 34.0$$

This tells us that the predicted value of education increases with age for Whites and for Blacks until age 65.3 and 51.9, respectively and then begins to decrease. But for Other Races, the expected value of education decreases with age until age 34.0 and then begins to increase.

We could also use the *margins* command to see this pattern and empirically find an approximate value for the sign change by calculating the moderated effect of age conditioned by race for values of age every 10 years from 20 to 90:

```
. margins , dydx(age) at(race=(1/3) age=(20(10)90) (means) _all)

Conditional marginal effects                    Number of obs    =    1,815
Model VCE    : OLS

Expression   : Linear prediction, predict()
dy/dx w.r.t. : age

1._at        : race          =            1
               age           =           20
               0.female      =    .4341598 (mean)
               1.female      =    .5658402 (mean)
               pared         =    12.10028 (mean)
```

```
 2._at         : race         =              1
                 age          =             30
                 0.female     =       .4341598 (mean)
                 1.female     =       .5658402 (mean)
                 pared        =      12.10028  (mean)


 3._at         : race         =              1
                 age          =             40
                 0.female     =       .4341598 (mean)
                 1.female     =       .5658402 (mean)
                 pared        =      12.10028  (mean)


 4._at         : race         =              1
                 age          =             50
                 0.female     =       .4341598 (mean)
                 1.female     =       .5658402 (mean)
                 pared        =      12.10028  (mean)


 5._at         : race         =              1
                 age          =             60
                 0.female     =       .4341598 (mean)
                 1.female     =       .5658402 (mean)
                 pared        =      12.10028  (mean)


 6._at         : race         =              1
                 age          =             70
                 0.female     =       .4341598 (mean)
                 1.female     =       .5658402 (mean)
                 pared        =      12.10028  (mean)


 7._at         : race         =              1
                 age          =             80
                 0.female     =       .4341598 (mean)
                 1.female     =       .5658402 (mean)
                 pared        =      12.10028  (mean)


 8._at         : race         =              1
                 age          =             90
                 0.female     =       .4341598 (mean)
                 1.female     =       .5658402 (mean)
                 pared        =      12.10028  (mean)


 9._at         : race         =              2
                 age          =             20
                 0.female     =       .4341598 (mean)
                 1.female     =       .5658402 (mean)
                 pared        =      12.10028  (mean)


10._at         : race         =              2
                 age          =             30
                 0.female     =       .4341598 (mean)
                 1.female     =       .5658402 (mean)
                 pared        =      12.10028  (mean)


11._at         : race         =              2
                 age          =             40
                 0.female     =       .4341598 (mean)
                 1.female     =       .5658402 (mean)
                 pared        =      12.10028  (mean)
```

```
12._at         : race         =               2
                 age          =              50
                 0.female     =       .4341598 (mean)
                 1.female     =       .5658402 (mean)
                 pared        =       12.10028 (mean)

13._at         : race         =               2
                 age          =              60
                 0.female     =       .4341598 (mean)
                 1.female     =       .5658402 (mean)
                 pared        =       12.10028 (mean)

14._at         : race         =               2
                 age          =              70
                 0.female     =       .4341598 (mean)
                 1.female     =       .5658402 (mean)
                 pared        =       12.10028 (mean)

15._at         : race         =               2
                 age          =              80
                 0.female     =       .4341598 (mean)
                 1.female     =       .5658402 (mean)
                 pared        =       12.10028 (mean)

16._at         : race         =               2
                 age          =              90
                 0.female     =       .4341598 (mean)
                 1.female     =       .5658402 (mean)
                 pared        =       12.10028 (mean)

17._at         : race         =               3
                 age          =              20
                 0.female     =       .4341598 (mean)
                 1.female     =       .5658402 (mean)
                 pared        =       12.10028 (mean)

18._at         : race         =               3
                 age          =              30
                 0.female     =       .4341598 (mean)
                 1.female     =       .5658402 (mean)
                 pared        =       12.10028 (mean)

19._at         : race         =               3
                 age          =              40
                 0.female     =       .4341598 (mean)
                 1.female     =       .5658402 (mean)
                 pared        =       12.10028 (mean)

20._at         : race         =               3
                 age          =              50
                 0.female     =       .4341598 (mean)
                 1.female     =       .5658402 (mean)
                 pared        =       12.10028 (mean)

21._at         : race         =               3
                 age          =              60
                 0.female     =       .4341598 (mean)
                 1.female     =       .5658402 (mean)
                 pared        =       12.10028 (mean)
```

```
22._at        : race            =               3
                age             =              70
                0.female        =       .4341598 (mean)
                1.female        =       .5658402 (mean)
                pared           =       12.10028 (mean)

23._at        : race            =               3
                age             =              80
                0.female        =       .4341598 (mean)
                1.female        =       .5658402 (mean)
                pared           =       12.10028 (mean)

24._at        : race            =               3
                age             =              90
                0.female        =       .4341598 (mean)
                1.female        =       .5658402 (mean)
                pared           =       12.10028 (mean)
```

		dy/dx	Delta-method Std. Err.	t	P>|t|	[95% Conf. Interval]	
age							
_at							
1		.0705406	.0147371	4.79	0.000	.041637	.0994443
2		.0548801	.0105475	5.20	0.000	.0341934	.0755668
3		.0392196	.006725	5.83	0.000	.0260299	.0524093
4		.0235591	.0043715	5.39	0.000	.0149854	.0321328
5		.0078986	.005736	1.38	0.169	-.0033512	.0191485
6		-.0077619	.009306	-0.83	0.404	-.0260135	.0104897
7		-.0234224	.0134242	-1.74	0.081	-.049751	.0029062
8		-.0390829	.0177123	-2.21	0.027	-.0738217	-.0043442
9		.1135337	.0316627	3.59	0.000	.0514343	.1756332
10		.0779391	.0219636	3.55	0.000	.0348624	.1210158
11		.0423446	.0135156	3.13	0.002	.0158366	.0688525
12		.00675	.0101661	0.66	0.507	-.0131886	.0266886
13		-.0288446	.0156343	-1.84	0.065	-.0595078	.0018186
14		-.0644392	.0246153	-2.62	0.009	-.1127167	-.0161616
15		-.1000338	.0344645	-2.90	0.004	-.1676283	-.0324393
16		-.1356284	.0446104	-3.04	0.002	-.2231219	-.0481348
17		-.002214	.0503628	-0.04	0.965	-.1009894	.0965615
18		-.000939	.033233	-0.03	0.977	-.0661182	.0642401
19		.0003359	.0186785	0.02	0.986	-.0362978	.0369697
20		.0016108	.0164249	0.10	0.922	-.030603	.0338246
21		.0028858	.0294466	0.10	0.922	-.0548672	.0606388
22		.0041607	.0462694	0.09	0.928	-.0865866	.094908
23		.0054356	.0639614	0.08	0.932	-.1200105	.1308818
24		.0067106	.0819615	0.08	0.935	-.1540388	.16746

This output is long and awkward to work with given that the values of race and age at which the effect of age is calculated are reported separately from the calculated effects of age. By looking at the list of _at values at the beginning of the output, we can determine that in the table at the end of the output each set of eight lines represents a race group (White, then Black, then Other Races) and that within a set the lines are for ages $(10, 20, \ldots, 90)$ $(10, 20, \ldots, 90)$. Thus, we can see that for Whites (Lines 1–8), the sign changes between Lines 5 and 6 (between ages 60 and 70); for Blacks (Lines 9–16), the sign changes between Lines 12 and 13 (between ages 50 and 60); and for Other Races (Lines 17–24), the sign changes between Lines 18 and 19 (between ages 30 and 40).

Tables and Plots of Predicted Values
of Education by Age and Race

To create plots of predicted values, you first use the *margins* command to calculate predicted education values for the display values of race and age, which as a by-product creates a table of predicted values, albeit one that is not well organized or formatted for reading or presenting. You then run the *marginsplot* command to create predicted value plots. For this example, I first create a scatterplot of predicted education against age, with a separate prediction line for each race group to highlight the effect of age, and then a bar chart of predicted education by race, with repeated bar charts for the display values of age.

To create the scatterplots of predicted education against age, I specify in the *margins* command the display (calculation) values for age as every 5 years from ages 20 to 90 to provide a smooth rendition of the prediction lines. In the *marginsplot* command, I set age to define the *x*-axis, xdim(age), and create separate prediction lines for race groups, plotd(race). I customize the appearance of each prediction line in Figure 12.1 with the plot#opts( ) option:

```
.  margins , at(race=(1/3) age=(20(5)90) (means) _all)
      ...
.  marginsplot , xdim(age) plotd(race)  noci name(AgebyRace, replace) ///
>    plot1opts(ms(i) lp(solid) lc(black) lw(*1.5)) plot2opts(ms(i) lp(dash) ///
>    lc(black) lw(*1.5)) plot3opts(ms(i) lp(longdash) lc(black) lw(*1.5) )
```

Given the length of the listing of predicted values from the *margins* command and the difficulty in reading it without extensive reorganization, I do not present and discuss this part of the output. The scatterplot in Figure 12.1 shows quite readily how race moderates the relationship between education and age. For the Other Races group, the prediction line is almost flat, indicating that there is little effect of age on education for this race group. Indeed, looking back at the prior output listing the moderated effect of age, the effect of age is never significantly different from 0 for Other Races. In contrast, there is a pronounced inverted U–shaped relationship between education and age for Whites and for Blacks. The predicted level of education increases with age through age 65.3 for Whites and through age 51.9 for Blacks and then decreases with age, particularly sharply for Blacks.

We can also see the change in the race effect by age from this figure relatively easily. The vertical distance between the curves at any age represents the difference in predicted education among race groups. The level of predicted education for Whites (*solid line*) is always greater than the predicted education for Blacks (*short dashed line*); the magnitude of the difference varies considerably at older ages and becomes increasingly large beyond age 52. The Other Race–White difference changes from positive (higher predicted education for Other Races) to negative (higher for Whites) between ages 30 and 35. The White advantage continues to increase until age 64 and then starts to diminish. Comparing Blacks and Other Races, predicted education is higher for Other Races at the young and old ends of the distribution (roughly ages 20–35 and 65 and older) but is greater for Blacks in the middle of the age range.

Alternatively, you could use *margins* and *marginsplot* to construct a bar chart focusing on the race differences by age. I specify fewer display (calculation) values for age—every 14 years from ages 20 to 90—in the *margins* command to provide a smaller and more manageable number of bar charts in Figure 12.2. In the *marginsplot* command, I set race to define the *x*-axis, xdim(race); create separate bar charts

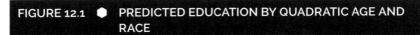

FIGURE 12.1 ● PREDICTED EDUCATION BY QUADRATIC AGE AND RACE

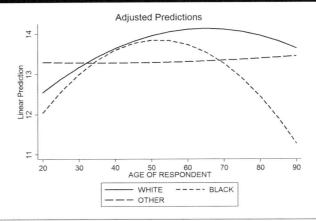

FIGURE 12.2 ● PREDICTED EDUCATION FOR RACE GROUPS BY PARABOLIC AGE

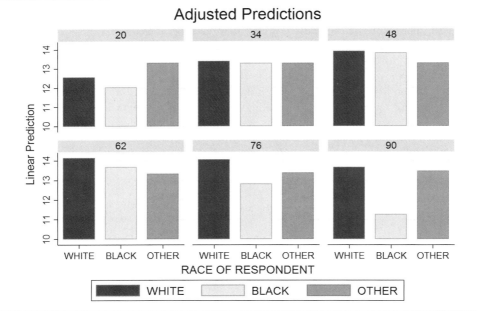

for the selected age values, byd(age); and change the plot type to a bar chart, recast (bar). I also specify plotd(race), which specifies different fill colors for the bars by race, and I control the fill colors with plot#opts( ).

```
. margins , at(race=(1/3) age=(20(14)90) (means) _all)
    ...
. marginsplot, byd(age) xdim(race) plot(race) recast(bar) noci ///
>   name(RacebyAge, replace) plot1opts(fc(black)) plot2opts(fc(gs13)) ///
>   plotopts(barwidth(.75) ysc(r(10)) leg( rows(1)))
```

The same patterns of race differences in predicted education that I discussed above are evident in the bar charts in Figure 12.2, so I will not repeat them. And with a bit of concentration, you can also see the varying parabolic age effect from the bar charts. But I think that both the age and the race effects are easier to discern in the scatterplots in Figure 12.1.

If you have an interaction of a polynomial function of a predictor with an interval rather than a nominal predictor, you could follow much the same approach as I used in this example. One difference is that you would select display (calculation) values for the interval predictor, which I would recommend equally spacing across the range of sample values. The resulting scatterplot of the outcome against the polynomial predictor would have a separate prediction line for each display value of the interval predictor. It is likely that this scatterplot would not show the effect of the interval predictor as readily as it did the effect of the nominal predictor. Thus, I would recommend that you create a second scatterplot of the outcome plotted against the interval predictor, with repeated prediction lines for selected values of the polynomial predictor.

Models With Censored (Selected) Outcomes

This set of models includes the Tobit model as well as sample selection models, the most well-known of which is the Heckman sample selection model, which I use for the example (Heckman, 1979). The starting premise is that the observed outcome is censored, meaning that for some cases in the sample the value of the outcome cannot be observed or is not known by definition but the independent variables are known for the full sample. For example, if the outcome is home equity, you cannot observe a value of home equity for non–home owners but you do have measures of the predictors for owners and non–home owners alike.

The modeling strategy is to estimate coefficients $(\underline{\beta})$ for the effect of the predictors X on a latent outcome y^* for both censored and uncensored cases while simultaneously estimating parameters for a mechanism selecting cases as censored or uncensored. The primary difference between the Tobit model and the Heckman sample selection model is that the Heckman model specifies an explicit selection mechanism; that is, the probability of selection is estimated as a function of coefficients $(\underline{\gamma})$ for a set of predictors W, which may or may not overlap with the predictors of y. In this example, I use the common choice of a probit model for the selection equation.

A key difference between these models and those covered in earlier chapters is that there are related but distinct predicted outcomes that refer to different definitions of the outcome and the populations of interest for your study. In this situation, there are three variations of the predicted outcome that you could study:

1. *The latent outcome y^* consisting of values of the predicted outcome for cases with both observed and unobserved values of y:* Effects on this outcome represent the effect of predictors if the outcome were observed for everyone in the sample—for example, if everyone owned a home, how a respondent's age would affect his or her home equity.

2. *The truncated outcome y^T consisting of the observed outcome:* That is, the population of interest are those cases that are not censored. For instance, you are interested in the effect of the predictors for home owners. Note that if you were to estimate a model based on only those cases with observed values it produces biased and inconsistent coefficient estimates (Greene, 2008, p. 868; Long, 1997, p. 189).

3. *The observed (censored) outcome y consisting of y* if the observation is not censored and a set value τ_y (usually zero) if the observation is censored:* Effects in this situation are on the actual values of the outcome observed in the sample. For example, the actual home equity value observed includes zero for non–home owners.

For the Tobit model and the Heckman selection model, Table 12.1 lists the formulas for the expected value of each outcome, $\varepsilon(\text{Outcome}_i \mid \underline{x}_i)$, and the marginal change in the outcome with respect to the focal variable F, $\dfrac{\partial \varepsilon(\text{Outcome}_i)}{\partial F}$. Calculate a centered discrete change in F as

$$\varepsilon\left(\text{Outcome}_i \mid \underline{x}_i, F = F_{ref} + \frac{1}{2}\right) - \varepsilon\left(\text{Outcome}_i \mid \underline{x}_i, F = F_{ref} - \frac{1}{2}\right) \quad (12.6)$$

There are no clear-cut guidelines concerning which of these should be the basis for interpretation, whether in terms of the effect of predictors or expected values of the outcome. In the main, it depends on the substantive question you wish to answer; there also appear to be some disciplinary differences (cf. Greene, 2008, pp. 871–872; Long, 1997, p. 206).

If you want to interpret the effects on the latent outcome y^*, you can use and interpret the results from the ICALC commands as you would for a linear regression model with an identity link (see Chapter 7). However, if you intend to interpret the results for either the observed (censored) outcome or the truncated outcome, these outcomes are not solely a linear function of the predictors (Greene, 2008, pp. 866–873, 884–886; Long, 1997, pp. 208–210, 216), and the ICALC commands do not currently perform the correct calculations. Moreover, because these outcomes are not a linear function of the prediction function, the results are subject to a similar issue of confounding the nonlinearity of the interaction effect with the nonlinearity of the mean function.

You can use the expression option in the *margins* command to calculate expected values for your chosen predicted outcome as it varies with the focal and moderating variables and then plot the results using the *marginsplot* command. You can also compute discrete change effects from calculating expected values according to Equation 12.6. I illustrate these calculations for a truncated and for a censored predicted outcome with results from a Heckman sample selection analysis of factors predicting home equity, which is observed only for owners. (For details on the data and measurement, see Krivo & Kaufman, 2004.) The analysis includes an interaction between education and age in the regression equation for home equity. Education, but not *age,* is also a predictor in the selection equation. Thus, $\beta_{F_{mod}}$ and $\gamma_{F_{mod}}$ are as defined in the notation description examples for Table 12.1. The syntax for the Heckman analysis is as follows:

```
.heckman eq1k i.hhraceth c.hhage##c.hheduc hhincome midwest south west cc ///
>    anywork numchild widow othntmar intermar ,///
>    sel(owner= c.hheduc c.pcfborn c.pcnhblk) vce(cluster fipscode)
    ...
. est store heck
. sca rho1=e(rho)
. sca sige=e(sigma)
```

The last three lines store the estimation results as a whole, as well as the values for ρ and σ_ε, which are needed for the calculations.

TABLE 12.1 ● EXPECTED OUTCOME VALUE AND MARGINAL CHANGE FORMULAS FOR TOBIT MODEL AND HECKMAN SELECTION MODEL

	Heckman	Tobit (Censored for $y^* \le 0$)
Latent outcome, y^*	$\varepsilon\left(y_i^* \mid \underline{x}_i\right) = \underline{x}_i'\underline{\beta}$ $\dfrac{\partial \varepsilon(y_i^* \mid \underline{x}_i)}{\partial F} = \beta_{F_{mod}}$	$\varepsilon\left(y_i^* \mid \underline{x}_i\right) = \underline{x}_i'\underline{\beta}$ $\dfrac{\partial \varepsilon(y_i^* \mid \underline{x}_i)}{\partial F} = \beta_{F_{mod}}$
Truncated outcome, y^T	$\varepsilon(y_i^T \mid \underline{x}_i) = \underline{x}_i'\underline{\beta} + \rho\sigma_\epsilon \dfrac{\phi(\underline{w}_i'\underline{\gamma})}{\Phi(\underline{w}_i'\underline{\gamma})}$ $\dfrac{\partial \varepsilon(y_i^T \mid \underline{x}_i)}{\partial F} = \beta_{F_{mod}} - \gamma_{F_{mod}}\rho\sigma_\epsilon \times$ $\left[\left(\dfrac{\phi(\underline{w}_i'\underline{\gamma})}{\Phi(\underline{w}_i'\underline{\gamma})}\right)^2 - \underline{w}_i'\underline{\gamma}\left(\dfrac{\phi(\underline{w}_i'\underline{\gamma})}{\Phi(\underline{w}_i'\underline{\gamma})}\right)\right]$	$\varepsilon\left(y_i^T \mid \underline{x}_i\right) = \underline{x}_i'\underline{\beta} + \sigma_\epsilon \dfrac{\phi\left(\dfrac{\underline{x}_i'\underline{\beta}}{\sigma_\epsilon}\right)}{\Phi\left(\dfrac{\underline{x}_i'\underline{\beta}}{\sigma_\epsilon}\right)}$ $\dfrac{\partial \varepsilon(y_i^T \mid \underline{x}_i)}{\partial F} = \beta_{F_{mod}} \times \left[1 - \left(\dfrac{\phi\left(\dfrac{\underline{x}_i'\underline{\beta}}{\sigma_\epsilon}\right)}{\Phi\left(\dfrac{\underline{x}_i'\underline{\beta}}{\sigma_\epsilon}\right)}\right)^2 - \dfrac{\underline{x}_i'\underline{\beta}}{\sigma_\epsilon}\left(\dfrac{\phi\left(\dfrac{\underline{x}_i'\underline{\beta}}{\sigma_\epsilon}\right)}{\Phi\left(\dfrac{\underline{x}_i'\underline{\beta}}{\sigma_\epsilon}\right)}\right)\right]$
Censored outcome, y	$\varepsilon\left(y_i \mid \underline{x}_i\right) = \left[\underline{x}_i'\underline{\beta} + \rho\sigma_\epsilon \dfrac{\phi\left(\underline{w}_i'\underline{\gamma}\right)}{\Phi\left(\underline{w}_i'\underline{\gamma}\right)}\right]\Phi\left(\underline{w}_i'\underline{\gamma}\right)$ $\dfrac{\partial \varepsilon\left(y_i \mid \underline{x}_i\right)}{\partial x_j} = \beta_{F_{mod}}\Phi\left(\underline{w}_i'\underline{\gamma}\right)$ $+\left[\gamma_{F_{mod}} \times \underline{x}_i'\underline{\beta} - \rho\sigma_\epsilon \underline{w}_i'\underline{\gamma}\right]\phi\left(\underline{w}_i'\underline{\gamma}\right)$	$\varepsilon\left(y_i \mid \underline{x}_i\right) = \left[\underline{x}_i'\underline{\beta} + \sigma_\epsilon \dfrac{\phi\left(\dfrac{\underline{x}_i'\underline{\beta}}{\sigma_\epsilon}\right)}{\Phi\left(\dfrac{\underline{x}_i'\underline{\beta}}{\sigma_\epsilon}\right)}\right]\Phi\left(\underline{x}_i'\underline{\beta}\right)$ $\dfrac{\partial \varepsilon\left(y_i \mid \underline{x}_i\right)}{\partial x_j} = \beta_{F_{mod}}\Phi\left(\underline{x}_i'\underline{\beta}\right)$

Notation:

$\beta_{F_{mod}}$ represents the GFI algebraic expression for the moderated effect of F on the outcome—for example, for an interaction between age and education with $F = education$, $\beta_{F_{mod}} = \beta_{education} + age \times \beta_{education \times age}$

$\gamma_{F_{mod}}$ represents the partial derivative with respect to F of the prediction function for the selection equation—for example, with $F = education$ having only a main effect in the selection equation $\gamma_{F_{mod}} = \gamma_{education}$

ρ is the correlation between the errors for the regression and selection equations in the Heckman model

σ_ϵ is the standard deviation of the error in the regression model/Tobit model

$\phi()$ and $\Phi()$ are the probability density function and cumulative distribution function of the standard normal distribution, respectively

The syntax in the following sections for the *margins*, the *marginsplot*, and other Stata commands is intended for advanced users who are knowledgeable about both commands, especially the expression( ) option for the *margins* command. The terms in the expression( ) option are specified in the same order as the terms in the corresponding formulas in Table 12.1 to facilitate understanding the Stata syntax equivalents. I first calculate a significance region table for discrete change effects and then a plot of the predicted outcome values by the interaction of education and age.

The discrete change calculation in the expression( ) option specifies computing the change between the following:

1. The expected outcome value evaluated at the reference values with the value for education (*hheduc*) increased by one-half unit

2. The expected outcome value evaluated at the reference values with the value for education (*hheduc*) decreased by one-half unit

That is, Quantity 2 is subtracted from Quantity 1. This discrete change value and its *p* value are extracted from the results stored by the *margins* command in r(table) and stored in a matrix (dctrun or dccen) to be then displayed in the Results window.

Syntax and Results for Truncated Home Equity, y*. Note that the margins calculations are limited to owners (if *Owner* = 1); otherwise, the reference values for the predictors (their means) would be based on non–home owners as well as owners, which is not appropriate for predicting the truncated outcome.

```
. mat dctrun=J(2,7,.)

. margins if owner==1 & e(sample), expression(((predict(xb) + _b[eq1k:hheduc] /// >   +
  _b[eq1k:c.hhage#c.hheduc]*hhage)*.5)+ rho1*sige*normalden(predict(xbsel) ///
>   + (_b[owner:hheduc]*.5))/normal(predict(xbsel)+ (_b[owner:hheduc]*.5))) - ///
>   ((predict(xb) + _b[eq1k:hheduc] + _b[eq1k:c.hhage#c.hheduc]*hhage)* /// > (-.5))+
  rho1*sige* normalden(predict(xbsel) + (_b[owner:hheduc]*(-.5)))/  ///
>   normal(predict(xbsel)+ (_b[owner:hheduc]*(-.5)))))) ///
>   at( hhage=(25(10)85) (means) _all) noatlegend

Adjusted predictions                      Number of obs     =     10,885
Model VCE    : Robust
       ...

------------------------------------------------------------------------------
             |            Delta-method
             |   Margin   Std. Err.      z    P>|z|     [95% Conf. Interval]
-------------+----------------------------------------------------------------
         _at |
          1  |  5.781163   1.050185    5.50   0.000     3.722839    7.839488
          2  |  6.137706   1.023078    6.00   0.000     4.132511    8.142901
          3  |  6.494249   1.009284    6.43   0.000     4.516089    8.472409
          4  |  6.850792    1.00935    6.79   0.000     4.872503    8.829081
          5  |  7.207335   1.023272    7.04   0.000     5.201758    9.212912
          6  |  7.563878   1.050501    7.20   0.000     5.504934    9.622823
          7  |  7.920421   1.090039    7.27   0.000     5.783983   10.05686
------------------------------------------------------------------------------

. mat gg =r(table)
. mat dctrun[1,1] = [gg[1,1..7] \ gg[4,1..7]]
. mat rownames dctrun = b_ed p
. mat colnames dctrun = age:25 35 45 55 65 75 85

. matlist dctrun, format(%8.3f) lines(one) nohead ///
>        tit("Significance Region Table for Effect of hheduc")

Significance Region Table for Effect of hheduc
             | age
             |     25       35       45       55       65       75       85
-------------+----------------------------------------------------------------
      b_ed  |   5.781    6.138    6.494    6.851    7.207    7.564    7.920
         p  |   0.000    0.000    0.000    0.000    0.000    0.000    0.000
```

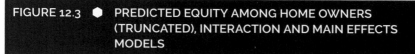

FIGURE 12.3 ● PREDICTED EQUITY AMONG HOME OWNERS (TRUNCATED), INTERACTION AND MAIN EFFECTS MODELS

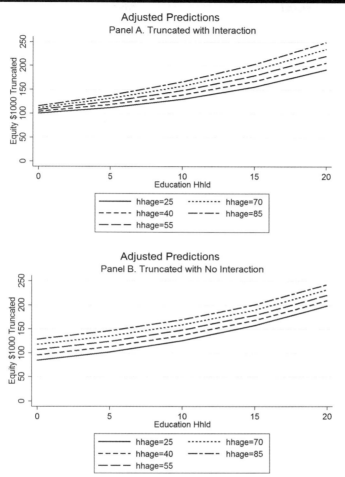

The discrete change effects in the significance region table are positive and significant across the range of age values. A 1-year difference in education predicts a change in home equity of $5,781 for a 25-year-old home owner, and this difference steadily rises with age up to a predicted increase of $7,920 for an 85-year-old home owner. The education effect increases by 37% between ages 25 and 85, suggesting that age moderates the education effect to a notable degree. We can assess this more readily by generating and comparing predicted value plots. I omit the table reporting the calculated margins from the following output in favor of the visual display produced by *marginsplot* in Figure 12.3, given that the table is not set up to easily show the relationship between equity, age, and education.

```
. margins if owner==1 & e(sample), expression(predict(xb)+ rho1*sige* ///
>   normalden(predict(xbsel))/normal(predict(xbsel))) at(hheduc=(0(5)20) ///
>   hhage=(25(15)85)(means) _all) noatlegend
```

```
Adjusted predictions                        Number of obs      =    10,885
  ...
. marginsplot , xdim(hheduc) plotd(hhage) noci  plotopts(leg(colfirst) ///
>   subtit("Panel A. Truncated with Interaction") ///
>   ytit("Equity $1000 Truncated" ) ylab(0(50)250)) plot1opts(ms(i) ///
>   lp(solid) lc(black) lw(*1.5)) plot2opts(ms(i) lp(dash) lc(black) lw(*1.5)) ///
>   plot3opts(ms(i) lp(longdash) lc(black) lw(*1.5)) plot4opts(ms(i) ///
>   lp(shortdash) lc(black) lw(*1.5)) plot5opts(ms(i) lp("_-") lc(black) ///
>   lw(*1.5) ) name(trunc, replace)
```

The prediction lines in Panel A of Figure 12.3 show that predicted equity increases with education, with a steeper slope for older versus younger home owners. Examining the vertical gaps between the prediction lines, you can see that age is also positively related to predicted equity and that the effect grows larger at higher levels of education. Moreover, the curvature of the prediction lines reveals the presence of nonlinearity created by sample selection and creates some uncertainty over the extent to which age moderates the education effect (and vice versa), as you see both here and in the significance region table. The plot in Panel B from a main effects model (syntax not shown) suggests that the sample selection creates the curvature, but the slopes of the prediction lines and the gaps between the lines are virtually constant. Thus, Panel A does in fact accurately portray the degree of moderation of the education effect by age.

Syntax and Results for Censored Home Equity, y. For this predicted outcome, you should include nonowners in the calculations and results, so the margins calculations are only limited to the estimation sample. I again start with a significance region table for the discrete change effects and then create a plot of the predicted censored equity values by age and education.

```
. mat dccen=J(2,7,.)

. margins if e(sample), expression( ///
>   normal(predict(xbsel)+_b[owner:hheduc]*(.5) )*((predict(xb) + ///
>     (_b[equity1k:hheduc] + _b[equity1k:c.hhage#c.hheduc]*hhage)*(.5))+ ///
>     rho1*sige* normalden(predict(xbsel) + (_b[owner:hheduc]*(.5)))/ ///
>         normal(predict(xbsel)+ (_b[owner:hheduc]*(.5)))) - ///
>   normal(predict(xbsel)+_b[owner:hheduc]*(-.5) )*((predict(xb) + ///
>     (_b[equity1k:hheduc] + _b[equity1k:c.hhage#c.hheduc]*hhage)*(-.5))+ ///
>         rho1*sige* normalden(predict(xbsel) + (_b[owner:hheduc]*(-.5)))/ ///
>         normal(predict(xbsel)+ (_b[owner:hheduc]*(-.5))))) ///
>   at( hhage=(25(10)85) (means) _all) noatlegend

Adjusted predictions                         Number of obs       =    18,576
Model VCE    : Robust

Expression    : normal(predict(xbsel)+_b[owner:hheduc]*(.5) )*((predict(xb) +
   (_b[equity1k:hheduc] + _b[equity1k:c.hhage#c.hheduc]*hhage)*(.5))+ rho1*sige*
   normalden(predict(xbsel) + (_b[owner:hheduc]*(.5)))/ normal(predict(xbsel)+
   (_b[owner:hheduc]*(.5)))) - normal(predict(xbsel)+_b[owner:hheduc]*(-.5) )*
   ((predict(xb) + (_b[equity1k:hheduc] + _b[equity1k:c.hhage#c.hheduc]*hhage)*
   (-.5))+ rho1*sige* normalden(predict(xbsel) + (_b[owner:hheduc]*(-.5)))/
   normal(predict(xbsel)+ (_b[owner:hheduc]*(-.5))))
```

		Delta-method				[95% Conf. Interval]	
		Margin	Std. Err.	z	P>|z|		
_at							
1		8.294661	.5603286	14.80	0.000	7.196437	9.392885
2		8.755892	.5485664	15.96	0.000	7.680721	9.831062
3		9.217122	.5476209	16.83	0.000	8.143805	10.29044
4		9.678353	.5575469	17.36	0.000	8.585581	10.77112
5		10.13958	.5777847	17.55	0.000	9.007146	11.27202
6		10.60081	.607304	17.46	0.000	9.41052	11.79111
7		11.06204	.6448315	17.15	0.000	9.798198	12.32589

```
. mat gg =r(table)
. mat dccen[1,1] = [gg[1,1..7] \ gg[4,1..7]]
. mat rownames dccen = b_ed p
. mat colnames dccen = age:25 35 45 55 65 75 85
.
. matlist dccen, format(%8.3f) lines(one) nohead ///
>     tit("Significance Region Table for Effect of hheduc")
```

```
Significance Region Table for Effect of hheduc
```

	age						
	25	35	45	55	65	75	85
b_ed	8.295	8.756	9.217	9.678	10.140	10.601	11.062
p	0.000	0.000	0.000	0.000	0.000	0.000	0.000

The discrete change effects of education in the significance region table suggest much the same story about the effect of education on actual home equity among all house-holders as they did among just home owners. The education effect is positive and significant at all ages and increases in magnitude by about 33% from age 25 to age 85. Note that the education effects on the censored outcome are about 40% larger than on the truncated outcome. Creating a predicted outcome plot will show this as well as the age effect quite handily.

```
. margins if e(sample), expression( (predict(xb)+ rho1*sige* ///
>     normalden(predict(xbsel))/normal(predict(xbsel)))*predict(psel))  ///
>     at(hheduc=(0(5)20) hhage=(25(15)85) (means) _all) noatlegend
```

```
Adjusted predictions                         Number of obs    =     18,576
Model VCE    : Robust
```

```
       ...
. marginsplot , xdim(hheduc) plotd(hhage) noci  plotopts(subti(Censored) ///
>     leg(colfirst) ytit("Equity $1000 Censored" )  ylab(0(50)250 )) ///
>     plot1opts(ms(i) lp(solid) lc(black) lw(*1.5)) plot2opts(ms(i) lp(dash) ///
>        lc(black) lw(*1.5)) ///
>     plot3opts(ms(i) lp(longdash) lc(black) lw(*1.5)) ///
>     plot4opts(ms(i) lp(shortdash) lc(black) lw(*1.5)) ///
>     plot5opts(ms(i) lp("_-") lc(black) lw(*1.5) ) name(censor, replace)
```

Both the similarities and the differences in the effect of education on the censored versus the truncated outcome are apparent in their respective scatterplots. To facilitate comparison, Figure 12.4 presents the plots one above the other. When education is at the bottom of its distribution, the predicted actual home equity is relatively close to zero (about $12K) but increases to almost $190K at the top of the education distribution. In contrast, the truncated distribution shows an increase in home equity across the range of education from around $100K to $250K. The steeper slope of education's effect on the censored outcome reflects in part the fact that education is a predictor in the selection model. Its positive effect on selection and ownership creates an indirect influence on home equity. Unlike education, age is not part of the selection model, and it has a smaller effect on the censored than on the truncated outcome. The vertical gaps between the education prediction lines, which represent the age effects, start smaller and end smaller in the censored outcome plot than in the truncated outcome plot.

It is hoped that this brief example with syntax provides a foundation for readers to adapt to their own analyses if they intend to interpret the truncated or the censored outcome from a Tobit or a Heckman-type selection model. Keep in mind that there

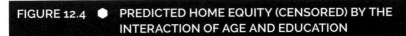

FIGURE 12.4 ● PREDICTED HOME EQUITY (CENSORED) BY THE INTERACTION OF AGE AND EDUCATION

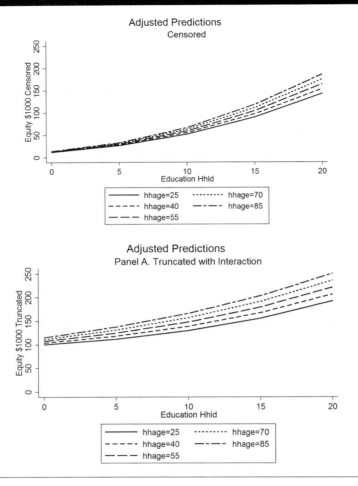

are a variety of selection models that differ from the Heckman model I used for this example, a regression model with a probit analysis of the selection process, as well as Tobit models with different forms of censoring. The formulas in Table 12.1 apply only to Tobit and Heckman models and would require adaptation for other variations.

Models for Survival Analysis (Cox Proportional Hazards Example)

What if the research questions you are asking concern the duration of time until an event occurs? For example, how quickly does a drug user relapse after treatment, and how is the speed of relapse predicted by the participant's age? To answer questions like that, your data must have information on the duration of time between the end of treatment and the occurrence of a relapse. It is common for such data to include cases for which the event does not occur before the end of data collection. That is, the duration outcome variable is censored because all you know is that the duration is greater than or equal to the duration at which the participant exited the study. All the models considered so far would not be appropriate to answer these questions or to analyze duration as an outcome variable (Greene, 2008, p. 933; Hosmer, Lemeshow, & May, 2008, p. 2). Instead, there are a variety of models for survival analysis—aka event history analysis—specifically designed for the analysis of event duration. For this brief example on interpreting interaction effects in survival models, I presume that interested readers are familiar with the basics of survival analysis. I would recommend Greene (2008, pp. 931–942) and/or the Institute for Digital Research and Education's (n.d.) online seminar on Survival Analysis with Stata for a brief but informative overview of survival models and Allison (2014) and/or Hosmer et al. (2008) for fuller treatments.

My example applies a Cox proportional hazards model—a very common model for survival analysis—to the data used in the Institute for Digital Research and Education's (n.d.) online seminar (see their website for details on the data and measures). The outcome is the elapsed time after treatment to drug relapse; the predictors are age, treatment site (A or B), the interaction of age and site, a dummy indicator of treatment length (long or short), and the number of prior drug treatments:

```
. stset time, failure(censor)
    ...
. stcox age ndrugtx i.treat i.site c.age#i.site, nohr

         failure _d:  censor
   analysis time _t:  time

Iteration 0:   log likelihood =  -2868.555
Iteration 1:   log likelihood =  -2851.487
Iteration 2:   log likelihood = -2850.8935
Iteration 3:   log likelihood = -2850.8915
Refining estimates:
Iteration 0:   log likelihood = -2850.8915

Cox regression -- Breslow method for ties

No. of subjects =            610          Number of obs    =          610
No. of failures =            495
Time at risk    =         142994
                                          LR chi2(5)       =        35.33
Log likelihood  =    -2850.8915           Prob > chi2      =       0.0000


------------------------------------------------------------------------------
         _t |      Coef.   Std. Err.      z    P>|z|     [95% Conf. Interval]
------------+-----------------------------------------------------------------
        age | -.0336943    .0092913    -3.63   0.000    -.051905    -.0154837
    ndrugtx |  .0364537    .0077012     4.73   0.000    .0213597     .0515478
            |
      treat |
       Long | -.2674113    .0912282    -2.93   0.003    -.4462153   -.0886073
            |
       site |
     Site_B | -1.245928    .5087349    -2.45   0.014     -2.24303   -.2488262
            |
 site#c.age |
     Site_B |  .0337728    .0155087     2.18   0.029     .0033764    .0641693
```

Interpretation of the effect of predictors for a Cox model is analogous to the interpretation of effects for binomial logistic regression.[2] Specifically, the coefficients represent effects on the logged hazard rate (akin to the log odds in a logistic), and by exponentiating the coefficients, you get factor effects on the hazard rate (called hazard ratios and akin to odds ratios). Thus, the ICALC commands can be used to probe and interpret interaction effects in the Cox model as effects on the log hazard rate, as factor changes, or as SPOST13 marginal effects (e.g., discrete changes). ICALC does not have the capability to create survival curves, which are commonly used for interpretation. I will demonstrate how to create them following the ICALC examples.

GFI **and** SIGREG **for the Effect of Age Moderated by Site**

```
. intspec focal(c.age) main((c.age, name(Age) range(20(9)56)) (i.site, ///
>   name(Site) range(0/1))) int2(c.age#i.site)  ndig(0)
  …


. gfi , factor

GFI Information from Interaction Specification of
Effect of Age on g(_T) from
-----------------------------------------------------------------------

Effect of Age =
   -0.0337 + 0.0338*Site_B


   Factor Change Effect (1 unit change in) Age =
      e^-0.0337 * e^( 0.0338*Site_B) =

   0.9669 * 1.0344^Site_B
  …

. sigreg , effect(factor(5))
  …

   Significance Region for Factor Change Effect of Age (5 unit difference)
   on _T at Selected Values of Site
---------------------------------------------------
                            At Site=
      Effect of    |    Site_A      Site_B
   --------------+----------------------------
            Age   |    0.8450*      1.0004
---------------------------------------------------
      Key: Plain font  = Pos, Not Sig    Bold font*  = Pos, Sig
           Italic font = Neg, Not Sig    Italic font* = Neg, Sig
```

Because site is a dummy indicator, the GFI results show that the baseline effect of age on the log hazard (−0.0337) is negative for Site A (*site* = 0) and changes sign to barely positive (0.0001) for Site B. You can get a better sense of the effect from the significance region table, which reports the moderated effect of age as a factor change for a 5-year difference in age. The effect of age in Site A is negative and significant and indicates that the rate of relapse decreases by 15% (100 × (1 − 0.845)) for a 5-year increase in age between two participants. Conversely, the effect of age in Site B is positive and not significant, and the relapse rate does not change meaningfully with age, a negligible 4/100 of a percent rise in the relapse rate.

GFI **and** EFFDISP **for the Effect of Site Moderated by Age**

```
. intspec focal(i.site) main( (c.age, name(Age) range(20(9)56)) (i.site, ///
>    name(Site) range(0/1))) int2( c.age#i.site )  ndig(0)
. gfi, factor

GFI Information from Interaction Specification of
Effect of Site on g(_T) from
---------------------------------------------------------------------

Effect of Site_B =
    -1.2459 + 0.0338*Age

    Factor Change Effect (1 unit change in) Site_B =
        e^-1.2459 * e^( 0.0338*Age) =

    0.2877 * 1.0344^Age
    ...


. effdisp, ndig(1) effect(factor) plot(name(effsite))
```

The GFI reports a negative baseline effect of site (−1.2459) and that the log hazard rate increases in the positive direction as age increases by 0.0338 per year of age. While the sign change analysis (not reported) would tell us about sign changes, it is more fruitful to look directly at the effects display in Figure 12.5, which portrays the sign and the significance of the site effect. A negative effect of site (coded 1 = *Site B*) would mean that the predicted rate of relapse is lower at Site B than at Site A, while a positive effect would show a higher relapse rate at Site B. Equivalently, a factor change value less than 1.0 indicates a lower relapse rate in Site B, while a value greater than 1.0 indicates a higher relapse rate. The effects display tells us that the effect of site is negative (slower relapse rate) and significant for *Age* ≤ 31 because the confidence bounds do not include the value of 1.0. The site effect turns positive around age 37 and continues to increase in magnitude with age but does not become significant.

FIGURE 12.5 ● EFFECTS DISPLAY FOR COX HAZARDS MODEL

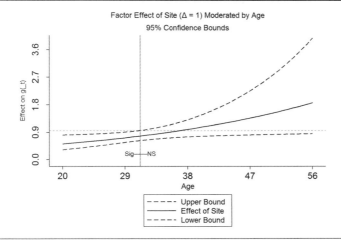

OUTDISP **for the Interaction of Site and Age**

You can choose which of your interacting variables is highlighted by your predicted outcome plot corresponding to what you declare as the focal variable in the *intspec* command. In this example, both default plots, produced by the syntax below, work equally well for interpreting the effects of both age and site on the relapse hazard rate.

```
** Age Focal
. intspec focal(c.age) main((c.age, name(Age) range(20(9)56)) ///
>   (i.site, name(Site) range(0/1))) int2( c.age#i.site ) ndig(0)
    ...
outdisp , tab(def) plot(def)
** Site Focal
. intspec focal(i.site) main((c.age, name(Age) range(20(9)56)) ///
>   (i.site, name(Site) range(0/1))) int2( c.age#i.site ) ndig(0)
    ...
. outdisp , tab(def) plot(def) pltopts(tit("Relapse  Hazard Rate (_t) by ///
>   Interaction of Site by Age") plotreg(ma(t+3))) ndig(3)
```

The scatterplot in Figure 12.6 shows the predicted relapse hazard rate plotted against age separately for Sites A and B. Clearly, age has no effect on the predicted relapse rate for participants treated at Site B (*dashed line*). Age has a negative effect on the predicted relapse at Site A; that is, younger participants are predicted to relapse into drug use more quickly than older participants. You can also discern how the effect of site changes with age by noting the vertical distance between the two prediction lines and how it changes with age. Namely, the predicted relapse rate at Site A is higher than at Site B for younger participants. This gap diminishes as age increases up to age 37, and then the relapse rate at Site A becomes increasingly lower than at Site B. But we know from the effects display results that the difference between sites is significant only when *Age* < 32.

The same patterns are evident from the bar charts in Figure 12.7. Starting with the effect of site on the hazard rate relapse, the *black bars* for Site A are higher than the *gray bars* for Site B in the first two bar charts (*Age* < 30). They are nearly identical at *Age* = 38 and become increasingly smaller than the gray bars at subsequent ages. You can also readily see that the height of the gray bars does not change appreciably with age, indicating no effect of age at Site B. In contrast, the height of the black bars declines from left to right, portraying a reduction in the relapse rate for older participants at Site A. If I were more interested in the age effects, I would present the scatterplot, which focuses attention on the prediction lines for the effect of age. Conversely, to highlight the site effects, I would present the bar charts.

Survival Curves for the Interaction of Site and Age

A survival curve is a commonly used alternative visual display of the relationship between age, site, and drug relapse. Instead of showing the risk (hazard) of the event occurring, it shows the predicted survival rates of not experiencing the event over time. To show an interaction effect, you need to create survival curves for the display values of your focal variable, with a separate survival curve for each display value of the moderator. You should then repeat the process by switching the roles of the focal and moderating variables. For example, to produce survival curves in Stata for the age effect conditional on treatment site, do the following:

1. Run the *margins* command with the at( ) option specifying age set to its display values and site set to its display values:

   margins, at(age=(20(9)56) site=(0/1))

FIGURE 12.6 ● RELAPSE HAZARD RATE, AGE FOCAL

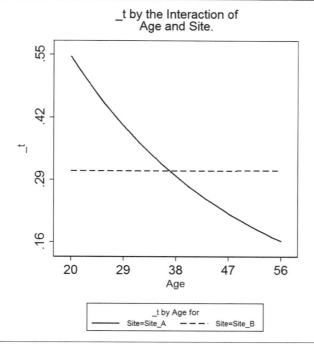

_t by the Interaction of
Age and Site.

FIGURE 12.7 ● RELAPSE HAZARD RATE, SITE FOCAL

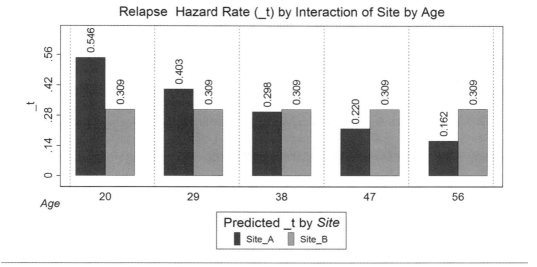

Relapse Hazard Rate (_t) by Interaction of Site by Age

2. Run the *stcurve* command with a set of at#( ) options specifying each display
    value of age paired with *site* = 0:

    ```
    stcurve, survival at1(age=20 site=0) at2(age=29 site=0) at3(age=38 ///
    site=0) at4(age=47 site=0) at5(age=56 site=0) name(site0)
    ```

3. Repeat specifying *site* = 1:

```
stcurve, survival at1(age=20 site=1) at2(age=29 site=1) at3(age=38 ///
site=1) at4(age=47 site=1) at5(age=56 site=1) name(site1)
```

To create survival curves for the site effect conditional on age, follow an analogous process. Create separate survival curves for the five display values of age by running *stcurve* five times—for example, for the first display value of age (20):

```
stcurve, survival at1(age=20 site=0) at2(age=20 site=1) name(age20)
```

Figure 12.8 shows the survival curve by age conditional on site. Like the outcome display for the predicted hazard rate in Figure 12.6, it is evident that age affects the survival rate at Site A (*site* = 0) but not at Site B (*site* = 1). And the survival rate (for not relapsing) in Site A is lower for younger ages than for older ages. Figure 12.9 portrays the survival rate by site conditional on age. Note how the survival curve for Site A (*solid line*) is higher than the survival curve for Site B (*dashed line*) for the youngest ages, but this reverses at around age 38. Just as Figure 12.7 showed, younger participants have a lower rate of survival (not relapsing) at Site A than at Site B, while participants older than age 37 have a higher rate of not relapsing at Site A than at Site B.

I produced the survival curves in Figures 12.8 and 12.9 using a set of Stata code to automate the creation of both sets of survival curves for a single-moderator interaction. This code is listed at the end of the do-file for this example (downloadable from www.icalcrlk.com), and it uses a file containing a graph editor recording (legendonly .grec) to create the legend (also downloadable). Substitute the name of one of your variables and its display values in Lines 5 and 6 of the code, substitute the name and display values for the other variable in Lines 9 and 10, and then highlight and run the code. This will produce two sets of plots: (1) survival curves by Variable 1 conditional on Variable 2 and (2) survival curves by Variable 2 conditional on Variable 1. Note that this will work properly for a nominal variable with any number of categories specified in factor-variable notation in the variable list for stcox, but only for a nominal variable with two categories if you are not using factor-variable notation.

FIGURE 12.8   ●   SURVIVAL CURVES FOR AGE CONDITIONAL ON SITE

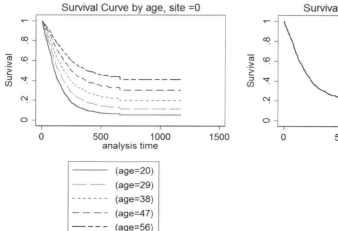
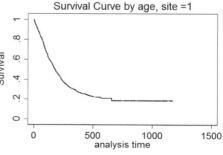

FIGURE 12.9 ● SURVIVAL CURVES FOR SITE CONDITIONAL ON AGE

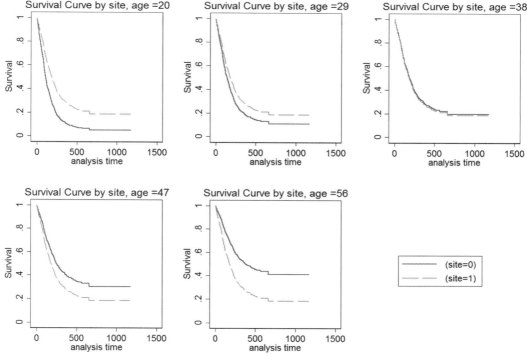

FINAL THOUGHTS: DOS, DON'TS, AND CAUTIONS

In conclusion, I want to concisely reiterate some important points about how you should work with and interpret interactions, what you should avoid, and when you should be cautious. I organize these final thoughts topically rather than as a list of dos and don'ts because some of the dos and don'ts correspond as opposites of each other.

Specifying Terms in the Prediction Function

It is rarely if ever appropriate to exclude lower order terms for an interaction specification because it changes the meaning of the higher order terms in the model. This is sometimes mistakenly done when the lower order term is not significant. This fails to recognize that a different but equivalent parameterization (e.g., centering) could easily lead to an opposite conclusion. Realistically, the only situation in which excluding a lower order term can be justified is when all of the predictors have true zeros (i.e., zero is not an arbitrary value), there is a good theoretical argument for why the lower order term should have a coefficient equal to 0, and the coefficient is not statistically different from 0 (Kam & Franzese, 2007, pp. 100–101).

On the other side of that spectrum, a recurrent argument—which recurrently gets debunked—is that it is not necessary to include product terms for interactions in GLMs with nonlinear link functions because the effects of all the predictors (including interactions if specified) depend on the values of all the predictors. This argument confuses the structured nonlinearity of an interaction effect, whose form can be tested, with the nonspecific nonlinearity created by models that specify that the observed outcome is a nonlinear function of the linear prediction function.

Centering your interval predictors in the prediction function is an option whose effect on the estimation process is sometimes misunderstood. Centering does not affect estimation problems due to collinearity among the interacting predictors and their product term(s), except in the very rare extreme case when the model cannot calculate estimates. So there is no reason to center if that is your reason. But there also is no problem with centering interval predictors, especially if it helps you interpret the effects. Many analysts like to center variables because the main effect coefficient is always interpretable. But many others, myself included, find it easier to work with uncentered predictors. If you interpret the results correctly, there is no difference between what you will find using centered versus uncentered predictors.

Interpreting Effects Versus Interpreting Coefficients

One of the common errors I discussed in the introductory chapter is interpreting the coefficient for a component term of an interaction specification as an unconditional effect—that is, without considering how it is conditioned by the other predictors constituting the interaction. The bigger-picture point is really what you should do. That is, you should describe what I have labeled as the moderated effect of a predictor. This expresses how the effect of a predictor on the outcome is conditional on the values of the other predictors with which it interacts.

This issue is sometimes misconstrued to mean that you should never discuss the meaning of the coefficient for an individual predictor such as a main effect term. Rather, the point is that if you do discuss the main effect coefficient, you need to be explicit about how it is conditioned by the other interacting predictors. There are situations in which the main effect coefficient is quite meaningful for helping understand the overall effect of the interaction specification. In particular, if your interaction is between a nominal predictor (a set of dummy variables) and an interval predictor, then the main effect coefficient for the interval predictor is the effect of the interval predictor when the nominal predictor equals its reference category. For example, if your interaction is between income and a three-category indicator of urban/suburban/rural location (reference category equals rural), then the main effect coefficient of income is the effect of income for those who live in rural areas.

More broadly, whether or not a predictor's main effect coefficient is meaningful depends partly on the coding of the other predictor(s) with which it interacts. Suppose zero is a valid value for the other predictors. Then the main effect of the first predictor can be meaningfully described as the first predictor's effect when the other predictors are all zero. Note that from a substantive point of view, this might or might not be a meaningful piece of information. However, if zero is not a valid value for the other predictors, then the main effect coefficient of the first predictor is not meaningful. It is for this reason that some analysts center the other predictors. To me, the interpretability (or not) of main effect coefficients is not a compelling reason to center the values of a predictor. I am more interested in the bigger picture of the whole effect, how the moderated effect of the predictor changes across the possible range of values of its moderators. Whether or not those possible values include zero is not necessarily important.

Consider the Totality of an Interaction Specification

You should explore the nature of a predictor's moderated effect across a wide if not the full range of its moderator's values to ensure that you fully understand the pattern and that any summary statement you make about the effect holds true. If you have only considered how the moderated effect varies between the moderator's mean ±1 standard deviation, then you may not know that it has changed sign or significance beyond those boundaries. Similarly, if you have multiple moderators, you should probe the moderated effect across the range of possible combinations of values of the different moderators.

Understanding the totality of the interaction specification more generally means that you should explore and interpret each of the component predictors that constitute the interaction. That is, treat each predictor in turn as the focal variable with whatever moderators apply to it, and explore and interpret how its moderated effect changes. Often, the initial rationale for why you are testing and including an interaction may be somewhat one-sided in the sense that you developed the idea that A's effect is conditional on B without necessarily having an explicit rationale for why B's effect should be conditional on A. You need both to develop an explicit conceptual rationale for B's effect being conditional on A and to explore empirically the nature of B's moderated effect.

Experiment with alternative ways to present results to tell the whole story of the interaction specification. There is no single right device that always works best. Sometimes, focusing on moderated effects in significance region tables or effect displays lets you tell the story most effectively. Sometimes, using a predicted values display works better than focusing on moderated effects, but it could be that predicted values tables work better than a visual display.

Comparing Effects

Interpreting interaction effects always involves a comparison of how the effect of the focal variable changes across the values of the moderating variable(s). A key premise of my interpretation approach is that when you make such comparisons, you must take into account the nonlinearity induced by the nonlinear link function of many GLMs. This can be confounded with the nonlinearity of the interaction specification itself, making it difficult to properly evaluate the meaning of the differences in the focal variable's effect as it varies with its moderators. Naively interpreting your results without considering the potential confounding can result in either an overstatement or an understatement of the degree of moderation actually present in your analysis. In some cases, most of the apparent pattern is due to the nonlinearity of the link function, while in others, the apparent pattern is nearly all due to the interaction specification. Remember that this concern is with accurately describing the degree of moderation of a predictor's effect by the variables with which it interacts; it does not apply to describing the relationship between the outcome and the interacting variables. The difference between the following two statements about Panel B in Figure 5.2 illustrates this distinction:

1. The probability of approval is consistently higher for Whites than for Blacks, but this gap is substantially moderated by education. The substantial

effect of race is reduced by higher levels of education, falling from a 63% difference at the lowest level of education to a 3% gap at the highest level.

2. The probability of approval is consistently higher for Whites than for Blacks, but this gap is substantially reduced at higher levels of education, falling from a 63% difference at the lowest level of education to a 3% gap at the highest level.

You can legitimately make the first statement about moderation only if it holds true after taking into account the nonlinearity due to the link function. The second statement is an accurate portrayal of the relationship, which makes no claims about moderation.

A different comparison problem can arise when using GLMs with a fixed value for the error variance of a latent outcome (e.g., logistic regression, probit analysis, ordinal regression models, etc.). In this situation, you cannot compare the as-estimated coefficients across nested models because the coefficients are scaled relative to the variance of the latent outcome, which changes across the nested models as predictors are added to the analysis. This applies to the estimated coefficients for all predictors, not just the interacting predictors. You can, however, compare moderated effects— or coefficients for noninteracting predictors—across nested models after standardizing them relative to the latent outcome's standard deviation, what Long and Freese (2014, p. 181) call y^*-standardized effects.

The prior statement raises an additional issue about standardized effects for interacting predictors. Can you meaningfully standardize them relative to the standard deviation of the predictors—that is, calculate an x-standardized effect? First of all, you cannot meaningfully calculate an x-standardized effect for a nominal focal variable. But if the focal variable F is interval level, you can standardize F's moderated effect by F's standard deviation (s_F), and in some circumstances, this calculation can be meaningful:

$$\text{Moderated effect of } F = \left(b_F + b_{F \times M} \times M\right)s_F \qquad (12.7)$$

The primary reason to calculate x-standardized effects is to compare within a model the magnitude of the effects of predictors measured on different scales to assess which predictor has the relatively largest effect. Because the effect of F changes with its moderators, you may not get a single or definitive description of how the magnitude of F's effect compares with that of the other predictors of the model. If F is moderated by a nominal predictor, then what you get is an x-standardized effect of F for each category of the nominal predictor, each of which can be separately compared with the effect of the other predictors. For example, it might be that F's effect for Category 1 is larger than the effect of any other predictor in the model, while F's effects for Categories 2 and 3 are the second largest effects compared with the effects of the other predictors. If F is moderated by another interval predictor, then you could calculate the range of the moderated x-standardized effect. If the values in the range are all positive or all negative, then you can compare that range of values with the other predictors' x-standardized effects, and you might be able to draw a useful conclusion. But if the values range from negative to positive, then the comparison would be inconclusive.

Model Misspecification and Diagnostic Testing

In Chapter 1, I briefly discussed the consequences of unspecified heterogeneity (heteroscedasticity) in the variance function of GLMs with nonlinear link functions. If the heterogeneity is associated with the model's predictors, then the estimated coefficients are biased and inconsistent. This is a topic of continuing research because there is no consensus about the proposed solutions. But there is agreement that diagnostic testing for the presence of heterogeneity should not be neglected.

In any analysis, two key concerns with model specification are the inclusion of unnecessary predictors in the model and the exclusion of predictors that should be in the model. The principal way to avoid these is to use a theory-guided specification of the model combined with significance and diagnostic testing of the inclusion/exclusion of predictors. Overreliance on data-mining approaches in this process can lead to the unneeded expansion of the model, while their avoidance can lead to an overly parsimonious model. With regard to interaction effects, this balance becomes a particular concern if testing for the inclusion of interaction effects does not yield unambiguous conclusions. For example, suppose a global test for the inclusion of a set of interaction terms is not significant but one or more of the individual interaction terms are significant, or vice versa. The question is whether it is worse to exclude an interaction that in reality should be included or to include an interaction that in reality should be excluded. The answer to this question at least is straightforward in terms of the severity of the consequences—it is worse to omit a relevant interaction. Excluding a needed interaction leads to biased and inconsistent coefficients, while including an extraneous interaction only leads to inefficient estimates (standard errors too large).

More broadly, the consequences of model misspecification reinforce the importance of doing thorough diagnostic testing for the appropriateness of the link function, the functional form of the prediction function, and the consistency of the data with the estimation model's assumptions. Although data limitations will inevitably raise concerns over model misspecification, you can and should try very hard to diagnose and correct for any avoidable problems.

CHAPTER 12 NOTES

1. The derivative of the quadratic expression $a + b\ Age + c\ Age^2$ with respect to Age is $b + 2 \times c \times Age$ where a, b, and c are constants (not functions of Age). More generally, the derivative of $d \times X^p$ is $p \times d \times X^{p-1}$.
2. Indeed, Allison (1984) demonstrates how to use logistic regression to estimate a Cox model.

APPENDIX: DATA FOR EXAMPLES

The data sets represent a variety of sources, and each has a subset of the variables from the original source. For each example used in the book, I report the Stata data set name (filename.dta) used in the do-file for the example, the size of the estimation sample, and a title describing the data source. For the variables used in the example, I list the Stata variable name, its descriptive label, descriptive statistics, and, for noninterval measures, value labels.

CHAPTER 2: ONE-MODERATOR EXAMPLE

Data Set: SIPP_Wealth.dta   Estimation Sample N=14237
    SIPP 1992: Subset Wealth data from Extract by Campbell and Kaufman
    (2006)from SIPP 1992 Panel (U.S. Bureau of the Census 1992)

Variables and Descriptives

netw10k "Networth in $10000"

  Mean =   11.87 S.D. =   20.14 Min =   -126.53 Max =   300.00

metro "Dummy for Metropolitan Residence"

  Mean =   0.72 S.D. =   0.45 Min =   0.00 Max =   1.00

kids "Number of Children"

  Mean =   0.77 S.D. =   1.08 Min =   0.00 Max =   10.00

age "Age in years"

  Mean =   49.89 S.D. =   17.12 Min =   17.00 Max =   83.00

agesq "Square of Age"

  Mean =   2781.96 S.D. =   1814.12 Min =   289.00 Max =   6889.00

retired "Dummy for Retired"

  Mean =   0.23 S.D. =   0.42 Min =   0.00 Max =   1.00

nilf "Dummy for Not in Labor Force"

  Mean =   0.04 S.D. =   0.20 Min =   0.00 Max =   1.00

hhld_income "Monthly Household Income in $1000s"

```
Mean =      3.26 S.D. =      2.52  Min =     -1.56  Max =     23.70
```

edcat "Education Degrees"

```
Mean =      1.84 S.D. =      1.26  Min =      0.00  Max =      4.00
```

```
        Labels:
          0 <HS
          1 HS
          2 SomeColl
          3 CollGrad
          4 PostGrad
```

hoh "Type of Household Head"

```
Mean =      0.69 S.D. =      0.87  Min =      0.00  Max =      2.00
```

```
        Labels:
          0 Couple
          1 Single_man
          2 Single_woman
```

CHAPTER 2: TWO-MODERATOR MIXED EXAMPLE

Data Set: GSS_1987.dta   Estimation Sample N=1372
   General Social Survey: Cumulative Data File, Year=1987 only (Smith, Marsden, Hout and Kim 2015)

Variables and Descriptives

ban "Approval of Ban on Racial Intermarriage"

```
Mean =      0.21 S.D. =      0.41  Min =      0.00  Max =      1.00
```

```
        Labels:
          0 Oppose
          1 Favor
```

ed "Highest year of school completed"

```
Mean =     12.56 S.D. =      3.07  Min =      0.00  Max =     20.00
```

age "Age of respondent"

```
Mean =     43.88 S.D. =     17.05  Min =     18.00  Max =     89.00
```

class "Social Class Rank"

```
Mean =      5.79 S.D. =      1.84  Min =      1.00  Max =     10.00
```

contact "Racial Contact (Count of Types of Contact)"

    Mean =   1.43 S.D. =   1.15 Min =   0.00 Max =   3.00

racew "Dummy for White-NonWhite"

    Mean =   0.72 S.D. =   0.45 Min =   0.00 Max =   1.00

        Labels:
          0 NonWhite
          1 White

region16 "Dummy for Southern Residence at Age 16"

    Mean =   0.35 S.D. =   0.48 Min =   0.00 Max =   1.00

        Labels:
          0 NonSouth
          1 South

CHAPTER 2: TWO-MODERATOR INTERVAL EXAMPLE

Data Set: GSS_2010.dta   Estimation Sample N=1085
    General Social Survey: Cumulative Data File, Year=2010 only (Smith et al.
    2015)

Variables and Descriptives

    childs "Number Of Children"

      Mean =   2.28 S.D. =   1.75 Min =   0.00 Max =   8.00

    faminc10k "Family Income In $10,000"

      Mean =   6.60 S.D. =   6.08 Min =   0.05 Max =   23.27

    educ "Highest Year Of School Completed"

      Mean =   13.62 S.D. =   3.20 Min =   0.00 Max =   20.00

    age "Age Of Respondent"

      Mean =   57.75 S.D. =   12.07 Min =   40.00 Max =   89.00

    sibs "Number Of Brothers And Sisters"

      Mean =   3.87 S.D. =   2.99 Min =   0.00 Max =   17.00

    religintens "Religious Intensity"

      Mean =   3.02 S.D. =   1.05 Min =   1.00 Max =   4.00

race "Race Of Respondent"

Mean = 1.24 S.D. = 0.53 Min = 1.00 Max = 3.00

Labels:
0 IAP
1 WHITE
2 BLACK
3 OTHER

CHAPTER 2: THREE-WAY INTERACTION EXAMPLE

Data Set: GSS_1987.dta   Estimation Sample N=1484
   General Social Survey: Cumulative Data File, Year=1987 only (Smith et al. 2015)

Variables and Descriptives

memnum "Number of memberships"

Mean = 1.71 S.D. = 1.92 Min = 0.00 Max = 16.00

ed "Highest year of school completed"

Mean = 12.55 S.D. = 3.13 Min = 0.00 Max = 20.00

age "Age of respondent"

Mean = 44.09 S.D. = 17.08 Min = 18.00 Max = 89.00

size "Size of Place of Residence in millions"

Mean = 0.42 S.D. = 1.26 Min = 0.00 Max = 7.07

class "Social Class Rank"

Mean = 5.79 S.D. = 1.88 Min = 1.00 Max = 10.00

sex "Dummy for Sex (female = 1)"

Mean = 0.57 S.D. = 0.49 Min = 0.00 Max = 1.00

Labels:
0 Men
1 Women

CHAPTER 3: ONE-MODERATOR EXAMPLE

See dataset information on SIPP_Wealth.dta for Chapter 2: One Moderator Example.

CHAPTER 3: TWO-MODERATOR EXAMPLE

See dataset information on GSS_1987.dta for Chapter 2: Two Moderators Example.

CHAPTER 3: THREE-WAY INTERACTION EXAMPLE

See dataset information on GSS_1987.dta for Chapter 2: Three-way Interaction Example.

CHAPTER 4: TABLES ONE-MODERATOR EXAMPLE AND FIGURES EXAMPLE 3

Data Set: GSS_2010.dta   Estimation Sample N=1429
   General Social Survey: Cumulative Data File, Year=2010 only (Smith et al. 2015)

Variables and Descriptives

sexfrqmonth "Freq of Sexual Intimacy Monthly"

   Mean =   4.03 S.D. =   5.17 Min =   0.00 Max =   19.50

age "Age Of Respondent"

   Mean =   49.27 S.D. =   15.80 Min =   25.00 Max =   89.00

ses "Ses Family"

   Mean =   53.21 S.D. =   19.84 Min =   17.10 Max =   97.20

female "Female"

   Mean =   0.55 S.D. =   0.50 Min =   0.00 Max =   1.00

        Labels:
           0 Male
           1 Female

nevermarr "Never Married"

   Mean =   0.23 S.D. =   0.42 Min =   0.00 Max =   1.00

        Labels:
           0 Ever Married
           1 Never Married

childs "Number Of Children"

   Mean =   1.91 S.D. =   1.63 Min =   0.00 Max =   8.00

attendmonth "Freq Attend Religious Services Monthly"

Mean =   1.37 S.D. =   1.74 Min =   0.00 Max =   4.33

CHAPTER 4: TABLES
TWO-MODERATOR EXAMPLE

See dataset information on GSS_2010.dta for Chapter 2: Two-Moderator
Interval Example.

CHAPTER 4: FIGURES EXAMPLES 1 AND 2

See dataset information on SIPP_Wealth.dta for Chapter 2: One Moderator
Example.

CHAPTER 4: FIGURES EXAMPLE 4

See dataset information on GSS_2010.dta for Chapter 2: Two Moderators
Interval Example

CHAPTER 4: TABLES THREE-WAY
INTERACTION EXAMPLE
AND FIGURES EXAMPLE 5

See dataset information on GSS_1987.dta for Chapter 2: Three-way Interaction Example.

CHAPTER 5: EXAMPLES 1 AND 2

See dataset information on GSS_1987.dta for Chapter 2: Two Moderators
Mixed Example.

CHAPTER 5: EXAMPLE 3

Data Set: GSS_2010.dta   Estimation Sample N=1115
   General Social Survey: Cumulative Data File, Year=2010 only (Smith et al.
   2015)

Variables and Descriptives

   pmhdays "Poor Mental Health Days"

   Mean =   3.81 S.D. =   7.29 Min =   0.00 Max =   30.00

         Labels:
          -1 IAP
          98 DONT KNOW
          99 NO ANSWER

sei   "Respondent Socioeconomic Index"

   Mean =   50.94 S.D. =   19.17 Min =   17.60 Max =   97.20

wfconflict   "Work Family Conflict Scale"

   Mean =   2.19 S.D. =   0.78 Min =   1.00 Max =   4.00

            Labels:
               1 Never
               2 Rarely
               3 Sometimes
               4 Often

age   "Age Of Respondent"

   Mean =   43.80 S.D. =   13.83 Min =   18.00 Max =   88.00

educ   "Highest Year Of School Completed"

   Mean =   14.00 S.D. =   3.03 Min =   0.00 Max =   20.00

childs   "Number Of Children"

   Mean =   1.67 S.D. =   1.52 Min =   0.00 Max =   8.00

female   "Female"

   Mean =   0.53 S.D. =   0.50 Min =   0.00 Max =   1.00

            Labels:
               0 Male
               1 Female

CHAPTER 5: EXAMPLE 4

See dataset information on GSS_1987.dta for Chapter 2: Three-way Interaction Example.

CHAPTER 6: ONE-MODERATOR EXAMPLE

See dataset information on GSS_2010.dta for Chapter 5: Example 3.

CHAPTER 6: TWO-MODERATOR EXAMPLE

See dataset information on GSS_2010.dta for Chapter 2: Two Moderators Interval Example.

CHAPTER 6: THREE-WAY INTERACTION EXAMPLE

See dataset information on GSS_1987.dta for Chapter 2: Three-way Interaction Example.

CHAPTER 7: ONE-MODERATOR EXAMPLE

See dataset information on GSS_2010.dta for Chapter 4: Tables One Moderator Example and Figures Example 3.

CHAPTER 7: TWO-MODERATOR EXAMPLE

See dataset information on GSS_2010.dta for Chapter 2: Two Moderators Interval Example.

CHAPTER 8: ONE-MODERATOR EXAMPLE

Data Set: GSS_1998.dta   Estimation Sample N=1707
   General Social Survey: Cumulative Data File, Year= 1998 only (Smith et al. 2015)

Variables and Descriptives

gun   "Gun Ownership"

  Mean =   0.25 S.D. =   0.43 Min =   0.00 Max =   1.00

        Labels:
           0 DontOwn
           1 Own

age   "Age in Years"

  Mean =   46.03 S.D. =   16.91 Min =   18.00 Max =   89.00

educ   "Education in Years"

  Mean =   13.34 S.D. =   2.97 Min =   0.00 Max =   20.00

sex   "Dummy for Male=1"

  Mean =   0.43 S.D. =   0.49 Min =   0.00 Max =   1.00

        Labels:
           0 Female
           1 Male

location   "Residential Location"

Mean =   1.85 S.D. =   0.57 Min =   1.00 Max =   3.00

       Labels:
         1 CentralCity
         2 Suburbs
         3 Rural

fearnbhd   "Fearful Walking in Neighborhood at Night"

Mean =   0.41 S.D. =   0.49 Min =   0.00 Max =   1.00

       Labels:
         0 NotFearful
         1 Fearful

race   "Dummy for Black=1"

Mean =   0.15 S.D. =   0.36 Min =   0.00 Max =   1.00

       Labels:
         0 White
         1 Black

CHAPTER 8: THREE-WAY INTERACTION EXAMPLE

See dataset information on GSS_1987.dta for Chapter 2: Two Moderators Mixed Example.

CHAPTER 9: ONE-MODERATOR EXAMPLE

Data Set: GSS_2010.dta   Estimation Sample N=1820
  General Social Survey: Cumulative Data File, Year=2010 only (Smith et al. 2015)

Variables and Descriptives

polviews3   "Political Views Recoded (1 2, 3 to 5, 6 7)"

Mean =   2.00 S.D. =   0.60 Min =   1.00 Max =   3.00

       Labels:
         1 Liberal
         2 Moderate
         3 Conservative

educ "Highest Year Of School Completed"

  Mean =   13.56 S.D. =   3.10 Min =   0.00 Max =   20.00

attendmonth "Freq Attend Religious Services Monthly"

  Mean =   1.39 S.D. =   1.74 Min =   0.00 Max =   4.33

age "Age Of Respondent"

  Mean =   47.60 S.D. =   17.53 Min =   18.00 Max =   89.00

class "Subjective Class Identification"

  Mean =   2.40 S.D. =   0.68 Min =   1.00 Max =   4.00

        Labels:
          0 IAP
          1 LOWER CLASS
          2 WORKING CLASS
          3 MIDDLE CLASS
          4 UPPER CLASS
          5 NO CLASS
          8 DK
          9 NA

female "Female"

  Mean =   0.56 S.D. =   0.50 Min =   0.00 Max =   1.00

        Labels:
          0 Male
          1 Female

race "Race Of Respondent"

  Mean =   1.32 S.D. =   0.63 Min =   1.00 Max =   3.00

        Labels:
          0 IAP
          1 WHITE
          2 BLACK
          3 OTHER

CHAPTER 9: TWO-MODERATOR EXAMPLE

Data Set: SIPP_OCC.dta   Estimation Sample N=10476
   SIPP 1992: Subset Occ Data from Extract by Campbell and Kaufman (2006)
   from SIPP 1992 Panel (U.S. Bureau of the Census 1992)

Variables and Descriptives

currocc "Occupation in 4 Classes"

 Mean =   2.35 S.D. =   1.21 Min =   1.00 Max =   4.00

            Labels:
              0 Other
              1 upperwc
              2 lowerwc
              3 upperbc
              4 lowerbc

educ "Education in Years"

 Mean =   13.44 S.D. =   2.81 Min =   0.00 Max =   18.00

age "Age in Years"

 Mean =   41.04 S.D. =   11.83 Min =   17.00 Max =   83.00

sex "Dummy for Female =1"

 Mean =   0.30 S.D. =   0.46 Min =   0.00 Max =   1.00

            Labels:
              0 Men
              1 Women

raceth "Race Ethnicity"

 Mean =   1.35 S.D. =   0.83 Min =   1.00 Max =   4.00

            Labels:
              1 White
              2 Black
              3 AsianAm
              4 Hispanic

married "Dummied for Married=1"

 Mean =   0.61 S.D. =   0.49 Min =   0.00 Max =   1.00

CHAPTER 10: ONE-MODERATOR EXAMPLE

Data Set: GSS_2010.dta   Estimation Sample N=1340
  General Social Survey: Cumulative Data File, Year=2010 only (Smith et al. 2015)

Variables and Descriptives

chemfree2 "Recoded Buy Pesticide Free Produce"

  Mean =   2.16 S.D. =   0.98 Min =   1.00 Max =   4.00

          Labels:
            1 Never
            2 Sometimes
            3 Often
            4 Always

educ "Highest Year Of School Completed"

  Mean =   13.57 S.D. =   3.06 Min =   0.00 Max =   20.00

age "Age Of Respondent"

  Mean =   47.90 S.D. =   17.62 Min =   18.00 Max =   89.00

size "Population In Millions"

  Mean =   0.36 S.D. =   1.24 Min =   0.00 Max =   8.01

          Labels:
           -1 NOT ASSIGNED

female "Female"

  Mean =   0.57 S.D. =   0.49 Min =   0.00 Max =   1.00

          Labels:
            0 Male
            1 Female

race "Race Of Respondent"

  Mean =   1.32 S.D. =   0.62 Min =   1.00 Max =   3.00

          Labels:
            0 IAP
            1 WHITE
            2 BLACK
            3 OTHER

polviews3 "Political Views Recoded (1 2, 3 to 5, 6 7)"

  Mean =   2.04 S.D. =   0.60 Min =   1.00 Max =   3.00

          Labels:
            1 Liberal
            2 Moderate
            3 Conservative

CHAPTER 10: TWO-MODERATOR EXAMPLE

Data Set: GSS_2010.dta   Estimation Sample N=1659
   General Social Survey: Cumulative Data File, Year=2010 only (Smith et al. 2015)

Variables and Descriptives

class "Subjective Class Identification"

Mean =   2.39 S.D. =   0.66 Min =   1.00 Max =   4.00

Labels:
   0 IAP
   1 LOWER CLASS
   2 WORKING CLASS
   3 MIDDLE CLASS
   4 UPPER CLASS
   5 NO CLASS
   8 DK
   9 NA

educ "Highest Year Of School Completed"

Mean =   13.70 S.D. =   3.05 Min =   0.00 Max =   20.00

loginc "Log Of Family Income In $1000 "

Mean =   1.34 S.D. =   1.14 Min =   -3.00 Max =   3.15

sei "Respondent Socioeconomic Index"

Mean =   49.39 S.D. =   19.23 Min =   17.10 Max =   97.20

age "Age Of Respondent"

Mean =   47.46 S.D. =   16.58 Min =   18.00 Max =   89.00

race "Race Of Respondent"

Mean =   1.31 S.D. =   0.62 Min =   1.00 Max =   3.00

Labels:
   0 IAP
   1 WHITE
   2 BLACK
   3 OTHER

female "Female"

Mean =   0.55 S.D. =   0.50 Min =   0.00 Max =   1.00

Labels:
   0 Male
   1 Female

CHAPTER 11: ONE-MODERATOR EXAMPLE

See dataset information on GSS_2010.dta for Chapter 5: Example 3.

CHAPTER 11: THREE-WAY INTERACTION EXAMPLE

See dataset information on GSS_1987.dta for Chapter 2: Three-way Interaction Example.

CHAPTER 12: POLYNOMIAL EXAMPLE

Data Set: GSS_2010.dta   Estimation Sample N=1815
    General Social Survey: Cumulative Data File, Year=2010 only (Smith et al. 2015)

Variables and Descriptives

educ "Highest Year Of School Completed"

Mean =   13.63 S.D. =   3.11 Min =   0.00 Max =   20.00

age "Age Of Respondent"

Mean =   48.45 S.D. =   16.95 Min =   18.00 Max =   89.00

pared "Parental Education"

Mean =   12.10 S.D. =   3.91 Min =   0.00 Max =   20.00

race "Race Of Respondent"

Mean =   1.30 S.D. =   0.60 Min =   1.00 Max =   3.00

            Labels:
                0 IAP
                1 WHITE
                2 BLACK
                3 OTHER

female "Female"

Mean =   0.57 S.D. =   0.50 Min =   0.00 Max =   1.00

            Labels:
                0 Male
                1 Female

CHAPTER 12: HECKMAN EXAMPLE

Data Set: AHS.dta   Estimation Sample N=18576
   AHS: Subset American Housing Survey data from Extract by Krivo and
   Kaufman (2004) from AHS 2001 Panel (U.S. Bureau of the Census 2001).

Variables and Descriptives

equity1k "Home Equity in $1000"

  Mean =   86.70 S.D. =   144.74 Min =   -35.43 Max =   681.01

owner "Home Ownership"

  Mean =   0.59 S.D. =   0.49 Min =   0.00 Max =   1.00

        Labels:
          0 Renter
          1 Owner

hhage "Age of Household (Mean Head & Spouse)"

  Mean =   48.74 S.D. =   17.39 Min =   15.00 Max =   93.00

hheduc "Education of Household (Max of Head and Spouse)"

  Mean =   13.62 S.D. =   3.15 Min =   0.00 Max =   20.00

hhincome "Income of Household"

  Mean =   66.18 S.D. =   86.66 Min =   -10.00 Max =   914.82

cc "Urban Location"

  Mean =   0.50 S.D. =   0.50 Min =   0.00 Max =   1.00

        Labels:
          0 Not Central City
          1 Central City

anywork "Head or Spouse Works"

  Mean =   0.68 S.D. =   0.47 Min =   0.00 Max =   1.00

        Labels:
          0 No Worker
          1 At least 1 Worker

numchild "# Children <18 in of Household"

  Mean =   0.60 S.D. =   1.02 Min =   0.00 Max =   9.00

intermar "Racial Intermarriage"

Mean =   0.03 S.D. =   0.17 Min =   0.00 Max =   1.00

 Labels:
  0 Not Intermarried
  1 Intermarried

hhraceth "Race Ethnicity of Household (Minority if Mixed)"

Mean =   1.73 S.D. =   1.11 Min =   1.00 Max =   4.00

 Labels:
  -9 White/Other Intermarr
  1 White Non-Hispanic
  2 Black Non-Hispanic
  3 Asian Non-Hispanic
  4 Hispanic
  5 Only Other Race/Eth
  6 Minority-Minority Intermarr

region "Region"

Mean =   1.47 S.D. =   1.12 Min =   0.00 Max =   3.00

 Labels:
  0 East
  1 South
  2 Midwest
  3 West

marital "Marital Status"

Mean =   0.91 S.D. =   0.94 Min =   0.00 Max =   2.00

 Labels:
  0 Married
  1 Widowed
  2 Other Not Married

owner "Home Ownership"

Mean =   0.59 S.D. =   0.49 Min =   0.00 Max =   1.00

 Labels:
  0 Renter
  1 Owner

pcfborn "MSA % Foreign Born"

Mean =   15.44 S.D. =   11.22 Min =   1.17 Max =   50.41

pcnhblk "Metro Area % Black Non-Hispanic"

Mean =   13.93 S.D. =   8.65 Min =   0.36 Max =   45.40

CHAPTER 12: SURVIVAL ANALYSIS EXAMPLE

Data Set: UIS.dta   Estimation Sample N=610
  UIS data (University of Massachusetts AIDS Research Unit Impact Study n.d.)
  downloaded "First-Edition Data.zip" on September 24, 2017 from ftp://
  ftp.wiley.com/public/sci_tech_med/survival/

Variables and Descriptives

  _t "Relapse time"

    Mean =   234.42 S.D. =   200.64 Min =   2.00 Max =   1172.00

  age "Age at Enrollment"

    Mean =   32.39 S.D. =   6.15 Min =   20.00 Max =   56.00

  ndrugtx "Number of Prior Drug Treatments"

    Mean =   4.58 S.D. =   5.50 Min =   0.00 Max =   40.00

  treat "Treatment Randomization Assignment"

    Mean =   0.49 S.D. =   0.50 Min =   0.00 Max =   1.00

        Labels:
          0 Short
          1 Long

  site "Treatment Site"

    Mean =   0.30 S.D. =   0.46 Min =   0.00 Max =   1.00

        Labels:
          0 Site_A
          1 Site_B

REFERENCES

Ai, C., & Norton, E. C. (2003). Interaction terms in logit and probit models. *Economics Letters*, *80*, 123–129.

Aiken, L. S., & West, S. G. (1991). *Multiple regression: Testing and interpreting interactions*. Newbury Park, CA: Sage.

Ailshire, J., Karraker, A., & Clarke, P. (2017). Neighborhood social stressors, fine particulate matter air pollution, and cognitive function among older U.S. adults. *Social Science and Medicine*, *172*, 56–63.

Aitken, M. A. (1973). Fixed-width confidence intervals in linear regression with applications to the Johnson-Neyman technique. *British Journal of Mathematical and Statistical Psychology*, *26*, 261–269.

Allison, P. D. (1977). Testing for interaction in multiple regression. *American Journal of Sociology*, *83*, 144–153.

Allison, P. D. (1999). Comparing logit and probit coefficients across groups. *Sociological Methods & Research*, *28*, 186–208.

Allison, P. D. (2014). *Event history and survival analysis* (2nd ed., Quantitative applications in the social sciences series, 07-46). Thousand Oaks, CA: Sage.

Auspurg, K., Hinz, T., & Sauer, C. (2017). Why should women get less? Evidence on the gender pay gap from multifactorial survey experiments. *American Sociological Review*, *82*, 179–210.

Bauer, D. J., & Curran, P. J. (2005). Probing interactions in fixed and multilevel regression: Inferential and graphical techniques. *Multivariate Behavioral Research*, *40*, 373–400.

Belsley, D. A., Kuh, E., & Welsh, R. E. (1980). *Regression diagnostics: Identifying influential data and sources of collinearity*. New York, NY: Wiley.

Berry, W. D., DeMeritt, J. H. R., & Esarey, J. (2010). Testing for interaction in binary logit and probit models: Is a product term essential? *American Journal of Political Science*, *54*, 248–266.

Berry, W. D., Golder, M., & Milton, D. (2012). Improving tests of theories positing interaction. *Journal of Politics*, *74*, 653–671.

Blau, P. M., & Duncan, O. D. (1967). *The American occupational structure*. New York, NY: Wiley.

Brambor, T., Clark, W. R., & Golder, M. (2006). Understanding interaction models: Improving empirical analysis. *Political Analysis*, *14*, 63–82.

Braumoeller, B. F. (2004). Hypothesis testing and multiplicative interaction terms. *International Organization*, *58*, 807–820.

Buis, M. L. (2010). Stata tip 87: Interpretation of interactions in nonlinear models. *Stata Journal*, *10*, 305–308.

Campbell, L. A., & Kaufman, R. L. (2006). Racial differences in household wealth: Beyond Black and White. *Research in Social Stratification and Mobility*, *24*, 131–152.

Cech, E., Rubineau, B., Silbey, S., & Seron, C. (2011). Professional role confidence and gendered persistence in engineering. *American Sociological Review*, *76*, 641–666.

Cortes, K. E., & Lincove, J. A. (2016). Can admissions percent plans lead to better collegiate fit for minority students? *American Economic Review: Papers & Proceedings*, *106*, 348–354.

Dawson, J. W. (2014). Moderation in management research: What, why, when, and how. *Journal of Business and Psychology*, *29*, 1–19.

Dobrzynska, A., & Blais, A. (2008). Testing Zaller's reception and acceptance model in an intense election campaign. *Political Behavior*, *30*, 259–276.

Easterlin, R. A. (1961). The American baby boom in historical perspective. *American Economic Review*, *51*, 869–911.

Eliot, T. S. (1925). *Eliot's poems: 1909–1925*. London, England: Faber & Gwyer.

Esposito, L., & Villaseñor, A. (2017). Relative deprivation: Measurement issues and predictive role for body image dissatisfaction. *Social Science and Medicine*, *192*, 49–57.

Feiveson, A. H. (2017). *Explanation of the delta method*. Retrieved from https://www.stata.com/support/faqs/statistics/delta-method/index.html/

Florescu, I., & Tudor, C. (Eds.). (2014). *Handbook of probability: Appendix B. Inequalities involving random variables and their expectations* [Online]. New York, NY: Wiley.

Fox, J. (2008). *Applied regression analysis and generalized linear models* (2nd ed.). Thousand Oaks, CA: Sage.

Fullerton, A. S., & Xu, J. (2016). *Ordered regression models: Parallel, partial, and non-parallel alternatives*. Boca Raton, FL: CRC Press.

Gafarian, A. V. (1964). Confidence bands in straight line regression. *Journal of the American Statistical Association*, 59, 182–213.

Gould, W. (2013). *Relationship between the chi-squared and F distributions*. Retrieved from www.stata.com/support/faqs/statistics/chi-squared-and-f-distributions

Greene, W. H. (2008). *Econometric analysis* (6th ed.). Upper Saddle River, NJ: Pearson.

Greene, W. H. (2010). Testing hypotheses about interaction terms in nonlinear models. *Economics Letters*, 107, 291–296.

Halli, S. (1990). *Ethnic demography: Canadian immigrant, racial and cultural variations*. Ottawa, Ontario, Canada: Carleton University Press.

Ham, J. C., & Reilly, K. T. (2002). Testing intertemporal substitution, implicit contracts, and hours. *American Economic Review*, 92, 905–927.

Hardin, J. W., & Hilbe, J. M. (2012). *Generalized linear models and extensions* (3rd ed.). College Station, TX: Stata Press.

Hayes, A. F. (2013). *Introduction to mediation, moderation, and conditional process analysis*. New York, NY: Guilford Press.

Hayes, A. F., Glynn, C. J., & Huge, M. E. (2012). Cautions regarding the interpretation of regression coefficients hypothesis tests in linear models with interactions. *Communication Methods and Measures*, 6, 1–11.

He, H. (2012). Effects of environmental policy on consumption: Lessons from the Chinese plastic bag regulation. *Environment and Development Economics*, 17, 407–431.

Heckman, J. (1979). Sample selection bias in the specification error. *Econometrica*, 47, 153–161.

Hosmer, D. W., Lemeshow, S., & May, S. (2008). *Applied survival analysis regression modeling of time-to-event data* (2nd ed.). Hoboken, NJ: Wiley.

Huitema, B. E. (2011). *The analysis of covariance and alternatives: Statistical methods for experiments, quasi-experiments, and single-case studies* (2nd ed.). New York, NY: Wiley.

Institute for Digital Research and Education. (n.d.). *Survival analysis with Stata*. Retrieved from https://stats.idre.ucla.edu/stata/seminars/stata-survival/

Jaccard, J. (1998). *Interaction effects in factorial analysis of variance* (Quantitative applications in the social sciences series, 07-118). Thousand Oaks, CA: Sage.

Jaccard, J. (2001). *Interaction effects in logistic regression* (Quantitative applications in the social sciences series, 07-135). Thousand Oaks, CA: Sage.

Jaccard, J., & Turisi, R. (2003). *Interaction effects in multiple regression* (2nd ed., Quantitative applications in the social sciences series, 07-072). Thousand Oaks, CA: Sage.

Jaccard, J., & Wan, C. K. (1996). *LISREL approaches to interaction effects in multiple regression* (Quantitative applications in the social sciences series, 07-114). Thousand Oaks, CA: Sage.

Johnson, P. O., & Fay, L. C. (1950). The Johnson-Neyman technique, its theory and application. *Psychometrika*, 15, 349–367.

Johnson, P. O., & Neyman, J. (1936). Tests of certain linear hypotheses and their applications to some educational problems. *Statistical Research Memoirs*, 1, 57–93.

Jones, A. M. (2017). When in Rome: Testing the moderating influence of neighborhood composition on the relationship between self-control and juvenile offending. *Crime and Delinquency*, 63, 759–785.

Kam, C. D., & Franzese, R. J., Jr. (2007). *Modeling and interpreting interactive hypotheses in regression analysis*. Ann Arbor: University of Michigan Press.

Karaman, K. K., & Pamuk, S. (2013). Different paths to the modern state in Europe: The interaction between warfare, economic structure, and political regime. *American Political Science Review*, 107, 603–626.

Kaufman, R. L. (1996). Comparing effects in dichotomous logistic regression: A variety of standardized coefficients. *Social Science Quarterly*, 77, 90–109.

Kaufman, R. L. (2013). *Heteroscedasticity in regression: Detection and correction* (Quantitative applications in the social sciences series, 07-172). Thousand Oaks, CA: Sage.

Keele, L., & Park, D. K. (2005, September). *Difficult choices: An evaluation of heterogeneous choice models*. Paper presented at the 2004 Meeting of the American Political Science Association, Chicago, IL.

Kim, I. S. (2017). Political cleavages within industry: Firm-level lobbying for trade liberalization. *American Political Science Review*, 111, 1–20.

Kleinbaum, D. G. (1992). *Logistic regression: A self-learning text*. New York, NY: Springer.

Kohler, U., Karlson, K. B., & Holm, A. (2011). Comparing coefficients of nested nonlinear probability models. *Stata Journal*, *11*, 420–438.

Krivo, L. J., & Kaufman, R. L. (2004). Housing and wealth inequality: Racial-ethnic differences in home equity in the United States. *Demography*, *41*, 585–605.

Kuha, J., & Mills, C. (2018). On group comparisons with logistic regression models. *Sociological Methods and Research*. Advance online publication.

Lazar, A. A., & Zerbe, G. O. (2011). Solutions for determining the significance region using the Johnson-Neyman type procedure in generalized linear (mixed) models. *Journal of Educational and Behavioral Statistics*, *36*, 699–719.

Liao, T. F. (1994). *Interpreting probability models* (Quantitative applications in the social sciences series, 07-101). Thousand Oaks, CA: Sage.

Liu, X. (2016). *Applied ordinal logistic regression using Stata*. Thousand Oaks, CA: Sage.

Long, J. S. (1997). *Regression models for categorical and independent variables* (Advanced quantitative techniques in the social sciences series). Thousand Oaks, CA: Sage.

Long, J. S. (2009). Group comparisons of logit and probit using predicted probabilities (Working paper). Retrieved from http://www.indiana.edu/~jslsoc/research_groupdif.htm

Long, J. S. (2016, May). *New methods of interpretation using marginal effects for nonlinear models*. Paper presented at the EUSMEX 2016: Mexican Stata Users Group. Retrieved from http://www.indiana.edu/~jslsoc/stata/eusmex2016/Versions/eusmex2016-effects-scott-long%202016-05-03.pdf

Long, J. S., & Freese, J. (2014). *Regression models for categorical dependent variables using Stata* (3rd ed.). College Station, TX: Stata Press.

Maddala, G. S. (1983). *Limited dependent and qualitative variables in econometrics*. New York, NY: Cambridge University Press.

Malmusi, D., Borrell, C., & Benach, J. (2010). Migration-related health inequalities: Showing the complex interactions between gender, social class and place of origin. *Social Science and Medicine*, *71*, 1610–1619.

Mare, R. D., & Winship, C. (1984). Regression models with ordinal variables. *American Sociological Review*, *49*, 512–525.

Matsuoka, R., & Maeda, T. (2015). Neighborhood and individual factors associated with survey response behavior: A multilevel multinomial regression analysis of a nationwide survey in Japan. *Social Science Japan Journal*, *18*, 217–232.

McVeigh, R., Beyerlein, K., Vann, B., Jr., & Trivedi, P. (2014). Educational segregation, Tea Party organizations, and battles over distributive justice. *American Sociological Review*, *79*, 630–652.

Mitchell, M. N. (2012). *A visual guide to Stata graphics* (3rd ed.). College Station, TX: Stata Press.

Mood, C. (2010). Logistic regression: Why we cannot do what we think we can do, and what we can do about it. *European Sociological Review*, *26*, 67–82.

Muenchen, R. A. (2017). *The popularity of data science software*. Retrieved from http://r4stats.com/articles/popularity

Nagler, J. (1991). The effect of registration laws and education on U.S. voter turnout. *American Political Science Review*, *85*, 1393–1405.

Nelder, J. A. (1977). A reformulation of linear models. *Journal of the Royal Statistical Society, Series A*, *140*, 48–77.

Nelder, J. A., & Wedderburn, R. W. M. (1972). Generalized linear models. *Journal of the Royal Statistical Society, Series A*, *135*, 370–384.

NIST/SEMATECH. (2017). *e-Handbook of statistical methods*. Retrieved from http://www.itl.nist.gov/div898/handbook/

Pisljar, T., van der Lippe, T., & den Dulk, L. (2011). Health among hospital employees in Europe: A cross-national study of the impact of work stress and work control. *Social Science and Medicine*, *72*, 899–906.

Potthoff, R. F. (1964). On the Johnson-Neyman technique and some extensions thereof. *Psychometrika*, *29*, 241–256.

Pregibon, D. (1981). Logistic regression diagnostics. *Annals of Statistics*, *9*, 705–724.

Rambotti, S. (2015). Recalibrating the spirit level: An analysis of the interaction of income inequality and poverty and its effect on health. *Social Science and Medicine*, *139*, 123–131.

Rogosa, D. (1977). Some results for the Johnson-Neyman technique. *Dissertation Abstracts International*, *38*(9), 5366A. (UMI No. AKf 7802225)

Rogosa, D. (1980). Comparing nonparallel regression lines. *Psychological Bulletin*, *88*, 307–321.

Rogosa, D. (1981). On the relationship between the Johnson-Neyman region of significance and statistical tests of parallel within group regressions. *Educational and Psychological Measurement*, *41*, 73–84.

Sakamoto, A., Goyette, K. A., & Kim, C. H. (2009). Socioeconomic attainments of Asian Americans. *Annual Review of Sociology*, *35*, 255–276.

Simon, J. L. (1975). Puzzles and further explorations in the interrelationships of successive births with husband's income, spouses' education and race. *Demography*, *12*, 259–274.

Southwood, K. E. (1978). Substantive theory and statistical interaction: Five models. *American Journal of Sociology*, *83*, 1154–1203.

Sribney, W., & Wiggins, V. (n.d.). *Standard errors, confidence intervals, and significance tests for ORs, HRs, IRRs, and RRRs*. Retrieved from https://www .stata.com/support/faqs/ statistics/delta-rule/

Visser, M., Mills, M., Heyse, L., Wittek, R., & Bollettino, V. (2016). Work–life balance among humanitarian aid workers. *Nonprofit and Voluntary Sector Quarterly*, *45*, 1191–1213.

Williams, R. (2009). Using heterogeneous choice models to compare logit and probit coefficients across groups. *Sociological Methods & Research*, *37*, 531–559.

Williams, R. (2010). Fitting heterogeneous choice models with OGLM. *Stata Journal*, *10*, 540–567.

Williams, R. (2012). Using the margins command to estimate and interpret adjusted predictions and marginal effects. *Stata Journal*, *12*(2), 308–331.

Williams, R. (2013, July 19). *Comparing logit and probit coefficients across groups: Problems, solutions, and problems with the solutions*. Paper presented at the 2013 Meetings of the European Survey Research Association, Ljubljana, Slovenia.

Williams, R. (2016). Understanding and interpreting generalized ordered logit models. *Journal of Mathematical Sociology*, *40*(1), 7–20.

Wilson, P. (2015). The misuse of the Vuong test for non-nested models to test for zero-inflation. *Economics Letters*, *127*, 51–53.

Yen, S. T., Coats, R. M., & Dalton, T. R. (1992). Brand-name investment of candidates and district homogeneity: An ordinal response model. *Southern Economic Journal*, *58*, 988–1001.

DATA SOURCES

Smith, T. W., Marsden, P., Hout, M., & Kim, J. (2015). *General Social Surveys, 1972–2014* [machine-readable data file]; Principal Investigator, T. W. Smith; Co-Principal Investigator, P. V. Marsden; Co-Principal Investigator, M. Hout; Sponsored by National Science Foundation (NORC ed.). Chicago, IL: NORC at the University of Chicago [producer]; Storrs, CT: The Roper Center for Public

Opinion Research, University of Connecticut [distributor], 2015. 1 data file (57,061 logical records) + 1 codebook (3,567 pp., National Data Program for the Social Sciences, No. 22).

University of Massachusetts AIDS Research Unit Impact Study. (1989–1994). [machine-readable data file downloaded "First-Edition Data.zip" on September 24, 2017 from

ftp://ftp.wiley.com/public/ sci_tech_med/survival/

U.S. Bureau of the Census. (1992). *Survey of Income and Program Participation, 1992 Panel*. [machine-readable data file]

U.S. Bureau of the Census. (2001). *American Housing Survey, 2001 Panel*. [machine-readable data file]

INDEX

Ai, C., 18
Aiken, L. S., 1, 3, 9, 11, 13, 19, 20, 50
Ailshire, J., 480
Aitken, M. A., 65
Akaike information criterion (AIC) statistics, 415
Allison, P. D., 13, 23, 24, 543, 553n2
As-observed reference values:
 linear (identity link) models, 103–105, 140–142
 nonlinear (nonidentity link) models, 157–159
Auspurg, K., 246

Bauer, D. J., 50, 63, 65, 66, 67, 97n4
Bayesian information criterion (BIC) statistics, 415
Belsley, D. A., 14, 22
Benach, J., 290
Berry, W. D., 15, 18
Beyerlein, K., 290
Blais, A., 346
Bollettino, V., 412
Bonferroni adjustment, 65–66
Borrell, C., 290
Boundary value analysis:
 ICALC, 73–74, 76–77
 Johnson-Neyman, 78–80
Brambor, T., 1, 3, 11, 13, 14, 18, 19
Braumoeller, B. F., 1, 3, 11, 13, 18
Buis, M. L., 19, 24

Campbell, R. A., 348
Categorical focal and moderator variables:
 education moderated by household headship type example, 122–123
 household headship type moderated by education and any children example, 123–127
 overview, 121
Categorical predictors, 7–8
Cech, E., 346, 412
Censored (selected) outcomes, models with:
 formulas for expected values of, 536–537
 home equity example, 538–542

modeling strategy for, 535–536
 Stata command syntax for, 537–538
Centering, myth of, 13–14
Central tendency option. See Display and reference values
Clark, W. R., 1
Clarke, P., 480
Coats, R. M., 412
Coefficients:
 centered versus uncentered, 13–14
 effects interpretation versus, 550–551
 estimated across nested models, 19–20
 factor change, 207–210
 main effect, 13–14
 scaled by $g(y)$'s standard deviation, 210–211
 unconditional marginal effects interpretation of, 13
Collinearity, problems of, 14
Concave functions, 181n
Concise option for SIGREG, 205
Conditional distribution functions, 245
Confidence bounds and error bar plots:
 definition of, 64
 EFFDISP tool output of, 257, 277
 one moderator, 85–87
 overview of, 84–85
 three-way interaction, 90–96
 two moderator, 88–90
Confidence intervals, estimating, 68
Confounded nonlinearity:
 interactive effect model predicted probabilities and, 472–473
 model metric outcome plot as solution to, 496–497
 sources of, 149–156
Convex functions, 181n
Cook's distance measure, 22
Cortes, K. E., 246
Count models, one-moderator example, 482–497
 data and testing, 482–483
 occupational status moderated by work-family conflict example, 489–492

OUTDISP for work–family conflict and occupational status simultaneously, 492–497
 work-family conflict moderated by occupational status example, 483–489
Count models, properties and use of, 479–482
Count models, three-way interaction example, 497–516
 data and testing, 497–498
 interpreting, 498–504
 OUTDISP for effects of age, education, and sex simultaneously, 504–516
Cox proportional hazards example. See Survival analysis models
Cumulative categories, ordinal regression model written as log odds of, 416, 441
Cumulative probabilities, ordinal regression model written as, 413–414
Curran, P. J., 50, 63, 65, 66, 67, 97n4

Dalton, T. R., 412
Dawson, J. W., 1, 3, 13
DeMerit, J. H. R., 18
den Dulk, L., 412
Derivations:
 equation 2.17, 60
 equation 5.13, 179–180
 equation 5.14, 180–181
Derivatives, partial, 34
Diagnostic testing:
 influential cases analysis, 22
 link function, 20, 245–246
 misspecification of models and, 20–24, 553
 residual plots, 21–22
Discrete change effect:
 central tendency reference values in calculation of, 398
 one-moderator example, multinomial logistic regression applications, 355–358
 one-moderator example, ordinal regression models, 419–420, 423–424

one-moderator example,
logistic regression and
probit applications,
297–299, 303–304
SPOST13 results on, 405
three-way interaction example,
count models, 502–504
two-moderator interaction
example, ordinal
regression models,
445–450
Disordinal interactions, ordinal
and, 7
Display and reference values:
discrete change effect
calculation and, 398
equivalence of as-observed and
central tendency options
for linear models, 141–142
focal variables, 100–101
linear (identity link) models,
103, 140–142
moderator variables, 100–102
nonidentity link functions,
156–159
other predictors, 102–105
Dobrzynska, A., 346
Dual-axis predicted value displays:
labeling of, 307–309
overview, 160
scatterplots customized in,
340–341
three-way interval-by-nominal
interaction in, 173–175
two-way interval-by-interval
interaction in, 168–170
two-way interval-by-nominal
interaction in, 165–166
two-way nominal-by-nominal
interaction in, 161–163

Easterlin, R. A., 262
EFFDISP (effect display) tool, in
ICALC:
linear regression model
applications, family income
modified by birth cohort
and education example,
273–277
linear regression model
applications, single
moderator example,
249–252, 257
linear regression model
applications, two
moderator example,
264–265
logistic regression and probit
applications, one-
moderator example,

nominal by nominal,
295–300
one-moderator example,
interval by nominal, count
models, 486–489, 491–492
overview, 27, 186–187, 213–227
survival analysis models, 545
Effect displays:
nonlinear models, 159–160
one-moderator example,
interval by nominal,
ordinal regression models,
420–421, 424–426
three-way interval-by-nominal
interaction, 172–173
two-way interval-by-interval
interaction, 167–168
two-way interval-by-nominal
interaction, 165
two-way nominal-by-nominal
interaction, 161
Effects:
comparing, 551–552
interpreting coefficients *versus*,
550–551
Empirically defined significance
regions:
one moderator, 81–82
overview of, 79–81
three-way interactions, 82–84
two moderators, 82
Equation 2.17 derivation, 60
Equation 5.13 derivation, 179–180
Equation 5.14 derivation, 180–181
Error bar plots. *See* Confidence
bounds and error bar plots
Esarey, J., 18
Esposito, L., 346
EXCEL:
stacked area chart created in,
407–408
table of significance regions
formatted in, 83, 276
Expected outcome value charts:
categorical focal and moderator
variables, 121–127
overview of, 119–121

Factor change:
coefficients of, 207–210
one-moderator example,
multinomial logistic
regression applications,
353–355
one-moderator example,
logistic regression and
probit applications,
296–297, 303
three-way interaction example,
count models, 500–502

three-way interaction example,
logistic regression and
probit applications,
316–320
two-moderator example,
multinomial logistic
regression (MNLR)
applications, 383–387,
392–396
two-moderator interaction
example, ordinal
regression models,
438–441, 473–476
Fay, L. C., 66
Focal variables *(F)*:
description of, 6
display and reference values for,
100–101
expected outcome value charts
for categorical, 121–127
isolating moderators' effects on,
113–114
limited range of moderator
values used to probe, 19
nature and shape of interaction
effect and, 7–8
path-style diagram including,
8–9
product terms of, 6–7
reversing roles with moderator
variables, 126–127
scatterplots for interval, 127–140
Focal variable's effect, varying
significance of, 63–98
confidence bounds and error
bar plots, overview of,
84–85
confidence bounds and
error bar plots for one
moderator, 85–87
confidence bounds and error
bar plots for three-way
interactions, 90–96
confidence bounds and
error bar plots for two
moderators, 88–90
empirically defined significance
region, overview of, 79–81
empirically defined significance
region for one moderator,
81–82
empirically defined significance
region for three-way
interactions, 82–84
empirically defined significance
region for two
moderators, 82
JN mathematically derived
significance region,
overview of, 66–69

JN mathematically derived
significance region for one
moderator, 69–74
JN mathematically derived
significance region for two
moderators, 74–79
overview, 63–64
specifications to estimate
standard errors,
significance regions, and
confidence intervals, 68–69
summary of how to explore,
96–97
test statistics for significance
levels, 64–66
Focal variable's effect in modeling
component:
equation 2.17 derivation, 60
GFI mathematical foundation,
34–37
one moderating variable,
algebraic regrouping and,
37–40
one-moderator plotting effects,
46–49
overview, 33–34
three-way interaction plotting
effects, 54–59
three-way interactions and,
42–46
two-moderator plotting effects,
49–54
two or more moderating
variables, algebraic
regrouping and, 40–42
Fox, J., 7, 8, 14, 17, 22, 23, 24, 246,
289, 409n1
Franzese, R. J., Jr., 1, 3, 11, 13, 14,
15, 18, 19, 549
Freese, J., 4, 25, 31n5, 64, 160,
161, 186, 289, 290, 291, 295,
298, 346, 347, 348, 355, 405,
411, 412, 414, 415, 433, 435,
482, 552
Fullerton, A. S., 411, 415

Gafarian, A. V., 65
Generalized least squares (GLS)
regression, 246
Generalized linear models (GLMs).
See GLMs (generalized linear
models)
GFI (gather, factor, and inspect)
process:
algebraic regrouping, point
estimates, and sign
change examples, 37–46
mathematical foundation for,
34–37
overview, 33–34

survival analysis models,
544–545
three-way interaction example,
logistic regression and
probit applications,
314–316
tool for, 25
GFI (gather, factor, and inspect)
tool, in ICALC:
linear regression model
applications, education
moderated by family
income example, 267–268
linear regression model
applications, family income
modified by birth cohort
and education example,
271–273
linear regression model
applications, single
moderator example,
248–249
linear regression model
applications, two
moderator example,
263–264
logistic regression and probit
applications, one-
moderator example,
nominal by nominal,
293–294, 301
multinomial logistic regression
(MNLR) applications,
one-moderator example,
interval by interval, 349–353
multinomial logistic regression
(MNLR) applications,
two-moderator example,
interval by two nominal,
371–373, 380–382, 391–392
overview, 185–186, 193–199
GLMs (generalized linear models):
equations defining, 15–16
errors in using interaction
effects in, 18–20
linear regression models as, 246
link and conditional distribution
functions in, 245
LR significance test preferred
for, 64–65
MNLR interpretation from,
346–347
modeling component interaction
effects in, 16
observed outcome interaction
effects in, 16–17
ordinal regression model
interpretation from,
413–414
rubric for choosing technique, 28

Global test of statistical
significance, 10
GLS (generalized least squares)
regression, 246
Glynn, C. J., 1
Golder, M., 1, 15
Greene, W. H., 5, 18, 20, 23, 64,
246, 247, 290, 535, 536, 543
Group comparisons, 24
Grouped data for logistic
regression, 343n–344n

Halli, S., 262
Ham, J. C., 290
Hardin, J. W., 15, 20, 22, 24
Hausman test, 370
Hayes, A. F., 1, 3, 5, 13, 19, 50, 65,
66, 101
He, H., 480
Heckman, J., 535
Heckman sample selection model,
535–537, 569–570
Heterogeneity, unspecified, 24
Heteroscedasticity, 5, 24
Heyse, R., 412
Hierarchically well-formulated
model, criteria for, 6
Highest order interaction
terms, tests of statistical
significance of, 10
Hilbe, J. M., 15, 20, 22, 24
Hinz, T., 246
Holm, A., 19
Hosmer, D. W., 543
Hosmer-Lemeshow statistic, 370
Huge, M. E., 1
Huitema, B. E., 65, 66

ICALC (Interaction CALCulator)
toolkit: syntax, options, and
examples, 185–244
EFFDISP tool (effect display),
186–187, 213–227
GFI tool (gather, factor, and
inspect), 185–186,
193–199
INTSPEC tool (interaction
specification), 187–193
linear regression model
applications, 254–261
OUTDISP tool (display of a
predicted outcome by the
interacting variables), 187,
227–243
overview of tools, 25–26
SIGREG tool (significance
regions), 25, 63, 186,
199–213
why Stata platform, 3–4
why toolkit is needed, 26–27

Identity link functions and Linear Regression Model, 246
Identity link models. *See* Linear (identity link) models
Influential cases, analysis of, 22
Institute for Digital Research and Education, 543
Interaction effects:
    description of, 4–6
    diagnostic testing of, 20–22
    errors in specifying and interpreting, 11–13
    GLMs (generalized linear models) and, 15–20
    inclusion in analysis, 5
    interpretive tools and techniques overview, 24–28
    model misspecifications, consequences of, 23–24
    outcome values from main effects model *versus*, 35–36
    overview, 1–4
    predicted value comparison, confounding in, 149–153
    slope comparison, confounding in, 153–156
    slope of logistic prediction function for, 180–181
    specifying, 6–9
    statistical significance of, 9–11
    interpret totality of specification, 551
    zero-inflated model component with, 521–522
INTSPEC (interaction specification) tool, in ICALC:
    linear regression model applications, education moderated by family income example, 267–268
    linear regression model applications, family income modified by birth cohort and education example, 271–273
    linear regression model applications, single moderator example, 248
    linear regression model applications, two moderator example, 263–264
    logistic regression and probit applications, one-moderator example, 293–294, 301

multinomial logistic regression (MNLR) applications, one-moderator example, 349–353
multinomial logistic regression (MNLR) applications, two-moderator example, 371–373, 380–382, 391–392
overview, 187–193

Jaccard, J., 1, 3, 5, 6, 7, 11, 13, 50
JN (Johnson-Neyman) mathematically derived significance region(*See also* Focal variable's effect, varying significance of):
    analysis of covariance models, 66
    boundary value plot for three-way interaction from, 80
    extension to GLMs, 63–74
    one moderator, 69–74
    overview of, 66–69
    two moderators, 74–79
JN (Johnson-Neyman) method, 63.
Johnson, P. O., 66
Jones, A. M., 480

Kam, C. D., 1, 3, 11, 13, 14, 15, 18, 19, 549
Karaman, K. K., 246
Karlson, K. B., 19
Karraker, A., 480
Kaufman, R. L., 7, 20, 40, 142n2, 246, 348, 536
Keele, L., 24
Kim, I. S., 290
Kleinbaum, D. G., 6
Kohler, U., 21, 26
Krivo, L. J., 536
Kuh, E., 14
Kuha, J., 24

Latent outcome:
    one-moderator example, ordinal regression models, 418–419, 422–423, 427–429
    one-moderator example, logistic regression and probit applications, 295–296, 302–303
    outcome displays to interpret, 324–330
    standard deviation and mean of, 405–407
    three-way interaction example, logistic regression and probit, 320

two-moderator examplemultinomial logistic regression (MNLR), 387–391
two-moderator interaction example ordinal regression models, 451–457, 476
Lazar, A. A., 66
Lemeshow, S., 543
Liao, T. F., 295
Limiting distribution, 97n
Lincove, J. A., 246
Linearizing (measurement) equation, 16–17, 33
Linear (identity link) models:
    as-observed and central tendency option equivalence for, 140–142
    expected outcome value charts, overview of, 119–121
    expected outcome value charts for categorical focal and moderator variables, 121–127
    focal variable display and reference values, 100–101
    moderator variable display and reference values, 100–102
    other predictors' reference values, 102–105
    overview of interpretation using predicted outcome, 99–100
    overview of linear regression model properties, 245–247predicted outcome value tables for one moderator, 105–109
    predicted outcome value tables for three-way interaction, 115–119
    predicted outcome value tables for two or more moderators, 109–114
    scatterplots for interval focal variables, 127–140
Linear regression model applications, single moderator example, 247–261
    data and testing, 247–248
    GFI analyses, 248–249, 255
    INTSPEC tool setup, 248, 255
    OUTDISP tool for, 252–253, 258–260
    SIGREG and EFFDISP tools for, 249–252, 256–258
    summary, 253–254, 260–261
Linear regression model applications, two moderator example, 261–283
    data and testing, 261–263

INTSPEC tool setup and GFI analyses, 263–264, 267–268, 271–273
OUTDISP tool for, 265–267, 268–271, 278–283
SIGREG and EFFDISP tools for, 264–265, 268, 273–277
Link functions for GLMs, 20, 245–246
Liu, X., 411
Log count for Poisson and negative binomial models, 522–523
Logistic regression and probit applications:
diagnostic test and procedures for, 292
properties and use of, 289–292
See also Multinomial logistic regression applications
Logistic regression and probit applications, one-moderator example, 292–312
data and testing, 292–293
OUTDISP tool for outcome displays, 304–309
probit *versus* logistic regression results, 310–313
residential location moderated by sex example, 300–304
sex moderated by residential location example, 293–300
what to present, 309–310
Logistic regression and probit applications, three-way interaction example, 313–339
additive and interaction model comparison, 341–343
bar chart customization, 341
data and testing, 313–314
dual-axis scatterplot customization, 340–341
factor change interpretation, 316–320
GFI results for, 314–316
predicted probabilities for interpretation, 330–339
significance region tables for interpretation, 321–330
standardized latent outcome interpretation, 320
what to present, 339–340
Log-linear analysis, 344n
Log odds of cumulative categories, ordinal regression model written as, 416
Long, J. S., 3, 19, 22, 24, 25, 29n5, 64, 160, 161, 186, 289, 290, 291, 295, 298, 310, 346, 347, 348, 355, 405, 406, 411, 411, 412, 414, 415, 433, 435, 482, 535, 536, 552

Lower order terms, errors from excluding, 11–13
LOWESS plots, 370
LR test of statistical significance, 10–11, 64–65

Maddala, G. S., 19
Maeda, T., 346
Main effect coefficients in model with interaction effects, interpreting, 13–14
Main effects model:
outcome values from, 35
predicted outcome category probabilities with, 431–433
slope comparison, confounding in, 153–156
slope of logistic prediction function for, 179–180
Malmusi, D., 290
Mare, R. D., 19
Marginality, principle of, 6, 526
Matsuoka, R., 346
May, S., 543
mchange command in SPOST13, 357, 405
McVeigh, R., 290, 480
Mediated effects, moderated effects *versus*, 20
Mills, C., 24
Mills, M., 412
Milton, D., 15
Misspecification of models. *See* model misspecication
Mitchell, M. N., 285
Model/fit departures, assessing, 20–21
Modeling component. *See* Focal variable's effect in modeling component
Modeling (structural) equation, in GLMs, 17, 33
Model misspecification:
consequences of, 23–24
diagnostic testing and, 20–24, 553
Moderator variables *(M)*:
description of, 7–9
display and reference values for, 100–102
expected outcome value charts for categorical, 121–127
nature and shape of interaction effect and, 7–8
path-style diagram including, 8–9
product terms of, 6–7
reversing roles with focal variables, 126–127
Mood, C., 19, 23, 24

Muenchen, R. A., 4
Multicategory nominal variables, path diagram for, 285
Multinomial logistic regression (MNLR) applications, one-moderator example, 348–369
data and testing, 348–353
discrete change interpretation, 355–358
factor change results, 353–355
predicted probabilities for interpretation, 359–365
predicted standardized latent outcomes for interpretation, 365–369
Multinomial logistic regression (MNLR) applications, overview of, 345–348. *See also* Logistic regression and probit applications
Multinomial logistic regression (MNLR) applications, two-moderator example, 369–404
base probability for discrete change effect from SPOST13, 405
data and testing, 369–371
education moderated by race/ethnicity and sex example, 391–405
latent outcome standard deviation and mean, 405–407
race/ethnicity moderated by education example, 379–391
sex moderated by education example, 371–379
stacked area chart, 407–409

Nagler, J., 18
Negative binomial models, log count for, 522–523
Nelder, J. A., 6, 15
Nested models, problems comparing estimated coefficients across, 19–20
Neyman, J., 66
Nonidentity link functions:
confounded sources of nonlinearity in , 149–156
display and reference value options for, 156–159
slope of logistic prediction function, interaction and main effects model, 179–181

Nonidentity link functions, issues interpreting interaction effects in:
confounded nonlinearity in comparing predicted outcomes, 149–153
confounded nonlinearity in comparing slopes, 153–156
identifying issues, 143–149
solutions for, overview, 159–160
examples of problems and solutions, 160-177
summary of recommendations, 177–179
Nonlinear effect of noninteracting predictors, 17, 153–156, 179–181
Norton, E. C., 18

Odds and probabilities, correspondence of, 387
One-moderator example, multinomial logistic regression applications:
data and testing, 348–353
discrete change interpretation, 355–358
factor change results, 353–355
predicted probabilities for interpretation, 359–365
predicted standardized latent outcomes for interpretation, 365–369
One-moderator example, count models:
data and testing, 482–483
occupational status moderated by work-family conflict example, 489–492
OUTDISP for work–family conflict and occupational status simultaneously, 492–497
work-family conflict moderated by occupational status example, 483–489
One-moderator example, ordinal regression models:
data and testing, 415–417
education moderated by sex, 417–421,
ordinal probit results, 433–435
OUTDISP for effects of education and sex simultaneously, 426–433
sex moderated by education, 421–426

One-moderator example, logistic regression and probit applications, 292–312
data and testing, 292–293
interaction effect, 309–310
OUTDISP tool for outcome displays, 304–309
probit *versus* logistic regression results, 310–313
residential location moderated by sex example, 300–304
sex moderated by residential location example, 293–300
Ordinal and disordinal interactions, 7
Ordinal regression models:
interpretation of, 414–415
properties and use of, 411–414
Ordinal regression models, one-moderator example:
data and testing, 415–417
education moderated by sex example, 417–421
ordinal probit results, 433–435
OUTDISP for effects of education and sex simultaneously, 426–433
sex moderated by education example, 421–426
Ordinal regression models, two-moderator interaction example:
data and testing, 435–436
factor change effects for different moderator values, 473–476
interpretation of, 436–437
latent outcome by race group, calculation of, 476
moderated effect of education by race, 437–438
moderated effect of log income by race, 438–441
moderated effect of race by education and race by log income, 441–450
OUTDISP for effects of race, education, and income simultaneously, 450
probability of outcome categories predicted values, 457–473
standardized latent outcome predicted values, 451–457
OUTDISP (display of a predicted outcome by the interacting variables) tool, in ICALC:

count model applications, one-moderator example, 492–497
count models, three-way interaction example, 504–516
linear regression model applications, education moderated by family income example, 268–271
linear regression model applications, family income modified by birth cohort and education example, 278–283
linear regression model applications, single moderator example, 252–254, 259
linear regression model applications, two moderator example, 265–267
logistic regression and probit applications, one-moderator example, 304–309
ordinal regression models, one-moderator example, 426–433
ordinal regression models, two-moderator interaction example, 450
overview, 25, 187, 227–244,
survival analysis models, 545, 546

Pamuk, S., 246
Park, D. K., 24
Partial residual-predictor plots, 21
Path-style diagram of interaction specification, 8, 274, 285
Perception bias, linear 145
Pisljar, T., 412
pltopts option to customize plots:
bar charts, 341
overview, 284–285
scatterplots, 340-341
Poisson and negative binomial models, log count for, 522–523
Polynomial function of a predictor interacting with another predictor, 525–535
moderated effect of age example, 528–532
moderated effect of race example, 527–528

overview, 525–526
predicted values of education
by race and age example,
533–535
Potthoff, R. F., 65
Predicted outcome value tables
for LRM:
one moderator, 105–109
three-way interaction, 115–119
two or more moderators,
109–114
Predicted probabilities:
one-moderator example, ,
multinomial logistic
regression, 359–365
plots of, 333–339
significance region tables for
discrete change effects,
330–333
three-way interaction example,
count models, 516–521
two-moderator example,
multinomial logistic
regression (MNLR)
applications, 396–404
two-moderator interaction
example, ordinal regression
models, 457–473
Prediction function, properly
specifying terms in, 549–550
Pregibon, D., 22
Pregibon's approximation of
Cook's distance measure, 22
Probabilities and odds,
correspondence of, 387
Probit applications. *See* Logistic
regression and probit
applications
Product terms (interaction), 7–8,
18–19

Rambotti, S., 246
Reilly, K. T., 290
Residual-omitted variable plots,
21–22
Residual-predictor plots, 21
Rogosa, D., 65, 97n4
Rubineau, B., 346

Sauer, C., 246
Scatterplots for interval focal
variables, 127–140
Seron, C., 346
Sidak adjustment, 65–66
Significance, statistical. *See*
Statistical significance
Significance regions:
boundary value calculation
example for, 75

Excel formatted table of, for
three-way interactions, 83
factor change effect of race on
class, 444
JN method for determining,
71–72
SIGREG (significance region)
tool, in ICALC:
factor change coefficients,
207–210
household headship type
example, 81–82
linear regression model
applications, education
moderated by family
income example, 268
linear regression model
applications, family income
modified by birth cohort
and education example,
273–277
linear regression model
applications, single
moderator example,
249–252
linear regression model
applications, two
moderator example,
264–265
logistic regression and probit
applications, one-
moderator example,
nominal by nominal,
295–302
one-moderator example of,
201–202
one-moderator example,
interval by nominal, count
models, 485–486,
490–491
overview, 25, 63, 186
scaled by g(y)'s standard
deviation, 210–211
SPOST13 marginal effects,
211–213
survival analysis models, 544
three-way interaction example,
how to use, 203–208
three-way interaction example,
logistic regression and
probit applications,
321–330
tool-specific options for,
199–201
Silbey, S., 346
Simon, J. L., 262
Simultaneous testing, significance
level adjustment for,
65–66

Slope of logistic prediction
function:
interaction model, 153–156,
180–181
main effects model, 153–156,
179–180
Small-Hsiao test, 370
Southwood, K. E., 1, 3
SPOST13:
discrete change effect, 298–299,
405marginal effects,
211–213
mchange command in, 25, 186,
200, 214, 357, 405
use by ICALC, 4
use to generate predicted
probabilities in count
models, 516–519
Stacked area chart, 407–409
Standard errors of moderated
effect, estimating, 68
Standardized log count value,
count value equivalent
for, 523
Statistical significance
of interaction effects, 9–11
test of predicted outcome
differences among
categories of a nominal
variable, 285–287
tests for, 12, 64–66
See also Focal variable's effect,
varying significance
ofSuperimposed displays:
alternative plot comparing,
341–343description of, 160
three-way interval-by-nominal
interaction example,
175–177
two-way interval-by-interval
interaction example,
170–172
two-way interval-by-nominal
interaction example, 166
two-way nominal-by-nominal
interaction example,
163–164
Survival analysis models:
example of, 571
GFI and SIGREG for effect
of age moderated by site,
544
GFI and SIGREG for effect of
site moderated by age, 545
OUTDISP for interaction of site
and age, 545, 546
overview, 542–544
survival curves for interaction of
site and age, 546–549

test command, in Stata, 473–475
testnl command, in Stata, 475–476
Test statistics for significance
  levels, 10, 64–66
3D surface plots, distortion in, 131
Three-way interaction example,
  count models:
  data and testing, 497–498
  interpreting, 498–504
  OUTDISP for effects of age,
    education, and sex
    simultaneously, 504–516
Three-way interaction example, ,
  logistic regression and probit
  applications:
  additive and interaction model
    comparison, 341–343
  bar chart customization, 341
  data and testing, 313–314
  dual-axis scatterplot
    customization, 340–341
  factor change interpretation,
    316–320
  GFI results for, 314–316
  predicted probabilities for
    interpretation, 330–339
  presentation of, 339–340
  significance region tables for
    interpretation, 321–330
  standardized latent outcome
    interpretation, 320
Tobit model, 535–537
Trivedi, P., 290
Turisi, R., 1, 3, 5, 7, 13, 50
Two-moderator interaction
  example, linear regression
  model applications:
  data and testing, 261–263
  INSPEC tool setup and GFI
    analysis, 263–264
  OUTDISP tool for, 265–267
  SIGREG and EFFDISP tools for,
    264–265

Two-moderator example,
  multinomial logistic
  regression applications:
  base probability for discrete
    change effect from
    SPOST13, 405
  data and testing, 369–371
  effect of education moderated
    by race/ethnicity and sex,
    391–405
  effect of race/ethnicity
    moderated by education,
    379–391
  effect of sex moderated by
    education, 371–379
  latent outcome standard
    deviation and mean,
    405–407
  stacked area chart, 407–409
Two-moderator interaction example,
  ordinal regression models:
  data and testing, 435–436
  factor change effects for
    different moderator
    values, 473–476
  interpretation of, 436–437
  latent outcome by race group,
    calculation of, 476
  moderated effect of education
    by race, 437–438
  moderated effect of log income
    by race, 438–441
  moderated effect of race by
    education and race by log
    income, 441–450
  OUTDISP for effects of race,
    education, and income
    simultaneously, 450
  probability of outcome
    categories predicted
    values, 457–473
  standardized latent outcome
    predicted values, 451–457

Uncentered coefficients, 14
Unconditional marginal
  effects, coefficients
  improperly interpreted
  as, 13

van der Lippe, T., 412
Vann, B., Jr., 290
Varying significance of focal
  variable's effect. *See* Focal
  variable's effect, varying
  significance of
Villasenor, A., 346
Visser, M., 412

Wald test of statistical
  significance:
  global nondirectional test,
    10–11
  LR test *versus*, 64–65
  significance regions from,
    79–80, 98n
  Stata *test* command for,
    98n
Wan, C. K., 3
Wedderburn, R. W. M., 15
Welsch, R. E., 16
West, S. G., 1, 3, 9, 11, 13,
  19, 20, 50
Williams, R., 19, 23, 24,
  102, 104, 142n1, 142n3, 298
Winship, C., 19
Wittek, L., 412

Xu, J., 411, 415

Yen, S. T., 412

Zerbe, G. O., 66
Zero-inflated model component,
  521–522
z test/*t* test of statistical
  significance, 10, 11